Conservation of Tropical Coral Reefs

Brian Joseph McFarland

Conservation of Tropical Coral Reefs

A Review of Financial and Strategic Solutions

Brian Joseph McFarland
Windham, NH, USA

ISBN 978-3-030-57014-9   ISBN 978-3-030-57012-5   (eBook)
https://doi.org/10.1007/978-3-030-57012-5

Cover illustration: Emma Holman/gettyimages

This Palgrave Macmillan imprint is published by the registered company Springer Nature Switzerland AG
The registered company address is: Gewerbestrasse 11, 6330 Cham, Switzerland

I would like to dedicate this book to my family and friends – particularly my son Attila, my wife and dive buddy Brigitta, my dad Joseph, my mom Tamara, my brother Trevor, my brother-in-law Dani, my sister-in-law Amanda, and my mother-in-law Eva - for all their love and support.

I would like to share a special dedication with my son Attila: when you are older, I hope that we, as humanity, have come up with the solutions to save tropical coral reefs and mitigate global climate change and so you can see, in person, all the wonderful animals that I used to read to you about. Likewise Attila, I began writing this book when you were three and just learning how to swim in our pool; jumping off the diving board and getting comfortable with swimming underwater with your eyes open and getting used to putting on a mask and snorkel; and how can we forget you "SCUBA diving" into pillow piles. I wish the same for my niece Adrian, my nephew Paul, and my goddaughter Olivia.

I would also like to thank my former teachers Dan Bisaccio, Christopher Brooks, and Joseph Domask and the School for Field Studies for teaching me about the natural world, about how to be an analytical thinker, and opening up my mind to a world of adventurous travels. A special thank you is reserved for Dan Bisaccio who provided my first opportunity to snorkel over coral reefs, in Puerto Morelos, Mexico, along with a special thank you to my dear friend Mike Edmonds who helped turn me on to SCUBA diving.

I would like to thank my hardworking colleagues at Carbonfund.org, particularly Jarett Emert, Linda Kelly, and Suzie Kaufman, and a particular thank you to Carbonfund.org's President and Founder Eric Carlson, along with

Carbonfund.org's Chairman Paul Rowland, for all the opportunities that have been provided to me over the last 13+ years at Carbonfund.org.
I would also like to say thank you to Gabriel Thoumi for all his hard work. I truly appreciate the time and effort of María José González for writing the foreword and of Peter Gash for writing the epilogue, along with the kind endorsements by Jennie Gilbert, Eric Carlson, and Cary Krosinsky.
In addition, thank you to the entire team at Palgrave Macmillan for their dedication and support.

Foreword

As the oceans go, so do we.

—David Doubilet

Coral reefs, especially those in the tropics, are possibly the world's most diverse ecosystems. Although they only cover about 284,000 km^2 of the world's surface, equivalent to about 5% of the world's rainforest cover, they are home to between one million and three million species depending on different projections.

Coral reefs provide many benefits to humans. Among others, they are a critical natural infrastructure that protect coastal communities from storm surges, beach erosion, and wave-induced damage. They reduce the threat to lives and property posed by hurricanes, and they are an important source of income, mainly through tourism and fisheries. For local communities, they provide livelihoods and food security.

Sadly, they are also one of the most degraded ecosystems and much of the information coming from scientists is telling us they have a bleak future. The biggest threat is global climate change, which is causing increased frequency and intensity of storms and hurricanes, as well as coral bleaching and seawater acidification. The bleaching events of 2016 and 2017 alone caused the mortality of half of the Great Barrier Reef of Australia, the largest reef on the planet. Climate change exacerbates anthropogenic threats to corals, such as overfishing, untreated sewage, solid waste and plastics, land use change and sedimentation, and coastal development. Human-caused threats reduce the resilience of coral reefs to climate change, making it more difficult for them to recover from climate-related events.

Coral reefs were symbols of time and permanence. For centuries, they were the barriers that protected/guarded tropical islands. They were the demise of adventurers and explorers; the end and the beginning of Robinson Crusoe and Lord Greystoke. It is an eerie sensation to realize this is changing dramatically during our lifetimes. Thanks to documentary films, families that have never seen the ocean now know of the wonders of coral reefs, but also that they are threatened and dying. For many scientists and recreational divers, it is happening before our eyes.

> *I can mention many moments that were unforgettable and revelatory. But the most single revelatory three minutes was the first time I put on scuba gear and dived into a coral reef. It's just the unbelievable fact that you can move in three dimensions.*
> *-Sir David Attenborough*

In results presented at a scientific meeting in early 2020, research indicates that climate change will be the main cause of the loss of the world's coral reefs.[1] By 2100, almost all of the coral habitats could be gone, and between 70–90% may disappear in the next 20 years. It seems to be a downward spiral with no return.

Is there hope? I think there is. There are reasons for optimism. An excellent program that operates in the Mesoamerican Reef (MAR) ecoregion, called the Healthy Reefs for Healthy People Initiative (HRI), measures the health of the MAR biennially and puts out a report card with a reef health index integrated by the following four variables: live coral cover, fleshy macroalgae cover, commercial fish biomass, and herbivorous fish biomass. In their recently published 2020 Report Card,[2] results indicate that over the whole MAR, live coral cover increased. This is good, very good given recent trends.

However, for the first time since their first Report Card in 2008, the overall reef health index dropped, mostly due to severe reductions in fish. This highlights another key aspect related to the health and potential permanence of reefs: they are intimately intertwined with other marine, coastal, and terrestrial ecosystems. So, a reduction in fish in the MAR may be a consequence of overfishing, but may also be tied to a reduction in mangrove forests that are critical nurseries for fish larvae, to lack of sewage treatment that pours untreated waters into the reefs, and of land use change upstream that washes sediment into the sea. How other ecosystems are managed—or not—will have an impact on coral reefs. It is important to protect and restore reefs, but also seagrass beds and mangroves, and other ecosystems along the watersheds that can ultimately affect reefs. To protect corals, we must have a ridge-to-reef approach.

Understanding this alone does not guarantee success. Effective protection and restoration of ecosystems requires funding. Actually, large amounts of funding. And obtaining continued, timely, and adequate funding is not an easy task. To provide insight into the possibilities for coral conservation finance, Brian McFarland skillfully provides an in-depth compilation on the sources of funding and strategies available for coral conservation, management and restoration today. This book begins with a careful account of coral reef degradation, ecology, and conservation policy that provide perspective on the conservation finance information provided.

The initial framework sets the stage for the array of thoroughly analyzed funding options for coral reefs. They encompass from philanthropic sources and government budgets, which are the more "traditional" sources—though by no means less important—to debt swaps, impact investing, payment for ecosystem services and innovative funding currently being piloted, such as blue bonds and parametric insurance for reef restoration. Critical to understanding these different financial mechanisms are the detailed case studies. They clearly explain how each mechanism or strategy has been established, how it operates, what challenges it has met with and lessons learned, among others. This is a very effective way for coral reef stewards to evaluate if a given mechanism is viable given their own specific circumstances. Overall, the author provides an extremely valuable toolbox for coral reef conservation finance.

Certainly, financial resources themselves will not protect nor restore corals. They are a means to an end. We need to work together, to share our experiences, to experiment with diverse strategies and to communicate successes, as much as failures, in order to move effective coral conservation and restoration forward. And we need to get involved with, and liaise outside, the strict coral conservation/restoration sphere, across professions and sectors, to sewage treatment, controlled and sustainable coastal development, sustainable fisheries, sustainable agriculture, sustainable tourism and, especially, to policies and actions that will contribute to the control of climate change if coral reefs are to survive.

We all need coral reefs, regardless of where we live and what we do. We cannot just stand by and continue to read the dismal news regarding the world's corals. We must take action from our neck of the woods, be it in the water out planting coral fragments grown in a nursery, teaching about recycling, supporting policies for conservation of resources and reduction of greenhouse gases, or by contributing financially for any of the range of financial mechanisms and strategies detailed in this book. Humans are the main cause of the current situation in which corals find themselves today, and yet we are also the hope for coral reefs and we need to act now.

March 2020

María José Gonzalez
Executive Director, Mesoamerican Reef Fund
Guatemala City, Guatemala

Notes

1. Yeung, Jessie. "Climate Change Could Kill All of Earth's Coral Reefs 102 by 2100, Scientists Warn." *CNN*. February 20, 2020. Accessed March 11, 2020. https://edition.cnn.com/2020/02/20/world/coral-reefs-2100-intl-hnk-scli-scn/index.html.
2. Healthy Reefs for Healthy People. "Mesoamerican Reef Report Card 2020." Accessed March 11, 2020. https://www.healthyreefs.org/cms/wp-content/uploads/2020/02/SmithReefs_RC19_Pages_BIL_f_E_LO.pdf.

Preface

The inspiration for this book comes from a lifetime of observing firsthand some of the world's most spectacular landscapes and wildlife, and the hope my son will be able to see the same seascapes, landscapes and wildlife when he is older. There are few things in life that I appreciate more than seeing wildlife in their natural landscapes. Over the years, I have snorkeled with whale sharks (*Rhincodon typus*) the size of a school bus, swam near Amazon river dolphins (*Inia geoffrensis*), come upon fresh jaguar (*Panthera onca*) tracks, watched colorful scarlet macaws (*Ara macao*) fly overhead, and stood face-to-face with a wild black rhinoceros (*Diceros bicornis*) (Fig. 1).

While growing up in the small town of Amherst, New Hampshire, my life was complemented by a close family, wildlife, and a well-financed public school system. I recall memories of my dad feeding hummingbirds out of his hand, my mom tending our vegetable garden, watching a moose (*Alces alces*) drink from the pond outside my bedroom, and fishing for largemouth bass (*Micropterus salmoides*) with my brother. Growing up relatively close to the ocean meant frequent trips to the seashore where I would spend time climbing over rocks to catch crabs, starfish, and sea urchins—so much climbing, that I often wound up with sprained ankles and crutches.

It was during the summer after my junior year at Souhegan High School that I traveled to Mexico's Yucatan Peninsula with our teacher Dan Bisaccio, the now-retired and former Director of Science Education at Brown University, for real-world fieldwork in conjunction with the Smithsonian Institution's Monitoring & Assessment of Biodiversity Program.[1] Although it

Fig. 1 Brigitta and Brian Diving the Great Barrier Reef (*Credit*: Brigitta Jozan)

took years working in a restaurant to earn the money to fund the trip, I became forever impassioned with conservation biology and our interconnected globe. For the first time, I saw poverty, I walked in a tropical rainforest, I snorkeled over the Mesoamerican Reef, and I watched Yucatan spider monkeys (*Ateles geoffroyi yucatanensis*) in the canopy. I started to understand the connections between poverty, commercial development, slash-and-burn agriculture, commercial agriculture—particularly the global cattle trade—and tropical deforestation, tropical degradation, catastrophic climate change, and its impact on coral reefs.

Since then, I traveled back to Mexico and shortly thereafter onto Costa Rica, Tanzania, and Kenya. During these additional education programs, I gained further insights into the interconnectivity between natural resource management, ecotourism, and sustainable economic development. Next, I chose to study under Joseph Domask at American University. During this time, I interned at the US Environmental Protection Agency and spent a month in Brazil studying tropical ecology and income inequality. I later enrolled in a dual Master degrees program in Global Environmental Policy and Business Administration at American University in Washington, DC.

I started as an intern at Carbonfund.org Foundation when I began my graduate studies and I still work today at Carbonfund.org and for its wholly-owned subsidiary CarbonCo, LLC. These two organizations fund climate change mitigation projects, along with the development and implementation of four, innovative forest conservation projects in the Brazilian Amazon. While my expertise is more on the conservation of tropical rainforests, tropical coral reefs are facing similar threats—particularly global climate change and habitat loss—and it is important to understand that healthy forests and responsible land use are related to healthy coastal areas, including tropical coral reefs. Furthermore, in my free time, I have snorkeled and/or dove throughout the Bahamas, Jamaica, Grand Cayman, Mexico, and the Philippines. For an applied research trip to help write this book, my wife Brigitta and I also had the remarkable opportunity to dive the Great Barrier Reef from September to October 2019. I also enjoy visiting aquariums, including ones in Baltimore, Boston, Chicago, New Orleans, Monterey, and Seattle, along with whale watching trips in California, Maine, and Washington State.

This book reflects my background. With travels to 30+ countries, 15+ years of conservation finance experience, 70+ semi-structured, in-depth interviews, and dozens of background readings, the book focuses on some of the financial instruments that can fund the conservation of the world's last remaining tropical coral reefs. The world's tropical coral reefs, from the Mesoamerican Reef to the Coral Triangle to the Great Barrier Reef, are under grave threat. These threats include, but are not limited to: global climate change and warming oceans; ocean acidification; plastics and other pollution such as runoff from cities and farms; overfishing; the spread of invasive species and disease; and the construction of megaprojects for infrastructure development such as ports, shipping lanes, and hotels.

The degradation and loss of tropical coral reefs are serious global environmental issues. For instance, conserving coastal ecosystems—including mangroves, seagrasses, and salt marshes - help mitigate global climate change by sequestering, at times, more carbon dioxide emissions per hectare than a primary tropical rainforest. Furthermore, conservation of tropical coral reefs can improve the livelihoods of local people who depend on reef fish for a large source of their protein. Finally, conservation can provide refuge for the world's highest levels of marine biodiversity, with coral reefs often being referenced as the "rainforests of the sea." However, conserving these tropical coral reefs is contingent upon raising both sufficient and long-term international financing that is structured to support scalable, seascape-level initiatives.

This book follows a chronological order. It briefly examines the historical development of conservation finance for tropical coral reefs. Conservation finance originally developed from governments in the form of domestic budgeting from taxation for the creation and maintenance of protected areas alongside philanthropy, followed by utilizing tax deductions created for conservation easements, and then by international aid to assist foreign countries with their own protected areas. Conservation finance then evolved to include impact investing, payments for ecosystem services, debt-for-nature conversions, and ecotourism. Today, conservation finance includes these approaches while also expanding into certified sustainable products and the issuance of blue bonds.

The book, in order to help an interdisciplinary audience, will begin with a context of tropical coral reef degradation and loss, followed by a review of coral reef ecology, global environmental policy, and international finance.

Each chapter will then be dedicated to a unique conservation financing instrument. Each of these chapters will follow a similar format, starting with a brief historical overview of the instrument, followed by an explanation of how the instrument works, then by providing background information on the size of the instrument, and concluding with a presentation of case studies, a financial analysis, a policy analysis, and a future outlook for the instrument.

All 30 case studies, which span 23 countries and 6 continents, are based off the following format:

I. The Problem

    i. Identify the Problem.
    ii. Explain Why the Problem Is Important.
    iii. How Was the Problem Identified?
    iv. Was the Process for Identifying the Problem Effective?

II. Steps Taken to Address the Problem
III. Results
IV. Challenges and How They Were Met
V. Beyond Results
VI. Lessons Learned.[2]

The financial analysis will examine the return versus risks of the financial instruments. The risk categories are:

- Business;
- Strategic;

- Reputation;
- Liquidity;
- Operational;
- Market;
- Legal and Regulatory; and
- Credit.

The policy analysis will utilize the following format:

I. Defining the Problem.
II. Establishing Goals.
III. Selecting a Policy.
IV. Implementing a Policy.
V. Evaluating the Policy.[3]

I would like to highlight, upfront, the following:

First, this book focuses on tropical coral reefs. While all biomes are important—from tropical rainforests to the deep ocean, cold water reefs—tropical coral reefs are the focus of this book given their unique issues.

Second, marine protected areas (MPAs) and conservation projects are often financed via complex mechanisms such as a combination of trust fund financing, domestic budgetary allocations, and revenue from ecotourism. This said, if a case study is categorized under domestic budgetary allocations, that is not to say that the MPA received exclusive financing from domestic budgetary allocations. Rather, projects and programs should seek diversified revenue models. For instance, ecotourism outfitters should adopt blue procurement models and impact investors should leverage matching funds from nonprofits and/or governments. Yet, a significant funding gap continues to exist.

Third, U.S.-based and EU-based conservation is different than conservation throughout the tropics. For example, Indigenous Peoples and local fishing communities are often living near tropical coral reefs and their buffer zones. In addition, there tends to be greater income inequality and less overall wealth in the host countries of tropical coral reefs. For example, the GINI coefficient—which is a leading economic indicator of wealth inequality among households—was 28.2 (in 2015) in the Netherlands, 32.7 (2015) for France, and 41.5 (2016) for the U.S., while the Philippines was 44.4 (2015), Seychelles was 46.8 (2013), and Mexico was 48.3 (2016).[4] Competing for government budgets are other pressing domestic issues such as poverty, energy access, education, and healthcare. Furthermore, corruption tends to be higher in such areas,[5] and the ease of doing business is often more difficult.[6]

Fourth, the requirements for effective tropical coral reef conservation include the rule of law, transparency, social justice, and long-term financing requirements. This said, the term "fisherfolk" will be used instead of "fishermen" because many "fishermen" around the world are in fact women.

Fifth, there are sources of financing, there are conservation outcomes, and there are financing instruments that bridge the gap between the sources of financing and these conservation outcomes. These three groups can often be confused. To clarify, sources of financing can come from individuals, foundations (corporate and family), businesses (small and medium enterprises and publicly traded corporations), and governments (domestic spending and through bilateral or multilateral channels). Conservation outcomes include: MPAs, international peace parks, conservation concessions, and conservation easements. The financing instruments that bridge the gap between the sources of financing and the conservation outcomes—which will be the focus of this book—include: debt-for-nature conversions, payments for ecosystem services, premiums for blue procurements, taxation, and tax deductions.

Sixth, unfortunately, developing long-term financing to conserve tropical coral reefs may not protect such reefs unless global climate change and ocean acidification are also mitigated.

Seventh, for every investment in the conservation of tropical coral reefs, there needs to be a root cause analysis to make sure funds are specifically targeting the underlying drivers and agents of coral reef degradation and loss.

Eighth, while tropical coral reefs are beautiful, they can be incredibly tough environments to work and live in. From remote atolls to hot, humid weather, work in the tropics—to say the least—is challenging and yet rewarding and absolutely necessary because of the need to conserve the world's remaining tropical coral reefs, halt the catastrophic collapse of biodiversity, and to improve the livelihoods of local communities.

Ninth, while I truly believe that tropical coral reefs have intrinsic value—due to their tremendous biodiversity and the resources they provide for local livelihoods—if there is no extrinsic value placed on these ecosystems, then decision-makers, whether it be political or business, will be more apt to undertake activities and policies that will lead to the conversion of these ecosystems to other uses.

Tenth, I tried to include as many case studies as possible, but some amazing places—such as Fiji, the Maldives, New Caledonia, and Vanuatu—were unfortunately excluded.

April 2020   Brian Joseph McFarland
                                                              Windham, NH, USA

Notes

1. UNESCO. "Man and the Biosphere Programme." Accessed December 30, 2019. http://www.unesco.org/new/en/natural-sciences/environment/eco logical-sciences/man-and-biosphere-programme/.
2. Pathfinder International. "Preparing a Case Study: A Guide for Designing and Conducting a Case Study for Evaluation Input." April 2006. Accessed August 1, 2019. http://www.pathfinder.org/publications-tools/pdfs/Preparing-a-Case-Study-A-Guide-for-Designing-and-Conducting-a-Case-Study-for-Evaluation-Input.pdf.
3. American University. "Tips for Writing a Policy Analysis." Accessed August 1, 2016.
4. Index Mundi. "GINI Index (World Bank estimate)—Country Ranking." Accessed March 17, 2020. http://www.indexmundi.com/facts/indicators/SI.POV.GINI/rankings.
5. Transparency International. "Corruption Perceptions Index 2019." Accessed March 17, 2020. https://www.transparency.org/cpi2019?/news/feature/cpi-2019.
6. World Bank Group. "Doing Business: Economy Rankings." Accessed March 17, 2020. http://www.doingbusiness.org/rankings.

Acknowledgments

The remarks that follow should not be attributed to any of the following interviewees unless otherwise noted. Nevertheless, I am very grateful for their comments, the thought-provoking conversations we had, and their dedication to the field:

- Abolade (Bola) Majekobaje
- Agustin (Tin Tin) Llanasa
- Alice Grainger
- Amy Gash
- Anabelle Plantilla
- Barrington Lewis
- Ben Kneppers
- Bill Crew
- Bryce Risley
- Chad Wiggins
- Charles Sheppard
- Charley Waters
- David White
- Denise Garcia
- Doug Rader
- Erik von Uexkull
- Fernanda F. C. Marques
- Fernando Secaira

- Gabriel Thoumi
- Gayatri Reksodihardjo-Lilley
- Gonzalo Merediz Alonso
- Guy Pinjuv
- Helen Pitman
- Helio Hara
- Helvig D. W. Cecilia-Thode
- Henk van de Velden
- James Bentley
- James Eaton
- James Schultz
- James Wright
- Jay Wink
- Jennie Gilbert
- John McManus
- Judith Denkinger
- Kelly Kryc
- Kenneth Johnson
- Kevin Kun He
- Leonard Sonnenschein
- Les Kaufman
- Marcel Bigue
- María Eugenia Arreola
- María José Gonzalez
- Melanie McField
- Michelle Portman
- Mike Berwick
- Nadia Bood
- Nick Zarlinga
- Nicola Bassett
- Nicolas Pascal
- Nigel Wenban-Smith
- Nilda S. Baling
- Nina Abalajon
- Paulina E. Martis-van Arneman
- Pervaze Sheikh
- Peter Gash
- Petra Lundgren
- Phil Townsing
- Rob Dunbar

- Rob Weary
- Sam Teicher
- Scott Dowd
- Scott Settelmyer
- Scott Winters
- Shannon Switzer Swanson
- Simon de Lestang
- Steve Box
- Ted Cheeseman
- Tim Fitzgerald
- Tim Miller-Morgan
- Tom Moore
- Trevor Jones
- Vic Ferguson
- Yael Teff-Seker

I would also like to acknowledge the companies, civil servants, local communities, volunteers, and dedicated professionals working to address tropical coral reef degradation and loss. Thank you for all that you do.

Timeline

This is a timeline outlining when conservation organizations and associations were established. Not every organization is listed, as this is illustrative. More detailed timelines are in each chapter.

Global Conservation Organizations

Nineteenth Century

- 1826: The Zoological Society of London (ZSL).[1]
- 1846: Smithsonian Institute.[2]
- 1865: Massachusetts State Fish & Wildlife Agency—the first agency in the United States.[3]
- 1873: Massachusetts Fish & Game Association (MFGA).[4]
- 1888: The National Geographic Society.[5]
- 1889: Royal Society for the Protection of Birds (RSPB).[6]
- 1891: The Trustees of Public Reservations, became Trustees of Reservations (TTOR), the world's first regional land trust.[7]
- 1892: The Sierra Club is founded by John Muir.[8]
- 1895: Wildlife Conservation Society (WCS), originally the New York Zoological Society.[9]

- 1896: Massachusetts Audubon Society, the world's oldest existing Audubon Society.[10]

Twentieth Century

- 1903: The Society for the Preservation of the Wild Fauna of the Empire, became Fauna and Flora International (FFI).[11]
- 1919: National Parks Conservation Association.[12]
- 1922: The Australian Coral Reef Society (ACRS), the world's oldest organization concerned with the study and conservation of coral reefs.[13]
- 1922: The International Council for Bird Preservation (ICBP), became BirdLife International in 1993.[14]
- 1926: Charles Stewart Mott Foundation.[15]
- 1930: The Woods Hole Oceanographic Institution (WHOI).[16]
- 1936: Ford Foundation.[17]
- 1945: Cooperative for Assistance and Relief Everywhere (CARE).[18]
- 1946: Barro Colorado Island joined the Smithsonian Institution in 1946, in 1966, became the Smithsonian Tropical Research Institute (STRI).[19]
- 1946: The Ecologists Union, became The Nature Conservancy (TNC) in 1950.[20]
- 1947: Defenders of Wildlife, formerly Defenders of Fur Bearers.[21]
- 1948: International Union for the Conservation of Nature (IUCN).[22]
- 1948: The Pew Charitable Trusts.[23]
- 1951: The Nature Conservancy (TNC) is incorporated.[24]
- 1951: Richard and Rhoda Goldman Foundation.[25]
- 1952: Resources for the Future (RFF).[26]
- 1959: The Charles Darwin Foundation.[27]
- 1959: Sea Turtle Conservancy (STC), formerly the Caribbean Conservation Corporation.[28]
- 1961: World Wildlife Fund (WWF).[29] WWF is known as World Wildlife Fund in the U.S. and Canada and known elsewhere as the World Wide Fund for Nature.
- 1963: Weeden Foundation.[30]
- 1964: David and Lucile Packard Foundation.[31]
- 1965: National Recreation and Park Association (NRPA).[32]
- 1967: Environmental Defense Fund (EDF).[33]
- 1967: American Cetacean Society.[34]
- 1969: The Oceanic Society.[35]
- 1970: Natural Resources Defense Council (NRDC).[36]
- 1970: MacArthur Foundation.[37]

- 1970: The Perry Institute for Marine Science (PIMS); in 1984, PIMS created the Caribbean Marine Research Center.[38]
- 1971: Greenpeace.[39]
- 1971: Pact International.[40]
- 1971: Earthwatch, formerly the Educational Expeditions International.[41]
- 1971: The Harbor Branch Oceanographic Institute (HBOI).[42]
- 1972: Trust for Public Land (TPL).[43]
- 1972: The Haribon Foundation.[44]
- 1972: The Delta Corporation, became the Ocean Conservancy.[45]
- 1973: The Cousteau Society is founded by Jacques-Yves Cousteau.[46]
- 1973: American Rivers.[47]
- 1973: The Endangered Wildlife Trust (EWT).[48]
- 1973: Rare.[49]
- 1974: The Worldwatch Institute.[50]
- 1975: Marine Mammal Center.[51]
- 1976: TRAFFIC.[52]
- 1977: Jane Goodall Institute for Wildlife Research, Education and Conservation (JGI) is founded by Jane Goodall and Princess Genevievedi San Faustino.[53]
- 1977: Sea Shepherd.[54]
- 1979: The Seychelles Island Foundation.[55]
- 1980: The School for Field Studies (SFS).[56]
- 1980: Pacific Whale Foundation.[57]
- 1981: Pronatura.[58]
- 1982: World Resources Institute (WRI).[59]
- 1982: The Equipe Cousteau, a sister organization to the Cousteau Society, was founded by Jacques-Yves Cousteau as Fondation Cousteau in 1982; became Equipe Cousteau in 1992.[60]
- 1983: The Fundación Defensores de la Naturaleza (FDN; Nature Defenders Foundation).[61]
- 1983: The Oak Foundation.[62]
- 1984: National Fish and Wildlife Foundation.[63]
- 1984: Environmental Investigation Agency.[64]
- 1985: Woods Hole Research Center (WHRC).[65]
- 1985: Winrock International is founded by the Rockefeller Family merging Winrock International Research and Training Center, the International Agricultural Development Service, and the Agricultural Development Council.[66]
- 1985: The Conservation Fund.[67]
- 1986: Amigos de Sian Ka'an (Friends of Sian Ka'an).[68]

- 1987: Conservation International (CI).[69]
- 1987: Curtis and Edith Munson Foundation.[70]
- 1987: Reef Relief began.[71]
- 1988: Ecosystem Survival Plan; Co-founders then founded The Center for Ecosystem Survival and SaveNature.org in 1993.[72]
- 1989: Stockholm Environmental Institute (SEI).[73]
- 1989: CERES (Coalition for Environmentally Responsible Economies).[74]
- 1989: Gapforce.[75]
- 1989: Frontier.[76]
- 1989: Sea Turtle Restoration Project, now the Sea Turtle Restoration Network.[77]
- 1990: International Institute for Sustainable Development (IISD).[78]
- 1990: Global Coral Reef Alliance.[79]
- 1990: Reef Environmental Education Foundation (REEF).[80]
- 1990: Tusk.[81]
- 1990: The Turner Foundation.[82]
- 1991: The Shark Research Institute.[83]
- 1991: Project Piaba.[84]
- 1992: EcoLogic Development Fund.[85]
- 1992: Project AWARE.[86]
- 1992: Foundation for the Philippine Environment (FPE).[87]
- 1993: The Waitt Foundation.[88]
- 1994: Yayasan Keanekaragaman Hayati Indonesia (KEHATI; The Indonesian Biodiversity Foundation).[89]
- 1994: The Mexican Fund for the Conservation of Nature.[90]
- 1994: Palau Conservation Society (PCS).[91]
- 1994: MarineLife Alliance.[92]
- 1995: The World Industry Council for the Environment (WICE) and the Business Council for Sustainable Development (BCSD) merge to form the World Business Council for Sustainable Development (WBCSD). In 1990, Stephan Schmidheiny created the first Business Council for Sustainable Development (BCSD).[93]
- 1995: Wildlife Alliance.[94]
- 1995: Operation Wallacea.[95]
- 1996: Nature and Culture International.[96]
- 1996: Fundo Brasileiro para a Biodiversidade (FUNBIO, The Brazilian Biodiversity Fund).[97]
- 1996: Reef Check Foundation.[98]
- 1996: Woodcock Foundation.[99]
- 1996: Doris Duke Charitable Foundation.[100]

- 1996: Marine Conservation Biology Institute; became the Marine Conservation Institute in 2011.[101]
- 1997: The Grantham Foundation for the Protection of the Environment.[102]
- 1998: Forest Trends.[103]
- 1998: Conservation Strategy Fund (CSF).[104]
- 1998: The United Nations Foundation.[105]
- 1998: Global Nature Fund.[106]
- 1998: The Leonardo DiCaprio Foundation.[107]
- 1999: Root Capital.[108]
- 1999: The Great Barrier Reef Foundation.[109]
- 1999: Mongabay.[110]
- 1999: Ocean Futures Society.[111]
- 1999: Algalita Marine Research Foundation.[112]

Twenty-First Century

- 2000: National Marine Sanctuary Foundation.[113]
- 2000: The Lighthouse Foundation.[114]
- 2000: Carbosur.[115]
- 2000: Trucost.[116]
- 2000: Gordon and Betty Moore Foundation.[117]
- 2000: Kemitraan Partnership.[118]
- 2000: Gili Eco Trust, or Yayasan Ekosistem Gili indah.[119]
- 2001: The Center for Global Development (CGD).[120]
- 2001: BlueOrchard.[121]
- 2001: Oceana.[122]
- 2002: SECORE (SExual COral REproduction) International, SECORE International formed in the U.S. in 2004.[123]
- 2003: The Ocean Foundation.[124]
- 2003: Save Our Seas Foundation.[125]
- 2003: Carbonfund.org Foundation.
- 2003: Blue Ventures.[126]
- 2004: Climate Focus.[127]
- 2004: Conservation Capital.[128]
- 2004: Green Fins.[129]
- 2005: Paso Pacifico.[130]
- 2005: Madagascar Biodiversity Fund.[131]
- 2006: Consilium Capital Limited.[132]

- 2007: EKO Asset Management Partners. EKO merged with Wolfensohn Fund Management in 2014 to form Encourage Capital.[133]
- 2007: ClientEarth.[134]
- 2007: Coral Restoration Foundation™ (CRF™).[135]
- 2008: Climate Advisers.[136]
- 2008: The World Federation for Coral Reef Conservation (WFCRC).[137]
- 2008: Southern Environmental Association (SEA-Belize) is created with the merger of Friends of Nature (FoN) and the Toledo Association for Sustainable Tourism and Empowerment (TASTE).[138]
- 2008: ClimateWorks Foundation.[139]
- 2009: The Carbon War Room.[140]
- 2009: Climate Bonds Initiative.[141]
- 2009: Corals for Conservation.[142]
- 2009: BACoMaB Trust Fund, Fonds Fiduciaire du Banc d'Arguin et de la Biodiversité Côtière et Marine (BACoMaB Trust Fund, Banc d'Arguin Trust Fund and Coastal and Marine Biodiversity).[143]
- 2010: Clarmondial.[144]
- 2010: Blue Marine Foundation.[145]
- 2010: Coral Triangle Center.[146]
- 2011: Althelia Ecosphere.[147]
- 2011: Oceans 5.[148]
- 2013: Bye Bye Plastics Bags is started by Melati and Isabel Wijsen at the ages of 10 and 12.[149]
- 2014: SeaLegacy.[150]
- 2016: Nature Trust Alliance.[151] The Nature Trust Alliance manages the Blue Action Fund.[152]
- 2017: Jr Ocean Guardians is formed by Shelby O'Neil for her 2017 Girl Scout of America Gold Award Project.[153]
- 2017: Blue Finance.[154]
- 2017: 4ocean.[155]

Associations/Networks

- 1870: The American Fisheries Society is established—"the world's oldest and largest organization dedicated to strengthening the fisheries profession, advancing fisheries science, and conserving fisheries resources."[156]
- 1973: Wild Oceans, formerly the National Coalition for Marine Conservation.[157]
- 1975: National Wildlife Refuge Association.[158]

- 1977: Coastal Conservation Association.[159]
- 1980: International Society for Reef Studies (ISRS).[160]
- 1985: Society for Conservation Biology (SCB).[161]
- 1993: The Marine Fish Conservation Network.[162]
- 1994: Coral Reef Alliance.[163]
- 1995: Alliance of Religions and Conservation (ARC).[164]
- 1995: The Global Coral Reef Monitoring Network (GCRMN) established by the International Coral Reef Initiative.[165]
- 1996: Tebtebba (Indigenous Peoples' International Centre for Policy Research and Education).[166]
- 1996: The WWF and Unilever convened a press conference to publish a Joint Statement of Intent, which would later result in the Marine Stewardship Council (MSC).[167]
- 1997: Environmental Markets Association (EMA).[168]
- 1997: Reef Conservation UK.[169]
- 1998: National Mitigation Banking Association (NMBA).[170]
- 1999: The International Emissions Trading Association (IETA).[171]
- 1999: Latin American and Caribbean Network of Environmental Funds (RedLAC).[172]
- 1999: The Partnership for Interdisciplinary Studies of Coastal Oceans (PISCO).[173]
- 2000: Carbon Disclosure Project, now CDP.[174]
- 2000: The International Coral Reef Action Network (ICRAN).[175]
- 2000: Locally-Managed Marine Area (LMMA) Network.[176]
- 2002: Conservation Finance Alliance.[177]
- 2002: Wildlife Conservation Network (WCN).[178]
- 2003: Global Footprint Network.[179]
- 2004: The Deep Sea Conservation Coalition.[180]
- 2004: The World Wildlife Fund (WWF) initiated the Aquaculture Dialogues, a multi-stakeholder roundtable launched in 2004; later in 2010, the Sustainable Trade Initiative IDH and WWF Netherlands created the Aquaculture Stewardship Council (ASC).[181]
- 2005: TNC launches the Reef Resilience Program, now the Reef Resilience Network.[182]
- 2006: Conservation Finance Network was "envisioned at Lincoln Institute of Land Policy in 2006 and held at Yale School of Forestry & Environmental Studies in 2007."[183]
- 2006: Sustainable Fisheries Partnership (SFP).[184]
- 2007: The Consortium for the Conservation of Coastal and Marine Ecosystems in the Western Indian Ocean (WIO-C).[185]

- 2008: The Alliance for Water Stewardship.[186]
- 2008: World Ocean Council (WOC).[187]
- 2008: International Carbon Reduction & Offset Alliance (ICROA).[188]
- 2008: The Carbon Markets and Investors Association (CMIA) is formed when the Carbon Markets Association (CMA) and International Carbon Investors and Services (INCIS) merge.[189] Rebranded in 2011 as the Climate Markets and Investment Association.[190]
- 2008: The World Federation for Coral Reef Conservation (WFCRC).[191]
- 2009: Global Impact Investing Network (GIIN).[192]
- 2009: Consumer Goods Forum.[193]
- 2010: The informal network Big Ocean.[194]
- 2015: The International Partnership for Blue Carbon launched by Australia at UNFCCC COP21.[195]
- 2016: The Conservation Finance Practitioner Roundtable.[196]
- 2016: Coalition for Private Investment in Conservation (CPIC).[197]
- 2016: Coral Restoration Consortium.[198]

Notes

1. ZSL. "Remember Sir Stamford Raffles, founder and first President of ZSL." Last modified July 1, 2015. https://www.zsl.org/blogs/artefact-of-the-month/remembering-sir-stamford-raffles-founder-and-first-president-of-zsl.
2. Smithsonian Libraries. "From Smithson to Smithsonian: The Birth of an Institution." Accessed November 23, 2016. http://www.sil.si.edu/Exhibitions/Smithson-to-Smithsonian/intro.html.
3. Foster, Dave. "Meeting the Conservation Challenge in New England." In *Conservation Capital in the Americas*, edited by James N. Levitt. 20–21.
4. Ibid.
5. National Geographic Society. "This Day in History: January 13, 1888: National Geographic Society Founded." Accessed November 23, 2016. http://nationalgeographic.org/thisday/jan13/national-geographic-society-founded/.
6. Royal Society for the Protection of Birds. "History of the RSPB." Accessed November 23, 2016. https://www.rspb.org.uk/whatwedo/history/.
7. Foster, Dave. "Meeting the Conservation Challenge in New England." In *Conservation Capital in the Americas*, edited by James N. Levitt. 20–21.
8. Sierra Club. "History: Sierra Club Timeline." Accessed November 23, 2016. http://vault.sierraclub.org/history/timeline.aspx.
9. WCS. "Zoos & Aquarium." Accessed November 23, 2016. https://www.wcs.org/parks.

10. National Audubon Society. "History of Audubon and Science-based Bird Conservation." Accessed November 23, 2016. http://www.audubon.org/content/history-audubon-and-waterbird-conservation.

11. FFI. "110 Years of Fauna & Flora International." Accessed November 23, 2016. http://www.fauna-flora.org/timeline/.

12. National Parks Conservation Association. "Our Story." Accessed November 30, 2016. https://www.npca.org/about/our-story.

13. Australia Coral Reef Society. "About." Accessed June 1, 2018. https://australiancoralreefsociety.org/about/.

14. BirdLife International. "Our History." Accessed November 23, 2016. http://www.birdlife.org/worldwide/partnership/our-history.

15. Charles Stewart Mott Foundation. "History and Founder." Accessed February 13, 2017. https://www.mott.org/about/history/.

16. Woods Hole Oceanographic Institution. "Home." Accessed December 2, 2018. https://www.whoi.edu/page.do?pid=9305.

17. Ford Foundation. "Our origins." Accessed February 13, 2017. https://www.fordfoundation.org/about-us/our-origins/.

18. CARE. "CARE Package." Accessed February 9, 2017. http://www.care.org/care-package.

19. Smithsonian Tropical Research Institute. "History." Accessed November 23, 2016. https://www.stri.si.edu/english/about_stri/history.php.

20. TNC. "Our History." Accessed November 23, 2016. http://www.nature.org/about-us/vision-mission/history/.

21. Defenders of Wildlife. "Frequently Asked Questions." Accessed November 23, 2016. http://www.defenders.org/frequently-asked-questions.

22. IUCN. "About." Accessed November 23, 2016. https://www.iucn.org/about.

23. Pew Charitable Trusts. "About: History." Accessed September 6, 2018. https://www.pewtrusts.org/en/about.

24. TNC. "Our History." Accessed November 23, 2016. http://www.nature.org/about-us/vision-mission/history/.

25. The Goldman Environmental Prize. "Founders." Accessed August 28, 2018. https://www.goldmanprize.org/about-us/founders/.

26. Resources for the Future. "About." Accessed April 5, 2017. http://www.rff.org/about.

27. Charles Darwin Foundation. "About Us." Accessed August 28, 2018. https://www.darwinfoundation.org/en/about.

28. Sea Turtle Conservancy. "About STC." Accessed September 6, 2018. https://conserveturtles.org/about-stc-organizational-background/.

29. WWF. "WWF from 1961 to 2006." Accessed November 23, 2016. http://wwf.panda.org/who_we_are/history/wwf_conservation_1961_2006/.

30. Weeden Foundation. "Weeden Foundation Mission Statement." Accessed February 13, 2017. http://www.weedenfdn.org/Weeden-Foundation-Mission-Statement.htm.

31. David and Lucile Packard Foundation. "Our History." Accessed February 13, 2017. https://www.packard.org/about-the-foundation/history/.
32. National Recreation and Park Association. "About NRPA." Accessed November 23, 2016. http://www.nrpa.org/About-National-Recreation-and-Park-Association/.
33. EDF. "Our Story: How EDF Got Started." Accessed November 23, 2016. https://www.edf.org/about/our-history.
34. American Cetacean Society. "Home." Accessed August 7, 2019. https://www.acsonline.org/.
35. Oceanic Society. "About Us." Accessed April 20, 2018. https://www.oceanicsociety.org/about.
36. Natural Resources Defense Council. "About Us." Accessed November 23, 2016. https://www.nrdc.org/about.
37. MacArthur Foundation. "Our History." Accessed February 13, 2017. https://www.macfound.org/about/our-history/.
38. Perry Institute for Marine Science. "Our History." Accessed December 2, 2018. http://www.perryinstitute.org/who-we-are/history-of-pims/.
39. Greenpeace. "About Us: History." Accessed November 23, 2016. http://www.greenpeace.org/canada/en/About-us/History/.
40. Pact International. "About Us." Accessed November 23, 2016. http://www.pactworld.org/our-promise.
41. Earthwatch. "History of Earthwatch." Accessed August 7, 2019. https://earthwatch.org/About/History-of-Earthwatch.
42. Harbor Branch Oceanographic Institute. "Who We Are." Accessed December 2, 2018. https://hboifoundation.org/who-we-are.
43. Trust for Public Land. "Our History." Accessed November 23, 2016. https://www.tpl.org/about/history.
44. Haribon Foundation. "About the Foundation." Accessed August 21, 2018. http://www.haribon.org.ph/index.php/haribon-foundation/about-haribon.
45. Ocean Conservancy. "About Us." Accessed April 20, 2018. https://oceanconservancy.org/about/history/.
46. Cousteau Society. "Who We Are?" Accessed April 20, 2018. https://www.cousteau.org/english/who.php.
47. American Rivers. "About American Rivers." Accessed April 20, 2018. https://www.americanrivers.org/about-us/who-we-are/.
48. Endangered Wildlife Trust. "Our Niche: History and Development." Accessed September 6, 2018. https://www.ewt.org.za/niche.html.
49. NGO Advisor. "Rare." Accessed March 30, 2020. https://www.ngoadvisor.net/ong/rare.
50. Worldwatch Institute. "History." Accessed November 23, 2016. http://www.worldwatch.org/mission.
51. Marine Mammal Center. "Fact Sheet." Accessed August 7, 2019. http://www.marinemammalcenter.org/about-us/media-center/fact-sheet.html.

52. TRAFFIC. "What We Do." Accessed September 19, 2019. http://www.traffic.org/overview/.
53. Jane Goodall Institute. "History of the Jane Goodall Institute." Accessed November 23, 2016. http://janegoodallug.org/about-us/history/.
54. Sea Shepherd. "Home." Accessed April 20, 2018. https://seashepherd.org/.
55. Seychelles Island Foundation. "About." Accessed August 28, 2018. http://www.sif.sc/about.
56. School for Field Studies. "Our History." Accessed November 23, 2016. http://www.fieldstudies.org/about/history.
57. Pacific Whale Foundation. "Mission & Vision." Accessed August 7, 2019. https://www.pacificwhale.org/about-us/mission-and-vision/.
58. San Diego Natural History Museum. "PRONATURA." Accessed September 19, 2019. http://sdnhm.org/oceanoasis/behindthescenes/pronatura.html.
59. WRI. "The WRI Story: 30 Years of Big Ideas." Accessed November 23, 2016. http://www.wri.org/wri-story-30-years-big-ideas.
60. Cousteau Foundation. "Who We Are?" Accessed April 20, 2018. https://www.cousteau.org/english/who.php.
61. Nature Defenders Foundation. "About Us." Accessed March 24, 2017. https://defensores.org.gt/en#conocenos.
62. Oak Foundation. "About Oak." Accessed September 6, 2018. http://www.oakfnd.org/about-oak.html.
63. National Fish and Wildlife Foundation. "About National Fish and Wildlife Foundation." Accessed February 15, 2017. http://www.nfwf.org/whoweare/Pages/home.aspx#.WKT9G39WfeQ.
64. Environmental Investigation Agency. "Our History." Accessed August 7, 2019. https://eia-international.org/about-us/our-history/.
65. Woods Hole Research Center. "About Us." Accessed November 23, 2016. http://whrc.org/about-whrc/.
66. Winrock International. "Winrock History." Accessed November 23, 2016. http://www.winrock.org/winrock-history/.
67. Conservation Fund. "History." Accessed September 19, 2019. http://www.conservationfund.org/about-us/history.
68. Amigos de Sian Ka'an. "¿What We Do and How We Work?" Accessed February 12, 2017. http://amigosdesiankaan.org/en/who-we-are/what-we-do-and-how-we-work.
69. CI. "About Conservation International." Accessed April 14, 2020. https://www.conservation.org/about.
70. Curtis and Edith Munson Foundation. "About the Curtis & Edith Munson Foundation." Accessed August 28, 2018. http://www.munsonfdn.org/about.html.
71. Reef Relief. "History." Accessed April 20, 2018. https://www.reefrelief.org/history/.

72. SaveNature.org. "History and Milestones." Accessed September 19, 2019. http://www.savenature.org/content/about/History.

73. Stockholm Environment Institute. "About SEI Asia." Accessed November 23, 2016. https://www.sei-international.org/asia/about-sei-asia.

74. Ceres. "Exxon Valdez Oil Spill Still Leaves a Painful Legacy." Last modified March 2014. http://www.ceres.org/about-us/our-history/exxon-valdez-oil-spill-still-leaves-a-painful-legacy.

75. Gapforce. "Gap Year Marine Conservation Volunteer Courses." Accessed September 19, 2019. https://gapforce.org/gb/gap-year-programs/marine-conservation.

76. Frontier. "About Frontier." Accessed August 21, 2018. https://frontier.ac.uk/AboutUs.aspx.

77. Sea Turtle Restoration Network. "Our History." Accessed August 7, 2019. https://seaturtles.org/about-us/history/.

78. International Institute for Sustainable Development. "Sustainable Development Timeline - 2012 – IISD." Accessed November 23, 2016. https://www.iisd.org/pdf/2002/sd_timeline2002.pdf.

79. Global Coral Reef Alliance. "About GCRA." Accessed August 21, 2018. http://www.globalcoral.org/about-gcra/.

80. REEF. "About REEF." Accessed May 6, 2018. http://www.reef.org/about.

81. Tusk. "Who We Are." Accessed September 6, 2018. https://www.tusk.org/about-us/who-we-are/.

82. Turner Foundation. "Our Story." Accessed February 13, 2017. http://www.turnerfoundation.org/our-story/.

83. Shark Research Institute. "Mission: Work." Accessed August 7, 2019. https://www.sharks.org/mission-work.

84. Project Piaba. "Background About the Fishery and History." Accessed April 10, 2017. http://projectpiaba.org/who-we-are/history/.

85. EcoLogic Development Fund. "History of EcoLogic." Accessed September 19, 2019. http://www.ecologic.org/about-us/history-of-ecologic/.

86. Project AWARE. "About Us." Accessed April 20, 2018. https://www.projectaware.org/aboutus.

87. Foundation for the Philippine Environment. "Organization: Our Beginnings: Birth." Accessed August 28, 2018. https://fpe.ph/about-fpe.html#our-beginnings.

88. Waitt Foundation. "History & Community." Accessed September 6, 2018. http://waittfoundation.org/history-community/.

89. Yayasan Keanekaragaman Hayati Indonesia. "History of KEHATI Foundation." Accessed March 26, 2017. http://www.kehati.or.id/about-us-2/.

90. Mexican Fund for the Conservation of Nature. "History." Accessed August 28, 2018. https://fmcn.org/history/?lang=en.

91. PCS. "History." Accessed August 26, 2019. https://www.palauconservation.org/about/history-and-mission/.

92. MarineLife Alliance. "Historical Background." Accessed March 12, 2020. https://www.marinelifealliance.org/historical-background/.

93. WBCSD. "History." Accessed November 23, 2016. http://www.wbcsd.org/Overview/About-us.

94. Wildlife Alliance. "Who We Are: History." Accessed September 19, 2019. https://www.wildlifealliance.org/history/.

95. Operation Wallacea. "Dr Tim Coles OBE: Project Director and Founder of Operation Wallacea." Accessed August 21, 2018. https://www.opwall.com/team/dr-tim-coles-obe/.

96. Nature and Culture International. "Celebrating 20 Years of conservation." Accessed November 23, 2016. https://natureandculture.org/20th-anniversary/.

97. FUNBIO. "About Us." Accessed November 28, 2016. http://www.funbio.org.br/en/o-funbio/quem-somos/.

98. Reef Check Foundation. "Who We Are." Accessed May 6, 2018. http://www.reefcheck.org/#services.

99. Inside Philanthropy. "Global Development: Funders: Woodcock Foundation." Accessed September 17, 2018. https://www.insidephilanthropy.com/grants-for-global-development/woodcock-foundation-grants-for-global-development.

100. Doris Duke Charitable Foundation. "Mission & History." Accessed February 13, 2017. http://www.ddcf.org/about-us/mission-and-history/.

101. Marine Conservation Institute. "Our History." Accessed December 2, 2018. https://marine-conservation.org/who-we-are/history/.

102. Grantham Foundation for the Protection of the Environment. "About." Accessed September 17, 2018. http://www.granthamfoundation.org/about.html.

103. Forest Trends. "Financial Information." Accessed November 23, 2016. http://www.forest-trends.org/page.php?id=189&name=Financial%20Information.

104. Conservation Strategy Fund. "About." Accessed November 30, 2016. http://conservation-strategy.org/en/page/about-conservation-strategy-fund.

105. UN Foundation. "Who We Are: Our Board: R. E. Turner (U.S.A.)." Accessed August 28, 2018. http://www.unfoundation.org/who-we-are/board/.

106. Global Nature Fund. "About Us." Accessed August 24, 2017. http://www.globalnature.org/34517/Home/About-us/resindex.aspx.

107. Leonardo DiCaprio Foundation. "History." Accessed September 6, 2018. https://www.leonardodicaprio.org/about/.

108. Root Capital. "About Us." Accessed November 23, 2016. https://www.rootcapital.org/about-us/our-team.

109. Great Barrier Reef Foundation "The Foundation." Accessed August 7, 2019. https://www.barrierreef.org/the-foundation.

110. Mongabay. "About Mongabay." Accessed March 27, 2017. https://www.mongabay.com/about/.
111. Ocean Futures Society. "Our Organization." Accessed April 20, 2018. http://www.oceanfutures.org/about.
112. Algalita Marine Research Foundation. "Home." Accessed August 7, 2019. https://algalita.org/.
113. National Marine Sanctuary Foundation. "About the Foundation." Accessed August 28, 2018. https://marinesanctuary.org/about-the-foundation/.
114. Lighthouse Foundation. "Our Organization." Accessed November 6, 2019. https://lighthouse-foundation.org/en/Lighthouse-Foundation-Foundation-for-the-seas-and-oceans.html.
115. Carbosur. "La Empresa." Accessed November 23, 2016. http://www.carbosur.com.uy/about/.
116. Trucost. "Company History." Accessed November 23, 2016. http://www.trucost.com/about-trucost/company-history/.
117. Foundation Directory Online. "Gordon and Betty Moore Foundation." Accessed February 15, 2017. https://fconline.foundationcenter.org/grantmaker-profile?key=MOOR151.
118. Kemitraan. "Our History." Accessed April 10, 2017. http://www.kemitraan.or.id/our-history.
119. Gili Eco Trust. "About: History." Accessed May 8, 2018. http://giliecotrust.com.cp-45.webhostbox.net/about/.
120. Center for Global Development. "About CGD." Accessed November 23, 2016. http://www.cgdev.org/page/about-cgd.
121. BlueOrchard. "About Us." Accessed November 30, 2016. http://www.blueorchard.com/about-us/blue-orchard/.
122. Oceana. "What We Do." Accessed April 20, 2018. http://oceana.org/what-we-do.
123. SECORE International. "About Us." Accessed April 20, 2018. http://www.secore.org/site/about-us.html.
124. Ocean Foundation. "History." Accessed April 20, 2018. https://www.oceanfdn.org/our-story/history.
125. Save Our Seas Foundation. "Founder: Abdulmohsen Abdulmalik Al-Sheikh." Accessed September 6, 2018. https://saveourseas.com/foundation/founder/.
126. Blue Ventures. "Rebuilding Tropical Fisheries with Coastal Communities: Blue Ventures Today and Tomorrow." Accessed September 6, 2018. https://bjyv3zhj902bwxa8106gk8x5-wpengine.netdna-ssl.com/wp-content/uploads/2018/05/BV-Today-Tomorrow-2018-web.pdf. 4.
127. Climate Focus. "About Us." Accessed November 23, 2016. http://www.climatefocus.com/about-us.
128. Conservation Capital. "Our Team: Giles Davies." Accessed August 7, 2019. https://www.conservation-capital.com/senior-team.

129. Green Fins. "History." Accessed April 20, 2018. http://greenfins.net/en/history.
130. Paso Pacifico. "Mission Statement." Accessed November 23, 2016. http://www.pasopacifico.org/our-mission.html.
131. Madagascar Biodiversity Fund. "About Us." Accessed August 28, 2018. http://www.fapbm.org/en.
132. Consilium Capital Limited. "Credentials." Accessed August 7, 2019. https://www.consiliumcapital.co.uk/credentials-2/.
133. Encourage Capital. "Our History." Accessed January 17, 2017. http://encouragecapital.com/our-firm/our-history/.
134. ClientEarth. "James Thornton." Accessed May 6, 2018. https://www.clientearth.org/people/thornton-james/.
135. Coral Restoration Foundation™. "The Coral Restoration Foundation Is Born…" Accessed April 20, 2018. https://coralrestoration.org/about/about/.
136. Thoumi, Gabriel. Interviewed by Brian McFarland. February 2017.
137. World Federation for Coral Reef Conservation. "About." Accessed February 12, 2020. https://www.wfcrc.org/about.html.
138. Southern Environmental Association. "Welcome to S.E.A. Belize." Accessed August 27, 2019. http://www.seabelize.org/.
139. ClimateWorks Foundation. "History." Accessed February 13, 2017. http://www.climateworks.org/about-us/our-history/.
140. Carbon War Room. "Frequently Asked Questions: When Did the Carbon War Room start?" Accessed February 13, 2017. http://carbonwarroom.com/what-we-do/faq.
141. LinkedIn. "Climate Bonds Initiative: About Us." Accessed April 11, 2017.https://www.linkedin.com/company-beta/1032198/.
142. Facebook. "Corals for Conservation." Accessed August 21, 2018. https://www.facebook.com/pg/C4Conservation/about/.
143. BACoMaB Trust Fund. "BACOMAB." Accessed May 8, 2018. http://bacomab.org/395-2/.
144. Clarmondial. "About Us." Accessed April 10, 2017. https://www.clarmondial.com/about-us/.
145. Blue Marine Foundation. "History." Accessed May 6, 2018. http://www.bluemarinefoundation.com/about/history/.
146. Coral Triangle Center. "Our Story." Accessed December 27, 2019. https://www.coraltrianglecenter.org/our-story/.
147. Ecosystem Marketplace. "Althelia Raises \$80 Million For REDD And Ecosystem Services." Accessed December 13, 2016. http://www.ecosystemmarketplace.com/articles/althelia-raises-80-million-for-br-redd-and-ecosystem-services/.
148. Oceans 5. "About Oceans 5." Accessed August 7, 2019. http://oceans5.org/results/introduction/.
149. Bye Bye Plastic Bags. "Our Team: Founders." Accessed June 1, 2018. http://www.byebyeplasticbags.org/team/.

150. SeaLegacy. "About Us: Founders." Accessed August 23, 2018. https://www.sealegacy.org/about-us.

151. Nature Trust Alliance. "Home." Accessed December 2, 2018. https://www.naturetrustalliance.org/.

152. Blue Action Fund. "About." Accessed December 2, 2018. https://www.blueactionfund.org/about-2/.

153. Jr Ocean Guardians. "About Us." Accessed June 1, 2018. https://www.jroceanguardians.org/about.

154. Pascal, Nicolas of Blue Finance. Email message to author. November 19, 2018.

155. PR Newswire. "4ocean Co-Founders Andrew Cooper and Alex Schulze Named Forbes 30 Under 30 Social Entrepreneurs." November 14, 2018. Accessed November 20, 2019. https://www.prnewswire.com/news-releases/4ocean-co-founders-andrew-cooper-and-alex-schulze-named-forbes-30-under-30-social-entrepreneurs-300750165.html.

156. American Fisheries Society. "About." Accessed April 20, 2018. https://fisheries.org/about/.

157. Wild Oceans. "Our Mission." Accessed August 7, 2019. https://wildoceans.org/about-us-2/our-mission/.

158. National Wildlife Refuge Association. "About the Refuge System and FWS." Accessed December 6, 2016. http://refugeassociation.org/about/about-the-refuge-system/.

159. Coastal Conservation Association. "Our Vision: Our Story." Accessed August 7, 2019. https://www.joincca.org/our-story/.

160. International Society for Reef Studies. "Home." Accessed April 20, 2018. http://coralreefs.org/.

161. Society for Conservation Biology. "What Is SCB." Accessed November 23, 2016. https://conbio.org/about-scb/who-we-are.

162. Marine Fish Conservation Network. "Our History." Accessed August 7, 2019. https://conservefish.org/about-us/our-history/.

163. Coral Reef Alliance. "Our History." Accessed April 20, 2018. https://coral.org/our-history/.

164. Alliance of Religions and Conservation. "About ARC." Accessed November 23, 2016. http://www.arcworld.org/about.asp?pageID=2.

165. Global Coral Reef Monitoring Network. "Background." Accessed February 26, 2020. https://gcrmn.net/about-gcrmn/background/.

166. Tebtebba. "Who We Are." Accessed November 23, 2016. http://tebtebba.org/index.php/content/who-we-are.

167. MSC. "The MSC Is Born." Accessed April 20, 2018. http://20-years.msc.org/.

168. Environmental Markets Association. "About EMA." Accessed November 23, 2016. http://www.emahq.org/about-us/about-ema.

169. Reef Conservation UK. "European Coral Reef Symposium: ECRS 2017 Was Brought to You By." Accessed June 1, 2018. https://www.reefconserva tionuk.co.uk/ecrs-2017.html.
170. National Mitigation Banking Association. "About." Accessed November 23, 2016. http://mitigationbanking.org/index.php/about/.
171. IETA. "Our History." Accessed November 23, 2016. http://www.ieta.org/Our-History.
172. Latin American and Caribbean Network of Environmental Funds. "Home." Accessed November 30, 2016. http://redlac.org/en/.
173. Partnership for Interdisciplinary Studies of Coastal Oceans. "About Us." Accessed April 20, 2018. http://www.piscoweb.org/about-us-0.
174. CDP. "Staff: Paul Dickinson." Accessed November 23, 2016. https://www.cdp.net/en/info/staff.
175. The International Coral Reef Action Network. "Home." Accessed January 14, 2020. http://www.icran.org/icran.html.
176. Locally-Managed Marine Area Network. "About the LMMA: History of the LMMA Network." Accessed September 6, 2018. http://lmmanetwork.org/who-we-are/vision/.
177. Conservation Finance Alliance. "About the CFA." Accessed November 23, 2016. http://www.conservationfinance.org/history.php.
178. Wildlife Conservation Network. "About WCN." Accessed September 6, 2018. https://wildnet.org/about/.
179. Global Footprint Network. "At a Glance." Accessed November 23, 2016. http://www.footprintnetwork.org/pt/index.php/GFN/page/at_a_glance/.
180. Deep Sea Conservation Coalition. "Overview." Accessed September 6, 2018. http://www.savethehighseas.org/about-us/.
181. Aquaculture Stewardship Council. "Our History." Accessed December 31, 2019. https://www.asc-aqua.org/about-us/history/.
182. Braverman. *Coral Whisperer*. 115.
183. Conservation Finance Network. "About Conservation Finance Network." Accessed March 12, 2020. http://www.conservationfinancenetwork.org/about-cfn.
184. Sustainable Fisheries Partnership. "About Us." Accessed November 6, 2019. https://www.sustainablefish.org/About-Us/About-Us.
185. Consortium for the Conservation of Coastal and Marine Ecosystems in the Western Indian Ocean. "About Us." Accessed August 7, 2019. https://wioc.org/about-us/.
186. Alliance for Water Stewardship. "History of The Alliance for Water Stewardship." Accessed June 27, 2018. http://a4ws.org/about/organisation/#history.
187. World Ocean Council. "About Us: The Organization." Accessed December 4, 2019. https://www.oceancouncil.org/about-us/.
188. ICROA. "About ICROA." Accessed November 23, 2016. http://www.icroa.wildapricot.org/About-ICROA.

189. MarketsWiki. "Carbon Markets and Investors Association." Accessed November 23, 2016. http://www.marketswiki.com/wiki/Carbon_Markets_and_Investors_Association.
190. MarketsWiki. "Climate Markets & Investment Association." Accessed November 23, 2016. http://www.marketswiki.com/wiki/Climate_Markets_%26_Investment_Association.
191. World Federation for Coral Reef Conservation. "About WFCRC." Accessed April 20, 2018. http://www.wfcrc.org/about-us/.
192. Global Impact Investing Network. "The GIIN History." Accessed December 6, 2016. https://thegiin.org/giin/history.
193. Consumer Goods Forum. "Our History." Accessed November 23, 2016. http://www.theconsumergoodsforum.com/about-the-forum/our-history.
194. Big Ocean. "Home." Accessed September 19, 2019. https://bigoceanmanagers.org/.
195. Wallington, Lucy of International Partnership for Blue Carbon. Email message to author. October 25, 2018.
196. USDA. "Conservation Finance Practitioner Roundtable." Accessed November 23, 2016. http://www.nrcs.usda.gov/wps/portal/nrcs/detail/national/technical/emkts/?cid=nrcseprd660806.
197. IUCN. "New Coalition Launches to Scale Private Conservation Investment at IUCN World Conservation Congress." Accessed December 12, 2016. https://www.iucn.org/newsa/new-coalition-launches-scale-private-conservation-investment-iucn-world-conservation-congress.
198. Moore, Tom. Interviewed by Brian McFarland. December 2019.

Contents

About the Author

Brian Joseph McFarland is currently the Senior Vice President of Carbonfund.org Foundation's Project Portfolio and the Senior Vice President of Project Origination at CarbonCo, the wholly owned subsidiary of Carbonfund.org.

At Carbonfund.org, he identifies climate change mitigation projects in the energy efficiency, renewable energy, and forestry sectors, conducts due diligence on such projects, and then structures the financial support and manages the project portfolio. This multimillion dollar project portfolio includes 215+ tree planting and carbon reduction projects across 40+ US states and 25+ countries.

At CarbonCo, Brian designed, financed, and now advises on the implementation of four Reducing Emissions from Deforestation and Degradation (REDD+) projects conserving ~300,000 hectares of tropical rainforests, while reducing 16 million tonnes of carbon dioxide emissions and benefitting 2000+ local community members. This includes spearheading the first-ever, dual Verified Carbon Standard and Climate, Community and Biodiversity Standards validated and verified REDD+ project in the State of Acre, Brazil.

Brian is a certified Project Management Professional (PMP) by the Project Management Institute (PMI), former certified Greenhouse Gas Inventory Quantifier (GHG-IQ) from CSA Standards, and a Certified Sustainability Professional from the International Society of Sustainability Professionals (ISSP).

Brian McFarland earned a dual graduate degree in Business Administration and Global Environmental Policy from American University. Brian's graduate thesis was entitled, *Origins, Development and Potential of the International REDD+ Market*. Brian has also published or co-published 20+ articles, along with a co-authored book chapter in *Sustainable Investing: Revolutions in Theory and Practice*. Brian authored two previous books entitled, *REDD+ and Business Sustainability: A Guide to Reversing Deforestation for Forward Thinking Companies* and *Conservation of Tropical Rainforests: A Review of Financial and Strategic Solutions*.

While finishing his Psychology and International Development undergraduate degree from Clark University, Brian conducted authentic environmental fieldwork in Mexico, Costa Rica, Kenya, and Brazil. Such fieldwork included addressing human-wildlife conflicts, working on sustainable community development projects and biodiversity monitoring. During graduate school, Brian also volunteered for projects with the Smithsonian Institution, the United Nations Global Compact, and the US Department of State. Today, Brian volunteers for a range of organizations including Code REDD, along with on his town's conservation commission and forestry committee.

The opinions expressed in this book are those of the author and do not necessarily represent the views of Carbonfund.org or CarbonCo.

Acronyms

ABNJ	Areas Beyond National Jurisdiction
ACES	A Community of Ecosystem Services
ACRS	Australia Coral Reef Society
ADB	Asian Development Bank
AfDB	African Development Bank
AIMS	Australia Institute of Marine Science
AOSIS	Alliance of Small Island States
APEC	Asia-Pacific Economic Cooperation
ARC	Alliance of Religions and Conservation
ASC	Aquaculture Stewardship Council
AUM	Assets Under Management
AUV	Autonomous Underwater Vehicle
BBOP	Business and Biodiversity Offsets Program
BCN	Biodiversity Conservation Network (USAID)
BCSD	Business Council for Sustainable Development
BIOFIN	Biodiversity Finance Initiative (UNDP)
BNMP	Bonaire National Marine Park
CBD	Convention on Biological Diversity (UN)
CCAD	Comisión Centroamericana de Ambiente y Desarrollo (Central American Commission for Environment and Development)
CCOM/JHC	Center for Coastal & Ocean Mapping/Joint Hydrographic Center
CDP	Carbon Disclosure Project (now known as CDP)
CEF	Closed-End Fund or Closed-Ended Fund
CFA	Conservation Finance Alliance
CGD	Center for Global Development

CHICOP	Chumbe Island Coral Park Ltd
CI	Conservation International
CITES	Convention on International Trade in Endangered Species of Wild Fauna and Flora
CLDP	Conservation and Limited Development Project
CMIA	Climate Markets and Investment Association
CO_2e	Carbon Dioxide Equivalent Emissions
CONABIO	Comisión Nacional para el Conocimiento y Uso de la Biodiversidad (Biodiversity National Council of Mexico)
CONANP	Comisión Nacional de Áreas Naturales Protegidas (National Commission of Protected Natural Areas)
CONAP	Consejo Nacional de Areas Protegidas (Guatemala, National Council for Protected Areas)
COP	Conference of the Parties
CORSIA	Carbon Offsetting and Reduction Scheme for International Aviation
CPIC	Coalition for Private Investment in Conservation
CPUE	Catch per Unit Effort
CRF™	Coral Restoration Foundation™
CSD	Commission on Sustainable Development (UN)
CSR	Corporate Social Responsibility
CTF	Conservation Trust Fund
DFI	Development Finance Institution
DFNS	Debt-for-Nature Swap
DIB	Development Impact Bond
DOD	Disbursed and Outstanding Debt
DOPA	Digital Observatory for Protected Areas
DWFN	Distant Water Fishing Nation
EAI	Enterprise for the Americas Initiative
EBRD	European Bank for Reconstruction and Development
EDF	Environmental Defense Fund
EEZ	Exclusive Economic Zone
EFJ	Environmental Foundation of Jamaica
EIB	Environmental Impact Bond
EIB	European Investment Bank
ENSO	El Niño-Southern Oscillation
EROS	Earth Resources Observation and Science
ESG	Environmental, Social and Governance
ETF	Exchange Traded Fund
EU	European Union
EU ETS	European Union Emissions Trading System
FAO	Food and Agriculture Organization of the United Nations
FASB	Financial Accounting Standards Board
FAST	Finance Alliance for Sustainable Trade

FCG	Fundación para la Conservación de los Recursos Naturales y de Ambiente en Guatemala (Foundation for the Conservation of Natural and Environmental Resources in Guatemala)
FFEM	Fonds Français pour l'Environnement Mondial (The French Facility for Global Environment)
FFI	Fauna and Flora International
FMAP	The Future of Marine Animals Populations Study
FMCN	Fondo Mexicano para la Conservación de la Naturaleza (The Mexican Fund for the Conservation of Nature)
FPE	Foundation for the Philippine Environment
FPIC	Free, Prior and Informed Consent
FRM	Financial Risk Manager
FSC	Forest Stewardship Council
FUNBIO	Fundo Brasileiro para a Biodiversidade (The Brazilian Biodiversity Fund)
GARP	Global Association of Risk Professionals
GATT	General Agreement on Tariffs and Trade
GBIF	Global Biodiversity Information Facility
GBRC	Great Barrier Reef Committee
GBRCA	Great Barrier Reef Catchment Area
GBRMP	Great Barrier Reef Marine Park
GBRWHA	Great Barrier Reef World Heritage Area
GCRA	Global Coral Reef Alliance
GCRMN	Global Coral Reef Monitoring Network
GDP	Gross Domestic Product
GEF	Global Environment Facility
GGGI	Global Ghost Gear Initiative
GHG	Greenhouse Gas
GIIN	Global Impact Investing Network
GMO	Genetically Modified Organism
GRID	Global Information Resource Database (UNEP)
GSTC	Global Sustainable Tourism Council
Gt	Gigaton
$GtCO_2eq$	Gigaton of Carbon Dioxide Equivalent
HDI	Human Development Index
HIPC	Highly Indebted Poor Country
IADB	Inter-American Development Bank
IBAT	Integrated Biodiversity Assessment Tool
IBRD	International Bank for Reconstruction and Development
ICAO	International Civil Aviation Organization
ICRAN	International Coral Reef Action Network
ICRI	International Coral Reef Initiative
IETA	International Emissions Trading Association
IFC	International Finance Corporation

IFI	International Financial Institution
IFQs	Individual Fishing Quotas
IFRS	International Financial Reporting Standards
IISD	International Institute for Sustainable Development
ILO	International Labor Organization
IMF	International Monetary Fund
IMO	International Maritime Organization
IP	Intellectual Property
IRR	Internal Rate of Return
IRS	Internal Revenue Service (U.S.)
ISO	International Organization for Standardization
ISRS	International Society for Reef Studies
ITMOs	Internationally Transferred Mitigation Outcomes
ITQs	Individual Transferable Quotas
IUCN	International Union for the Conservation of Nature
IUU	Illegal, Unregulated, and Unreported (Fishing)
IYOR	International Year of the Reef
KBAs	Key Biodiversity Areas
LIBOR	London Interbank Offered Rate
LLC	Limited Liability Company
LLP	Limited Liability Partnership
LMMA	Locally-Managed Marine Area
LMPA	Large Marine Protected Area
LSE	London Stock Exchange
LWCF	Land and Water Conservation Fund
MAR	Mesoamerican Reef Fund
MDG	Millennium Development Goal
MDRI	Multilateral Debt Relief Initiative
MEA	Multilateral Environmental Agreement
MOU	Memorandum of Understanding
MPA	Marine Protected Area
MRV	Monitoring (or Measuring), Reporting and Verification
MSC	Marine Stewardship Council
MSP	Marine Spatial Planning
MSY	Maximum Sustainable Yield
MT	Megaton
Mt	Metric Tonne or Metric Ton
mtCO$_2$e	Metric Ton(s) of Carbon Dioxide Equivalent Emissions
NASA	National Aeronautics and Space Administration
NBSAPs	National Biodiversity Strategies and Action Plans
NDC	Nationally Determined Contributions
NGO	Nongovernmental Organization
NMBA	National Mitigation Banking Association
NOAA	National Oceanic and Atmospheric Administration

NPV	Net Present Value
NTZ	No-Take Zone
OBIS	Ocean Biogeographic Information System
OCTO	Open Communications for the Ocean
ODA	Official Development Assistance
OECD	Organization for Economic Co-operation and Development
OPIC	Overseas Private Investment Corporation
OSB	Ocean Studies Board (U.S.)
PACT	Protected Areas Conservation Trust (Belize)
PADI	Professional Association of Diving Instructors
PBF	Performance-Based Financing
PCS	Palau Conservation Society
PES	Payments for Ecosystem Services
PICRC	Palau International Coral Reef Center
PISCO	Partnership for Interdisciplinary Studies of Coastal Oceans
PMCES	Payments for Marine and Coastal Ecosystem Services
PNMS	Palau National Marine Sanctuary
PPP	Public-Private Partnership
PRI	Program-Related Investment
RBF	Results-Based Financing
REA	Rapid Ecological Assessment
RedLAC	Latin American and Caribbean Network of Environmental Funds
ROV	Remotely Operated Vehicle
RRTTLLU	Risk, Return, Time (Horizon), Taxes, Liquidity, Legal and Unique
SDGs	Sustainable Development Goals (UN)
SeyCCAT	Seychelles Conservation and Climate Adaptation Trust
SFP	Sustainable Fisheries Partnership
SIB	Social Impact Bond
SICA	Sistema de la Integración Centroamericana (Central American Integration System)
SIDS	Small Island Developing States
SIMAB	Smithsonian Institution's Monitoring & Assessment of Biodiversity Program
SINAC	Sistema Nacional de Áreas de Conservación (Costa Rica's National System of Conservation Areas)
SLOSS	Single Large or Several Small
SMART	Spatial Monitoring and Reporting Tool
SPC	Secretariat for the Pacific Community
SPOTT	Sustainability Policy Transparency Toolkit
SPREP	Secretariat for the Pacific Regional Environmental Programme
SPV	Special Purpose Vehicle
SSI	Scuba Schools International
SST	Sea Surface Temperature
STC	Sea Turtle Conservancy

TAC	Total Allowable Catches
TAE	Total Allowable Effort
TAF	Technical Assistance Facility
TBPA	Transboundary Protected Area
TEEB	The Economics of Ecosystems and Biodiversity
TFCA	Transfrontier Conservation Area
TIDE	Toledo Institute for Development and Environment
TIES	The International Ecotourism Society
TNC	The Nature Conservancy
TPL	Trust for Public Land
TRASE	Transparency for Sustainable Economies
(U)HNWI	(Ultra-) High-Net-Worth Individual
U.S.	United States of America
UNCED	United Nations Conference on Environment and Development
UNCLOS	United Nations Convention on the Law of the Sea
UNDP	United Nations Development Programme
UNDRIP	United Nations Declaration on the Rights of Indigenous Peoples
UNEP	United Nations Environment Programme
UNESCO	United Nations Educational, Scientific and Cultural Organization
UNFCCC	United Nations Framework Convention on Climate Change
UNWTO	United Nations World Tourism Organization
USAID	United States Agency for International Development
USCRTF	The United States Coral Reef Task Force
USD	United States Dollar
USEPA	United States Environmental Protection Agency
USFWS	United States Fish and Wildlife Service
VCS	Verified Carbon Standard
VERs	Verified Emission Reductions
VPA	Voluntary Partnership Agreements
WBCSD	World Business Council for Sustainable Development
WCED	World Commission on Environment and Development
WCS	Wildlife Conservation Society
WFCRC	World Federation for Coral Reef Conservation
WHIP	Wildlife Habitat Incentives Program
WHOI	Woods Hole Oceanographic Institution
WOC	World Ocean Council
WOCAT	World Overview of Conservation Approaches and Technologies
WoRMS	World Register of Marine Species
WRI	World Resources Institute
WTO	World Trade Organization
WWF	World Wildlife Fund (also known as World Wide Fund for Nature)

List of Figures

List of Tables

1

Executive Summary

Tropical coral reefs are the most biologically diverse marine ecosystems on Earth and provide a tremendous amount of ecosystem services. This includes providing food to millions of people and generating billions of dollars in revenue from ecotourism and fisheries.

> The annual 'gross marine product' (GMP) – equivalent to a country's annual gross domestic product – is at least US$2.5 trillion; the total 'asset' base of the ocean is at least US$24 trillion. Underpinning this value are direct outputs (fishing, aquaculture), services enabled (tourism, education), trade and transportation (coastal and oceanic shipping) and adjacent benefits (carbon sequestration, biotechnology). Putting it into an international context, if the ocean were a country it would have the seventh largest economy in the world.[1]

However, there is a funding gap when it comes to conserving the world's tropical coral reefs. For instance, as noted by the International Coral Reef Initiative (ICRI):

> One of the main barriers to conservation enhancement is that the funds currently needed to achieve effective, lasting conservation significantly exceed the available funds. An average US$270 million commitment per year for marine conservation was recorded between 2010 and 2016. Meanwhile, the funds necessary to achieve the United Nations Convention on Biological Diversity target of having 20% of the ocean protected are estimated at between US$4 and $8 billion per year in management costs only.[2]

© The Author(s) 2021
B. J. McFarland, *Conservation of Tropical Coral Reefs*,
https://doi.org/10.1007/978-3-030-57012-5_1

In addition to a shortage of funding, there is "significant variation in the amount of funding provided for coral reef management in different regions, including relatively low investment per unit area of reef in some regions and countries where reefs are extensive, highly biodiverse or hold a large proportion of climate refugia."[3]

Yet, there are solutions and progress is being made. Dr. Sylvia Earle, one of the foremost oceanographers, suggests we should:

1. Treat the 60% of the world ocean outside of national Exclusive Economic Zones (EEZs) as a World Ocean Public Trust, with legal approaches concerning the use of the high seas, including fisheries, under coordinated, international, multiuse zoning regimes.
2. Reform fisheries using market-based mechanisms, subsidy changes, and sustainable practices.
3. Implement global and regional communications plans to educate the public.
4. Create, consolidate, and strengthen marine protected areas into a globally representative network.
5. Develop an expanded research program focused on top priority marine environments high in endemism and biodiversity.[4]

Furthermore, Dr. Earle's recommendations include "developing EEZs for long-range sustainability; removing perverse subsidies (i.e., estimated at nearly $34 billion in the fishing sector); restoration of species and systems; along with investments in aquaculture, cultivation of some species lower on the food chain, and the practice of closed-system cultivation."[5]

Similarly, Dr. Hoegh-Guldberg et al. suggest that the following eight actions are required:

1. Ensure ocean recovery features strongly in the UN Post-2015 Agenda, including the Sustainable Development Goals;
2. Take global action to avoid dangerous climate change and further damage to the ocean;
3. Conserve and effectively manage 10% of representative coastal and marine areas by 2020, increasing coverage to 30% by 2030;
4. Rebuild fish stocks to ecologically sustainable harvest levels;
5. Drive new global cooperation and investment for the ocean;
6. Reinvent public/private partnerships;
7. Build transparent accounting of the value of ocean assets to improve decision-making; and

8. Share knowledge more effectively and drive institutional collaboration.[6]

Specifically related to financing, the focus of this book, the ICRI-commissioned report, *Innovations for Coral Finance*, suggests the key activities that need financing are the:

- Creation and development of marine protected areas (MPAs);
- Suppression of external pressure factors (marine pollution and land threats);
- Sustainable fisheries;
- Sustainable tourism; and
- Restoration of coral ecosystems.[7]

The ICRI report also suggests we need to: favor new financing approaches for coral conservation in the future; adapt business models to conservation to make projects more attractive to investors; and strengthen the development of conservation trust funds to better address global issues.[8]

Furthermore, "marine 'no-take' reserves, areas closed to fishing and all other extractive activities, are among the most essential tools required to protect and restore the health of our oceans from multiple stressors."[9] Likewise, "a recent meta analysis of 87 MPAs found that conservation benefits increased exponentially with the accumulation of five attributes: no take, well enforced, old (>10 years), large (<100 km^2) and isolated by deep water or sand."[10]

There are many examples of effective financing and implementation of coral reef conservation and this book explores the diversity of these tools.

Notes

1. Hoegh-Guldberg, O. et al. *Reviving the Ocean Economy*. 7.
2. Vertigo Lab. "Innovations for Coral Finance, ICRI Publication." 2017. http://vertigolab.eu/wp-content/uploads/2018/04/rapport-innovation-for-coral-finance.pdf. 1.
3. UN Environment et al. "Analysis of International Funding for the Sustainable Management of Coral Reefs and Associated Coastal Ecosystems." 11.
4. Earle. *The World is Blue*. 217–218.
5. Ibid. 237–238.
6. Hoegh-Guldberg, O. et al. *Reviving the Ocean Economy*. 32.

7. Vertigo Lab. "Innovations for Coral Finance, ICRI Publication." 2017. http://vertigolab.eu/wp-content/uploads/2018/04/rapport-innovation-for-coral-finance.pdf. 14.
8. Ibid. 73.
9. Sobel and Dahlgren. *Marine Reserves.* 19.
10. Rotjan et al. "Establishment, Management, and Maintenance of the Phoenix Islands Protected Area." 298.

2

The Context of Coral Reef Degradation and Loss

Where Are the World's Coral Reefs?

Planet Earth is a blue planet. The "seven seas"—being the North and South Atlantic Ocean, the North and South Pacific Ocean, along with the Arctic, Indian and Southern Ocean—account for "97 percent of Earth's water, and also make up more than 95 percent of the biosphere {…}."[1]

> To us, the oceans are immense. They cover 140 million square miles and fill a volume of 324 million cubic miles. It is hard to imagine a cubic mile. If you flooded all of New York's Central Park with water to the height of a thirty-story building, the volume would be a little greater than one eighth of a cubic mile; it would take 2.66 billion Central Parks of water to fill the oceans, an almost unthinkable volume of water. Yet at the scale of the planet, the oceans form a layer only as thick as the skin of an apple.[2]

Approximately 42% of the world's oceans are within a nation's Exclusive Economic Zone (EEZ), while approximately 58% of the world's oceans are outside of these EEZs and considered to be the high seas.[3] This said:

> Under the Law of the Sea, nations may claim jurisdiction over an Exclusive Economic Zone of the ocean, a region extending seaward 200 nautical miles from the edge of their territorial sea, 12 nautical miles from shore. Foreign nations have the freedom of navigation and overflight, subject to regulation by the coastal states, but in general, the states have sole exploitation rights over the natural resources within the region.[4]

© The Author(s) 2021
B. J. McFarland, *Conservation of Tropical Coral Reefs*,
https://doi.org/10.1007/978-3-030-57012-5_2

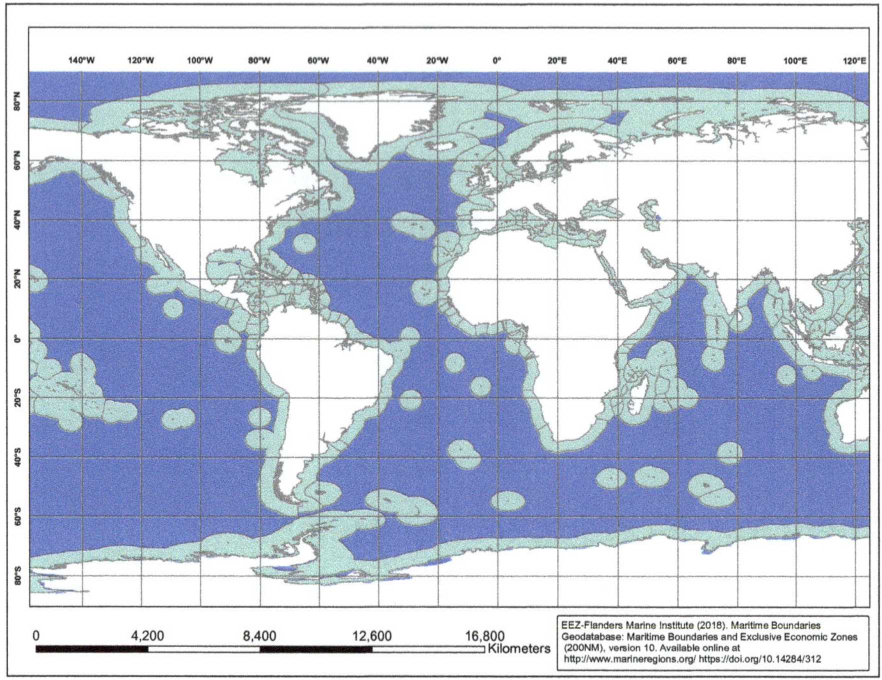

Fig. 2.1 Map of EEZ boundaries (*Credit* James Eaton)

As shown Fig. 2.1, most of the world's oceans are high seas.

The following Table 2.1 compares the largest countries on Earth as determined by their land area,[5] and the largest countries as determined by their EEZs.[6]

Specific to tropical coral reefs:

> The living coral coasts of the world are confined to waters in which the temperature seldom falls below 70°F. (and never for prolonged periods), for the massive structures of the reefs can be built only where the coral animals are bathed by waters warm enough to favor the secretion of their calcareous skeletons. Reefs and all the associated structures of a coral coast are therefore restricted to the area bound by the Tropics of Cancer and Capricorn. Moreover, they occur only on the eastern shores of continents, where currents of tropical water are carried toward the poles in a pattern determined by the earth's rotation and the direction of the winds. Western shores are inhospitable to corals because they are the site of upwellings of deep, cold water, with cold coastwise currents running toward the equator.[7]

This said, "some of the world's most northerly reefs are found in the seas off Kuwait, where temperatures swing wildly,"[8] and Bermuda, due to the Gulf Stream, also has some of the most northerly reefs. Furthermore, "all tropical coral reefs inhabit waters that are less than three hundred feet deep {…}."[9]

Table 2.1 Countries with the largest landmass and with the largest EEZs

	Country	Land area (km^2)		Country	EEZs (km^2)
1	Russia	17,098,242	1	France	11,691,000
2	Canada	9,984,670	2	United States	11,351,000
3	United States	9,826,675	3	Australia	8,505,348
4	China	9,596,960	4	Russia	7,566,673
5	Brazil	8,514,877	5	United Kingdom	6,805,586
6	Australia	7,741,220	6	Indonesia	6,159,032
7	India	3,287,263	7	Canada	5,599,077
8	Argentina	2,780,400	8	Japan	4,479,388
9	Kazakhstan	2,724,900	9	New Zealand	4,083,744
10	Algeria	2,381,741	10	Chile	3,681,989
11	Democratic Republic of the Congo	2,344,858	11	Brazil	3,660,955
12	Greenland (Denmark)	2,166,086	12	Kiribati	3,441,810
13	Saudi Arabia	2,149,690	13	Mexico	3,269,386
14	Mexico	1,964,375	14	Micronesia	2,996,419
15	Indonesia	1,904,569	15	Denmark	2,551,238
16	Sudan	1,861,484	16	Papua New Guinea	2,402,288
17	Libya	1,759,540	17	Norway	2,385,178
18	Iran	1,648,195	18	India	2,305,143
19	Mongolia	1,564,116	19	Marshall Islands	1,990,530
20	Peru	1,285,216	20	Portugal	1,727,408

The following map shows where warm water tropical coral reefs exist (Fig. 2.2).

According to the World Wildlife Fund (WWF), "coral reefs only occupy 0.1% of the area of the ocean but they support 25% of all marine species on the planet."[10] The largest tropical coral reefs are:

- The Great Barrier Reef of Australia;
- The Coral Triangle of Indonesia, Malaysia, Papua New Guinea, the Philippines, the Solomon Islands and Timor Leste;
- The Mesoamerican Reef of Mexico, Belize, Guatemala, and Honduras;
- East Africa, including Madagascar;
- The Indian Ocean;
- The Red Sea; and
- The Caribbean (Figs. 2.3, 2.4, 2.5, 2.6, and 2.7).

Approximately 80 countries on Earth host tropical coral reefs. The Table 2.2 describes the global distribution of coral reefs across these 80 countries, according to the 2001 seminal report, *World Atlas of Coral Reefs*.[11]

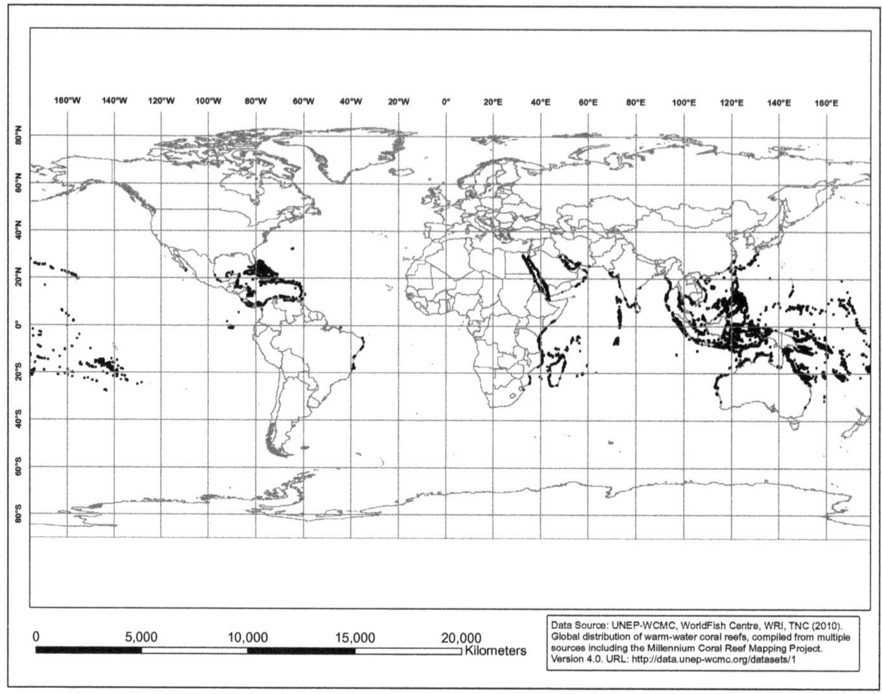

Fig. 2.2 Global distribution of warm water coral reefs (*Credit* James Eaton)

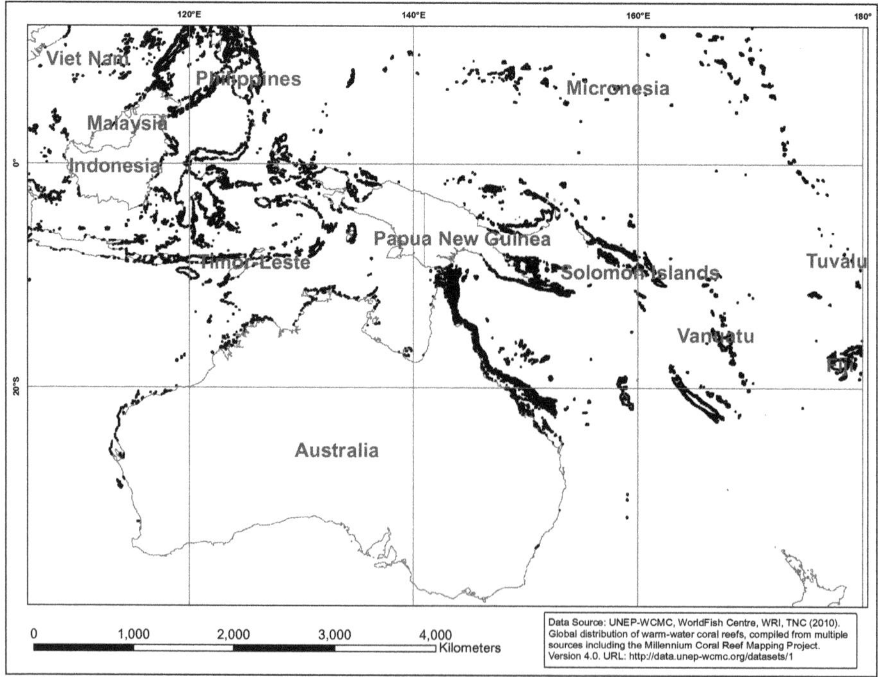

Fig. 2.3 Warm water coral reefs in Australia and South Pacific (*Credit* James Eaton)

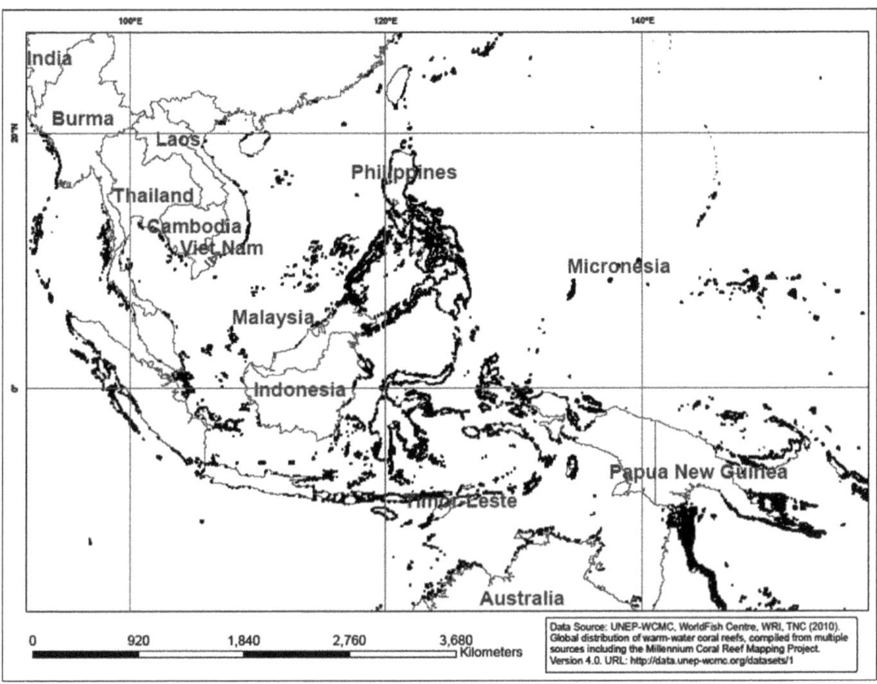

Fig. 2.4 Warm water coral reefs in Southeast Asia (*Credit* James Eaton)

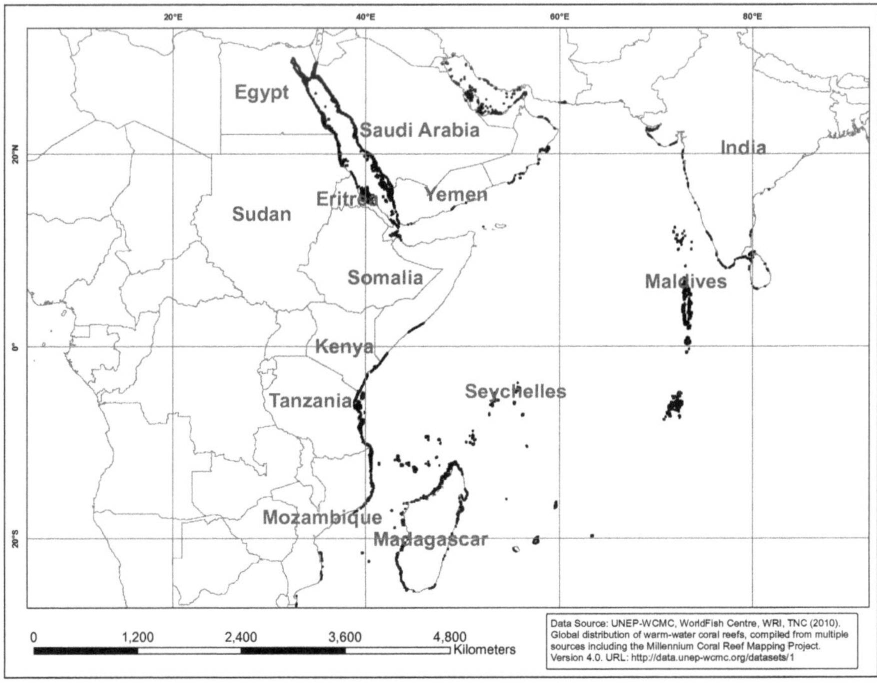

Fig. 2.5 Warm water coral reefs in East Africa and the Indian Ocean (*Credit* James Eaton)

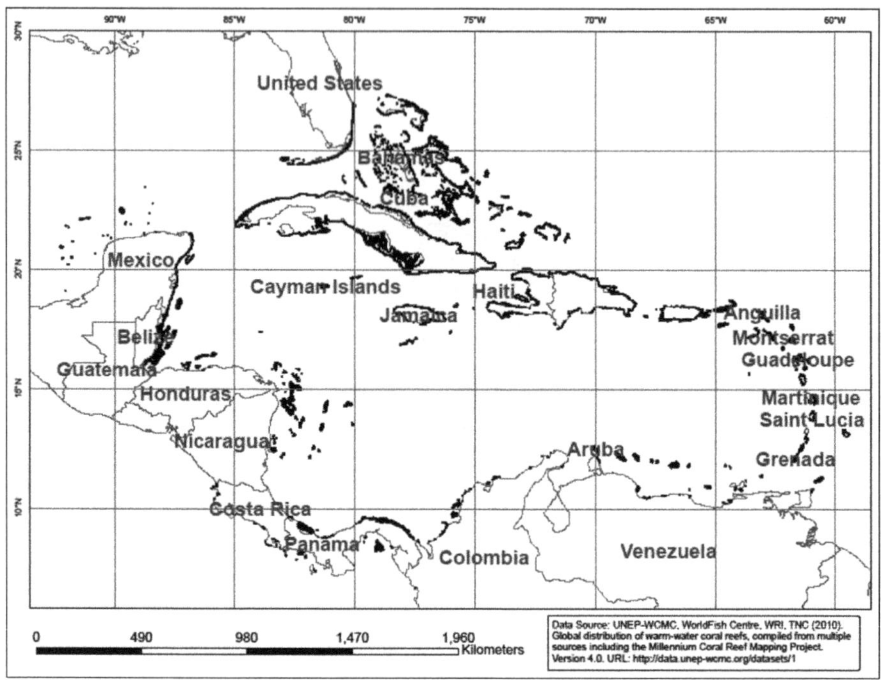

Fig. 2.6 Warm water coral reefs in and around the Caribbean Sea (*Credit* James Eaton)

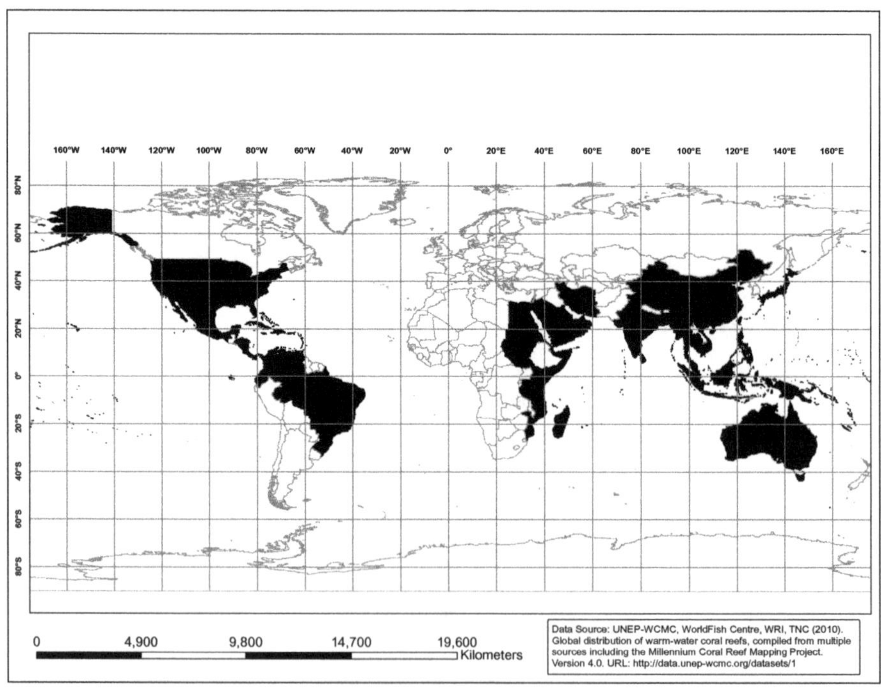

Fig. 2.7 Countries with warm water coral reefs (*Credit* James Eaton)

Table 2.2 Countries with tropical coral reefs

	Country	Total area of coral reefs (km^2)	Percentage of global total (%)		Country	Total area of coral reefs (km^2)	Percentage of global total (%)
1	Indonesia	51,020	17.95	41	Costa Rica	970	0.34
2	Australia	48,960	17.22	42	Colombia	940	0.33
3	Philippines	25,060	8.81	43	Taiwan	940	0.33
4	France (i.e., Clipperton, Mayotte, Réunion, Guadeloupe, Martinique, New Caledonia, French Polynesia, Wallis, and Futuna Islands)	14,280	5.02	44	Mauritius	870	0.31
5	Papua New Guinea	13,840	4.87	45	Honduras	810	0.28
6	Fiji	10,020	3.52	46	Panama	720	0.25
7	Maldives	8920	3.14	47	Nicaragua	710	0.25
8	Saudi Arabia	6660	2.34	48	Somalia	710	0.25
9	Marshall Islands	6110	2.15	49	Tuvalu	710	0.25
10	India	5790	2.04	50	Iran	700	0.25
11	Solomon Islands	5750	2.02	51	Qatar	700	0.25

(continued)

Table 2.2 (continued)

	Country	Total area of coral reefs (km^2)	Percentage of global total (%)		Country	Total area of coral reefs (km^2)	Percentage of global total (%)
12	UK (i.e., British Indian Ocean Territory, Anguilla, Bermuda, Cayman Islands, Pitcairn, Turks and Caicos Islands, and the British Virgin Islands)	5510	1.94	52	Yemen	700	0.25
13	Micronesia, Federated States of	4340	1.53	53	Sri Lanka	680	0.24
14	Vanuatu	4110	1.45	54	Kenya	630	0.22
15	Egypt	3800	1.34	55	Dominican Republic	610	0.21

	Country	Total area of coral reefs (km^2)	Percentage of global total (%)		Country	Total area of coral reefs (km^2)	Percentage of global total (%)
16	United States (i.e., Florida and Gulf of Mexico, Hawaii, United States Minor Outlying Islands, American Samoa, Puerto Rico, US Virgin Islands, and Guam)	3770	1.33	56	Bahrain	570	0.20
17	Malaysia	3600	1.27	57	Oman	530	0.19
18	Tanzania	3580	1.26	58	Western Samoa	490	0.17
19	Eritrea	3260	1.15	59	Venezuela	480	0.17
20	Bahamas	3150	1.11	60	Netherlands (i.e., Aruba and Netherlands Antilles)	470	0.17
21	Cuba	3020	1.06	61	Djibouti	450	0.16
22	Kiribati	2940	1.03	62	Haiti	450	0.16
23	Japan	2900	1.02	63	Comoros	430	0.15
24	Sudan	2720	0.96	64	Antigua and Barbuda	240	0.08
25	Madagascar	2230	0.78	65	Brunei Darussalam	210	0.07
26	Thailand	2130	0.75	66	Saint Kitts and Nevis	180	0.06

(continued)

Table 2.2 (continued)

	Country	Total area of coral reefs (km^2)	Percentage of global total (%)		Country	Total area of coral reefs (km^2)	Percentage of global total (%)
27	Myanmar	1870	0.66	67	Saint Lucia	160	0.06
28	Mozambique	1860	0.65	68	Grenada	150	0.05
29	Mexico	1780	0.63	69	Saint Vincent and the Grenadines	140	0.05
30	Seychelles	1690	0.59	70	Kuwait	110	0.04
31	China	1510	0.53	71	Barbados	<100	
32	Tonga	1500	0.53	72	Dominica	<100	
33	Belize	1330	0.47	73	Singapore	<100	
34	New Zealand (i.e., Cook Islands, Niue, and Tokelau)	1310	0.46	74	Trinidad and Tobago	<100	
35	Vietnam	1270	0.45	75	Bangladesh	<50	
36	Jamaica	1240	0.44	76	Cambodia	<50	
37	Brazil	1200	0.42	77	Ecuador	<50	
38	United Arab Emirates	1190	0.42	78	Jordan	<50	
39	Palau	1150	0.40	79	Nauru	<50	
40	Spratly Islands	1150	0.40	80	Israel	<10	

There exist "around 5,000 MPAs {marine protected areas} around the world, with 10 alone accounting for 74% of the total area of MPAs."[12] As of 2018:

> MPAs cover 7% of the ocean, but the majority of these are 'paper parks' – meaning that the rules for the reserve are not being enforced. The Malta Declaration calculates that because many MPAs do not offer full protection, a more realistic estimate would be that only 2% of the global ocean is strongly protected.[13]

As of 2011, there were "an estimated 2,679 marine protected areas (MPAs) in coral reef areas worldwide, encompassing approximately 27 percent of the world's coral reefs."[14] Further, coral reefs, particularly reefs within MPAs, are important spawning grounds for fish. In fact, there are approximately 4800 fisheries worldwide.

Yet, besides MPAs, other solutions exist to conserving coral reefs. For instance, "'MPAs can't finance water treatment or sewage plants for example.' he {Nicolas Pascal, founder of Blue Finance} says 'And they can't prevent the use of plastics, nor encourage recycling efforts and clean up. We can't also counteract decisions to sell coastal land for development.'"[15]

Thus, a variety of conservation financing mechanisms and integrated management plans are required to conserve coral reefs. Further, it is imperative to prioritize conservation investment decisions. In the seminal paper, "Risk-sensitive planning for conserving coral reefs under rapid climate change," the researchers:

> {...} conducted a global scale analysis to identify a portfolio of regions in which long-term coral reef conservation investment might be least subject to impacts from climate change (e.g., thermal stress and coral bleaching), while including reefs with a capacity, to repopulate other reefs over time, and that are not likely to be frequently devastated by cyclones {...}.[16]
>
> Regions that are projected to experience relatively lower levels of threat impact, as measured by their suitability metric, include portions of central and western portions of southeast Asia (the Philippines, Borneo, Indonesia), Australia's Great Barrier Reef, French Polynesia, East Africa, the Red Sea, and the Caribbean. {...} Several regions containing reefs of high ecological and social value that are projected to suffer higher levels of impacts include, for example, Hawaii, Meso-American Reef, and Western Australia. {...} Our study identified a global portfolio of 50 BCUs {bioclimatic units} for conservation investment that maximizes the chance these reefs are secure in the future. {...}.[17]

This balanced solution includes reefs in 31 of the 87 countries that have more than 500 km^2 of tropical coral reefs, with multiple BCUs in countries such as Australia, Cuba, French Polynesia, Philippines, Bahamas, and Malaysia. This solution also includes 36 of the 150 Corals of the World ecoregions, between them containing 95% of the documented species of corals and representatives in 31 of the 232 Marine Ecoregions of the World, which contain approximately 68% of all coral reefs.[18]

These BCUs, based on the balanced solution, representing the optimal portfolio and the maximum return solution are as follows[19]: (Table 2.3).

Applying the World Resources Institute's (WRI) Reefs at Risk Revisited dataset[20] to these BCUs,[21] risks can be mapped. Using this tool, the Coral Reef Rescue Initiative, led in part by WWF, is focusing on seven key countries with "regeneration reefs," which are essentially lower exposed reefs:

1. Cuba;
2. Madagascar;
3. Tanzania;
4. Indonesia;
5. Solomon Islands;
6. Fiji; and
7. The Philippines.[22]

Thus, not all coral reefs are declining at the same rate, as some areas have slower rates of heating and/or fewer storms.[23]

In another seminal paper, "Global Marine Biodiversity Conservation Priorities," researchers found that:

Peaks in marine species richness and endemism generally occur in the tropics, although temperate areas were also identified, particularly for proportional range rarity. We found high species richness values (range of top 5% = 1618 − 5099 species) in the EEZ waters of several Southeast Asian nations in the region known as the Coral Triangle. We also found higher richness in less well-known places in the Indian Ocean along the coasts of Madagascar and the Chagos, Maldives, and Lakshadweep archipelagos. In contrast, higher range rarity values within EEZs (range of 5% = 301 − 680) were most prevalent in the Coral Triangle, the Bahamas, and along the Pacific Central American coast. High proportional range rarity values (range of top 5% = 376 − 500) were found along Arctic and Antarctic coasts and within semi-enclosed seas such as the Mediterranean, Baltic, and Black Seas. {...}

The designation of priority areas was driven both by spatial patterns in diversity and human impact. When we considered the degree of human impact on these places of high biodiversity, several large areas emerged as priority areas for marine conservation. Areas of high biodiversity − high impact were identified around India, South Africa, Sri Lanka, Fiji, southeastern Australia, the South China Sea, the Mediterranean Sea, the Baltic Sea, and the coasts of

Table 2.3 Top 50 Bioclimatic units

	Bioclimatic unit	Region	Balanced solution/Optimal portfolio	Maximum return solution
1	Cuba & Bahamas (Grand Bahama, Bimini, Abacos & Eastern Cuba)	Caribbean and Bahamas	X	
2	Cuba		X	X
3	Greater Antilles (E. Cuba, Dominican Republic & Haiti)		X	X
4	Northern Cuba			X
5	Southern Cuba		X	
6	Bahamas (2 BCUs)		X	X
7	Far Northern Great Barrier Reef (2 BCUs)	Coral Sea	X	X
8	Fiji (Vanua levu)		X	
9	Fiji (Viti levu & Somo somo Strait)		X	
10	Papua New Guinea		X	
11	Solomon Islands		X	X
12	Torres Strait		X	X
13	Whitsundays (Great Barrier Reef) (3 BCUs)		X	X
14	Cenderawasih Bay	Coral Triangle		X
15	Central Sulawesi		X	X
16	East Timor & East Nusa Tenggara		X	X
17	Northern & Central Sulawesi		X	X
18	North Maluku		X	X
19	Southeast Sulawesi			X
20	South Sulawesi & Flores Sea (2 BCUs)		X	X
21	Taka Bonerate		X	

(continued)

Table 2.3 (continued)

	Bioclimatic unit	Region	Balanced solution/Optimal portfolio	Maximum return solution
22	West Papua & Maluku		X	X
23	Comoros	East Africa		X
24	Kenya & Tanzania		X	X
25	Madagascar			X
26	Mozambique			X
27	Somalia & Kenya		X	X
28	Tanzania		X	
29	Tanzania		X	X
30	India & Sri Lanka (Gulf of Mannar & Palk Bay)	Indian Ocean	X	
31	Lakshadweep		X	X
32	Maldives (2 BCUs)			X
33	Myanmar		X	
34	Egyptian Red Sea	Middle East & North Africa	X	X
35	Eritrea & Saudi (Dahlaks & Farasan Islands)			X
36	Hala'ib Triangle			X
37	Saudi Arabia		X	X
38	Southern Red Sea (Eritrea & Djibouti)		X	X
39	Sudan			X
40	Yemen		X	X
41	Bangka-Belitung	Southeast Asia	X	X
42	Mindanao & Negros (2 BCUs)		X	X
43	North Sumatra & Aceh		X	X
44	Palawan (Philippines)		X	X
45	Riau Islands (2 BCUs)		X	X
46	Sabah (Malaysia)		X	X
47	West Nusa Tenggara		X	X

(continued)

Table 2.3   (continued)

	Bioclimatic unit	Region	Balanced solution/Optimal portfolio	Maximum return solution
48	West Sumatra		X	X
49	Abrolhos Archipelago (Brazil)	South Atlantic and South Pacific	X	
50	Rangiroa (Palliser Group)		X	
51	Tuamotu Atoll (2 BCUs)		X	X
52	Society Islands		X	X

Credit Beyer, HL et al.

Southeast Asia among others. High biodiversity-low impact areas were identified within the Pacific EEZs of Mexico, Colombia, and Honduras as well as the Bahamas, the Galapagos (Ecuador), Madagascar, Mozambique, West Papua (Indonesia), Papua New Guinea, the north and west coasts of Australia, southern Kalimantan (Indonesia), the Solomon Islands, and in the Arctic and Antarctic Oceans.[24]

Similarly, the seminal paper entitled, "Bright spots among the world's coral reefs," researchers identified several "bright spots" and "dark spots"[25]: (Table 2.4).

Table 2.4   Bright and dark spots

	Bright spots	Nation/State	Location	Populated or unpopulated	Level of protection
1	Bright	British Indian Ocean Territory	Chagos	Unpopulated	Unfished (high compliance)
2	Bright	Commonwealth of the Northern Mariana Islands	Agrihan	Unpopulated	Fished
			Guguan	Unpopulated	Fished
3	Bright	Indonesia	Raja Ampat 1 & 2	Populated	Restricted
			Kalimantan	Populated	Restricted
4	Bright	Kiribati	Tabueran 1 & 2	Populated	Fished
5	Bright	Papua New Guinea	Karkar	Populated	Restricted

(continued)

Table 2.4 (continued)

	Bright spots	Nation/State	Location	Populated or unpopulated	Level of protection
6	Bright	Pacific Remote Islands Area	Baker	Unpopulated	Restricted
			Jarvis Island	Unpopulated	Restricted
7	Bright	Solomon Islands	Choiseul	Populated	Fished
			Isabel	Populated	Fished
			Makira	Populated	Fished
			New Georgia	Populated	Fished
	Dark Spots	**Nation/State**	**Location**	**Populated or Unpopulated**	**Level of Protection**
8	Dark	Australia	Lord Howe	Populated	Unfished (high compliance)
9	Dark	Hawaii	Hawaii	Populated	Fished
			Kauai 1 & 2	Populated	Fished
			Lanai	Populated	Fished
			Maui 1 & 2	Populated	Fished
			Molokai	Populated	Fished
			Oahu 1–6	Populated	Fished
10	Dark	Indonesia	Karimunjawa 1	Populated	Fished
			Karimunjawa 2 & 3	Populated	Unfished (low compliance)
			Pulau Aceh	Populated	Fished
11	Dark	Jamaica	Montego Bay 1	Populated	Unfished (low compliance)
			Montego Bay 2	Populated	Fished
			Rio Bueno	Populated	Fished
12	Dark	Kenya	Diani	Populated	Fished
13	Dark	Madagascar	Toliara	Populated	Fished
14	Dark	Mauritius	Anse Raie	Populated	Fished
			Grand Sable	Populated	Fished
15	Dark	NW Hawaii	Lisianski	Unpopulated	Unfished (high compliance)
			Pearl & Holmes 1 & 2	Unpopulated	Unfished (high compliance)
16	Dark	Reunion	Reunion	Populated	Fished
17	Dark	Seychelles	Bel Ombre	Populated	Restricted
18	Dark	Tanzania	Bongoyo	Populated	Unfished (high compliance)
			Chapwani	Populated	Fished
			Mtwara	Populated	Fished
			Stone Town, Zanzibar	Populated	Fished
19	Dark	Venezuela	Chuspa	Populated	Fished

Adapted from Table 5 from Cinner et al.

As of November 2019, the World Heritage Marine Programme had fifty sites listed[26]: (Table 2.5).

Table 2.5 Marine world heritage sites

	Location	Country	Status
1	Aldabra Atoll	Seychelles	Green
2	Archipiélago de Revillagigedo	Mexico	Green
3	Area de Conservación Guanacaste	Costa Rica	Green
4	Banc d'Arguin National Park	Mauritania	Green
5	Belize Barrier Reef Reserve System	Belize	Green
6	Brazilian Atlantic Islands: Fernando de Noronha and Atol das Rocas Reserves	Brazil	Green
7	Cocos Island National Park	Costa Rica	Green
8	Coiba National Park and its Special Zone of Marine Protection	Panama	Green
9	East Rennell	Solomon Islands	Red
10	Everglades National Park	United States (Florida)	Red
11	French Austral Lands and Seas	France	Green
12	Galápagos Islands	Ecuador	Green
13	Gough and Inaccessible Islands	United Kingdom	Green
14	Great Barrier Reef	Australia	Green
15	Gulf of Porto: Calanche of Piana, Gulf of Girolata, Scandola Reserve	France	Green
16	Ha Long Bay	Vietnam	Green
17	Heard and McDonald Islands	Australia	Green
18	High Coast/Kvarken Archipelago	Finland and Sweden	Green
19	Ibiza, Biodiversity and Culture	Spain	Yellow, Green
20	iSimangaliso Wetland Park	South Africa	Green
21	Islands and Protected Areas of the Gulf of California	Mexico	Red
22	Kluane/Wrangell-St. Elias/Glacier Bay/Tatshenshini-Alsek	Canada, United States	Green
23	Komodo National Park	Indonesia	Green
24	Lagoons of New Caledonia: Reef Diversity and Associated Ecosystems	France	Green
25	Lord Howe Island Group	Australia	Green

(continued)

Table 2.5 (continued)

	Location	Country	Status
26	Macquarie Island	Australia	Green
27	Malpelo Fauna and Flora Sanctuary	Colombia	Green
28	Natural System of Wrangel Island Reserve	Russia	Green
29	New Zealand Sub-Antarctic Islands	New Zealand	Green
30	Ningaloo Coast	Australia	Green
31	Ogasawara Islands	Japan	Green
32	Papahānaumokuākea	United States (Hawaii)	Yellow, Green
33	Península Valdés	Argentina	Green
34	Phoenix Islands Protected Area	Kiribati	Green
35	Puerto-Princesa Subterranean River National Park	Philippines	Green
36	Rock Islands Southern Lagoon	Palau	Yellow, Green
37	Sanganeb Marine National Park and Dungonab Bay—Mukkawar Island Marine National Park	Sudan	Green
38	Shark Bay, Western Australia	Australia	Green
39	Shiretoko	Japan	Green
40	Sian Ka'an	Mexico	Green
41	Socotra Archipelago	Yemen	Green
42	St Kilda	United Kingdom	Yellow, Green
43	Sundarbans National Park	India	Green
44	Surtsey	Iceland	Green
45	The Sundarbans	Bangladesh	Green
46	Tubbataha Reefs Natural Park	Philippines	Green
47	Ujung Kulon National Park	Indonesia	Green
48	Wadden Sea	Denmark, Germany, Netherlands	Green
49	West Norwegian Fjords—Geirangerfjord and Nærøyfjord	Norway	Green
50	Whale Sanctuary of El Vizcaino	Mexico	Green

Credit World Heritage Marine Programme

This book presents case studies on several of these priority marine areas such as the Belize Barrier Reef Reserve System, the Great Barrier Reef, and the Phoenix Islands Protected Area.

Where Is Coral Reef Degradation and Loss Occurring Today?

In the seminal paper, "A Global Map of Human Impact on Marine Ecosystems," the researchers found that "{…} no area is unaffected by human influence and that a large fraction (41%) is strongly affected by multiple drivers. However, large areas of relatively little human impact remain, particularly near the poles."[27]

The future of tropical coral reefs is grim and there is debate as to whether coral reefs will entirely disappear or just be severely degraded by mid-century. For instance, "even if they are met {the Paris Agreement's emission reduction goals}, an estimated 70-90% of the world's corals could disappear by mid-century."[28] Then again, the saying that "'all corals are going to be dead by 2050.' It's a lie, a dirty lie. They're not going to be dead by 2050. Massively degraded, perhaps."[29]

Global, regional, and local factors contribute to the degradation and loss of coral reefs. In addition, these factors can be categorized as direct and indirect impacts.

Global factors contributing to degradation and loss are:

- Global climate change; and
- Ocean acidification.

Regional factors include:

- Natural disasters;
- Pollution runoff;
- Plastics; and
- Overfishing.

Local factors include:

- Noise pollution;
- Disease;
- Invasive species; and
- High-human traffic.

Global Climate Change

Global climate change, says the World Economic Forum's Global Risks Report 2019, is one of the world's greatest risks due to biodiversity loss and ecosystem collapse, natural disasters, and extreme weather events.[30]

Since the Industrial Revolution, an increasingly rapid volume of greenhouse gas (GHG) emissions has been released, primarily due to the fossil fuel combustion for power generation and transportation, along with tropical deforestation and land-use change. Increasing anthropogenic GHG emissions results in global climate change, which increases terrestrial surface and sea surface temperatures. In turn, these increasingly cause coral bleaching:

> Warm water corals live on a knife edge; wherever they occur, they adapt to local temperatures. If temperatures rise more than a degree or two above the typical maximum, they bleach. Bleaching happens when there is a breakdown in the relationship between corals and the microscopic plants that live within their tissues and provide them with much of their food. At higher temperatures these microscopic plants, or zooxanthellae, as they are called, begin to harm their hosts, at which point the corals kick them out. Since it is the zooxanthellae that give corals their color, the tissue becomes transparent and colonies turn deathly white, as the chalky skeleton beneath is revealed. Bleached corals are starving; they die within weeks unless temperatures revert to normal.[31]

Thus, "when disturbed or stressed by water too warm or too cold, by salinity too high or too low, by pollutants, disease, high turbidity, or even increased ultraviolet radiation, the coral may expel the algae. Often, a combination of factors seems to set off the process {…}."[32]

Heat waves also cause coral bleaching events. The US National Oceanic and Atmospheric Administration (NOAA) describes global bleaching as "mass bleaching of at least one hundred square kilometers in all three ocean basins."[33] Early global mass bleaching occurred in 1981–1982 with the next major bleaching event occurring in 1997–1998, which "hammered reefs in more than fifty countries, through the Indo-Pacific, the Red Sea, the Caribbean, and even among the hot-water corals of the Arabian Sea."[34] Similarly, in 1998:

> the first major global bleaching event was declared when a huge underwater heatwave killed 16 percent of the corals on reefs around the world's oceans. The second global bleaching event was caused by the El Niño of 2010. In 2015, NOAA announced the third global bleaching event, the longest on record, which impacted reefs around the world between June 2014 and May 2017. It

hit the Great Barrier Reef particularly hard, with 93 percent of the surveyed reefs bleaching between 2015 and 2017"[35]

Global climate change also results in sea-level rise as the runoff from onshore glaciers increases and because of the expanding molecules in warming waters.

According to recent estimates, even the most modest sea-level rise predicted by the IPCC, a rise of up to twenty-four inches by 2100 could displace hundreds of millions of people and inundate four hundred thousand square miles of the world's agricultural land and coastal cities. Ten percent of the world population lives near coasts on land less than thirty-three feet above the present sea level. Eleven of the world's sixteen megacities, each home to more than fifteen million people, are built on coasts or estuaries: Tokyo, Guangzhou, Shanghai, Mumbai, New York, Manila, Jakarta, Los Angeles, Karachi, Osaka, and Kolkata. That number will grow as the world's population increases, with a further four coastal megacities predicted by 2025. There are three possible ways we can respond when rising tides threaten: retreat, adapt, or defend.[36]

This said:

Nowhere is the battle to protect land more pressing than on low-lying oceanic islands such as the Maldives, Phoenix Islands, and Tuvalu. {…} The combination of faltering coral growth and rising seas means that nations whose fortunes are founded on coral have begun to contemplate the unthinkable. Tuvalu in the South Pacific has made an arrangement to relocate its people to New Zealand when sea levels rise beyond their capacity to adapt. Already the highest tides wash over islands in Tuvalu that were dry a generation ago.[37]

Furthermore, climate change impacts coral reef production:

Climate change is hampering the birth of new corals in Australia's Great Barrier Reef, according to new research. A study published this week in the journal *Nature* found that new corals on the reef have plummeted by 89% since two major back-to-back bleaching events in 2016 and 2017, demonstrating how climate change is pushing coral ecosystems towards 'ecological collapse'. 'We never thought we'd see this happen,' lead author Andrew Baird told *The New York Times*. 'We thought the Barrier Reef was too big to fail, but it's not.'[38]

Climate change will also likely lead to increased storm frequency and intensity with more hurricanes in the Caribbean and more typhoons in the Pacific. According to the Global Climate Risk Index, coral reef countries such as the Philippines, Madagascar, India, Sri Lanka, Kenya and Fiji are the most affected by climate change.[39]

Ocean Acidification

Global climate change, because the ocean stores carbon dioxide, causes increased ocean acidification:

> Only about half of the annual release of six billion metric tons of CO_2 from burning fossil fuels increases CO_2 in the atmosphere. The remainder goes into the ocean and is giving rise to an unexpected suite of additional concerns. Of greatest worry is the steady rise in ocean acidification caused by the conversion of the excess carbon dioxide into carbonic acid: As the amount of carbon dioxide in the ocean increases, so does the water's acidity.
>
> Some think first of the consequences to coral reefs, where increasing acidity can interfere with the ability of corals to build and maintain their stony skeletons; past a certain level, they dissolve. The structure of coral reefs also depends on red and green species of coralline algae that may make up as much as 90 percent of the mass of a coral reef. They, too, dissolve when the surrounding water becomes too acidic. Everything with a calcium carbonate shell is vulnerable – oysters, clams, snails, pteropods (planktonic swimming mollusks), many sponges, sea stars, sea cucumbers, sea urchins – the list is long.[40]

Sadly, "scientists are suddenly waking up to the possibility that acidification, in combination with coral bleaching due to global warming, will cause the destruction of reefs worldwide. {...} Ocean acidification has been dubbed 'osteoporosis for reefs' because of this skeletal weakening."[41]

Natural Disasters

Natural disasters such as hurricanes, cyclones, typhoons, and earthquakes impact coral reefs. In September 2018, a 7.5 magnitude earthquake hit Indonesia causing more than 4400 deaths, "triggered tsunamis, landslides and nightmarish liquefaction, where moist soil turns into quicksand, swallowing up houses, cars and people. Hundreds of thousands of people were displaced, and the government estimates rebuilding will cost $2.6 billion." In addition, "'the loss {of coral cover off Palu, Central Sulawesi} was extraordinary,' says Dwi Sulistiawati, a fisheries researcher at Palu's Tadulako University. 'We went down to observe the coral after the earthquake and saw that only 1 to 9.5 per cent of the ocean floor had coral. Before the earthquake there was maybe 35 to 50 per cent coverage.'"[42]

Tropical storms, such as Cyclone Yasi in February 2011 that hit Australia and Hurricane Dorian that hit the Bahamas in August 2019, severely impacted coastal communities and adjacent coral reefs. Runoff after storms, such as debris, household chemicals, and sewage, also destroy coral reefs.

Pollution Runoff

Pollution runoff includes toxic chemicals, untreated sewage, petro-based fertilizers, siltation, discharge of bilge water, and other chemicals such as sunscreen. Pollution runoff relates to coastal land use including farming, manufacturing facilities, tourism, and the location of the world's megacities. The following map and Fig. 2.8 and Table 2.6 identify the world's largest cities in 2018 that are located near coral reefs.[43]

Runoff from farms, specifically animal manure and fertilizers containing nitrogen and phosphorus, are contributing to algae blooms and dead zones. In addition, "our coastal waters are poisoned not only from land-based runoff but also from airborne deposits of nitrogen oxide, a tailpipe pollutant that may account for more than 25 percent of the nitrogen buildup in offshore waters."[44] Runoff also comes from driveways, lawns, and gardens. Likewise, "low-oxygen dead zones have formed in many coastal zones, largely as a

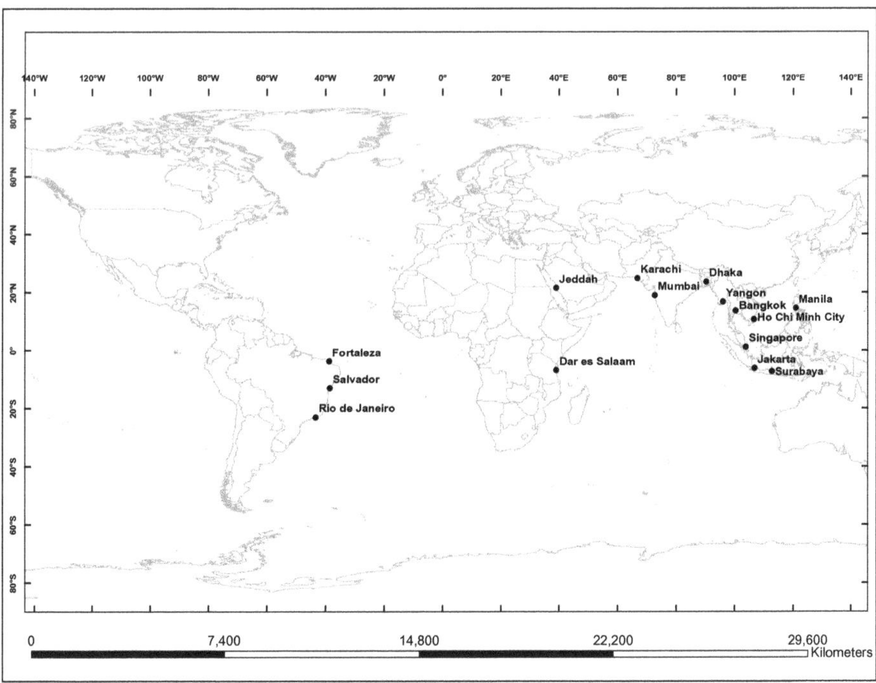

Fig. 2.8 World's largest cities near tropical coral reefs (*Credit* James Eaton)

Table 2.6 World's largest cities located near tropical coral reefs

World rank	City	State/Country	City's population	Metro's population
3	Karachi	Pakistan	18,000,000	27,506,000
5	Dhaka	Bangladesh	14,543,000	18,898,000
8	Manila	Philippines	12,877,000	22,710,000
10	Mumbai (Bombay)	India	12,400,000	27,750,000
21	Jakarta	Indonesia	9,608,000	30,214,000
31	Ho Chi Minh City	Vietnam	8,426,000	10,050,000
32	Bangkok	Thailand	8,281,000	14,566,000
41	Rio de Janeiro	Brazil	6,454,000	12,281,000
47	Singapore	Singapore	5,607,000	5,607,000
54	Yangon	Myanmar (Burma)	5,160,000	7,360,000
64	Dar Es Salaam	Tanzania	4,364,000	4,364,000
66	Jeddah	Saudi Arabia	4,276,000	4,276,000
104	Salvador	Brazil	2,913,000	3,920,000
108	Surabaya	Indonesia	2,865,000	6,485,000
119	Fortaleza	Brazil	2,610,000	4,019,000

consequence of excess fertilizers and toxic chemicals flowing from upstream fields, farms, and backyards."[45] Untreated waste is dumped into storm drains; because "people often think that storm drains lead to their local water treatment plant, but most do not."[46] Likewise:

> By the time these rivers reach the sea they are loaded with sewage, toxins, and trash from cities, as well as industrial and agricultural runoff. The problem increased sharply after World War II, when farms became dependent on chemical fertilizers to boost yields. Fertilizers now fuel coastal plankton blooms that can cover thousands of square miles and are often visible from space. {...} When plankton die their decay takes oxygen from the water. As a result, beneath pollution-fueled plankton blooms the water often becomes anoxic, spreading a shroud of death. Dead zones are now permanent or seasonal features at over four hundred places worldwide, all of them coastal or in enclosed seas, like the Baltic.[47]

Dead zones are dangerous to animals that are unable to escape fast enough from the lack of oxygen. This includes coral reefs, crustaceans (e.g., crabs, lobsters and prawns) and mollusks (e.g., clams, mussels, and oysters). Algae blooms can lead to anoxia, hypoxia, and dead zones, and create "red tide" (Fig. 2.9).

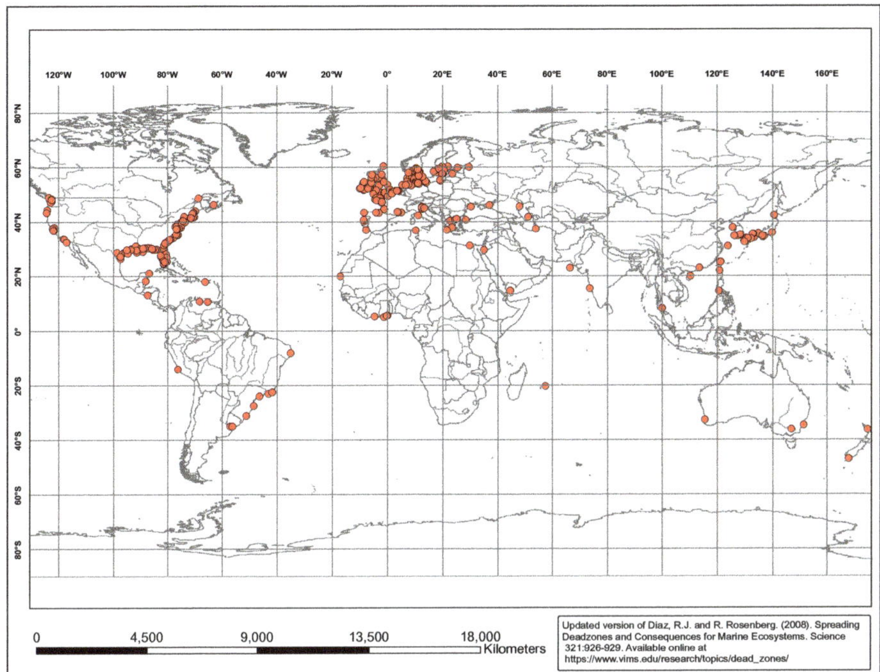

Fig. 2.9 World's documented dead zones and major river systems (*Credit* James Eaton)

Life in red tide hot spots, such as Tampa Bay, Naples, and Fort Myers, has become bad enough that there is concern it could drive away tourists and depress coastal property values. There are many other toxic phytoplankton out there able to make trouble for us in places that are polluted by excess nutrients. {…} The problem, as with Florida's red tides, is most severe when storms churn the sea into a spray that can be inhaled, causing lung inflammation and itchy skin. The same algae can be found in Hawaii, where it is eaten by a type of mullet that has become known locally as the 'nightmare weke' – when eaten, toxic fish cause hallucinations and nightmares. The toxins aren't just responsible for short-term health problems. Some promote tumor growth, or produce birth defects. It may be no coincidence that Hawaiians who eat seaweeds have high rates of stomach and intestinal cancers.[48]

While rivers are a source of healthy nutrients for coastal ecosystems, many rivers dump excessive levels of nutrients and pollution into the sea and other rivers are dammed up:

Today rivers are again running thick with soil, but less of it reaches the coast than in the past. Dams, which have been constructed along almost all of the world's large rivers and countless smaller ones, hold back mud that once would have reached the sea. Crop irrigation in arid regions often diverts water and sediment that would have replenished wetlands. Australia's legendary Murray and Darling rivers, for instance, lose so much water to irrigation that they now reach the coast as mere trickles.[49]

In addition, "nearly half of the world's largest rivers are dammed, and over 40 percent of river flows to the sea are intercepted by reservoirs. Collectively they exert a growing influence over life in the sea."[50] Researchers identified ten "rivers that carry 93 percent of that trash {which} are the Yangtze, Yellow, Hai, Pearl, Amur, Mekong, Indus and Ganges Delta in Asia, and the Niger and Nile in Africa."[51]

Other sources of toxic chemicals include pharmaceuticals, flame retardants, and heavy metals:

Throughout the world human populations are growing, aging, and increasingly dosing themselves with drugs to combat health problems. Lax controls on pharmaceutical manufacturers in developing countries mean there is an increased potential for release into the environment. {...} Downstream of sewage treatment works, male fish have been feminized by synthetic estrogen from contraceptive pills and hormone replacement therapy.[52]

It is easy to be indignant about the ubiquity of chemical pollution and to rail at the rapacious corporations that peddle these products. But we shouldn't forget the thousands of lives they save. Flame retardants have spared many people from being torched in their beds and DDT has prevented countless deaths from malaria. But a balance must be struck between safety and danger. Once chemicals get into the sea it is very difficult to remove them.[53]

Heavy metals like copper, lead, zinc, and mercury also contaminate the oceans and have proven far more difficult to control than PCBs. One of the most toxic is methyl mercury, because it is easily absorbed and can cross the blood-brain barrier. Much of the mercury we are exposed to today comes from emissions by coal-fired power stations that countries seem hell-bent on building more of, despite their contribute to global warming. Asian power stations produce over half of the world's mercury pollution, and much of it blows straight over the Pacific, where it combines with particles of organic matter and is converted to methyl mercury by microbes.[54]

Oil spills are also a less frequent source of marine pollution. The following table outlines some of history's largest spills[55,56] (Table 2.7).

While these spills received extensive media coverage, many smaller oil spills do not, yet their cumulative risks add up as "one gallon of oil can pollute 250,000 gallons of seawater."[57] This includes leaks in oil infrastructure, along

Table 2.7 Some of the largest oil spills in world history

	Name of oil spill	Year of oil spill	Gallons of oil spilled	Location of oil spill
1	Persian Gulf War oil spill	1991	240–520 million	Persian Gulf
2	BP's Deepwater Horizon	2010	134–206 million	Gulf of Mexico, United Kingdom
3	Ixtoc 1 oil well	1979–1980	126–140 million	Bay of Campeche, Mexico
4	Atlantic Empress	1979	88–90 million	Near Atlantic Ocean islands of Trinidad and Tobago
5	The Mingbulak (or Fergana Valley) oil spill	1992	~88 million	Uzbekistan
6	The Kolva River spill	1994	~84 million	The Russian Arctic
7	Nowruz Field platform	1983	~80 million	Northern Persian Gulf
8	The Amoco Cadiz	1978	~69 million	Near Brittany, France
9	The Castillo de Bellver	1983	53–79 million	Near Cape Town, South Africa
10	ABT Summer	1991	51 million	Off Coast of Angola
11	MT Haven	1991	~45 million	Off Coast of Genoa, Italy
12	Odyssey oil spill	1988	~43 million	Off Coast of Nova Scotia, Canada
13	Exxon Valdez oil spill	1989	~11 million	Prince William Sound, Alaska

Credit Encyclopedia Britannica and Marine Insight

with leaks and lubricants from both automobiles and boats. As Dr. Callum Roberts explains:

> I don't want to leave the impression that the oceans are toxic. In much of the sea, especially far from human habitation, pollution is very low. Dangerous chemicals are concentrated where inputs are high, around estuaries and cities, ports, and shipping lanes. Only a few places are contaminated enough to be dangerous, like New Bedford Harbor in Massachusetts. This site received industrial wastes for so many years that sediments there are loaded with a cocktail of chemical nasties. Fishing is banned and a cleanup is underway to remove contaminated mud.[58]

As described by Dr. Curtis Ebbesmeyer "it's surprising, and a little shocking, to think how much of my life has been intertwined with oil – the best and worst of substances, which powers ships, produces plastics, fouls the ocean, and enriches and ravages human life like nothing else."[59] Yet:

> {…} I often shock people when I tell them that, as marine pollutants go, oil is relatively innocuous stuff. Certainly, it looks awful as it coats the water, shore, and helpless seabirds, and it can cause grievous short-term damage. But oil dissipates and breaks down, becoming food for microorganisms. {…} When we turn petroleum into plastic, however, we make it far more persistent and, I fear, more deadly. Animals take up the plastic nurdles and dust but cannot digest them, and slowly the toxins and hormone mimics leach out.[60]

Marine plastic pollution is becoming a significant problem.

Plastics

Marine plastic pollution is becoming such a problem that plastic trash has been found inside the Mariana Trench, the deepest point on Earth.[61] According to a 2016 report from the Ellen MacArthur Foundation, "at least 8 million tons of plastic end up in the ocean each year. That amount is equivalent to unloading 'one garbage truck into the ocean every minute'."[62] According to Sir David Attenborough, plastics kill up to one million people a year.[63] Whales,[64] dolphins, and seals digest plastics.[65] Plastics are also having a demonstrable impact on corals.[66] Further, marine plastic waste is costing humanity billions of dollars. The Plymouth Marine Laboratory found:

> {…} microplastic pollution will result in a 1-5 per cent decline in marine ecosystem service delivery. This equates to an annual loss of $500-$2,500 billion in the value of benefits derived from marine ecosystem services, globally. The study broke this figure down to find that, based on 2011 values, every tonne of plastic that enters the ocean has a cost of between $3,300 and $33,000 of reduced environmental value.[67]

Historically, "plastics came into widespread use after World War II, so pretty much all of the plastic at-large on the high seas got there after 1945."[68] In 1975, 30 years after the introduction of plastics, a:

> National Academy of Sciences study found that 14 billion pounds of garbage, much of it consisting of long-lasting plastic goods, were being deliberately dumped into the sea – *at* sea- every year, merchant ships contributing the

most (close to 12 ½ billion pounds), then commercial fishing (almost a billion pounds) followed by recreational boating, military interests, passenger vessels, oil-drilling rigs and platforms, and various other sources.[69]

Despite the disastrous consequences of marine plastic pollution and its horrible impact on both human and marine health, it is important to recognize that:

It is difficult not to be impressed with the characteristics of plastics. They are light, strong, durable, corrosion-resistant, versatile – and inexpensive, compared to alternatives. It is also difficult to get rid of them, once used. Appealing qualities translate to appalling quantities of ocean debris: cups, lids, straws, spoons, boxes, toothbrushes, toys, rope, bags, packing material, fishing gear, and more. Much, much more. Of the many billions of pounds of plastics produced in recent decades, most if it is still around.[70]

Despite its benefits, there is so much accumulated marine plastic pollution that there are literally great gyres of plastic trash floating in the oceans off the coast of Chile, off the western coast of Antarctica, in the northwestern Gulf of Mexico, in the Sargasso Sea, and in the North Pacific Ocean.[71] Likewise:

{...} the so-called Great Pacific Garbage Patch, which swirls in the currents of the North Pacific Ocean. It is connected to the entire ocean, the planet's circulatory system, and all of that ocean, Pole to Pole, is clogged with large, medium, and especially, small pieces of plastic and other goods discarded by people worldwide. Like modern archaeological confetti, tiny bits of colorful plastic get caught up with bottles, shoes, plates, buckets, forks, spoons, cups, straws, toys, toothbrushes, packing straps, shavers, lawn chairs, bottle caps, coolers, crates, bags, and much more, carried by ocean currents across the face of the planet but collected in especially high concentrations in a few favored places.[72]

According to Dr. Curtis Ebbesmeyer (Fig. 2.10):

Only a few garbage patches have been documented. I have tallied eight of them, four in the Pacific, three in the Atlantic, and one in the Indian Ocean. The Turtle Gyre (like the Heyerdahl and Columbus) contains two, which I've dubbed the Great Eastern and Western Garbage Patches. The former, stretching between California and Hawaii, is the largest of all the patches, about half the size of the United States; together the eight patches cover an area more than twice the size of the fifty states.[73]

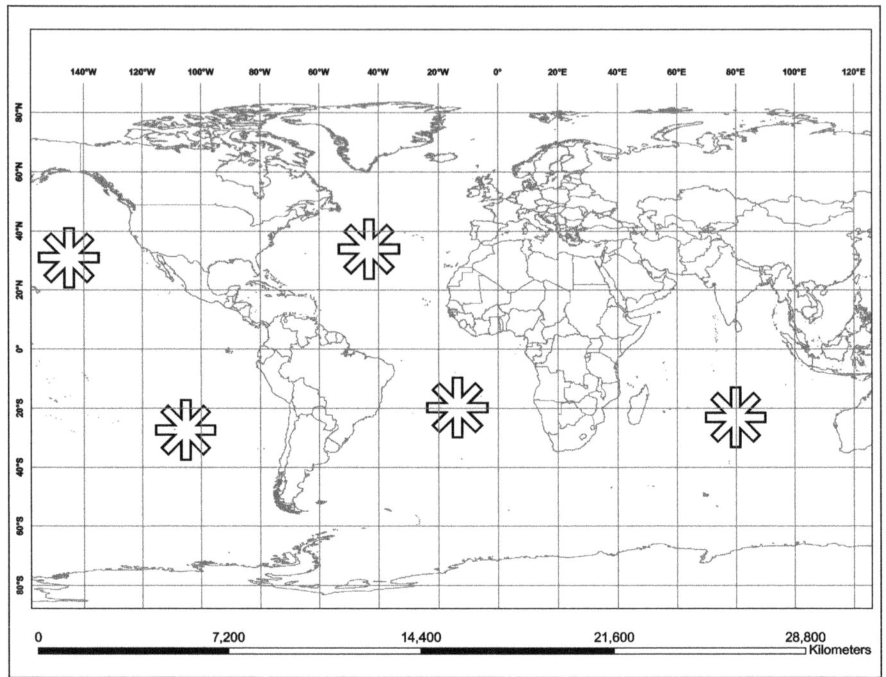

Fig. 2.10 General location of world's major plastic gyres (*Credit* James Eaton)

Today, some of the largest sources of plastic waste include: fishing gear; balloons; cigarette butts; straws; and other single use-plastics, such as bags.[74] The top 20 countries contributing to mismanaged plastic waste and plastic marine debris (estimated for 2010) are[75]: (Table 2.8).

Trade is global so that mismanaged waste in Indonesia, for instance, could be the result of plastic products purchased from the United States, or plastics mismanaged in China could have been destined to fulfill consumer demand in the EU.

Another way to understand where marine plastic pollution comes from is to consider sources beyond the country level, such as supermarkets, hotels or restaurants. A survey by the Environmental Investigation Agency and Greenpeace UK showed "the extent of plastic use in UK supermarkets. {…} Seven of them {UK supermarkets} are putting in circulation the equivalent of some 59 billion pieces of plastic packaging, over 2000 items for every household in the country."[76]

Similar to other pollution, plastic pollution also originates from land-based sources such as overturned trash bins, uncovered dump trucks, and littering, while some also originate from water-based sources. For instance,

Table 2.8 Largest contributors to marine plastic pollution

	Country	Mismanaged plastic waste (Millions of Metric Tonnes per Year)	Plastic Marine Debris (Millions of Metric Tonnes per Year)
1	China	8.82	1.32–3.53
2	Indonesia	3.22	0.48–1.29
3	Philippines	1.88	0.28–0.75
4	Vietnam	1.83	0.28–0.73
5	Sri Lanka	1.59	0.24–0.64
6	Thailand	1.03	0.15– 0.41
7	Egypt	0.97	0.15–0.39
8	Malaysia	0.94	0.14 –0.37
9	Nigeria	0.85	0.13–0.34
10	Bangladesh	0.79	0.12–0.31
11	South Africa	0.63	0.09–0.25
12	India	0.60	0.09–0.24
13	Algeria	0.52	0.08–0.21
14	Turkey	0.49	0.07–0.19
15	Pakistan	0.48	0.07–0.19
16	Brazil	0.47	0.07–0.19
17	Burma	0.46	0.07–0.19
18	Morocco	0.31	0.05–0.12
19	North Korea	0.30	0.05–0.12
20	United States	0.28	0.04–0.11

"cargo practices have since improved, but in the 1990s as many as ten thousand containers may have gone overboard each year. The largest known spill, during the 1998 typhoon, dropped three to four hundred containers into the mid-Pacific."[77] Furthermore, "the *Alguita's* net trawls suggest that fully 80 percent of marine debris comes from land. Greenpeace estimates that 10 percent of the 100 million tons of plastic produced each year worldwide ends up in the sea. That global production includes, by various estimates, 500 billion to 1 trillion plastic bags. It takes just one bag to choke a hungry sea turtle."[78]

Marine plastic pollution disastrously impacts ocean and human health. For instance, "those who consume seafood should be asking another question: How far *up* the food chain do ingested plastics go?"[79]

Large pieces of trash are eyesores, and some kinds, especially plastic bags, are lethal to sea turtles, whales, and whale sharks when the indigestible material is engulfed and jams their digestive system. A whale, washed ashore in California in 2008, died of 'unknown causes' but had 181 kilograms (400 pounds) of plastic in its stomach. Lost and discarded fishing gear causes major problems by entangling and killing marine mammals, birds, fish, and other marine life, as well as by endangering boat propellers and submarines. Even submersibles can be brought to a halt when snagged with masses of old fishing nets, such as the Russian sub trapped in the Bering Sea in August 2005. The seven sailors aboard were rescued after their encounter with the derelict gear; most marine mammals, turtles, birds and fish are not.[80]

Microplastics and nurdles, called "mermaid tears,"[81] are a devastating pollutant. It is said that "many of the particles at-large in the oceans range from a few hundredths to a few thousandths of an inch in size. They now fall into the size range of planktonic, food for a huge variety of life at the bottom of the food web."[82] These microplastics are:

Less obvious, much harder to retrieve, and more insidious {…} fragments of larger objects, as well as countless tiny pale pebbles dubbed nurdles, the pre-production plastic spheres that are later melted and molded to produce thousands of products, from juice jugs and action figures to lamp shades and chairs. Over 113 billion kilograms (250 billion pounds) of nurdles are created from petrochemicals every year and transported by trucks and tankers, loaded onto containerships, and carried to global destinations. The pellets are light, bouncy, and hard to restrain and readily wash down drains, blow into rivers, or spill directly into the sea. Some are deliberately incorporated into cosmetic 'scrubs' and when washed off, slide down pipes and eventually flow to the sea.[83]

As stated by Dr. Ebbesmeyer, "imagine salting your food with plastic ground to dust. That is what's happening to the oceanic food chain, all the way up to us."[84]

Through a process called *photo degradation*, sunlight slowly breaks down plastic polymers into pellets and fine dust. A study by the Algalita Marine Research Foundation found that plastic dust in parts of the North Pacific Ocean weighs six times more than zooplankton, the tiny animals that form the base of the marine food web. Marine plastic also acts as a toxic sponge, absorbing pollutants in the water such as PCBs and DDT. It concentrates these poisons tens of thousands of times more than seawater can. When consumed by fish, this poisonous plastic dust becomes part of the food web, increasing the toxic load in the flesh of tuna, billfish, sharks, and other top predators that humans then consume. Studies are now underway to see how these toxic loads may affect the development and reproduction of marine animals.[85]

Sadly, "the oceans are choking on plastics and will continue to do so for hundreds of years, even if we were to stop dumping plastics today. But it is never too late to start the cleanup."[86] Perhaps most awful, plastics may fossilize and remain in geologic history for millions of years.[87] As explained by Dr. Sylvia Earle, "thoughtfully, I disengaged a starkly barren polystyrene cup (projected lifetime, five hundred years) from where it languished in folds of soft gray-brown mud. Had a cup such as this been tossed overboard five centuries before, when Christopher Columbus was making his way across the Atlantic, it might still be around."[88]

Overfishing

Fishing, whether directly or indirectly, provides a livelihood for millions of people around the world. Some researchers suggest that 6.3 million people specifically fish coral reefs.[89]

According to the Food and Agriculture Organization of the United Nations' (FAO) State of World Fisheries and Aquaculture 2018 report, based off 2015 data, 33.1% of fish stocks were overfished, 59.9% of stocks were "maximally sustainably fished," (i.e., 90% of stocks are fully exploited, over-exploited or depleted) and only 7% of stocks were classified as "underfished stocks."[90]

It is important to make a distinction between legal and illegal fishing, along with between subsistence, artisanal, and commercial fishing, and whether it is nearshore or distant water fishing. There are also different fishing methods such as longline, seine purse, and trawling. Further, fishing access can be categorized as: (1) sole ownership; (2) open access; (3) limited entry, and; (4) individual transferable quotas.[91]

Reducing overfishing includes "stopping illegal, unreported and unregulated fishing and limiting perverse fisheries subsidies." Fishing is a vitally important livelihood for millions of people around the world. For instance, "developing countries make more money from seafood exports than from exports of coffee, cocoa, rubber, tea, meat, and rice combined. Worldwide, around 35 million people depend on fishing for their living."[92] As described by Jack Sobel and Craig Dahlgren, there are four types of overfishing:

If fishing is intense enough, fish are removed from the population before having the chance to grow enough to maximize their value to the fishery, a condition referred to as *growth overfishing*. {…} *Recruitment overfishing* occurs when a population produces too few offspring to sustain itself and is generally considered more serious because it threatens the population's future viability. *Serial overfishing* (*depletion*) describes the frequent situation in which a fishery sequentially targets and overfishes one species or species complex after another repeating the overfishing cycle. *Genetic overfishing* describes fishing-induced genetic changes that can result from targeting older, larger individuals or from other means {…}.[93]

Overfishing leads to when fishing effort increases (i.e., "race to fish" or "effort creep"), but the catch per unit effort (CPUE) decreases, and the species of fish composition changes.[94] When overfishing impacts grazers, algae can come to dominate coral reefs:

> In the case of the Caribbean, which is home to 9% of the world's coral reefs, overfishing (not climate change) has caused the most damage. When fishing is stopped, reefs can regenerate in just a few years to produce greater stocks that can compensate the years of income loss, but fishermen must be supported in the interim and so investments in managed MPAs must take this into consideration.[95]

Arguably, overfishing even led to the initial pirate attacks from Somalia. As noted in Jay Bahadur's book, *The Pirates of Somalia*:

> In 1994, he {Boyah} still worked as an artisanal lobster diver in Eyl {Somalia} – 'one of the best,' he said. {…} Since then, the lobster population off the coast of Eyl has been devastated by foreign fishing fleets – mostly Chinese, Taiwanese, and Korean ships, Boyah said. Using steel-pronged drag fishing nets, these foreign trawlers did not bother with nimble explorations of the reefs: they uprooted them, netting the future livelihood of the nearby coastal people along with the day's catch. Through their rapacious destruction of the reefs, foreign drag-fishers wiped out the lobster breeding grounds. Today, according to Boyah, there are no more lobsters to be found in the waters off Eyl.[96]

Many factors contribute to overfishing.

First, today's fishing fleets are larger and more technologically advanced:

> In the decade since the 1982 convention, advances in fishing technology, from the deployment of thousands of miles of lightweight, inexpensive drift nets to the use of sophisticated sonar and even satellite observation techniques to locate populations of fish and squid, have led to swift and devastating reductions in what once seemed to be 'limitless' populations.[97]

This said, "Japanese fishing fleets, so large and conspicuous as to be visible to space shuttle astronauts, illuminate the night sea to attract for capture millions of tons of squid."[98] Further:

High overall catches had been maintained in the latter part of the 20th century by the advent of industrial fishing. Fishermen were now using increasing numbers of large, fast boats, employing greatly improved methods for finding and catching various species, moving farther offshore, working in deeper water, and developing new markets for fish previously regarded ignominiously as trash fish. These included most kinds of sharks, rays, and toothy anglerfish, now a popular culinary specialty marketed as monkfish. Deep-sea fish from Antarctic waters known to scientists as Patagonia toothfish became 'Chilean sea bass,' slimeheads emerged as 'orange roughy,' beautiful, dark-eyed brotulid rat-tails were magically repacked as 'hoki,' and hundred-year-old fish that can with-stand freezing temperatures owing to an unusual kind of 'antifreeze' in their blood came to market as Antarctic cod.[99]

Such overfishing has led to depleted fisheries around the world.

> Intensive exploitation over long timescales leads to the progressive loss of larger animals. This phenomenon is so pervasive that it has been given a name: 'fishing down the food web.' It is brother to the process of sequential over-exploitation – the removal of species in sequence from large to small, predator to prey, high to low value – as they tend to go hand in hand.[100]

Second, government subsidies, including low or no interest loans, fuel or equipment subsidies, or direct payments to fishers, contributes to overfishing as many fleets would not be profitable without subsidies. It is estimated that governments annually spend USD$20 billion on such subsidies.[101]

Third, bycatch caught while fishing for the targeted species is a form overfishing:

> From 1950 to the 1990s, more than six million spotted, common, and spinner dolphins died in the eastern tropical Pacific alone as 'incidental bycatch' by fishermen using purse seines to capture yellowfin tuna. The dolphins associate closely with the tuna and are swept up with them when the seine is brought to the surface.[102]

WWF estimates "{…} more than 300,000 marine mammals, hundreds of thousands of sea turtles and seabirds, and millions of tons of fish and invertebrates are taken as bycatch every year."[103] Similarly:

> For every 10 pounds of Gulf of Mexico shrimp scraped from the sea floor, 80 to 90 pounds of 'trash fish' – rays, eels, flounder, butterfish, redfish, batfish, and more, including juveniles of many species – are mangled and discarded, in addition to tons of plants and animals not even considered worth reporting

as 'bycatch,' i.e., starfish, sand dollars, urchins, crabs, turtle-grass, seaweed, sponges, coral, sea hares, sea squirts, polychaete worms, horse conchs, and whatever else constitutes the sea-floor communities that are in the path of the nets.[104]

Fourth, there are destructive fishing practices including the use of poisons like cyanide, the use of dynamite, and bottom trawlers.

> Every year, estimates suggest, trawlers sweep an area of seabed equivalent to half of the world's continental shelves. Together with dredgers like the Clyde's scallop boats, they have transformed life on the seabed, converting three-dimensionally complex habitats rich in coral, sponge, sea fan, and seaweed into endless, monotonous expanses of shifting gravel, sand, and mud. While trawls and dredges decimate the ocean floor, the hand lines of old have evolved into longlines. The length of hook-studded longlines set every night in our seas is enough to wrap around the world five hundred times over.[105]

Restricting fish harvests can have a negative impact on local and artisanal fisheries. Fair compensation is often warranted. Also, closing fisheries can be done at certain times of the year, for instance, to protect spawning grounds. Over-fishing and destructive fishing practices have in part contributed to the rise of aquaculture and particularly "smart aquaculture" along with mariculture.

> In early forms of agriculture, people 'managed' wild plants and animals, giving naturally occurring populations special protection, care, or even modest enhancement, such as providing food or inhibiting predators. Protecting spawning areas or natural oyster beds, or taking sea creatures only in certain seasons, are comparable mild methods of aquaculture management.
>
> The second phase of agriculture builds on the first but goes farther with manipulation of the environment and the species. Examples in aquaculture include providing artificial structures for oysters or open-sea cage cultivation of enclosed species. Coastal shrimp and oyster farms are in this category. So is salmon farming. The life history of the fish is controlled, but its environment is not. This is true even of the futuristic submersible self-propelled current-riding geodesic-sphere fish pen recently launched in the Caribbean, filled with baby cobia. {…}
>
> The third kind of aquaculture, the approach most likely to yield enduring benefits, involves true domestication: control of the life cycle of the desired species and of the environment in closed systems.[106]

With this in mind, "many now talk of a blue revolution in aquaculture to match the much-vaunted green revolution in agriculture. Farmed fish and shellfish already make up around 46 percent of the fish sold for human consumption worldwide."[107] Yet:

> Suffice to say that many farm-raised fish are certainly not guilt-free, and they may come loaded with contaminant chemicals used to prevent disease in over-crowded pens. Some fish farms are stocked with fry or young caught from the wild, like some prawns and bluefin tuna. Aquaculture also has direct impacts on the environment including pollution from chemicals, excess feed and wastes produced, alien species introductions, and habitat destruction to make way for ponds or cages, not to mention the human rights abuses associated with unscrupulous businesses in some parts of the world. Shrimp farming has caused vast areas of mangrove forest to be cleared along tropical coasts. {…} However, not all aquiculture is bad.[108]

Similarly,

> Wild-caught shrimp? Too many problems with bycatch, habitat destruction, and overexploitation. Farmed shrimp? Some are better than others, but it is difficult to identify the source when buying in large quantities, and most involve large-scale loss of coastal mangroves, marshes, reefs, and other critically important coastal systems.[109]

Fifth, overfishing occurs by the targeting of rare, endemic or charismatic species, such as killing sharks for shark fin soup. In 2004, "a single large whale shark fin could sell for over US$50,000."[110] As noted by Dr. Sylvia Earle, "whatever name is given to the pieces of wild fish sold as fish sticks, a fish sandwich, fish and chips, or catch of the day, what you think you are eating and what is actually being served may be like asking for chicken salad and getting chopped pelican."[111]

Challenging questions arise. For instance, how should a new no-take zone compensate the former local, fisherfolk who previously fished its waters? How does one prevent the density of fishing boats increasing in another area now that a no-take zone has been established? This said,

> {…} it is generally accepted and even legally mandated that some regulation is necessary to protect fished populations from overfishing. We regularly use a suite of tools to achieve this objective, including catch quotas, effort quotas, size limits, gear restrictions, closed seasons, temporary closed areas, and even stocking of fish by artificial means. {However,} none of these techniques have more scientific validation than marine reserves.[112]

For an excellent book on overfishing, see Mr. Charles Clover's *The End of the Line: How Overfishing Is Changing the World and What We Eat.*

Noise Pollution

Due to increasing global commerce and tourism, along with the exploration and mining of the world's seabed, there has been an increase in marine noise pollution.

> Noise levels have grown to a roar since the 1950s as the world's globalized economy has launched tens of thousands of new merchant ships. The sound of a supertanker is ear-splitting, and because sound goes further underwater than in air, it can be heard by a whale a day before the ship arrives.[113]

Similarly, "oil spillage has fallen considerable as a result of the imposition of double hulls, the compartmentalization of oil in the hold, pinpoint satellite navigation, and a raft of better sailing practices. It is high time for them to tackle the twin problems of noise and fuel efficiency."[114] Further, "acoustic or environmental impacts may be more insidious. Seismic exploration, military exercises, shipping, or drilling may have far-reaching acoustic impacts that cause species to leave an area, to become temporarily unable to forage, or even to sustain physical damage."[115]

Disease

Disease is another factor that is harming the world's coral reefs:

> Caribbean coral reefs possess a handful of coral species that do the heavy lifting of reef construction; West Pacific reefs have dozens. Disease epidemics in the Caribbean have knocked out two of the main reef builders and are now hacking away at the third. As a result, growth of reefs in the Caribbean at best has juddered to a halt, and in most places is in reverse as erosion takes hold. It is easy to see the results underwater.[116]

Disease is also related to pollution. For instance, the "cause of the elkhorn epidemic was later discovered to be a human gut bacterium, carried in sewage effluent. This outbreak was one of a collection of pathogens that has contributed to the loss of 80 percent of living coral cover in the Caribbean since 1977."[117] Similarly, white-band disease, "distinguishable by the white band of dead coral tissue it forms {...} has been identified as the central reason for the mass decline in the two dominant acroporid coral species in the Caribbean over the past forty years."[118]

Invasive Species

Invasive species – whether marine or terrestrial species – have a negative impact on coral reefs:

> Most invasive creatures are released from commercial ships' ballast waters. Large ships take in water for stability before a voyage and then release this ballast water when they reach their port of destination. Every hour of every day some 2 million gallons of ballast water are released in United States ports, harbors, and bays. {...} But commercial ships aren't the only contributors to this problem; small boat operators, fishermen, divers, pet owners, aquarium owners, travelers, and even seafood lovers also introduce invasive species into our marine waters.[119]

In the Caribbean, one infamous invasive species is the lionfish:

> Enough lionfish were released or escaped from aquaria in the late twentieth century in south Florida that a breeding population became established. Consequently, lionfish have few Caribbean predators. Their lacy fins conceal sharp spines that deliver a potent shot of venom to the unwary. The Caribbean was poised for invasion. {...} Lionfish have explored across the Caribbean in less than twenty years. They can now be found from Rhode Island south to Colombia, and from Belize east to the Windward Islands, and may ultimately range along the entire coast of Brazil.[120]

In addition to the lionfish, the most notorious invasive species "is a tropical seaweed, *Caulerpa taxifolia*, a dark green weed that scrambles over the seabed in mats of twisted stems and has been dubbed the 'killer algae'."[121] Other examples of invasive marine species include the freshwater Zebra mussel (*Dreissena polymorpha*) and the Asian swamp eel (*Monopterus albus*).

Terrestrial invasive species—rats, feral dogs and cats—also hurt the marine environment:

> One solution designed to protect coral reefs threatened with climate change is to eliminate rats from nearby tropical islands, according to a new study. Usually, birds inhabiting the islands provide natural fertiliser to the reefs in the form of their droppings, but rats prey on the birds, leading to lost nutrients for the coral. 'There is no example that is as clear cut and effective in enhancing the functioning of coral reefs in the face of climate change,' said contributing

researcher Dr. Aaron MacNeil, a fisheries ecologist at Dalhousie University in Nova Scotia.[122]

While invasive species are a major issue around the world, there are also introduced, alien, and/or exotic species which are foreign species that may not necessarily be as detrimental as invasive species.

High-Human Traffic

High-human traffic, whether it be commerce to global ports or cruise liners and tourists, is impacting the world's oceans through the spread of invasive species, the generation of excessive noise, their contribution to marine pollution, and the physical damage to coral reefs for infrastructure development such as using anchors, mining, dredging, ship wrecks, and port construction.

Some of the world's largest port terminals, which host cruise liners and which are adjacent to tropical coral reefs, include the cities of Cozumel (Mexico), George Town (Grand Cayman, Cayman Islands), and Nassau (Bahamas).[123] According to Cruise Market Watch, "total worldwide ocean cruise capacity at the end of 2018 will be 537,000 passengers and 314 ships. Annualized total passengers carried worldwide will be 26.0 million (a 3.3% increase over 2017)."[124]

Why Are Coral Reefs Valuable?

Tropical coral reefs are one of the most stunning ecosystems on Earth that host the greatest amount of marine biodiversity, provide livelihoods to millions of people, and generate billions of dollars in value each year from fisheries, storm protection, and tourism. In fact:

> The annual 'gross marine product' (GMP) – equivalent to a country's annual gross domestic product – is at least US$2.5 trillion; the total 'asset' base of the ocean is at least US$24 trillion. Underpinning this value are direct outputs (fishing, aquaculture), services enabled (tourism, education), trade and transportation (coastal and oceanic shipping) and adjacent benefits (carbon sequestration, biotechnology). Putting it into an international context, if the ocean were a country it would have the seventh largest economy in the world.[125]

Direct outputs from marine fisheries, mangroves, coral reefs, and seagrass are estimated at USD$6.9 trillion, the use of shipping lanes for trade and transportation are estimated at USD$5.2 trillion, and the adjacent assets of productive coastlines and carbon sequestration are estimated at USD$7.8 trillion and USD$4.3 trillion, respectively.[126]

In the seminal article, entitled "The value of the world's ecosystem services and natural capital," the global extrinsic value of the biosphere was estimated at USD$16–$54 trillion per year, with an average value of USD$33 trillion. This average value is more than 1.5 times the size of the entire U.S. economy or nearly three times the size of the China economy in 2018.[127] Coral reefs, due to their role in providing recreation, disturbance regulation, and food production, were valued at USD$6,075 per hectare per year for a global annual estimate of USD$375 billion in ecosystem services.[128]

As summarized by Dr. Sylvia Earle:

The ocean drives climate and weather, regulates temperature, absorbs much of the carbon dioxide from the atmosphere, holds 97 percent of Earth's water, and embraces 97 percent of the biosphere. Far and away the greatest abundance and diversity of life occurs in the ocean, occupying liquid space from the sunlit surface to the greatest depths.[129]

According to The Economics of Ecosystems and Biodiversity (TEEB), ecosystem services can be categorized as provisioning services, regulating services, habitat or supporting services, and cultural services.[130] TEEB was initiated as a global study in 2007 by the G8+5 countries to analyze "the global economic benefit of biological diversity, the costs of the loss of biodiversity, and the failure to take protective measures versus the costs of effective conservation."[131]

Provisioning Services

Provisioning services include raw materials, freshwater, and medicinal resources. The oceans, and particularly tropical coral reefs, provide a substantial amount of provisioning services from seafood to medicines. Consider:

One kind of blue-green bacteria, *Prochlorococcus*, is so abundant – about 100 octillion (1 octillion = 10^{27}) are alive at any given moment – that it alone is responsible for about 20 percent of the oxygen in the atmosphere. Put another way, this nearly invisible form of life generates the oxygen in one of every five breaths you take, no matter where on the planet you live. Other ocean photosynthesizers contribute an additional 50 percent of our atmospheric oxygen. Clearly, we would all have a hard time breathing if we had to rely on trees, grass, and other terrestrial plants alone for oxygen.[132]

With respect to medicine, "pharmaceutical companies are studying the toxins in nudibranchs in search of new compounds that might yield drugs to combat cancer and other diseases."[133] Furthermore:

> Two widely used drugs – Acyclovir, which treats herpes, and AZT, which fights the HIV-AIDS virus – are derived from compounds first identified in marine sponges. Many other potentially useful drugs are now being developed from marine organisms, including cancer-fighting compounds from soft corals, anti-inflammation chemicals from sea feathers, virus-killing proteins from sea grass molds, and painkillers from cone snails.[134]

In addition, healthy reefs provide food security. For instance, "a billion people rely on seafood as their main source of animal protein today, most of them in the developing world. The continued decline of wild fisheries threatens malnourishment for many more by the middle of the century."[135] More specifically, "US$6.8 billion is the annual vale of reef fisheries worldwide."[136] Similarly, "{…} it has been estimated that fish catch from reefs is 6 million metric tonnes. On top of this, an estimated 9 million metric tonnes of shellfish and other molluscs are taken per year in and around coral reefs."[137]

Mangroves "store carbon and provide over 100 million people with a variety of goods and services, such as fisheries and forest products, clean water, and protection against erosion and extreme weather events. {Yet} The rate of deforestation of the planet's mangroves is three to five times greater than even the average forest global forest loss."[138] Such coastal deforestation, particularly of mangroves, has occurred due to aquaculture (i.e., shrimp farming) and agriculture, along with coastal developments for infrastructure, housing, and tourism facilities.

Furthermore:

> The global population in the Low-Elevation Coastal Zones (LECZ) is expected to grow from 600 million people in 2000 to an average of 1.2 billion in 2060 (Neumann et al., 2015). Therefore, both the number of people benefitting from those SDGs and the external pressures on those ecosystems will increase, reinforcing the importance of protecting both. {Further}, there are 850 million people worldwide {who} live within 100 kilometers of reefs and 275 million people live within 30 kilometers of reefs and less than 10 kilometers from the coast.[139]

Another provisioning service is freshwater. This said, "but all of us dwell near the sea, even if we live a thousand miles inland – the sea falls from the sky when it rains; every drop of water we use eventually finds its way into the ocean."[140]

Regulating Services

Regulating services described by TEEB include the regulation of local climate and air quality, carbon sequestration and storage, moderation of extreme events, erosion prevention and maintenance of soil fertility, nutrient recycling, pollination, and biological control.

> For the globe as a whole, the ocean is the great regulator, the great stabilizer of temperatures. It has been described as 'a savings bank for solar energy, receiving deposits in seasons of excessive insolation and paying them back in seasons of want.' Without the ocean, our world would be visited by unthinkably harsh extremes of temperature. For the water that covers three-fourths of the earth's surface with an enveloping mantle is a substance of remarkable qualities. It is an excellent absorber and radiator of heat. {...} The redistribution of heat for the whole earth is accomplished about half by the ocean currents, and half by the winds.[141]

In fact, "the ocean produces half the oxygen we breathe, and absorbs 30 per cent of the anthropogenic emissions of carbon dioxide and around 93 percent of the added heat arising from human-driven changes to the atmosphere."[142]

Coastal ecosystems can sequester more carbon dioxide per hectare than some tropical rainforests as "intact coastal habitats store hundreds to thousands of tonnes of carbon beneath each hectare: for example, seagrasses 500 tonnes CO_2e per ha, salt marshes 917 tonnes CO_2e per ha and 1800 tonnes CO_2e per ha for oceanic mangroves."[143] Estimates suggest that "{...} corals remove about 700 billion kg {700 million metric tonnes} of carbon per year."[144]

Regarding the regulation of extreme events, "healthy coral reefs are naturally self-repairing breakwaters, better at protecting coasts than any concrete wall we can throw up to defend cities, resorts, or agricultural land, and far cheaper."[145] In fact, "up to 90% of wave energy {is} reduced by coral conservation {and} US$6 billion is the value of built capital protected by coral ecosystems."[146]

> Like coral reefs, wetlands and dunes are living barriers between land and sea. {...} Mangroves in Indonesia and Sri Lanka likewise seem to have helped dissipate the destructive force of the terrible tsunami of 2004. (The correlation between reduced wave impacts and the presence of mangroves has been challenged, because mangroves colonize sheltered shores that were already likely to be less affected by the tsunami). But where would you prefer to be if a tsunami struck: on a shore fronted by open sea or behind a dense natural stand of mangrove trees?[147]

Coral reefs also provide the regulating services of filtration and purification of water runoff. Similarly, seagrasses help reduce turbidity in the water and essentially allow more light to hit the seafloor, which allows greater photosynthesis for corals.[148]

Habitat or Supporting Services

Habitat or supporting services include habitat for species, along with the maintenance of genetic diversity. There are important linkages between habitats, biodiversity, and the health of coral reefs. For instance, mangroves provide an important fish nursery, while coral reefs host the greatest abundance of marine biodiversity on Earth. One such species that use mangroves as a nursery are parrotfish, which are an important algae grazer and which help to keep a balance between corals and algae – especially after bleaching events.[149]

In fact, "fewer than 1,000 reef-building coral species exist, but they host a massive diversity of marine life. 35,000 to 60,000 reef dwelling species have been described, but between 1 to 9 million species are estimated live on coral reefs."[150] Yet,

> Biodiversity – short for biological diversity, the amazing variety of life on Earth – is being reduced to the lowest levels in human history. *Homo sapiens* has already wiped out one-quarter of all bird species, and an estimated 11 percent more are on the path to extinction, along with 24 percent of mammal and 11 percent of plant species. One-quarter of the world's coral reefs have been destroyed, with many others undergoing serious decline.[151]

Similarly, "throughout the world, with every passing hour an estimated two to three species are committed to extinction, mostly as victims of habitat destruction and the spread of non-native species, which kill or compete with natives."[152] Sadly, we are currently in the sixth great extinction period known as the Holocene extinction period. The last mass extinction prior to the arrival of humans was known as the Cretaceous-Tertiary, which took place approximately 65 million years ago and resulted in the end of the dinosaurs.[153] As proclaimed by Dr. E. O. Wilson, "this mass extinction is the folly our descendants will be least likely to forgive."[154]

As described by WWF, "an ecoregion is defined as a large area of land or water that contains a geographically distinct assemblage of natural communities that:

a. share a large majority of their species and ecological dynamics;
b. share similar environmental conditions, and;
c. interact ecologically in ways that are critical for their long-term persistence."[155]

These ecoregions include the world's tropical coral reefs, and similarly, Conservation International (CI) has designated 36 biodiversity hotspots, including many regions adjacent to coral reefs. Dr. Callum Roberts et al., identified the following eighteen marine biodiversity hotspots which are the eighteen "richest multitaxon centers of endemism. They include 35.2% of the world's coral reefs and cover only 0.028% of the world's oceans, but include between 58.6 and 68.7% of restricted-range species from the four taxa."[156]

- 1. South Japan;
- 2. Western Australia;
- 3. Gulf of Guinea;

- 4. Great Barrier Reef;
- 5. Hawaiian Islands;
- 6. Gulf of California;

- 7. Lord Howe Island;
- 8. North Indian Ocean;
- 9. New Caledonia;

- 10. Eastern South Africa;
- 11. Cape Verde Islands;
- 12. West Caribbean;

- 13. Red Sea;
- 14. Philippines;
- 15. South Mascarene Islands;

- 16. St. Helena and Ascension Islands;
- 17. Sunda Islands; and
- 18. Easter Island.

Cultural Services

Cultural services include recreation, mental, and physical health; tourism; aesthetic appreciation and inspiration for culture, art, and design; spiritual experiences; and sense of place. Reefs have inherent value and reefs are also extrinsically valued by people around the world as demonstrated by the money spent on tourism, movies (e.g., Finding Nemo), and books, along with school visits to zoos and aquariums. This said, coastlines and offshore coral reefs are some of the most beautiful places on Earth. Yet, the ocean is full of mysteries:

- "Fish, amphibian, and reptile, warm-blooded bird and mammal – each of us carries in our veins a salty stream in which the elements sodium, potassium, and calcium are combined in almost the same proportions as in seawater."[157]
- "{…} jellyfish and sharks and crustaceans, have survived, more or less intact, through hundreds of millions of years."[158]

- Blue whales are the largest animals on Earth, even larger than the iconic dinosaurs of the past like Tyrannosaurus Rex or Apatosaurus.
- The fish known as the "orange roughy {*Hoplostethus atlanticus*} mate at about 30 years and may live more than 150 years {or even up to 200 years old}."[159] Even more astonishing, a 400-year old Greenland shark (*Somniosus microcephalus*) was discovered.[160] Regarding corals, "like the ancient trees of the High Sierras in the United States, some corals live for hundreds of years, while a few reach a thousand or more."[161]
- Dolphins and whales have some of the highest levels of intelligence in the animal kingdom, and sperm whales have the largest brain ever discovered;
- Albatross are {…} winged giants that can stay aloft above the open sea for a year {…}."[162]
- "Other treasures will forever be lost if bluefins and other tunas continue on their current slide into oblivion, not the least being answers to questions we do not yet know enough to ask. The knowledge we could gain would seem to be of greater worth than can be measured in pounds or ounces of meant. How is it possible for them to power their way across entire ocean basins to specific locations, repeatedly, without a road map? What senses enable them to locate sources of food? How do they maintain position in amazing V-shaped formations as they move near the ocean's surface? Who leads, and why? How do they rest?"[163]
- "The arctic jellyfish Cyanea, on the other hand, is so hardy that it continues to pulsate when half its bell is imprisoned in ice, and may revive even after being solidly frozen for hours. The horseshoe crab is an example of an animal that is very tolerant of temperature change. It has a wide range as a species, and its northern forms can survive being frozen into ice in New England, while its southern representatives thrive in tropical waters of Florida and southward to Yucatán."[164]
- The blood of bristle worms is "red in some, green in others."[165]

The mysteries are endless. Consider salmon—"But why return to the exact same spawning ground of your parents? Why not just any stream? To me, that is a profound mystery with no provable solution."[166]

Other Resources on the Context of Coral Reef Degradation and Loss

Resources on Global Climate Change, Bleaching and Ocean Acidification

A Reef Manager's Guide to Coral Bleaching

- https://www.coris.noaa.gov/activities/reef_managers_guide/

Climate Adaptation Toolkit for Marine and Coastal Protected Areas

- https://www.cakex.org/MPAToolkit

Climate Watch

- https://www.climatewatchdata.org/

Global Carbon Atlas

- www.globalcarbonatlas.org/

Global Carbon Project

- https://www.globalcarbonproject.org/

Global Climate Risk Index by Germanwatch

- https://germanwatch.org/en/cri

IPCC Fifth Assessment Report

- Assessment Report: https://www.ipcc.ch/assessment-report/ar5/; and
- Special Report: https://www.ipcc.ch/sr15/

NOAA Coral Reef Watch

- https://coralreefwatch.noaa.gov/satellite/index.php

NOAA's National Centers for Coastal Ocean Science (NCCOS) Biogeography Projects

- https://maps.coastalscience.noaa.gov/biogeography/

Ocean Health Index

- http://www.oceanhealthindex.org/; and
- http://www.oceanhealthindex.org/methodology/components/fisheries-status

Project Drawdown

- https://www.drawdown.org/

Surging Seas: Sea Level Rise Analysis by Climate Central

- http://sealevel.climatecentral.org/

Resources on Fishing

Environmental Performance Index: Fisheries

- https://epi.envirocenter.yale.edu/2018-epi-report/fisheries

FAO: The State of World Fisheries and Aquaculture

- http://www.fao.org/fishery/sofia/en

Near Shore Artisanal Fishers Enforcement Guide by TNC and WildAid

- http://wildaid.org/wp-content/uploads/2017/09/Nearshore-Artisanal-Fisheries-Enforcement-Guide_0.pdf

Resources on Plastics and Pollution

Beachcombers' Alert

- http://beachcombersalert.org/

Break Free from Plastic

- https://www.breakfreefromplastic.org/

For a Strawless Ocean

- http://strawlessocean.org/alternatives

GRID-Arendal's Waste and Marine Litter

- http://www.grida.no/activities/27

Indonesian Waste Platform

- http://www.indonesianwaste.org/en/home/

Mapotic's Zero Plastics, Anti-Plastic's Mapping the Movement

- https://www.mapotic.com/zero-plastics

Plastic Free Challenge

- http://www.plasticfreechallenge.org/

Plastic Pollution Coalition

- http://www.plasticpollutioncoalition.org/

Plastic Soup Foundation

- https://www.plasticsoupfoundation.org/en/

Plastic Soup: An Atlas of Ocean Pollution by Michiel Roscam Abbing

- https://islandpress.org/books/plastic-soup

Plastisphere

- http://anjakrieger.com/plastisphere/

Santa Aguila Foundation's Plastic Pollution Site

- http://plastic-pollution.org/

The MARLISCO Project

- http://www.marlisco.eu/

The Story of Plastic

- https://www.storyofplastic.org/watch

World Oceans Day's Plastic Pollution Resources

- https://www.worldoceansday.org/plastic-pollution-resources-1

Resources on Related Topics

Allen Coral Atlas

- https://allencoralatlas.org/

Conservation Biology Institute's Data Basin

- https://databasin.org/

Global Footprint Network

- http://data.footprintnetwork.org/

How to Recycle

- http://www.how2recycle.info/about

Ocean Optimism Movement

- http://www.oceanoptimism.org/

Reef Base

- http://www.reefbase.org/gis_maps/default.aspx

Regional State of the Coral Triangle

- http://www.coraltriangleinitiative.org/sites/default/files/resources/regional-state-coral-triangle_0.pdf

Tools for Landscapes for People, Food and Nature

- http://peoplefoodandnature.org/learning-network/find-tools/

UN's Atlas of the Oceans

- http://www.oceansatlas.org

UNEP's Global Environment Outlook

- https://www.unenvironment.org/global-environment-outlook

UNEP's Global Information Resource Database (GIRD)

- https://na.unep.net/

WRI's Reefs at Risk Revisited

- https://www.wri.org/resources/data-sets/reefs-risk-revisited

WWF's Water Risk Filter

- http://waterriskfilter.panda.org/

Notes

1. Ibid. XIV.
2. Roberts. *The Ocean of Life*. 15.
3. Cutlip, Kimbra. "Who Owns the Fish: High Seas and the EEZs." *Global Fishing Watch*. July 1, 2016. Accessed March 19, 2020. https://globalfishin gwatch.org/fisheries/who-owns-the-fish-high-seas-and-the-eezs/.
4. Earle. *The World is Blue*. 143.

5. World Atlas. "The Largest Countries in the World." Accessed October 5, 2018. https://www.worldatlas.com/articles/the-largest-countries-in-the-world-the-biggest-nations-as-determined-by-total-land-area.html.

6. World Atlas. "Countries with the Largest Exclusive Economic Zones." Accessed October 5, 2018. https://www.worldatlas.com/articles/countries-with-the-largest-exclusive-economic-zones.html.

7. Carson. *The Edge of the Sea.* 191–192.

8. Roberts. *The Ocean of Life.* 85.

9. Ibid. 110.

10. World Wildlife Fund. "Coral Reefs." Accessed December 4, 2019. https://wwf.panda.org/our_work/oceans/coasts/coral_reefs/.

11. UNEP. "UNEP-WCMC *World Atlas of Coral Reefs*: The Most Detailed Assessment Ever of the Status and Distribution of the World's Coral Reefs." September 11, 2011. Accessed July 23, 2019. http://coral.unep.ch/atlaspr.htm.

12. Avery, Helen. "Blue finance: Why Marine PPPs Could be a Win-Win-Win." *Euromoney.* June 5, 2018. Accessed December 4, 2019. https://www.euromoney.com/article/b18hg7vjwmy9hn/blue-finance-why-marine-ppps-could-be-a-win-win-win.

13. Ibid.

14. WRI. "Marine Protected Areas in Coral Reef Regions Classified According to Management Effectiveness Rating." February 2011. Accessed December 4, 2019. https://www.wri.org/resource/marine-protected-areas-coral-reef-regions-classified-according-management-effectiveness.

15. Avery, Helen. "Blue finance: Why Marine PPPs Could be a Win-Win-Win." *Euromoney.* June 5, 2018. Accessed December 4, 2019. https://www.euromoney.com/article/b18hg7vjwmy9hn/blue-finance-why-marine-ppps-could-be-a-win-win-win.

16. Beyer et al. "Risk-Sensitive Planning for Conserving Coral Reefs Under Rapid Climate Change." 3.

17. Ibid. 4.

18. Ibid. 5.

19. Ibid. 8.

20. WRI "Reefs at Risk Revisited." Accessed December 20, 2019. https://www.wri.org/resources/data-sets/reefs-risk-revisited.

21. Hoegh-Guldberg, Ove. "Coral Reef Rescue: Resilient People, Coral Reefs, and Their Future in a 1.5° World." Side Event at WWF Pavilion—The Ambition Emergency Room—at the Conference of the Parties (COP25). Madrid, Spain. December 11, 2019.

22. World Wildlife Fund. "Coral Reef Rescue: Resilient People, Coral Reefs, and Their Future in a 1.5° World." Side Event at WWF Pavilion—The Ambition Emergency Room—at the Conference of the Parties (COP25). Madrid, Spain. December 11, 2019.

23. Hoegh-Guldberg, Ove. "Coral Reef Rescue: Resilient People, Coral Reefs, and Their Future in a 1.5° World." Side Event at WWF Pavilion—The Ambition

Emergency Room—at the Conference of the Parties (COP25). Madrid, Spain. December 11, 2019.

24. Selig et al. (2014). "Global Priorities for Marine Biodiversity Conservation." 4, 6.

25. Cinner, Joshua E., et al. "Bright Spots Among the World's Coral Reefs." https://doi.org/10.1038/nature18607.

26. World Heritage Marine Programme. "World Heritage Sites (50)." Accessed November 23, 2019. http://whc.unesco.org/en/marine-programme.

27. Halpern, Benjamin S., et al. "A Global Map of Human Impact on Marine Ecosystems." https://doi.org/10.1126/science.1149345.

28. Beyer et al. "Risk-Sensitive Planning for Conserving Coral Reefs Under Rapid Climate Change." 2 of 10.

29. Braverman. *Coral Whisperers*. 244.

30. World Economic Forum. "The Global Risks Report 2019, 14th Edition." 2019. Accessed December 4, 2019. http://www3.weforum.org/docs/WEF_Global_Risks_Report_2019.pdf. 5.

31. Roberts. *The Ocean of Life*. 85.

32. Earle. *Sea Change*. 199.

33. Braverman. *Coral Whisperers*. 9.

34. McCalman. *The Reef*. 270.

35. Braverman. *Coral Whisperers*. 8–9.

36. Roberts. *The Ocean of Life*. 95.

37. Ibid. 96–98.

38. Carbon Pulse. "CP Daily: Friday April 5, 2019." April 5, 2019. Accessed December 4, 2019. http://carbon-pulse.com/72580/.

39. Germanwatch. "Global Climate Risk Index 2020." Accessed February 26, 2020. https://germanwatch.org/en/17307.

40. Earle. *The World Is Blue*. 178–179.

41. Roberts. *The Ocean of Life*. 108.

42. Morse, Ian. "In Indonesia, an Earthquake Leaves Devastation on Land and Under the Sea." *Eco-Business*. April 10, 2019. Accessed December 4, 2019. https://www.eco-business.com/news/in-indonesia-an-earthquake-leaves-devastation-on-land-and-under-the-sea/.

43. City Mayors Statistics. "Largest Cities in the World." Accessed July 30, 2019. http://www.citymayors.com/statistics/largest-cities-population-125.html.

44. Ibid. 69.

45. Earle. *The World Is Blue*. 25.

46. Helvarg. *50 Ways to Save the Ocean*. 71.

47. Roberts. *The Ocean of Life*. 120.

48. Ibid. 128.

49. Ibid. 91.

50. Ibid. 122–123.

51. Patel, Prachi. "Stemming the Plastic Tide: 10 Rivers Contribute Most of the Plastic in the Oceans." *Scientific American*. February 1, 2018. Accessed

April 14, 2020. https://www.scientificamerican.com/article/stemming-the-plastic-tide-10-rivers-contribute-most-of-the-plastic-in-the-oceans/.
52. Roberts. *The Ocean of Life*. 146.
53. Ibid. 148.
54. Ibid. 143.
55. Rafferty. "9 of the Biggest Oil Spills in History." *Encyclopedia Britannica*. Accessed September 10, 2019. https://www.britannica.com/list/9-of-the-biggest-oil-spills-in-history.
56. Kaushik, Mohit. "11 Major Oil Spills of the Maritime World." *Marine Insight*. February 15, 2019. Accessed September 10, 2019. https://www.marineinsight.com/environment/11-major-oil-spills-of-the-maritime-world/.
57. Helvarg. *50 Ways to Save the Ocean*. 72.
58. Roberts. *The Ocean of Life*. 147.
59. Ebbesmeyer and Scigliano. *Flotsametrics and the Floating World*. 6.
60. Ibid. 219.
61. Morelle, Rebecca. "Mariana Trench: Deepest-Ever Sub Dive Finds Plastic Bag." *BBC News*. May 13, 2019. Accessed December 4, 2019. https://www.bbc.com/news/science-environment-48230157.
62. Zhang, Amanda. "Financing the Battle Against Marine Plastic Pollution." *Conservation Finance Network*. September 25, 2019. Accessed April 13, 2020. https://www.conservationfinancenetwork.org/2019/09/25/financing-the-battle-against-marine-plastic-pollution.
63. Cooper, Rachel. "Plastic Killing Up to One Million People a Year, Says Sir David Attenborough." *Climate Action*. May 14, 2019. Accessed November 6, 2019. http://www.climateaction.org/news/plastic-killing-up-to-one-million-people-a-year-says-sir-david-attenborough.
64. Cooper, Rachel. "Dead Whale Found with 40 kg of Plastic in Its Stomach." *Climate Action*. March 18, 2019. Accessed November 6, 2019. http://www.climateaction.org/news/dead-whale-found-with-40kg-of-plastic-in-its-stomach.
65. Cooper, Rachel. "Microplastics Found in All Dead Dolphins, Seals and Whales Tested in British Waters." *Climate Action*. January 31, 2019. Accessed November 6, 2019. http://www.climateaction.org/news/microplastics-found-in-all-dead-dolphins-seals-and-whales-tested-in-british.
66. Lamb, Joleah B., et al. "Plastic Waste Associated with Disease on Coral Reefs." *Science*. 26 January 2018: Vol. 359, Issue 6374, pp. 460–462. https://doi.org/10.1126/science.aar3320.
67. Cooper, Rachel. "Marine Plastic Pollution Is Costing the World Billions, Report Finds." *Climate Action*. April 4, 2019. Accessed November 6, 2019. http://www.climateaction.org/news/marine-plastic-pollution-is-costing-the-world-billions-report-finds.
68. Roberts. *The Ocean of Life*. 161.
69. Earle. *Sea Change*. 256.
70. Ibid. 255.
71. Earle. *The World Is Blue*. 102.

72. Ibid. 101.
73. Ebbesmeyer and Scigliano. *Flotsametrics and the Floating World*. 189–190.
74. UN Environment. "Our planet Is Drowning in Plastic Pollution." Accessed April 10, 2020. https://www.unenvironment.org/interactive/beat-plastic-pollution/; Also see: Ritchie, Hannah Ritchie and Max Roser. "Plastic Pollution." Our World in Data. September 2018. Accessed April 10, 2020. https://ourworldindata.org/plastic-pollution.
75. Jambeck et al. "Plastic Waste Inputs from Land Into Ocean." *Science*. 13 February 2015. Vol 347. Issue 6223. https://doi.org/10.1126/science.1260352.
76. Cooper, Rachel. "Supermarkets Put Out 59 Billion Items of Plastic Packaging Every Year." *Climate Action*. November 15, 2018. Accessed December 4, 2019. www.climateaction.org/news/supermarkets-put-out-59-billion-items-of-plastic-packaging-every-year.
77. Ebbesmeyer and Scigliano. *Flotsametrics and the Floating World*. 70.
78. Ibid. 205.
79. Earle. *The World Is Blue*. 114.
80. Ibid. 105–106.
81. Ibid. 108.
82. Roberts. *The Ocean of Life*. 163.
83. Earle. *The World Is Blue*. 107.
84. Ebbesmeyer and Scigliano. *Flotsametrics and the Floating World*. 211.
85. Helvarg. *50 Ways to Save the Ocean*. 55.
86. Roberts. *The Ocean of Life*. 164.
87. Cooper, Rachel. "The Plastic Age, Fossilised Plastic Could Remain for 10 Million Years." *Climate Action*. September 14, 2018. Accessed October 16, 2018. www.climateaction.org/news/the-plastic-age-fossilised-plastic-could-remain-for-10-million-years.
88. Earle. *Sea Change*. 254.
89. Teh, Louise S. L. et al. "A Global Estimate of the Number of Coral Reef Fishers." *PLOS ONE*. June 19, 2013. Accessed December 4, 2019. https://doi.org/10.1371/journal.pone.0065397.
90. FAO. 2018. "The State of World Fisheries and Aquaculture 2018—Meeting the Sustainable Development Goals." Rome. Licence: CC BY-NC-SA 3.0 IGO. http://www.fao.org/3/I9540EN/i9540en.pdf. 39–40.
91. Deacon, Robert T. and Dominic P. Parker. "Encumbering Harvest Rights to Protect Marine Environments: A Model of Marine Conservation Easements." 39.
92. Balmford. *Wild Hope*. 163.
93. Sobel and Dahlgren. *Marine Reserves*. 64–65.
94. Ibid. 45.

95. Avery, Helen. "Blue Finance: Why Marine PPPs Could be a Win-Win-Win." *Euromoney*. June 5, 2018. Accessed December 4, 2019. https://www.euromoney.com/article/b18hg7vjwmy9hn/blue-finance-why-marine-ppps-could-be-a-win-win-win.

96. Bahadur. *The Pirates of Somalia*. 16.

97. Earle. *Sea Change*. 162.

98. Ibid. 185.

99. Earle. *The World Is Blue*. 66–67.

100. Roberts. *The Ocean of Life*. 52.

101. Wilson, Elizabeth. "Fishing Subsidies Are Speeding the Decline of Ocean Health." *Pew Charitable Trusts*. July 19, 2018. Accessed April 10, 2020. https://www.pewtrusts.org/en/research-and-analysis/articles/2018/07/19/fishing-subsidies-are-speeding-the-decline-of-ocean-health.

102. Earle. *The World Is Blue*. 48.

103. Ibid. 69.

104. Earle. *Sea Change*. 173.

105. Roberts. *The Ocean of Life*. 49.

106. Earle. *The World Is Blue*. 227–229.

107. Roberts. *The Ocean of Life*. 244.

108. Ibid. 352.

109. Earle. *The World Is Blue*. 234.

110. Roberts. *The Ocean of Life*. 217.

111. Earle. *The World Is Blue*. 67.

112. Nowlis and Friedlander. "Research Priorities and Techniques." In: Sobel and Dahlgren. *Marine Reserves*. 188.

113. Roberts. *The Ocean of Life*. 166.

114. Ibid. 180.

115. Hooker, Sascha K. and Leak R. Gerber. "Marine Reserves as a Tool for Ecosystem-Based Management: The Potential Importance of Megafauna." 30.

116. Roberts. *The Ocean of Life*. 343.

117. Ibid. 200.

118. Braverman. *Coral Whisperers*. 35.

119. Helvarg. *50 Ways to Save the Ocean*. 78–79.

120. Roberts. *The Ocean of Life*. 183–184.

121 Ibid. 186.

122. Carbon Pulse. "CP Daily: Thursday July 12, 2018." July 12, 2018. Accessed December 5, 2019. www.carbon-pulse.com/55421/; Also see: Gabbatiss, Josh. "Eradicating Rats from Tropical Islands Could Protect Coral Reefs from Effects of Climate Change, Scientists Find." *The Independent*. July 11, 2018. Accessed December 5, 2019. https://www.independent.co.uk/environment/rats-coral-reefs-eradicate-tropical-island-seabirds-fish-chagos-archipelago-indian-ocean-a8442371.html.

123. World Atlas. "The 10 Busiest Cruise Ports in The World." 2019. Accessed December 5, 2019. https://www.worldatlas.com/articles/the-10-busiest-cruise-ports-in-the-world.html.

124. Cruise Market Watch. "2018 Worldwide Cruise Line Passenger Capacity." 2018. Accessed December 5, 2019. https://cruisemarketwatch.com/capacity/.

125. Hoegh-Guldberg, O. et al. *Reviving the Ocean Economy*. 7.

126. Ibid. 14.

127. World Bank. "Gross Domestic Product 2018." Accessed July 13, 2019. https://databank.worldbank.org/data/download/GDP.pdf.

128. Costanza, Robert et al. "The Value of the World's Ecosystem Services and Natural Capital." *Nature*. Volume 387. 15 May 1997. http://www.esd.ornl.gov/benefits_conference/nature_paper.pdf. 253–260.

129. Earle. *The World Is Blue*. 17.

130. The Economics of Ecosystems and Biodiversity. "Ecosystem Services." Accessed November 23, 2016. http://www.teebweb.org/resources/ecosystem-services/.

131. TEEB. "The Initiative." Accessed February 20, 2017. http://www.teebweb.org/about/the-initiative/.

132. Earle. *The World Is Blue*. 62.

133. Adams, Mary Jane. "Beautiful Aliens: The Invertebrates." In *Underwater Eden*, edited by Gregory S. Stone and David Obura. VII–IX.

134. Helvarg. *50 Ways to Save the Ocean*. 114.

135. Roberts. *The Ocean of Life*. 216–217.

136. Vertigo Lab. "Innovations for Coral Finance, ICRI Publication." 12.

137. Austin (n 26), 7; Souter and Linden (n 41), 659. In Goodwin. *International Environmental Law and the Conservation of Coral Reefs*. 12.

138. Hoegh-Guldberg, O. et al. *Reviving the Ocean* Economy. 4.

139. Vertigo Lab. "Innovations for Coral Finance, ICRI Publication." 12.

140. McKibben, Bill. "Foreword." In Earle, Sylvia. *The World Is Blue*. 7.

141. Carson. *The Sea Around Us*. 170.

142. Hoegh-Guldberg, O. et al. *Reviving the Ocean Economy*. 12.

143. Mohammed (Ed.). *Economic Incentives for Marine and Coastal Conservation*. 5.

144. Nybakken and Bertness (n 13), 407. In Goodwin. *International Environmental Law and the Conservation of Coral Reefs*. 7.

145. Roberts. *The Ocean of Life*. 97.

146. Vertigo Lab. "Innovations for Coral Finance, ICRI Publication." 12.

147. Roberts. *The Ocean of Life*. 102.

148. Settelmyer, Scott. Interviewed by Brian McFarland. March 2020.

149. Settelmyer, Scott. Interviewed by Brian McFarland. March 2020.

150. Stanford University. "Microdocs: Species on Coral Reefs." April 11, 2012. Accessed December 5, 2019. https://web.stanford.edu/group/microdocs/species.html.

151. Daily, Gretchen C. and Katherine Ellison. *The New Economy of Nature*. 7–8.

152. Ibid. 143.

153. BBC Nature. "Big Five Mass Extinction Events." Accessed March 29, 2017. http://www.bbc.co.uk/nature/extinction_events.

154. Daily, Gretchen C. and Katherine Ellison. *The New Economy of Nature.* 157.

155. WWF. "Ecoregion Conservation." Accessed March 12, 2020. https://wwf.panda.org/knowledge_hub/where_we_work/alps/ealp2/ecoregion_conservation_plan2222/.

156. Roberts, Callum M. et al. "Marine Biodiversity Hotspots and Conservation Priorities for Tropical Reefs." 1283.

157. Carson. *The Edge of the Sea.* 13.

158. Earle. *Sea Change.* 6.

159. Earle. *The World Is Blue.* 59.

160. Nielsen, Julius et al. "Eye Lens Radiocarbon Reveals Centuries of Longevity in the Greenland Shark (*Somniosus microcephalus*)." *Science.* Volume 353. Issue 6300. 12 August 2016. http://science.sciencemag.org/content/353/6300/702.

161. Roberts. *The Ocean of Life.* 86.

162. Earle. *The World Is Blue.* 91.

163. Ibid. 74.

164. Carson. *The Edge of the Sea.* 19.

165. Ibid. 259.

166. Levinton, Jeffrey S. "Afterword." In Carson, *The Sea Around Us.* 232.

3

Coral Reef Ecology

Introduction

Biological diversity was first coined by conservation biologist Raymond F. Dasmann in 1968.[1] Biological diversity, also known as biodiversity, is one of the central components of ecology. There are three types of diversity: genetic diversity, species diversity, and ecosystem diversity. To begin, consider:

> The coral animal is an exercize in biological minimalism. It is usually much less than an inch across and consists of a soft cup, open upwards and topped with a ring of tiny tentacles. The tentacles trap minute zooplankton and bring them into the cup, which is stomach, intestine, and excretory organ all in one. Corals are usually colonial, and hundreds of thousands of individuals, or polyps, often compromise the colony. Each can secrete a limy skeleton beneath, and the whole colony produces layers upon layers of lime, which build up eventually to form a large skeleton. The skeleton can be massive or delicately branched.
>
> Corals grow slowly; a mound shaped coral will grow one half an inch upward a year. The polyps contain large numbers of tiny single-celled algae called zooxanthellae, which live within the coral tissues. The zooxanthellae still function like plants. They capture light from the sun and manufacture sugars and other vital constituents of life. A good deal of the photosynthetic sugar is transported from the zooxanthellae to the animal, which benefits by growing faster. But the animal protects the zooxanthellae by harboring them within the coral tissues. Thus, the interdependency benefits both partners. Without the algae, the coral would not grow nearly as fast. Without this symbiotic relationship, the reef would never grow.[2]

© The Author(s) 2021
B. J. McFarland, *Conservation of Tropical Coral Reefs*,
https://doi.org/10.1007/978-3-030-57012-5_3

While some coral reefs, such as the Great Barrier Reef, are relatively new, older reefs have been discovered. For instance, "in 1952, a group of scientists drilled a borehole in Eniwetak Atoll, in the western Pacific {…}. Using modern dating techniques, the geologists could determine that the reef had been growing for over 40 million years – little tentacled animals, reaching for the sun."[3]

Ecosystem Diversity

Oceans, and its ecosystem diversity, are driven by wind, water currents, tides, and upwellings:

> The surface of the sea is unequally heated by the sun; as the water is warmed it expands and becomes lighter, while the cold water becomes heavier and more dense. Probably a slow exchange of polar and equatorial waters is brought about by these differences, the heated water of the tropics moving poleward in the upper layers, and polar water creeping toward the equator along the floor of the sea. But these movements are obscured and largely lost in the far greater sweep of the wind-driven currents. The steadiest winds are the trades, blowing diagonally toward the equator from the northeast and southeast. It is the trades that drive the equatorial currents around the globe. {…} the spinning earth exerts a deflecting force, turning all moving objects to the right in the Northern Hemisphere and to the left in the Southern.[4]
>
> There is, then, no water that is wholly of the Pacific, or wholly of the Atlantic, or of the Indian or the Antarctic. The surf that we find exhilarating at Virginia Beach or at La Jolla today may have lapped at the base of Antarctic icebergs or sparkled in the Mediterranean sun, years ago, before it moved through dark and unseen waterways to the place we find it now. It is by the deep, hidden currents that the oceans are made one.[5]

Ocean ecosystem diversity refers to the variety of ecosystems and micro-ecosystems that exist. This includes underground seamounts, which are the longest and largest mountain chains on Earth, along with the Kuril waters, "which are home to the largest kelp forests in the world."[6] Oceanic ecosystems also include both tropical coral reefs and deep, cold-water coral reefs. Furthermore, there are barrier, fringing, and ribbon reefs. An atoll is "a ring-shaped reef rising from exceptionally deep water," while coral cays "consist basically of a layer of coral sand atop a varied collection of flotsam. Some also have a base of hard limestone rock."[7] In addition, "the surrounding reefs of a coral cay are all submerged at high tide. However, when the tide recedes many hectares of rough coral surface are exposed. This exposed seascape is called the reef flat."[8]

In addition, a coral reef may provide a variety of micro-ecosystems from the sea anemones' protective home for a clownfish (*Amphiprioninae*), to a crevice providing a home for a Moray eel (*Muraenidae*) and a rock lobster (*Palinuridae*), to further down, below the subsoil where a stingray may reside.

Other coastal ecosystems include seagrasses, salt marshes, and mangroves, along with pelagic zones, which may play an interactive role with coral reefs. Ecosystems are complex, and as a result of coral reef degradation and loss, these ecosystems become isolated islands surrounded by changing landscapes. This may lead to habitat fragmentation which reduces the ability of animals to migrate and adapt to climate change.

Species Diversity

About 8.7 million ± 1.3 million species call Earth home.[9] Species diversity in the marine environment, including plants and animals, is unsurpassed in tropical coral reefs. This diversity ranges from blue-green algae "which are among the simplest and oldest forms of life and are the most ancient plants that still exist,"[10] to the largest living species on Earth, the blue whale. Furthermore, "most of the major distinctive categories of animal and plant life that ultimately evolved are still represented in the ocean, including many that are found only there; only about half have ever become established on the land."[11] Take for example, that "sixty thousand or so kinds of crustaceans are included within the most diverse group of animals on Earth, the Arthropoda."[12]

Regarding corals, there are both hard corals and soft corals. One tends to think of corals that build reef systems, like brain and elkhorn corals, which are known as hard corals. In contrast, corals such as sea fans, are considered soft corals. From yellow-colored lettuce (*Agaricia agaricites*) and purple-colored finger corals (*Porites divaricate*) to staghorn (*Acropora cervicornis*) and elkhorn corals (*Acropora palmata*), there are more than 800 species of reef-building corals.[13] Interestingly "{…} if you were to add them all up, there are more coral species [living] deeper than fifty meters than [there are living] shallower than fifty meters. The weird thing is that although the world focuses on shallow tropical corals, for all kinds of good reasons, more than half the diversity, really, is in deep waters"[14] (Fig. 3.1).

Fig. 3.1 Picture of Soft Coral at Opal Reef off Coast of Port Douglas, Australia (*Credit* Brigitta Jozan)

This diversity contributes to ecosystem stability and can help contribute to food security. This said, there are a wide variety of roles or niches which diversity helps to fulfill.

Species diversity also relates to endemic species, threatened and endangered species, indicator species, and keystone species, along with charismatic or flagship species.

Endemic species are species found in a particular location and nowhere else. For example:

We found that many coral reef species were not in fact widespread, and some were limited to remarkably small fragments of habitat. The dark indigo Limbaugh's angelfish, to take one example, is known only from Clipperton Reef, a pinnacle of coral in the eastern Pacific that is isolated from neighboring reefs by nearly six hundred miles of blue water. Or there is the pale dottyback, a shy and ghostly fish found only in the Gulf of Aqaba, a narrow finger of water in the northern Red Sea. Or the splendid toadfish, whose only known haunts are the reefs of the resort island Cozumel in Mexico.[15]

In contrast to endemic species, there are other marine species such as humpback whales (*Megaptera novaeangliae*), sea turtles, and whale sharks (*Rhincodon typus*), that travel great distances.

Threatened, endangered, and critically endangered species are species whose population has historically been in decline and is now facing possible extinction. The International Union for the Conservation of Nature (IUCN), established in 1948, provides a comprehensive Red List for species and a Red List for habitats that are threatened by extinction.

According to WWF, 60% of all wildlife has been wiped out by humans since 1970,[16] and another one million species are threatened with extinction by humans.[17] Countless species within tropical coral reefs are now extinct or face extinction. In 2007, "corals were added to the IUCN Red List of threatened species for the first time – not because these colonial organisms were not threatened earlier, but because there had been a common misperception that marine species are not as vulnerable to extinction as terrestrial ones."[18] As of December 2019, the IUCN Red List indicated there were 21 critically endangered, 66 endangered, and 346 vulnerable species within the "marine neritic - coral reef" zone.[19] For example, this includes the vulnerable leatherback turtles (*Dermochelys coriacea*),[20] the largest turtles on Earth, along with the Kemp's ridley sea turtle (*Lepidochelys kempii*)[21] and Nassau grouper (*Epinephelus striatus*),[22] both of which are critically endangered.

Indicator species—like canaries in a coal mine—are species that indicate an ecosystem's health. Sea turtles, as described by Jennie Gilbert, are an indicator species because a sick turtle—whether due to marine debris or poor water quality and disease—equals a sick ocean.[23] Similarly, corals can indicate the underlying health of the local marine environment.

Keystone species are species that are particularly important for an ecosystem. While all species are important for a given ecosystem, some species, such as coral reefs and zooxanthellae algae, have a multiplying effect on other species. For example, a keystone species in the Caribbean is the black or long-spined sea urchin (*Diadema antillarum*). However, "the massive Caribbean-wide epidemic that killed off over 95 percent of the previously ubiquitous black-spined sea urchin, *Diadema antillarum*, illustrates the key role that coral reef grazers play {...}."[24]

Charismatic species represent an ecosystem and can raise both awareness and financing for the preservation of the entire ecosystem and thus, conserve a multitude of lesser-known species.

Similarly, a flagship species "is a species selected to act as an ambassador, icon, or symbol for a defined habitat, issue, campaign, or environmental cause. A mascot species has many of the same attributes as a flagship species, but is selected for its communications value instead of its ecological value."[25] For instance, there are Important Marine Mammal Areas (IMMAs).[26] Charismatic marine species include blue whales (*Balaenoptera musculus*), common bottlenose dolphins (*Tursiops truncatus*), Ocellaris clownfish (also known as the common clownfish, *Amphiprion ocellaris*) and the West Indian manatee (*Trichechus manatus*). All seven species of sea turtles are charismatic species as they are one of the species that tourists most often want to see.

Similarly, umbrella species are "wide-ranging species {e.g., manatees}, the protection of whose habitat will encompass several other species within their ecosystem"[27] (Fig. 3.2).

Sharks, which play a critical role in the ecosystem, are also charismatic, umbrella species that are often sought out by divers (Fig. 3.3).

There are complex, symbiotic interactions among all species—whether they are more common species, indicator species or endangered species—throughout tropical coral reefs. These symbiotic interactions include neutralism, competition, mutualism, protocooperation, commensalism, amensalism, parasitism, and predation.[28]

Fig. 3.2 Picture of Green Sea Turtles at Heron Bommie, Heron Island, Australia (*Credit* Brigitta Jozan)

Fig. 3.3 Picture of Blacktip Reef Sharks at Shark Bay, Heron Island, Australia (*Credit* Brigitta Jozan)

Genetic Diversity

Genetic diversity refers to the number of individuals within a given species and the diversity of their genes. A sufficient number of breeding pairs must exist to maintain a healthy, sustainable population. Overfishing, degraded reefs, and dead zones at the mouths of rivers, for instance, can decrease populations and their genetic diversity.

Connectivity to Seascapes

Alongside MPAs and conservation projects, it is important to establish large-scale, oceanscape- or seascape-level conservation initiatives which cover numerous terrestrial and marine ecosystems, ranging from seagrass and mangrove forests to offshore tropical coral reefs and pelagic zones. Thus, healthy forests contribute to healthy rivers, and healthy rivers can contribute downstream to healthy oceans. This is important because:

In a few cases, we understand the linkages among habitat types. Shallow, nearshore, tropical ocean environments, for example, are characterized by the connectivity among coral reefs, mangrove lagoons, and sea grass beds (Ogden 1988). These connections can result in different fish assemblages on a coral reef depending on whether there are mangroves and seagrass beds nearby.[29]

For instance, "mangroves provide corals with shade during hot weather events and can limit bleaching {…}."[30]

The creation of marine reserves is an important tool to protect coral reef seascapes:

Marine reserves, regardless of their size, and with few exceptions, lead to increases in density, biomass, individual size, and diversity in all functional groups. The diversity of communities and the average size of the organisms within a reserve are between 20 and 30 percent higher relative to unprotected areas. The density of organisms is roughly double in reserves, while the biomass of organisms is nearly triple. These results are robust despite the many potential sources of error in the individual studies.[31]

Marine spatial planning or zoning is another key tool that supports conservation.

SLOSS, which stands for single large or several small, is a way to frame conservation. It is best to have a single large protected area with interconnected biological or wildlife corridors and to have all of them surrounded by low-impact activities.[32] (Fig. 3.4).

There has been much head scratching over how big and how close {marine} protected areas need to be to function optimally. Based on how far animals and plants move and disperse, the consensus is that they should typically be about six to twelve miles across and no more than twenty-five to fifty miles apart. At its most risk prone – six-mile reserves fifty miles apart – such a network would cover one ninth of the sea. At its most risk averse – twelve-mile reserves twenty-five miles apart – it would cover a third.[33]

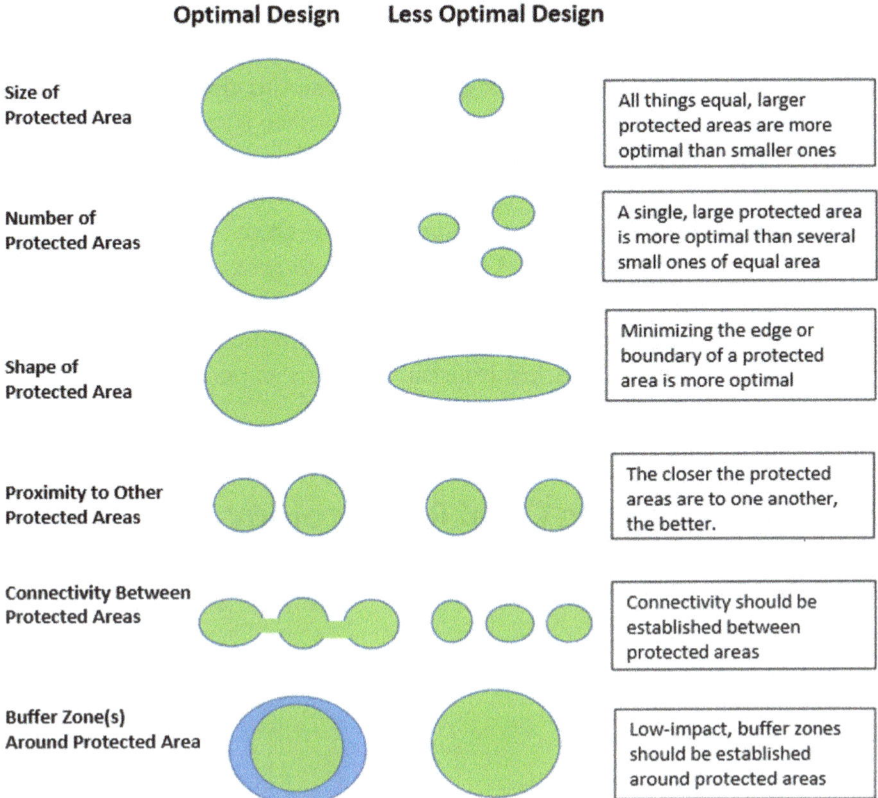

Fig. 3.4 Optimal design of protected areas (*Credit* Brian McFarland)

The IUCN classifies protected areas as:

- Category Ia: Strict Nature Reserve[34];
- Category Ib: Wilderness Area[35];
- Category II: National Park[36];
- Category III: Natural Monument or Feature[37];
- Category IV: Habitat / Species Management Area[38];
- Category V: Protected Landscape / Seascape[39]; and
- Category VI: Protected Area with Sustainable Use of Natural Resources.[40]

There are important distinctions between a marine reserve, a marine protected area (MPA), and a marine wilderness. A marine reserve is the same as a marine "no-take" reserve, which is "an area of the sea in which all consumptive or extractive uses, including fishing, are effectively prohibited and other human interference is minimized to the extent practicable."[41]

A MPA is "any area of intertidal or subtidal terrain, together with its over-lying water and associated flora, fauna, historical, and cultural features, which has been reserved by law or other effective means to protect part or all of the enclosed environment."[42] A marine wilderness is an "area of the sea … along with coastal land where appropriate, that has been protected to preserve or restore its natural character, condition, vistas, living communities, and habitats for present and future generations to enjoy, experience, explore, and study, but leave unaltered. Ocean wilderness areas are large, generally at least 100 square miles, closed to all extractive activities, including all forms of fishing, and to other damaging human activities as needed to ensure that natural communities within flourish, as much as possible unaffected by human activities."[43]

Other Resources on Coral Reef Ecology

5 Gyres Institute

- https://www.5gyres.org/

AquaMaps

- www.aquamaps.org

BirdLife International's Data Zone

- http://datazone.birdlife.org/home

Census of Marine Life; Census of Marine Life's History of Marine Animal Populations Project; and Census of Marine Life's The Future of Marine Animals Populations Study

- http://www.coml.org/;
- http://www.coml.org/history-marine-animal-populations-hmap/; and
- http://www.comlmaps.org/mcintyre/ch16/the-future-of-marine-animal-populations

Conservation Evidence

- http://www.conservationevidence.com/

Corals of the World Database

- http://www.coralsoftheworld.org/page/home/

Coral Trait Database

- https://coraltraits.org/

Critical Ecosystem Partnership Fund

- http://www.cepf.net/Pages/default.aspx

Earth Observatory

- https://earthobservatory.nasa.gov/

Encyclopedia of Earth

- http://eol.org/

Global Biodiversity Information Facility

- http://www.gbif.org/

Global Biodiversity Outlook

- https://www.cbd.int/gbo/

Global Invasive Species Database

- http://www.iucngisd.org/gisd/

Global Ocean Observing System

- https://www.goosocean.org/

Google Ocean / Underwater Street View

- https://www.google.com/streetview/;

- https://www.google.com/maps/about/behind-the-scenes/streetview/treks/oceans/; and
- https://theocean.agency/google-underwater-street-view/

Guidelines for Applying the IUCN Protected Area Management Categories to Marine Protected Areas

- https://portals.iucn.org/library/node/10201

IUCN Red List of Threatened Species

- http://www.iucnredlist.org/

IUCN Species Programme Marine Biodiversity Unit's Global Marine Species Assessment

- https://sites.wp.odu.edu/GMSA/

Joint Research Centre's Digital Observatory for Protected Areas

- http://dopa-explorer.jrc.ec.europa.eu/

Living Planet Index

- http://www.livingplanetindex.org/home/index

Lophelia.org (Resource on Deep Sea Corals)

- http://lophelia.org/

MarXiv

- https://marxiv.org/

Millennium Ecosystem Assessment

- http://www.millenniumassessment.org/en/index.html

National Biodiversity Strategies and Action Plans' Forum

- http://nbsapforum.net/#

Network of Regional Governments for Sustainable Development's Biodiversity Learning Platform

- http://www.nrg4sd.org/biodiversity/regions-biodiversity-learning-platform/

NOAA's Ocean Explorer

- http://www.oceanexplorer.noaa.gov

Ocean Biogeographic Information System (OBIS)

- http://www.iobis.org/

Ocean Data Viewer

- https://data.unep-wcmc.org/

Ocean + Library

- https://library.oceanplus.org/

Open Communications for the Ocean (OCTO)

- https://www.octogroup.org/;
- OCTO's OpenChannels: https://www.openchannels.org/;
- Ecosystem-Based Management Tools Network: http://www.natureserve.org/conservation-tools/ecosystem-based-management-tools-network;
- Marine Ecosystem and Management: https://meam.openchannels.org/meam/archives;
- MPA News: https://mpanews.openchannels.org/mpanews/archives;
- MarineDebris.Info: https://marinedebris.openchannels.org/;
- MarXiv: https://marxiv.org/ and https://www.marxivinfo.org/; and
- NatureServe: http://www.natureserve.org/

Rosenstiel School of Marine and Atmospheric Sciences' Digital Atlas of Marine Species and Locations

- www.damsl.org

SeaLifeBase

- https://www.sealifebase.ca/

TRAFFIC, The Wildlife Trade Monitoring Network

- http://www.traffic.org/

Tree of Life Web Project

- http://www.tolweb.org/tree

Tropicos®

- http://www.tropicos.org/

UNEP's World Conservation Monitoring Centre

- https://www.unep-wcmc.org/

University of New Hampshire's Center for Coastal & Ocean Mapping Joint Hydrographic Center

- http://ccom.unh.edu/

World Ocean Observatory and World Ocean Radio

- http://www.worldoceanobservatory.org/; and
- http://www.worldoceanobservatory.org/world-ocean-radio

World Register of Marine Species (WoRMS)

- http://www.marinespecies.org

WRI's Resource Watch

- https://resourcewatch.org/dashboards

WWF's Global Observation and Biodiversity Information Portal

- http://panda.maps.arcgis.com/home/index.html

WWF's Wildlife and Climate Change

- https://www.worldwildlife.org/initiatives/wildlife-and-climate-change

XL Catlin Global Reef Record
  http://globalreefrecord.org/home_scientific

Notes

1. Martin. *On the Edge*. 135.
2. Levinton, Jeffrey S. "Afterword." In *The Sea Around Us*, edited by R. Carson. 234–235.
3. Ibid. 235.
4. Carson. *The Sea Around Us*. 132.
5. Ibid. 147.
6. Lambacher, Jason. "Nesting Cranes: Envisioning a Russo-Japanese Peace Park in the Kuril Islands." In *Peace Parks*, edited by Saleem H. Ali. 264.
7. Walsh. *Lady Elliot Island*. 11.
8. Ibid. 14.
9. Martin. *On the Edge*. 137–138.
10. Carson. *The Edge of the* Sea. 252.
11. Earle. *Sea Change*. 9.
12. Earle. *The World is Blue*. 93.
13. UN Atlas of the Oceans. "Biodiversity and Coral Reefs." Accessed April 3, 2020. http://www.oceansatlas.org/subtopic/en/c/765/.
14. Braverman. *Coral Whisperers*. 185.
15. Roberts. *The Ocean of Life*. 81.
16. Cooper, Rachel. "60% of Wildlife Has Been Wiped Out by Humans Since 1970, Finds WWF." *Climate Action*. October 31, 2018. Accessed March 12, 2020. http://www.climateaction.org/news/60-of-wildlife-has-been-wiped-out-by-humans-since-1970-finds-wwf.

17. Cooper, Rachel. "Humans Threaten 1 Million Species with Extinction." *Climate Action*. May 7, 2019. Accessed March 12, 2020. http://www.climat eaction.org/news/humans-threaten-1-million-species-with-extinction.

18. Earle. *The World Is Blue*. 125.

19. IUCN 2019. The IUCN Red List of Threatened Species. Version 2019-3. https://www.iucnredlist.org.

20. Wallace et al. 2013. *Dermochelys coriacea. The IUCN Red List of Threatened Species* 2013: e.T6494A43526147. http://dx.doi.org/10.2305/IUCN.UK. 2013-2.RLTS.T6494A43526147.en. Downloaded on 13 December 2019.

21. Wibbels, T. and E. Bevan. 2019. *Lepidochelys kempii* (errata version published in 2019). *The IUCN Red List of Threatened Species* 2019: e.T11533A155057916. Downloaded on 13 December 2019.

22. Sadovy et al. 2018. *Epinephelus striatus. The IUCN Red List of Threatened Species* 2018: e.T7862A46909843. http://dx.doi.org/10.2305/IUCN.UK. 2018-2.RLTS.T7862A46909843.en. Downloaded on 26 November 2019.

23. Gilbert, Jennie. Interviewed by Brian McFarland. September 2019; Also see: Gilbert, Jennie. "Turtles on Track." *TEDxJCUCairns*. November 18, 2015. Accessed October 22, 2019. https://www.youtube.com/watch?v=IzClvb CbDVo.

24. Sobel and Dahlgren. *Marine Reserves*. 41, 43.

25. Hayden, Daniel and Benjamin Dills. "Smokey the Bear Should Come to the Beach." 1.

26. Hoyt, Erich, and Giuseppe Notarbartolo di Sciara. "Perspective: Thirty More Marine Mammal Habitats Awarded Status as Important Marine Mammal Areas." *MPA News*. February 28, 2019. Accessed March 12, 2020. https://mpa news.openchannels.org/news/mpa-news/perspective-thirty-more-marine-mam mal-habitats-awarded-status-important-marine-mammal.

27. Hooker, Sascha K. and Leak R. Gerber. "Marine Reserves as a Tool for Ecosystem-Based Management: The Potential Importance of Megafauna." 35.

28. Bisaccio, Daniel J. of Brown University. Interviewed by Brian McFarland. July 2016.

29. Nowlis and Friedlander. "Research Priorities and Techniques." In Sobel and Dahlgren. *Marine Reserves*. 218–219.

30. Iyer, Venkat, et al. *Finance Tools for Coral Reef Conservation*. 28.

31. Halpern 2003, p. S129. In Sobel and Dahlgren. *Marine Reserves*. 92.

32. Diamond, Jared. "The Island Dilemma: Lessons of Modern Biogeographic Studies of the Design of Natural Reserves." *Biological Conservation* (7) (1975). England: Applied Science Publishers Ltd. Accessed February 20, 2017. http://jareddiamond.org/Jared_Diamond/Further_Reading_files/Dia mond%201975.pdf. 129–146.

33. Roberts. *The Ocean of Life*. 344–345.

34. IUCN. "Protected Areas: Category Ia: Strict Nature Reserve." Accessed February 20, 2017. https://www.iucn.org/theme/protected-areas/about/protec ted-areas-categories/category-ia-strict-nature-reserve.

35. IUCN. "Protected Areas: Category Ib: Wilderness Area." Accessed February 20, 2017. https://www.iucn.org/theme/protected-areas/about/protected-areas-categories/category-ib-wilderness-area.
36. IUCN. "Protected Areas: Category II: National Park." Accessed February 20, 2017. https://www.iucn.org/theme/protected-areas/about/protected-areas-categories/category-ii-national-park.
37. IUCN. "Protected Areas: Category III: Natural Monument or Feature." Accessed February 20, 2017. https://www.iucn.org/theme/protected-areas/about/protected-areas-categories/category-iii-natural-monument-or-feature.
38. IUCN. "Protected Areas: Category IV: Habitat/Species Management Area." Accessed February 20, 2017. https://www.iucn.org/theme/protected-areas/about/protected-areas-categories/category-iv-habitatspecies-management-area.
39. IUCN. "Protected Areas: Category V: Protected Landscape/Seascape." Accessed February 20, 2017. https://www.iucn.org/theme/protected-areas/about/protected-areas-categories/category-v-protected-landscapeseascape.
40. IUCN. "Protected Areas: Category VI: Protected Area with Sustainable Use of Natural Resources." Accessed February 20, 2017. https://www.iucn.org/theme/protected-areas/about/protected-areas-categories/category-vi-protected-area-sustainable-use-natural-resources.
41. Sobel and Dahlgren. *Marine Reserves*. 21.
42. Kelleher 1999. In Sobel and Dahlgren. *Marine Reserves*. 22.
43. The Ocean Conservancy 2000. In Sobel and Dahlgren. *Marine Reserves*. 22.

4

Global Environmental Policy

Introduction

Global environmental policy focused on tropical coral reef conservation is primarily done through local and national policy. Thus, there is "no single dedicated {international} agreement" specific to tropical coral reefs.[1]

One way to assess global environmental policy is to map policy against the major global, regional, and local factors contributing to coral reef degradation and loss. International coral reef policy should primarily, but not exclusively, focus on climate change and ocean acidification; regional policies should primarily focus on pollution runoff, plastics, and overfishing; and local policies should primarily focus on noise pollution, disease, invasive species, and high-human traffic.

Policy broadly includes two main legal systems, English common law and Napoleonic code (also referred to as civil law or Napoleonic law).

Common law is "generally *uncodified*. This means that there is no comprehensive compilation of legal rules and statutes. While common law does rely on some scattered statutes, which are legislative decisions, it is largely based on *precedent*, meaning the judicial decisions that have already been made in similar cases."[2]

Civil law "is *codified*. Countries with civil law systems have comprehensive, continuously updated legal codes that specify all matters capable of being brought before a court, the applicable procedure, and the appropriate punishment for each offense."[3]

© The Author(s) 2021
B. J. McFarland, *Conservation of Tropical Coral Reefs*,
https://doi.org/10.1007/978-3-030-57012-5_4

Countries with tropical coral reefs, such as Brazil and Indonesia, follow the civil law system, while other coral reef countries such as Belize and Jamaica are based on common law.

It is also important to note that designing and implementing policy is challenging, especially in developing countries which at times lack transparency, have increased levels of corruption, and lack adequate funding.

Local Policies

Local policies can have a direct and/or indirect impact on the marine environment. This could include local zoning ordinances, such as stipulating the types and locations of construction. For instance, the province of Bali in Indonesia announced in 2019, "plans to implement a tourist tax which targets plastic pollution,"[4] while Hawaii, in 2018, became the first U.S. "state to ban sale of sunscreens with coral-harming chemicals."[5]

Federal Policies

Similar to local policies, there are numerous federal policies that have direct and/or indirect impacts on the marine environment.

While policy at the local level is critical, tropical coral reefs often transcend state or provincial borders—not to mention country borders—and thus, policy at the federal level is vital to the conservation of tropical coral reefs at a landscape level. However, federal policy—and similarly on the local level—can often conflict. For example, are "{…} the comparatively weaker powers of the coastal state in the EEZ actually a significant issue if the vast majority of the world's coral reefs lie within internal and territorial waters?"[6]

Many countries, such as Belize, have developed integrated coastal zone management plans:

> Small-island developing states are also becoming more and more aware of their rights. Ninety-eight percent of Belizeans voted against off-shore drilling and banned it. And they don't trawl, either. Belize cut the cord, realizing that should an accident ever occur, their tourism would be negatively impacted.[7]

The Seychelles, with a population of less than 100,000 people,[8] has both wealth and beauty. The country issued the world's first sovereign blue bond,[9] and one of the first marine debt-for-nature conversions was completed in the Seychelles in 2015.[10] This said, Seychelles is quite remote and off the normal tourist bucket list, so there are less impacts than, for example, the Galápagos Islands.

An example at a different scale occurred in the United States:

George W. Bush made a huge splash with his Pacific protected areas and single-handedly added over 30 percent to the total area under protection in the world's oceans. That makes him one of the world's great marine conservationists! That expression might stick in some people's throats given his rather less glorious environmental record on land. Whatever your feelings about Bush, he left a wonderful legacy in the sea.[11]

{Years later} US President Barack Obama expanded a national monument off Hawaii, creating the world's largest marine reserve, the White House says. His announcement on Friday quadruples in size a monument originally created by President George W Bush in 2006. The Papahānaumokuākea Marine National Monument will now span 1.5m sq km (582,578 sq miles), more than twice the size of Texas.[12]

Further, a growing number of countries, from Belize[13] to the Cook Islands,[14] have increased the size of their MPAs; yet, the effectiveness of some of these MPAs is still in question.

The United States adopted the Lacey Act of 1900, and then in 2008, the US Congress "followed suit and passed an amendment to the Lacey Act of 1900, which prohibits trafficking in illegally harvested plants and wildlife."[15] The U.S. Endangered Species Act of 1973 is also an important, complementary piece of legislation. In addition, a "growing number of countries, such as Mexico, Cuba, and Costa Rica, have introduced turtle excluder devices (TEDs) on shrimp trawl nets."[16]

It is important to note that policy matters within countries that host tropical coral reefs as well as in those countries with large economies that import products from these host countries. This includes policies surrounding the import of reef fish for consumption or the aquarium trade, as well as importing sugar from countries with adjacent, onshore sugarcane plantations.

Bilateral, Multilateral, and Regional Policies

There are several bilateral, multilateral, and regional policies and initiatives impacting the marine environment including the Alliance of Small Island States (AOSIS) and the Small Island Developing States (SIDS).

The Pacific Islands region is best described as a sea of islands. At 38 million square kilometers (sq km), amounting to more than four times the size of the continental United States, the region lies in the heart of the world's largest ocean. The islands, of which there are more than 30,000, are like jewels in a blue crown, renowned for their unique biodiversity which support an unparalleled diversity of Pacific Island cultures and peoples who are connected by their ocean heritage.

The Pacific Islands are comprised of 23 island nations and territories, with generally small populations and economies, but with management responsibilities under the UN Convention on the Law of the Sea (UNCLOS) for an area larger than the surface of the moon. Their Exclusive Economic Zones are amongst the largest in the world making their small island states in reality large ocean stewards. {...}

Pacific Island nations bear a disproportionate burden of the effects of global climate change, and are highly vulnerable to the impacts of sea level rise, warming sea temperatures, and ocean acidification, whilst also being nations that contribute relatively little to carbon dioxide emissions.[17]

The Pacific Islands' 23 island nations and territories are:

- American Samoa;
- Australia;
- Cooks Islands;
- Federated States of Micronesia;
- Fiji;
- French Polynesia;
- Guam;
- Kiribati;
- Marshall Islands;
- Nauru;
- New Caledonia;
- New Zealand and Tokelau;
- Niue;
- Northern Mariana Islands;
- Palau;
- Papua New Guinea;
- Samoa;
- Solomon Islands;
- Tonga;
- Tuvalu;
- Wallis and Futuna; and
- Vanuatu.

The UN currently identifies 37 member-state as SIDS organized into three regional commissions (i.e., Atlantic, Indian Ocean and South China Sea; the Caribbean; and the Pacific). There are also an additional twenty non-UN members/associate members of regional commissions.[18]

The Pacific Islands Forum Secretariat and the Secretariat of the Pacific Regional Environment Program are a couple of intergovernmental entities. There is also the Coral Triangle Initiative on Coral Reefs, Fisheries, and Food Security (CTI-CFF), a collaborative of six countries.[19] Further, there are several "challenges" such as the Western Indian Ocean Coastal Challenge,[20] Micronesia Challenge,[21] the Caribbean Challenge Initiative.[22]

One difficulty with these forums is that most, if not all, countries strongly hold onto national sovereignty. Yet, because tropical coral reefs often transcend national borders, there is a need to work on a regional level.

An important "{…} global conference was held in Washington, DC in late 1995. This conference established the Global Programme of Action for the Protection of the Marine Environment from Land-Based Activities (GPA). The GPA remains to this day the principal global multilateral framework for addressing LBSP {land-based sources of pollution)."[23]

In addition, there is the:

- Action Plan for the Caribbean Environment Programme and the Cartagena Convention for the Protection and Development of the Marine Environment of the Wider Caribbean Region (the Cartagena Convention);
- Action Plan for the Protection and Management of the Marine and Coastal Environment of the South Asian Seas Region; and
- Convention for the Protection, Management and Development of the Marine and Coastal Environment of the Eastern African Region.

With respect to land-based sources of pollution, "the most significant action plan, however, has been formulated for the Red Sea and Gulf of Aden. The action plan was adopted in March 2003 and is dedicated to coral reefs."[24] However, it is important to note that, "only two protocols on LBSP have been negotiated and entered into force for maritime regions containing coral reefs."[25] These are:

- 1983 Protocol for the Protection of the South-East Pacific Against Pollution from Land-Based Sources; and
- 1990 Protocol for the Protection of the Marine Environment Against Pollution from Land-Based Sources.

In addition, "three further protocols on LBSP have been adopted but are awaiting entry into force."[26] These are:

- 1990 Protocol Concerning Pollution from Land-Based Sources and Activities to the Convention for the Protection and Development of the Marine Environment of the Wider Caribbean Region;
- 2005 Protocol Concerning Land-Based Activities: A Regional Protocol, to the Jeddah Convention, concerning Protection of the Marine Environment from Land-Based Activities in the Red Sea and Gulf of Aden; and
- 2010 Protocol for the Protection of the Marine and Coastal Environment of the Western Indian Ocean from Land-Based Sources and Activities.

Asides from protocols to address LBSP, there are "protocols focused upon protected areas {which} are in force in three of the regions of relevance {…}."[27] These are the:

- 1985 Protocol Concerning Protected Areas and Wild Fauna and Flora in the Eastern African Region (Nairobi Protocol);
- 1989 Protocol for the Conservation and Management of Protected Marine and Coastal Areas of the South-East Pacific (Paipa Protocol); and
- 1990 Protocol Concerning Specially Protected Areas and Wildlife to the Convention for the Protection and Development of the Marine Environment of the Wider Caribbean Region (the SPAW Protocol).

Mr. David Freestone, "who acted as a member of the delegation for Antigua and Barbuda at all three of the meetings convened to negotiate the protocol, has described the SPAW as:

Arguably the most comprehensive regional wildlife protection treaty in the world – it is certainly the most comprehensive of its kind… {reflecting} much of the best in modern thinking on wildlife protection and management.[28]

Another interesting regional initiative seeks to tackle climate change:

A coalition of Pacific island nations wants to raise $500 mln to make all shipping in the Pacific Ocean zero carbon by the middle of the century. The Pacific Blue Shipping Partnership, announced on Tuesday by the governments of Fiji, the Marshall Islands, Samoa, Vanuatu, the Solomon Islands and Tuvalu, has set an emissions reduction target of 40% by 2030, and full decarbonisation by 2050. The partnership intends to raise money through grants from multinational institutions, concessional loans, direct private sector investment, and through issuing regional "blue bonds". The money would be used to retrofit existing passenger and cargo ferries with low-carbon technologies, and to buy new zero-emissions vessels.[29]

Furthermore, there are regional fishery management organizations such as the:

- Agreement on the International Dolphin Conservation Program;
- Indian Ocean South-East Asian Marine Turtle Memorandum of Understanding;
- Indian Ocean Tuna Commission;
- Inter-American Tropical Tuna Commission; and
- Western and Central Pacific Fisheries Commission.[30]

International Policies

Some of the main international policies related to the marine environment are the Law of the Sea Treaty, the United Nations Framework Convention on Climate Change's Paris Agreement, and the Convention on Biological Diversity. An interesting question arises: "which takes precedence in situations where a state finds itself a party to two conventions seeking to conserve the same habitat or species?"[31]

For an excellent resource on international coral reef legislation, see Dr. Edward J. Goodwin's book *International Environmental Law and the Conservation of Coral Reefs.*[32]

Law of the Sea Treaty

One of the most significant international policies is the Law of the Sea Treaty of 1982. Further:

> In 1992 the United Nations imposed an unprecedented constraint on the 'freedom to fish' concept by banning the use of large drift nets on the high seas, largely because of the outrage generated over indiscriminate killing of incidentally caught victims – birds, turtles, whales, dolphins, seals, and numerous nontarget kinds of fish.[33]

The Law of the Sea Treaty outlines the sovereign rights of coastal states, the allowable use of another country's EEZs (i.e., for "innocent passage"), and allowable uses of the high seas.[34]

Paris Agreement

The Rio Convention of 1992 established the United Nations Framework Convention on Climate Change (UNFCCC). Nearly every country on Earth submitted Intended Nationally Determined Contributions (INDCs) prior to the UNFCCC Conference of the Parties (COP) 21 event in Paris. Then, almost 200 countries signed onto the Paris Agreement in December 2015, which now commits countries to carry out their Nationally Determined Contributions (NDCs). The threshold for the Paris Agreement to come into force was met on October 4, 2016, and then the Paris Agreement officially came into force one month later on November 4, 2016.[35]

Prior to COP21, many countries were beginning to regulate GHG emissions. This includes:

- The European Union Emissions Trading Scheme (EU ETS);
- China's developing national carbon market with pilot provinces and cities (Beijing, Chongqing, Guangdong, Hunan, Shanghai, Shenzhen, and Tianjin);
- The Western Climate Initiative (WCI) and the Regional Greenhouse Gas Initiative (RGGI) in the United States; and
- South Korea's cap-and-trade system.

It is important to note that "present climate change policies focus on the atmosphere, largely neglecting the ocean, despite ample evidence that the ocean drives and regulates planetary climate, weather, temperature, and chemistry."[36] However, the UNFCCC COP25, held in Madrid, Spain, was coined "the Blue COP."

Convention on Biological Diversity

The Rio Convention of 1992 also established the Convention on Biological Diversity (CBD). The CBD has a target for MPAs, currently set at 10% "global" coverage by 2020 and this was also adopted as a target under Sustainable Development Goal 14.

According to Dr. Edward J. Goodwin, the CBD:

has succeeded in bringing a large number of states within a single MEA {multilateral environmental agreement} regime charged with stimulating policy development across a comprehensive agenda for conserving biodiversity, ensuring the sustainable use of its components, and the fair and equitable sharing of the benefits arising from the use of genetic resources. In turn, this throws up particular problems, namely focusing the agenda in a detailed and meaningful way, managing external relations with other MEAs and garnering consensus among so many contracting parties in what has historically been a highly politicised negotiating environment.[37]

Furthermore:

{...} all of the eligible states endowed with coral reefs are contracting parties, with the sole exception of the United States. This total represents nearly 98 per cent of global coral reefs. This places the CBD in the strongest position for influencing coral reef conservation in comparison to the other MEAs {...}. Of course, bringing the United States within the regime would offer a significant increase in coverage (1.3 per cent). This may not be a realistic possibility in the light of that country's position with regards to the convention and its provisions on biotechnology, IP rights and access to technology. Fortunately, the United

States still takes something of an active role with regard to biological diversity as it follows developments under the CBD and often attends COPs.[38]

This said, many countries are pushing for marine protection from 20% to 30% by 2030.

Ramsar Convention

The Ramsar Convention on Wetlands of International Importance especially as Waterfowl Habitat (Ramsar Convention) includes coral reefs and "was concluded as the first global agreement to deal with a particular habitat."[39]

Sustainable Development Goals

The Sustainable Development Goals (SDGS) are also important:

Before doctors begin their practice, they pledge to do no harm and remember to 'remain a member of society, with special obligations to all my fellow human beings.' This principle underpins the U.N. SDGs as well. Yet, the 2030 global goals may need their own Hippocratic Oath when it comes to environmental conservation and sustainability.

The most acute risk of unintended consequences may arise in SDG No. 14: ocean conservation. Almost 50% of the world's population lives within 150 km of a coastline. Many of those people depend on the seas for food security and their livelihoods; most of the world's fisheries are small-scale or community run. Ocean conservation therefore needs a 'social code of conduct,' argue Rebecca Jarvis and Nathan Bennett, research fellows at Auckland University of Technology and the University of Washington, respectively. 'There is a risk that some actions taken will undermine the rights and needs of local people.' Such a code could include:

Fair conservation governance and decision-making processes to ensure respect for and participation from local communities, indigenous people, traditional users, and marginalized populations.

Socially-just conservation action that recognizes tenure and rights of indigenous communities and local traditions, accounts for food security and social welfare, and incorporates enforceable 'benefit sharing' arrangements.

Accountable conservation organizations that adapt best practices and commit to transparency dand conflict remediation.

The recommendations are broad, but are the starting point for a more comprehensive framework.[40]

Many of the book's case studies will explore these issues.

Other International Policies

There are numerous international policies that either directly aim to address the degradation and loss of tropical coral reefs or which aim to address related aspects such as climate change, biodiversity, and human rights. This includes, but is not limited to:

- Convention on International Trade in Endangered Species of Wild Fauna and Flora (CITES);
- Universal Declaration of Human Rights;
- United Nations Declaration on the Rights of Indigenous Peoples (UNDRIP);
- International Labor Organization (ILO) Convention;
- World Trade Organization (WTO) and other international trade agreements such as the Asia-Pacific Economic Cooperation (APEC), the North American Free Trade Agreement (NAFTA; now known as the United States-Mexico-Canada Agreement or USMCA), and Mercosur; and
- 2016 UNEA Resolution 2/12 on sustainable coral reefs management.[41]

Despite good intentions, such international policies appear to be less effective. As described in *The Economy of Nature*:

> Although governments have negotiated a wide array of global and regional agreements to protect certain ecosystems from degradation and extinction – such as the Ramsar Conventions on Wetlands, the Convention on Biological Diversity, and the Convention on the Law of the Sea – these agreements are mostly weak, lacking the participation, resources and systems of incentives and enforcement they need to be effective.[42]

As succinctly explained by Dr. Edward J. Goodwin:

> {…} the principal needs are: driving the formulation of further regional protocols on land-based sources of pollution in coral reef regions and helping participants access financial support and technology so as to implement commitments; helping the Ramsar Convention to promote greater participation of states and coverage of reefs as wetlands of international importance; encouraging the promotion of more reefs to World Heritage status; exploring ways to drive the creation of new coral reef enclaves whilst also improving capacities for implementing MPAs and conservation policies; ensuring that

replication of effort is not occurring (or perpetuated by inter-agency competition) where conventions' mandates overlap; and, influencing negotiations for appropriate mitigation targets at climate change conferences. Creating and financing such a unit or group that can deliver on these objectives, would do much for ensuring international law is providing a suitable framework to support the vital work needed to conserve coral reefs.[43]

As proposed by Dr. Curtis Ebbesmeyer, "we might start by requiring that shippers and vessel operators report anything they lose or jettison in the sea. As it stands, they aren't even obliged to report the contents of lost shipping containers. They need only report the loss of eight or more containers at a time, because of the hazard to shipping."[44]

Company Policies

In addition to government policies, it is important to recognize that many companies around the world are adopting internal company policies that are designed to reduce the company's environmental impact. For example, Puma implemented an environmental profit-loss statement,[45] and Microsoft implemented an internal carbon price.[46] In 1987, "Anheuser-Busch, the world's largest brewer, voluntarily changed to photodegradable loop carriers, and other major users have followed."[47]

More recently, there has been the emergence of single-use plastic free commitments from many of the world's largest companies including American Airlines,[48] Marriott International,[49] Starbucks,[50] and McDonald's.[51]

The following timeline is an overview of some important global environmental policies as well as the dates certain governmental institutions were established.

Dates of Global Environmental Policy

Nineteenth Century

- 1892: "As long ago as the 1890s there was concern in the Bahamas over the destruction of coral reefs, as evident in the Sea Gardens Protection Act of 1892, which prohibited dredging or the removal of coral, sea fans, or other organisms from the seabed."[52]

Twentieth Century

- 1900: Lacey Act is passed, amended in 2008.[53]
- 1911: North Pacific Fur Seal Treaty.[54]
- 1931: The "Convention for the Regulation of Whaling was signed by twenty-two nations."[55]
- 1949–1955: "…maximum sustainable yield became the goal of international fishing management policies."[56]
- 1958: "The first appearance of such a measure in a legislative framework seems to be in the United States in 1958, in the Fish and Wildlife Coordination Act, to mitigate damages to wetlands."[57]
- 1971: Convention on Wetlands of International Importance Especially as Waterfowl Habitat (Ramsar Convention).[58]
- 1972: U.S. National Marine Sanctuaries Act.[59]
- 1972: The UN Conference on the Human Environment is held in Stockholm.[60]
- 1972: The Convention Concerning the Protection of the World Cultural and Natural Heritage is signed, came into force in 1975.[61]
- 1972: The "U.S. Congress passed legislation authorizing the first underwater equivalent of national parks, national marine sanctuaries, to provide a measure of security for natural, historic, and cultural values under the sea {…}."[62]
- 1972: U.S. Marine Mammal Protection Act.[63]
- 1973: U.S. Endangered Species Act.[64]
- 1973: CITES is passed, enters into force on July 1, 1975.[65]
- 1974: Convention on the Prevention of Marine Pollution from Land-Based Sources is adopted.[66]
- 1975: The Convention on the Prevention of Marine Pollution by Dumping of Wastes and Other Matter of 1972 goes into force.[67]
- 1976: The U.S. "Congress made an attempt in 1976 to control overfishing in the nation's coastal waters by the passage of the Magnuson Fishery, Conservation, and Management Act (MFCMA). This legislation protects U.S. fishermen by prohibiting foreign fleets from fishing within the Exclusive Economic Zone (EEZ) {…}."[68]
- 1976: Convention on Conservation of Nature in the South Pacific (the Apia Convention).[69]
- 1978: Australia ceases commercial hunting of whales.[70]
- 1982: UN Convention on the Law of the Sea.[71]
- 1982: "Tuna purse seine licensing is currently regulated under the Nauru Agreement, a subregional agreement spanning eight countries including

Kiribati that make up the Parties to the Nauru Agreement (PNA) established in February 1982."[72]

- 1982: Regional Convention for the Conservation of the Red Sea and the Gulf of Aden (Jeddah Convention).[73]
- 1983: Convention for the Protection and Development of the Marine Environment of the Wider Caribbean Region (Cartagena Convention).[74]
- 1987: World Commission on Environment and Development is held, and the *Our Common Future*, known as the Brundtland Report, is published.[75]
- 1988: MARPOL Annex V came into force which regulated marine debris, particularly garbage from ships.[76]
- 1990: U.S. Clean Air Act Amendments are signed into law; establishing a market-based trading system for SOx and NOx emissions.[77]
- 1990: U.S. Oil Pollution Act is passed.[78]
- 1991: "{…} the 'mother of all oil spills,' as some called the deliberate release into the Persian Gulf in 1991 of more than 45 times the amount of oil lost from the *Exxon Valdez*. {…} the majority of the oil flowed onto the beaches, reefs, and marshes of the upper Gulf, from Saudi Arabia to Kuwait {…}."[79]
- 1992: The UN Conference on Environment and Development (UNCED, also known as the Rio "Earth Summit") is held:

  - Agenda 21 adopted;
  - The Rio Declaration on Environment and Development adopted;
  - The Convention on Biological Diversity adopted;
  - UNFCCC adopted.[80]
  - Furthermore, "The SIDS were first recognized as a distinct group of developing countries at the UN Conference on Environment and Development in 1992."[81]

- 1992: UN Commission on Sustainable Development (CSD) is created "to ensure effective follow-up of the United Nations Conference on Environment and Development (UNCED), also known as the Earth Summit."[82]
- 1992: "{…} since 1992, high seas drift netting has been banned by the United Nations."[83]
- 1993: The CBD enters into force.[84]
- 1994: The International Coral Reef Initiative founded.[85]
- 1994: The first UN Global Conference on the Sustainable Development of SIDS held in Barbados; the Barbados Programme of Action for the Sustainable Development of SIDS is adopted.[86]
- 1997: The Kyoto Protocol adopted in Kyoto, Japan.[87]
- 1998: The U.S. Coral Reef Task Force (USCRTF) established "by Presidential Executive Order to lead U.S. efforts to preserve and protect coral

reef ecosystems. The USCRTF includes leaders of 12 Federal agencies, seven U.S. States, Territories, Commonwealths, and three Freely Associated States."[88]

Twenty-First Century

- 2000: Cartagena Protocol on Biosafety is adopted and entered into force on September 11, 2003.[89]
- 2000: U.S. Coral Reef Conservation Act of 2000.[90]
- 2000: The UN Millennium Summit was held.[91] The UN Millennium Declaration (later known as the Millennium Development Goals or MDGs) is first adopted.[92] The MDGs are now known as the Sustainable Development Goals (SDGs), adopted in September 2015.[93]
- 2000: U.S. Oceans Act of 2000.[94]
- 2001: The Stockholm Convention on Persistent Organic Pollutants adopted, later enters into force in 2004.[95]
- 2002: World Summit on Sustainable Development in Johannesburg, South Africa.[96]
- 2003: "The establishment of a High Seas Marine Protected Area (MPA) Executive Committee, which brings together the International Union for Conservation of Nature (IUCN), the World Wildlife Fund (WWF), the World Commission on Protected Areas (WCPA), and some governments in the causes of high-seas MPAs, in 2003."[97]
- 2003: World Parks Congress in Durban, South Africa is held.[98]
- 2004: Marine Turtle Conservation Act of 2004 is passed in the U.S.[99]
- 2005: The Kyoto Protocol enters into force.[100]
- 2005: The first International Marine Protected Areas Congress is held in Geelong, Australia.[101]
- 2005: Tommy Remengesau, then-President of Palau, launched the Micronesia Challenge.[102]
- 2006: Assembly Bill 32 (AB 32), the California Global Warming Solutions Act of 2006, is passed.[103]
- 2006: The U.S. Congress enacts the Marine Debris Research, Prevention, and Reduction Act.[104]
- 2008: In 2008, the "US Congress followed suit and passed an amendment to the Lacey Act of 1900, which prohibits trafficking in illegally harvested plants and wildlife."[105]
- 2009: The second International Marine Protected Areas Congress is held in Washington, DC.[106]
- 2009: The World Ocean Conference is held in Manado, Indonesia.[107]

- 2010: The Pacific Oceanscape was "unanimously endorsed by Pacific Islands Forum Leaders."[108]
- 2012: The UN Conference on Sustainable Development (Rio + 20) took place in Rio de Janeiro, Brazil.[109]
- 2015: UNFCCC's Paris Agreement adopted.[110]
- 2018: Agreement is reached that states "commercial fishing in the high seas portion of the Central Arctic Ocean will be banned, until scientists can confirm that it can be done sustainably."[111]
- 2018: U.S. Save Our Seas Act of 2018.[112]

Dates When Government Institutions Were Established

Twentieth Century

- 1920: League of Nations.[113]
- 1944: The Articles of Agreement of the International Bank for Reconstruction and Development (i.e., the World Bank) and the International Monetary Fund were drawn up and adopted; articles became effective on December 31, 1945.[114]
- 1945: FAO.[115]
- 1945: The UN officially comes into existence.[116]
- 1946: International Whaling Commission.[117]
- 1948: Germany's KfW Bank.[118]
- 1958: The U.S. NASA.[119]
- 1959: Inter-American Development Bank (IADB).[120]
- 1961: The Organisation for Economic Co-operation and Development (OECD).[121]
- 1961: USAID.[122]
- 1964: African Development Bank (AfDB).[123]
- 1965: United Nations Development Programme. "UNDP is based on the merging of the United Nations Expanded Programme of Technical Assistance, created in 1949, and the United Nations Special Fund, established in 1958."[124]
- 1966: Asian Development Bank (ADB).[125]
- 1970: The U.S. NOAA.[126]
- 1972: UNEP is created by the UN General Assembly.[127]

- 1973: The World Heritage Trust is "established through UNESCO's 1973 Convention for the Protection of the World Cultural and Natural Heritage."[128]
- 1991: European Bank for Reconstruction and Development (EBRD).[129]
- 1991: Global Environment Facility (GEF).[130]
- 1994: Uruguay Round of the General Agreement on Tariffs and Trade (GATT) concluded.[131]
- 1994: World Trade Organization (WTO).[132]

Twenty-First Century

- 2000: The National Oceanic and Atmospheric Administration's (NOAA) Coral Reef Conservation Program is established by the Coral Reef Conservation Act of 2000.[133]

Other Resources on Global Environmental Policy

American Bar Association's Section of Environment, Energy, and Resources

- https://www.americanbar.org/groups/environment_energy_resources/

Assessing Progress on Ocean and Climate Action, 2019: A Report of the Roadmap to Oceans and Climate Action (ROCA) Initiative

- https://rocainitiative.files.wordpress.com/2019/12/roca-2019-progress-report-1.pdf

Caribbean Challenge Initiative

- https://www.caribbeanchallengeinitiative.org/

Climate Change Litigation Databases

- http://climatecasechart.com/;
- http://www.lse.ac.uk/GranthamInstitute/climate-change-laws-of-the-world/; and
- http://www.lse.ac.uk/GranthamInstitute/countries/

Convention on Biological Diversity

- https://www.cbd.int/; and
- https://www.cbd.int/blg/

<u>Coral Digest</u>

- http://coraldigest.org/index.php/ReefRegulations

<u>ECOLEX</u>

- http://www.ecolex.org/

<u>Environmental Law Institute</u>

- www.elr.info

<u>FAO's FAOLEX Database</u>

- http://www.fao.org/faolex/en/

<u>GovTrack, A Project of Civic Impulse, LLC</u>

- https://www.govtrack.us/

<u>International Coral Reef Initiative and UN Documents</u>

- https://www.icriforum.org/icri-documents/ICRI-and-UN-Documents

<u>INTERPOL's Environmental Crime Division</u>

- https://www.interpol.int/Crimes/Environmental-crime; and
- https://www.interpol.int/Crimes/Environmental-crime/Fisheries-crime

<u>Micronesia Challenge</u>

- http://themicronesiachallenge.blogspot.com/

<u>National Biodiversity Strategies and Action Plans' (NBSAP) Policy and Legislation Page</u>

- http://nbsapforum.net/#categories/293

NDC Partnership

- http://ndcpartnership.org/

Ramsar Convention and The List of Wetlands of International Importance (the Ramsar List)

- https://www.ramsar.org/about-the-ramsar-convention; and
- https://www.ramsar.org/document/the-list-of-wetlands-of-international-importance-the-ramsar-list

The Last Beach Cleanup's Country Laws on Plastic Products

- https://www.lastbeachcleanup.org/countrylaws

UN's Audiovisual Library on International Law

- http://legal.un.org/avl/intro/introduction.html?tab=2

UNEP Coral Reef Unit and WWF Coral Reefs Advocacy Initiative's *Conventions and Coral Reefs*

- https://www.icriforum.org/sites/default/files/Conventions_CoralReefs.pdf

UNEP's Publication Single-Use Plastics: A Roadmap for Sustainability

- https://wedocs.unep.org/bitstream/handle/20.500.11822/25496/single UsePlastic_sustainability.pdf

UNFCCC's Paris Agreement

- https://unfccc.int/process-and-meetings/the-paris-agreement/the-paris-agreement

Yale Center for Environmental Law and Policy

- http://envirocenter.yale.edu/

Notes

1. Goodwin. *International Environmental Law and the Conservation of Coral Reefs.* 28.
2. University of California at Berkeley's School of Law. "The Common Law and Civil Law Traditions." Accessed April 5, 2017. https://www.law.berkeley.edu/library/robbins/CommonLawCivilLawTraditions.html.
3. Ibid.
4. Cooper, Rachel. "Bali to Impose 'Tourist Tax' for Plastic Pollution." *Climate Action.* January 28, 2019. Accessed April 10, 2020. http://www.climateaction.org/news/bali-to-impose-tourist-tax-for-plastic-pollution.
5. Wu, Nina. "Hawaii Becomes First State to Ban Sale of Sunscreens with Coral-Harming Chemicals." *Honolulu Star-Advertiser.* July 3, 2018. Accessed April 10, 2020. https://www.staradvertiser.com/2018/07/03/breaking-news/hawaii-becomes-1st-state-to-ban-sale-of-sunscreens-with-coral-harming-chemicals/.
6. Goodwin. *International Environmental Law and the Conservation of Coral Reefs.* 54.
7. Braverman. *Coral Whisperers.* 195.
8. Worldometers. "Seychelles Population." Accessed December 13, 2019. https://www.worldometers.info/world-population/seychelles-population/.
9. World Bank. "Seychelles Launches World's First Sovereign Blue Bond." October 29, 2018. Accessed December 13, 2019. https://www.worldbank.org/en/news/press-release/2018/10/29/seychelles-launches-worlds-first-sovereign-blue-bond.
10. SeyCCAT. "Seychelles' Conservation and Climate Adaptation Trust (SeyCCAT): Achieving Conservation Through Innovative Finance and Creative Collaborations." Accessed August 7, 2019. https://seyccat.org/wp-content/uploads/2017/09/infographic_talking_points_bat_web-1.pdf.
11. Roberts. *The Ocean of Life.* 344.
12. BBC News. "World's Largest Marine Reserve Created Off Hawaii." August 27, 2016. Accessed December 13, 2019. http://www.bbc.com/news/world-us-canada-37202045.
13. Dasgupta, Shreya. "Belize to Nearly Triple Area Under Strict Marine Protected Areas." *Mongabay.* April 8, 2019. Accessed April 10, 2020. https://news.mongabay.com/2019/04/belize-to-nearly-triple-area-under-strict-marine-protected-areas/.
14. Evans, Monica. "Will a Massive Marine Protected Area Safeguard Cook Islands' Ocean?" *Mongabay.* September 19, 2019. Accessed April 10, 2020. https://news.mongabay.com/2019/09/will-a-massive-marine-protected-area-safeguard-cook-islands-ocean/.
15. Martin. *On the Edge.* 82.
16. Helvarg. *50 Ways to Save the Ocean.* 52.

17. CI. "CI Pacific Islands and the Pacific Oceanscape Program." Accessed October 9, 2018. https://www.conservation.org/publications/Documents/Pacific_Islands_Program_Factsheet.pdf. 1.

18. Sustainable Development Goals Knowledge Platform. "Small Island Developing States." Accessed April 10, 2020. https://sustainabledevelopment.un.org/topics/sids/list.

19. Coral Triangle Initiative. "About." Accessed April 10, 2020. http://www.coraltriangleinitiative.org/about.

20. Global Island Partnership. "Western Indian Ocean Coastal Challenge (WIOCC)." Accessed December 13, 2019. http://www.glispa.org/11-commitments/32-western-indian-ocean-coastal-challenge-wio-cc.

21. Micronesia Challenge. "Home." Accessed December 13, 2019. http://themicronesiachallenge.blogspot.com/.

22. Caribbean Challenge Initiative. "Home." Accessed December 13, 2019. https://www.caribbeanchallengeinitiative.org/.

23. Goodwin. *International Environmental Law and the Conservation of Coral Reefs*. 75.

24. Ibid. 85.

25. Ibid. 98.

26. Ibid. 98.

27. Ibid. 89.

28. D. Freestone, 'Specially Protected Areas and Wildlife in the Caribbean—The 1990 Kingston Protocol to the Cartagena Convention' (1990) 5(4) *International Journal of Estuarine and Coastal Law* 362, 368. In: Goodwin. *International Environmental Law and the Conservation of Coral Reefs*. 90.

29. Doherty, Ben. "Pacific Islands Seek $500m to Make Ocean's Shipping Zero Carbon." *The Guardian*. September 24, 2019. Accessed December 13, 2019. https://www.theguardian.com/environment/2019/sep/24/pacific-islands-seek-500m-ocean-shipping-zero-carbon.

30. NOAA. "International Fisheries Organizations." Accessed February 19, 2020. https://www.fisheries.noaa.gov/international-affairs/international-fisheries-organizations.

31. Goodwin, Edward J. *International Environmental Law and the Conservation of Coral Reefs*. 118.

32. Goodwin, Edward J. *International Environmental Law and the Conservation of Coral Reefs*. New York: Routledge of Taylor & Francis Group, 2011.

33. Earle. *Sea Change: A Message of the Oceans*. 162–163.

34. UN's Office of Legal Affairs. "United Nations Convention on the Law of the Sea of 10 December 1982 Overview and full text." 2018. Accessed April 14, 2020. https://www.un.org/depts/los/convention_agreements/convention_overview_convention.htm.

35. UNFCCC. "Paris Agreement." Accessed April 14, 2020. https://unfccc.int/process-and-meetings/the-paris-agreement/the-paris-agreement.

36. Earle. *The World is Blue*. 164.

37. Goodwin. *International Environmental Law and the Conservation of Coral Reefs.* 135.
38. Ibid. 114.
39. Ibid. 138.
40. Pothering, Jessica. "A Hippocratic Oath for the Sustainable Development Goals." *ImpactAlpha.* July 5, 2017. Accessed January 9, 2020. https://impact alpha.com/a-hippocratic-oath-for-the-sustainable-development-goals-5563b6 bbe160/.
41. International Coral Reef Initiative. "Resolution 2/12 on Sustainable Coral Reefs Management (UNEA 2)." Accessed February 17, 2020. https://www. icriforum.org/news/2016/10/resolution-212-sustainable-coral-reefs-manage ment-unea-2.
42. Daily, Gretchen C. and Katherine Ellison. *The New Economy of Nature.* 7.
43. Goodwin. *International Environmental Law and the Conservation of Coral Reefs.* 272.
44. Ebbesmeyer and Scigliano. *Flotsametrics and the Floating World.* 221.
45. Puma. "Environmental Profit and Loss Account." Accessed February 9, 2017. http://about.puma.com/en/sustainability/environment/environmental-profit-and-loss-account.
46. Microsoft. "Internal Carbon Fee." Accessed February 9, 2017. Available: http://download.microsoft.com/download/0/A/B/0AB2FDD7-BDD9-4E23-AF6B-9417A8691CF5/Microsoft%20Carbon%20Fee%20Impact.pdf.
47. Earle. *Sea Change: A Message of the Oceans.* 260.
48. Wentworth, Adam. "American Airlines Switches Away from Plastic Straws to Cut Waste." *Climate Action.* July 12, 2018. Accessed December 6, 2018. www.climateaction.org/news/american-airlines-switches-away-from-pla stic-straws-to-cut-waste.
49. Danigelis, Alyssa. "Marriott International Eliminating Plastic Straws Next Year." *Environmental Leader.* July 18, 2018. Accessed October 16, 2018. https://www.environmentalleader.com/2018/07/marriott-international-plastic-straws/.
50. Danigelis, Alyssa. "Starbucks Vows to Ditch Plastic Straws by 2020." *Environmental Leader.* July 13, 2018. Accessed October 16, 2018. https://www.env ironmentalleader.com/2018/07/starbucks-plastic-straws/.
51. Danigelis, Alyssa. "McDonald's Begins Transition Away From Plastic Straws." *Environmental Leader.* June 15, 2018. Accessed October 16, 2018. https:// www.environmentalleader.com/2018/06/mcdonalds-transition-plastic-straws/.
52. Dahlgren. "Bahamian Marine Reserves—Past Experience and Future Plans." In: Sobel and Dahlgren. *Marine Reserves.* 269.
53. USFWS. "Lacey Act." Accessed November 23, 2016. https://www.fws.gov/int ernational/laws-treaties-agreements/us-conservation-laws/lacey-act.html.
54. National Ocean Service. "North Pacific Fur Seal Treaty of 1911." Accessed August 12, 2019. https://celebrating200years.noaa.gov/events/furseltreaty/ welcome.html.

55. Earle. *Sea Change*. 209.

56. Earle. *The World Is Blue*. 56.

57. Vertigo Lab. "Innovations for Coral Finance, ICRI Publication." 24; Also see: USFWS. "Digest of Federal Resource Laws of Interest to the U.S. Fish and Wildlife Service: Fish and Wildlife Coordination Act." Accessed January 9, 2020. https://www.fws.gov/laws/lawsdigest/fwcoord.html.

58. Ramsar Convention. "About the Ramsar Convention." Accessed December 10, 2019. https://www.ramsar.org/about-the-ramsar-convention.

59. NOAA National Marine Sanctuaries. "Legislation: The National Marine Sanctuaries Act." Accessed December 13, 2019. https://sanctuaries.noaa.gov/about/legislation/.

60. UN. "United Nations Conference on the Human Environment (Stockholm Conference)." Accessed November 23, 2016. https://sustainabledevelopment.un.org/milestones/humanenvironment.

61. UNESCO. "Convention Concerning the Protection of the World Cultural and Natural Heritage." Accessed November 29, 2016. http://whc.unesco.org/en/conventiontext/.

62. Earle. *The World is Blue*. 244.

63. USFWS. "Marine Mammal Protection Act." Accessed August 12, 2019. https://www.fws.gov/international/laws-treaties-agreements/us-conservation-laws/marine-mammal-protection-act.html.

64. USFWS. "A History of the Endangered Species Act of 1973." Accessed November 23, 2016. https://www.fws.gov/endangered/laws-policies/esa-history.html.

65. CITES. "What Is CITES?" Accessed November 29, 2016. https://cites.org/eng/disc/what.php.

66. NOAA Office of General Counsel. "Land-Based Sources of Marine Pollution." Accessed December 10, 2019. https://www.gc.noaa.gov/gcil_land_based_pollution.html.

67. International Maritime Organization. "The London Convention and Protocol." Accessed August 12, 2019. http://www.imo.org/en/KnowledgeCentre/ReferencesAndArchives/IMO_Conferences_and_Meetings/London_Convention/Pages/default.aspx.

68. Earle. *Sea Change*. 194.

69. Secretariat of the Pacific Regional Environment Programme. "Apia Convention." Accessed December 12, 2019. https://www.sprep.org/convention-secretariat/apia-convention.

70. Australia Government's Department of the Environment and Energy. "Whaling," Accessed August 12, 2019. https://www.environment.gov.au/marine/marine-species/cetaceans/whaling.

71. UN. "United Nations Convention on the Law of the Sea of 10 December 1982: Overview and full text." Accessed June 6, 2018. http://www.un.org/depts/los/convention_agreements/convention_overview_convention.htm.

72. Rotjan et al. "Establishment, Management, and Maintenance of the Phoenix Islands Protected Area." 308.

73. Regional Organization for the Conservation of the Environment of the Red Sea & Gulf of Aden. "Consolidated Jeddah Conventions (1982–2009)." Accessed December 10, 2019. http://www.persga.org/inner.php?id=32.

74. UNEP's Caribbean Environment Programme. "Learn about the Cartagena Convention and Its Protocols." Accessed December 10, 2019. http://cep.unep.org/cartagena-convention.

75. World Commission on Environment and Development. "Our Common Future." Accessed November 29, 2016. http://www.un-documents.net/our-common-future.pdf.

76. International Maritime Organization. "Prevention of Pollution by Garbage from Ships." Accessed December 13, 2019. www.imo.org/en/OurWork/Environment/PollutionPrevention/Garbage/Pages/Default.aspx.

77. USEPA. "1990 Clean Air Act Amendment Summary." Accessed November 23, 2016. https://www.epa.gov/clean-air-act-overview/1990-clean-air-act-amendment-summary.

78. USEPA. "Laws & Regulations: Summary of the Oil Pollution Act." Accessed December 28, 2018. https://www.epa.gov/laws-regulations/summary-oil-pollution-act.

79. Earle. *The World is Blue*. 155.

80. UN. "UN Conference on Environment and Development (1992)." Accessed November 29, 2016. http://www.un.org/geninfo/bp/enviro.html.

81. Vertigo Lab. "Innovations for Coral Finance, ICRI Publication." 28.

82. UN. "Commission on Sustainable Development (CSD)." Accessed November 29, 2016. https://sustainabledevelopment.un.org/csd.html.

83. Balmford. *Wild Hope*. 170.

84. CBD. "Introduction." Accessed November 29, 2016. https://www.cbd.int/intro/.

85. ICRI. "About Us." Accessed December 13, 2019. https://www.icriforum.org/about-icri.

86. Sustainable Development Goals Knowledge Platform. "BPOA (1994)—Barbados Programme of Action." Accessed April 10, 2020. https://sustainabledevelopment.un.org/conferences/bpoa1994.

87. UNFCCC. "Kyoto Protocol." Accessed November 23, 2016. http://unfccc.int/kyoto_protocol/items/2830.php.

88. U.S. Coral Reef Task Force. "About the U.S. Coral Reef Task Force." Accessed June 6, 2018. https://www.coralreef.gov/about/.

89. CBD. "The Cartagena Protocol on Biosafety." Accessed November 23, 2016. https://bch.cbd.int/protocol.

90. NOAA. "Appendix A: Coral Reef Conservation Act of 2000." Accessed December 13, 2019. https://www.coris.noaa.gov/activities/actionstrategy/08_cons_act.pdf.

91. UN. "Millennium Summit (6–8 September 2000)." Accessed December 13, 2019. https://www.un.org/en/events/pastevents/millennium_summit.shtml.

92. UN Millennium Project. "What They Are." Accessed November 23, 2016. http://www.unmillenniumproject.org/goals/.

93. Division for Sustainable Development. "Mission Statement." Accessed November 23, 2016. https://sustainabledevelopment.un.org/about.

94. GovTrack. "S. 2327 (106th): Oceans Act of 2000." Accessed June 6, 2018. https://www.govtrack.us/congress/bills/106/s2327.

95. Stockholm Convention. "History of the Negotiations of the Stockholm Convention." Accessed December 13, 2019. http://www.pops.int/TheConvention/Overview/History/Overview/tabid/3549/Default.aspx.

96. UN. "World Summit on Sustainable Development." Accessed November 23, 2016. http://www.un.org/events/wssd/summaries/envdevj1.htm.

97. Earle. *The World Is Blue*. 214.

98. IUCN. "2003 Durban World Parks Congress." Accessed December 13, 2019. https://www.iucn.org/content/2003-durban-world-parks-congress.

99. Congress.gov. "H.R.3378—Marine Turtle Conservation Act of 2004." Accessed August 28, 2018. https://www.congress.gov/bill/108th-congress/house-bill/3378.

100. UNFCCC. "Kyoto Protocol." Accessed November 23, 2016. http://unfccc.int/kyoto_protocol/items/2830.php.

101. International Marine Protected Areas Congress. "Background." Accessed June 6, 2018. http://www.impac3.org/en/congress/about/background.

102. MPA News. "The Micronesia Challenge: Assessing Progress over the Past Two Years." Accessed December 13, 2019. https://mpanews.openchannels.org/news/mpa-news/micronesia-challenge-assessing-progress-over-past-two-years.

103. California Environmental Protection Agency. "Assembly Bill 32 Overview." Accessed November 23, 2016. https://www.arb.ca.gov/cc/ab32/ab32.htm.

104. GovTrack. "S. 362 (109th): Marine Debris Research, Prevention, and Reduction Act." Accessed June 6, 2018. https://www.govtrack.us/congress/bills/109/s362.

105. Martin. *On the Edge*. 82.

106. International Marine Protected Areas Congress. "Background." Accessed June 6, 2018. http://www.impac3.org/en/congress/about/background.

107. IISD Reporting Services. "World Ocean Conference Bulletin." Accessed December 13, 2019. http://enb.iisd.org/oceans/woc2009/html/ymbvol162num5e.html.

108. CI. "CI Pacific Islands and the Pacific Oceanscape Program." Accessed October 9, 2018. https://www.conservation.org/publications/Documents/Pacific_Islands_Program_Factsheet.pdf. 2.

109. UN. "United Nations Conference on Sustainable Development. Rio + 20." Accessed November 23, 2016. https://sustainabledevelopment.un.org/rio20.

110. UNFCCC. "Historic Paris Agreement on Climate Change." Accessed November 23, 2016. http://newsroom.unfccc.int/unfccc-newsroom/finale-cop21/.

111. Cooper, Rachel. "New Global Agreement Results in Fishing Ban in the Arctic Ocean." *Climate Action*. October 5, 2018. Accessed December 13, 2019. http://www.climateaction.org/news/new-global-agreement-results-in-fishing-ban-in-the-arctic-ocean.

112. Congress.gov. "S.3508—Save Our Seas Act of 2018." Accessed August 12, 2019. https://www.congress.gov/bill/115th-congress/senate-bill/3508/text.

113. U.S. Department of State's Office of the Historian. "The League of Nations, 1920." Accessed November 23, 2016. https://history.state.gov/milestones/1914-1920/league.

114. World Bank Group. "World Bank Group Archivists' Chronology 1944-2013." Accessed November 23, 2016. http://pubdocs.worldbank.org/en/186241442500110286/PDF-World-Bank-Group-Archivists-Chronology-1944-2013.pdf.

115. FAO. "History." November 23, 2016. http://www.fao.org/world-food-day/2016/history/en/.

116. UN. "History of the United Nations." Accessed November 23, 2016. http://www.un.org/en/sections/history/history-united-nations/index.html.

117. International Whaling Commission. "Key Documents: The Convention." Accessed June 6, 2018. https://iwc.int/convention.

118. KfW. "KfW Turns 65." Last modified November 11, 2013. https://www.kfw.de/KfW-Group/Newsroom/Aktuelles/Pressemitteilungen/Pressemitteilungen-Details_171713.html.

119. NASA. "History Home." Accessed November 28, 2016. http://history.nasa.gov/.

120. IADB. "History of the Inter-American Development Bank." Accessed November 23, 2016. http://www.iadb.org/en/about-us/history-of-the-inter-american-development-bank,5999.html.

121. OECD. "History." Accessed November 23, 2016. http://www.oecd.org/about/history/.

122. USAID. "USAID History." Accessed November 23, 2016. https://www.usaid.gov/who-we-are/usaid-history.

123. African Development Bank. "History." Accessed November 23, 2016. http://www.afdb.org/en/about-us/corporate-information/history/.

124. UNDP. "Frequently Asked Questions." Accessed November 23, 2016. http://www.undp.org/content/undp/en/home/operations/about_us/frequently_askedquestions.html#being.

125. Asian Development Bank. "Origins." Accessed November 23, 2016. http://www.adb.org/about/history.

126. NOAA. "Our history." Accessed June 6, 2018. http://www.noaa.gov/our-history.

127. UNEP. "Milestones." Accessed November 23, 2016. http://www.unep.org/Documents.multilingual/Default.asp?DocumentID=287.

128. Myers. *The Shrinking Ark*. 246.
129. European Bank for Reconstruction and Development. "History of the EBRD." Accessed November 23, 2016. http://www.ebrd.com/who-we-are/history-of-the-ebrd.html.
130. World Bank Group. "The World Bank Group's Global Environment Facility Program." Last modified May 21, 2014. http://www.worldbank.org/en/topic/climatechange/brief/gef.
131. WTO. "The Uruguay Round." Accessed November 23, 2016. https://www.wto.org/english/thewto_e/whatis_e/tif_e/fact5_e.htm.
132. WTO. "20th Anniversary of the WTO." Accessed November 23, 2016. https://www.wto.org/english/thewto_e/20y_e/20y_e.htm.
133. NOAA Coral Reef Conservation Program. "Who We Are." Accessed June 6, 2018. https://coralreef.noaa.gov/about/welcome.html.

5

International Finance

Introduction

According to the World Bank, the size of the world's gross domestic product (GDP), in nominal terms in 2018, was USD\$85.8 trillion.[1] In 2018, the largest national economies by GDP were[2]: (Table 5.1).

Global GDP is primarily derived, based on 2017 estimates, from household consumption (56.4%), government consumption (16.1%), fixed capital investments (25.7%) and investments in inventories (1.4%). While the services (63%) and industry sectors (30%) are the largest sectors contributing to GDP, followed by agriculture (6.4%), it is important to understand that the agricultural sector is a key employer (31%) when compared to industry (23.5%). The services sector, as of 2014, is estimated to be the largest employment sector (45.5%).[3]

This said:

There's \$500 trillion of wealth on planet Earth, give or take - maybe \$230 trillion in land and property, \$200 trillion in debt and \$70 trillion in equity. All of that wealth comes, ultimately, from the planet and from a stable climate. Last week's Nobel Prize winner William Nordhaus points out in his 2013 book 'The Climate Casino' that 'the last 7,000 years have been the most stable climatic period in more than 100,000 years,' and they have also seen the rise of civilization and the creation of that wealth. This is not a coincidence. According to Axios, last week's IPCC puts the cost of a 1.5C increase at \$54 trillion, in today's money. That number comes from research that also says that a 2.0°C increase will cause \$69 trillion of damage, and a 3.7°C increase will cause a

© The Author(s) 2021
B. J. McFarland, *Conservation of Tropical Coral Reefs*,
https://doi.org/10.1007/978-3-030-57012-5_5

Table 5.1 Largest national economies in the world

	Country	GDP (trillion USD$)		Country	GDP (trillion USD$)
1	U.S.	$20.494	11	Russia	$1.658
2	China	$13.608	12	South Korea	$1.619
3	Japan	$4.971	13	Australia	$1.432
4	Germany	$3.997	14	Spain	$1.426
5	UK	$2.825	15	Mexico	$1.224
6	France	$2.778	16	Indonesia	$1.042
7	India	$2.726	17	Netherlands	$0.913
8	Italy	$2.074	18	Saudi Arabia	$0.782
9	Brazil	$1.869	19	Turkey	$0.767
10	Canada	$1.709	20	Switzerland	$0.706

Source The World Bank

stunning $551 trillion in damage - more than all the wealth currently existing in the world, which gives an indication of just how much richer humanity could become if we don't first destroy our planet.[4]

With this context in mind, as of January 2016, only "about $52 billion per year flows to conservation projects {not just coral reef conservation projects}, the bulk of it is public and philanthropic funds. The best estimates are that $300 to $400 billion per year is needed to preserve healthy ecosystems on land and in the oceans, and with them the earth's natural capital stock of clean air, fresh water and species biodiversity."[5] In return, consider:

The annual 'gross marine product' (GMP) – equivalent to a country's annual gross domestic product – is at least US$2.5 trillion; the total 'asset' base of the ocean is at least US$24 trillion. Underpinning this value are direct outputs (fishing, aquaculture), services enabled (tourism, education), trade and transportation (coastal and oceanic shipping) and adjacent benefits (carbon sequestration, biotechnology). Putting it into an international context, if the ocean were a country it would have the seventh largest economy in the world.[6]

There is a clear funding gap when it comes to conserving the tropical coral reefs. The International Coral Reef Initiative states:

One of the main barriers to conservation enhancement is that the funds currently needed to achieve effective, lasting conservation significantly exceed the available funds. An average US$270 million commitment per year for marine conservation was recorded between 2010 and 2016. Meanwhile, the funds necessary to achieve the United Nations Convention on Biological

Diversity target of having 20% of the ocean protected are estimated at between US$4 and $8 billion per year in management costs only.[7]

Besides this funding gap, there is "significant variation in the amount of funding provided for coral reef management in different regions, including relatively low investment per unit area of reef in some regions and countries where reefs are extensive, highly biodiverse or hold a large proportion of climate refugia."[8] This said, "research does show that government sources are a dominant source of funding for MPAs in developed countries whereas in developing countries foreign assistance and park entry fees provide relatively more resources."[9]

Going forward, "{…} to finance the preservation of the world's precious ecosystems will require $200 billion to $300 billion in additional capital, and private investment capital may be the main source of additional capital. Attracting such a level of private capital will require attractive risk-adjusted rates of return, in addition to clear and measurable conservation impacts."[10] While this USD$200 billion to USD$300 billion is not exclusively for coral reefs, it is important to understand that all such ecosystems need to be conserved and humanity is falling short of these financial targets. This additional USD$200 billion to USD$300 billion "only" represents 0.23–0.35% of global GDP.

Regarding the private sector, the *State of Private Investment in Conservation 2016* report states USD$8.2 billion was invested by the private sector between 2004 and 2015 in the following three conservation-oriented investments:

- 1. Sustainable food and fiber production (e.g., forestry, agriculture, and fisheries);
- 2. Habitat conservation (e.g., mitigation banking and forest carbon trading); and
- 3. Water quality and quantity protection (e.g., watershed protection and water rights trading).[11]

Yet, the overall "travel and tourism industry is the largest business sector in the world, accounting for more than 230 million jobs and 10 percent of the global economy. The $6.5 trillion industry is one of the top five export earners in 80 percent of the world's countries and the top export earner in 60 of them."[12] The size of the U.S.' pet supplies industry was USD$72.56 billion in 2018.[13] Zoos and aquariums accredited by the Association of Zoos & Aquariums "contributed more than $22.5 billion to the US economy in 2018."[14] In addition, a significant amount of money is raised from recreational fishing licenses and SCUBA divers.

Although outdated, "the Melbourne, Australia based Economists at Large & Associates, shows more than 13 million people took whale watching tours last year {2008} in 119 countries worldwide, generating a whopping $2.1 billion in total expenditures during 2008."[15]

Between domestic and international box offices, the combined gross revenue generated from the movies Finding Nemo (2003)[16] and its sequel Finding Dory (2016)[17]—two animated movies starring marine animals and coral reefs—was more than USD$1.95 billion.

Before explaining the financial framework to be used to analyze the various conservation financing instruments, the following will briefly review some key terms.

Capital Markets

Capital markets, whether via the primary or secondary market, are where equities and bonds are traded in order to raise financing.

Equities and Bonds

Equities are shares in the stock issued by publicly traded companies or shares in a private company, whereas bonds are debt instruments. Bonds can be issued by governments or companies. Typically, a bond's principal is paid back at the end of the bond's maturity, and intermittent coupon payments are paid to the investor (lender). With respect to bonds issued by a government, there are general obligation bonds and special revenue bonds. A general obligation bond is paid back from the government's overall revenue (e.g., sales and/or property taxes collected), while a special revenue bond is paid back from a specific government revenue (e.g., a specific tax on tourism).

Short-Term and Long-Term Debt

In addition to bonds, there are other forms of debt such as traditional bank loans and notes. Oftentimes, conservation organizations will utilize short-term loans, as opposed to long-term loans:

> Short-term loans can also pay for the up-front costs of a transaction, such as options, earnest money, or down payments, or to cover legal, appraisal, or

planning costs. Restoration projects often have retention requirements (withholding of money until the project is completed), which hinder cash flow for the project.[18]

Long-term debt could be raised to finance education centers or offices,[19] along with establishing ecotourism operations.

Many conservation organizations have established internal and external revolving loan funds, because it can be challenging to get a loan from a traditional bank that is unfamiliar with conservation.

> An internal fund, such as that of The Nature Conservancy (TNC), recirculates money within a large organization. OSI {Open Space Institute}, on the other hand, uses external revolving funds, which lend money to outside organizations seeking funding for conservation projects.[20]

With respect to debt:

> Debt reschedulings involve the deferral of debt service payments falling due to some future date as a means of providing interim cash relief or, effectively, a reduction in debt in present value terms. Debt refinancing involves the exchange of an old loan for a new loan on improved terms. Debt forgiveness involves the extinguishing of all of part of a debt obligation. A less common form of debt restructuring is a debt exchange.[21]

This debt exchange can take the form of a debt-for-nature swap (see Chapter 13) or a debt-for-development initiative.

Insurance

There are different types of insurance, including "traditional" insurance, parametric insurance, and catastrophe bonds.

> Catastrophe bonds or 'cat' bonds are a possible means by which small, disaster-prone developing countries can insure themselves against natural disasters. Cat bonds are typically structured as floating rate securities which pay an attractive yield to investors but waive some or all of the interest and principal repayments when a specific, predetermined event such as a natural disaster occurs. Cat bonds therefore transfer some of the risk to a country's public finances from a natural disaster to the purchasers of the bond.[22]

For a more detailed discussion of parametric insurance, see Chapter 10's case study.

Conservation Trust Funds

There are three main types of conservation trust funds: an endowment fund; a sinking fund; and a revolving fund. An endowment fund is a fund where the project or organization employs an endowment to obtain a rate of return to support conservation activities. A sinking fund uses both the endowment's principle and returns to support conservation activities until the endowment's principle is exhausted. A revolving fund is similar to a sinking fund, except instead of having a fixed time period until the endowment is exhausted, the revolving fund is continuously replenished with new funds.

Currencies

The value of a currency relative to another currency is constantly changing, and this can be particularly challenging in developing countries where exotic currencies exhibit more volatility relative to currencies in developed countries. Currency volatility can impact investment decisions and directly impact conservation activities.

Interest Rates

Interest rates are the rates charged to borrow money. Interest rates are often pegged to a published bank lending rate and then adjusted based on the borrower's creditworthiness. Interest rates can be a variable rate (which can fluctuate over time) and a fixed rate (a set rate for a predetermined period of time), or combination of the two. Due to increased risk, some countries with coral reefs and the companies that operate within these countries tend to have higher interest rates and thus a higher cost to borrow.

For instance, USD$1 million on March 23, 2016, was equivalent to J$121,604,633 (Jamaican Dollar).[23] This same USD$1 million, three years later, on March 22, 2019, was equivalent to J$125,088,665 (2.87% increase).[24] Similarly, €1 million on October 21, 2014, was equivalent to Ar3,416,979,020 (Malagasy Ariary, currency used in Madagascar).[25] Four years later, this €1 million on October 22, 2018, was equivalent to Ar3,982,606,171 (16.55% increase).[26] Another example, taking £1 million on March 18, 2014, was equivalent to B$1,656,954 (Bahamian Dollar).[27] Five years later, this £1 million was equivalent to B$1,320,715 (20.29% decrease) on March 18, 2019.[28] Furthermore, A$1 million (Australian

Dollar) on May 22, 2009, was equivalent to Rp8,009,611,625 (Indonesian Rupiah).[29] Ten years later, this A$1 million was equivalent to Rp9,986,695,778 (24.68% increase) on May 22, 2019.[30]

Currency movements, such as these examples, can impact investment decisions and directly impact conservation activities. While there might be less implications in the examples about with a U.S.-based investor with U.S. dollars investing in a coral reef conservation project in Jamaica (i.e., due to a modest 2.87% appreciation) there could more significant implications—depending on the investment structure—with the British Pound's 20.29% depreciation vis-à-vis the Bahamian Dollar.

Project Finance

Project finance involves structuring the finance of a project where the project's cash flows are used to pay back investors and where there is no recourse against the project sponsor's corporate assets. Project finance is often used for large infrastructure projects.

Public Finance

Public finance is funding provided by a government. Public finance can be raised via taxation to pay for the establishment and maintenance of a MPA or the issuance of a bond by a city to upgrade the public water supply and restore an upstream watershed.

Trade Finance

In addition to direct financing for trade, there is also specialized contracting and insurance products for trade. Trade finance is often provided for the international trade of soft commodities such as cattle, soy, palm oil, timber, and pulp paper and for hard commodities such as diamonds, gold, and petroleum. For example, farmed salmon in Chile is eligible for trade finance.

Financial Intermediaries

Financial intermediaries are market makers that connect borrowers with lenders including banks, traders, and brokers.

Cash Flows

Cash flows involve the sources, quantity, and timing of both revenue and expenses related to a particular project, program, or portfolio. Challenges can arise when there is a mismatch between cash outflows and cash inflows.

Conservation Finance

Conservation finance is a subset field of finance that involves deploying funds for the conservation and restoration of ecosystems. Conservation finance can come from public or private sources and can be structured in a variety of ways including project finance or through a debt offering.

Financial Frameworks

There are three financial frameworks that will be used to analyze each of the conservation financing instruments that will be discussed throughout the book. These financial frameworks are: the efficient frontier; the Global Association of Risk Professionals' Framework for Financial Risk Management; and the Risk, Return, Time (Horizon), Taxes, Liquidity, Legal and Unique (RRTTLLU) framework for Portfolio Management.

A common financial calculation that is conducted on investments is a basic review of expected revenue and expenses. Additional financial calculations, including Net Present Value (NPV), Internal Rate of Return (IRR), payback periods, and sensitivity analyses, can also be performed.

These calculations are important when making investment decisions. For instance: If you need to raise USD$1 million in five years from now to launch a new MPA, with a 5% annual interest rate, how much do you need to raise each year starting at the end of year one until the end of year five? The following assumes at the end of each year, you deposit an equal amount into a fund that returns 5% a year, compounded annually, and then you will liquidate the full USD$1 million starting day one of year six. The answer is approximately USD$180,975. Similarly, if you need USD$10 million each year, fixed for perpetuity, to manage MPAs throughout the Coral Triangle, what size of an endowment—with 7% annual returns—do you need to have? Assuming you will not be raising additional funds, the answer is approximately USD$142,857,143.

Efficient Frontier

The efficient frontier is a conceptual framework to align the expected return of an investment with an appropriate level of risk. In contrast, an inefficient investment is an investment with a higher level of risk than the corresponding expected return warrants. Thus, to make up for this deficiency, an investment can seek ways to reduce risk (e.g., through internally mitigating the risk or purchasing insurance) and/or increase the expected return (e.g., reducing costs, diversifying revenue, expanding sales, etc.) to reach the efficient frontier (Fig. 5.1).

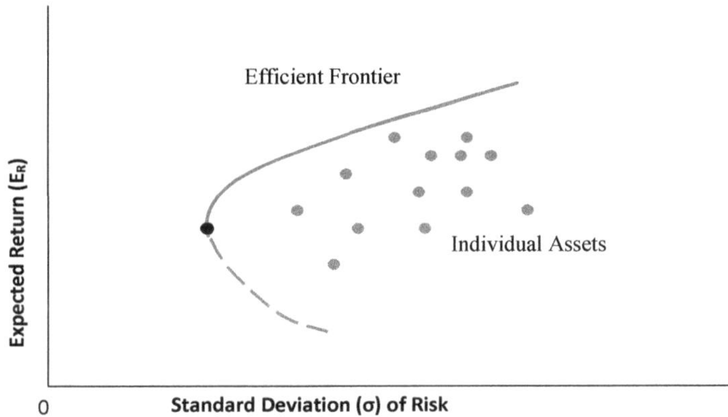

Fig. 5.1   Efficient frontier (*Image Credit* Brian Joseph McFarland)

Financial Risk Management

The Global Association of Risk Professionals' (GARP) Framework for Financial Risk Managers (FRM) is one of the leading frameworks for identifying and managing risk. Risks are inserted into thematic categories. These categories are business, credit, reputation, liquidity, operational, strategic, market, and regulatory or legal risks.

This risk management framework allows analysts to assess financial instruments and their possible impacts over time. In addition, this assessment can inform decision-making for project design. Similar to the RRTTLLU framework below, this risk model allows for the integration of characteristics such as land-use change, biodiversity loss, and hydrological systems, along with human and labor rights, into financial terms that may fit at times within financial models.

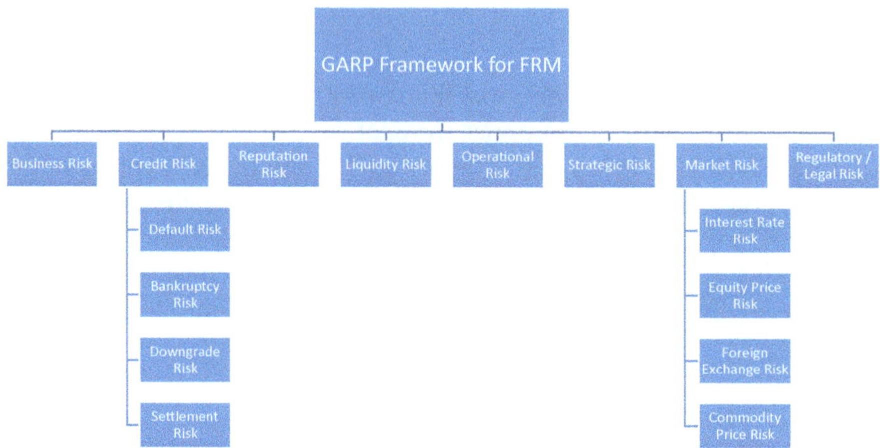

Fig. 5.2 GARP framework for FRM (*Image Credit* Brian Joseph McFarland and Gabriel Thoumi)

For example, the GARP thematic model shown below feeds into enterprise risk management tools that can be used to manage and model the overall risk of an organization. In this case, Environmental, Social and Governance (ESG) characteristics of an investment—including deforestation, human rights, and other risks—can be included within each risk category. These ESG risks often have direct operational risks which then result in indirect business, credit, reputation, liquidity, strategic, market, and regulatory or legal risks (Fig. 5.2).

Portfolio Management: Risk, Return, Time (Horizon), Tax, Liquidity, Legal and Unique

To develop an Investment Policy Statement (IPS), the Risk, Return, Time (Horizon), Taxes, Liquidity, Legal and Unique (RRTTLLU) framework can be utilized.[31] This RRTTLLU complements the efficient frontier and the GARP Framework. In the RRTTLLU framework, analysts review five constraints (time horizon, taxes, liquidity, legal, and unique) to inform both return and risk objectives. Conservation finance is informed by this RRTTLLU process of categorizing information, even though most in the field might not use this acronym. When the possible financial structure is described using risk and return objectives informed by the five constraints, it then fits inside the efficient frontier model.

Other Resources on International Finance

2° Investing Initiative

- https://2degrees-investing.org/; and
- Also see Paris Agreement Capital Transition Assessment: https://www.transitionmonitor.com/

Biodiversity Finance Initiative from UNDP

- http://www.biodiversityfinance.net/

Ceres' Investor Water Toolkit

- https://www.ceres.org/resources/toolkits/investor-water-toolkit

Earth Genome

- https://www.earthgenome.org/overview/

Fisheries Economics Research Unit, University of British Columbia's Institute for the Oceans and Fisheries

- http://feru.oceans.ubc.ca/

Global Alliance for Banking on Values

- http://www.gabv.org/

Global Footprint Network's Ecological Wealth of Nations

- http://www.footprintnetwork.org/content/documents/ecological_footprint_nations/

Global Green Economy Index 2018

- https://dualcitizeninc.com/global-green-economy-index/

Impact Alpha's Conservation Finance

- http://impactalpha.com/category/conservation/; and
- http://impactalpha.com/new-products-attract-investors-to-the-conservat ion-finance-marketplace/.

Instrumentl

- https://www.instrumentl.com/

International Finance Corporation's (IFC) Presentations and Factsheets; IFC's Performance Standard 6

- http://www.ifc.org/wps/wcm/connect/corp_ext_content/ifc_external_cor porate_site/about+ifc_new/ifc+governance/investor+relations/presentat ions+and+factsheets; and
- https://www.ifc.org/wps/wcm/connect/3baf2a6a-2bc5-4174-96c5-eec808 5c455f/PS6_English_2012.pdf?MOD=AJPERES&CVID=jxNbLC0

Journal of Environmental Investing

- http://www.thejei.com/

Ocean Finance: Definition and Actions

- https://www.icriforum.org/sites/default/files/Ocean_Finance_Definition_ Paper_Walsh_June_2018_1_.pdf

Sustainable Blue Economy Finance Principles

- https://www.wwf.org.uk/updates/sustainable-blue-economy-finance-princi ples

The Conservation Planning Database

- http://database.conservationplanning.org/

The Little Biodiversity Finance Book (3rd Edition)

- http://d6.globalcanopy.org/materials/little-biodiversity-finance-book; and
- http://d6.globalcanopy.org/materials/little-biodiversity-finance-book-met hodology-appendix

<u>The Little Climate Finance Book</u>

- http://d6.globalcanopy.org/materials/little-climate-finance-book

<u>TNC's Conservation Gateway and Marine Sector</u>

- http://www.conservationgateway.org/ConservationPractices/Marine/Pages/marine.aspx

<u>Trust for Public Land, What Conservation Finance is Really About</u>

- https://www.youtube.com/watch?v=PSVZsXEA8tw&feature=youtu.be

<u>Undercurrent News</u>

- https://www.undercurrentnews.com/

<u>UNEP's Finance Initiative</u>

- http://www.unepfi.org/

<u>Yale University's Environmental Performance Index</u>
  https://epi.envirocenter.yale.edu/

Notes

1. World Bank. "Gross Domestic Product 2018." Accessed July 13, 2019. https://databank.worldbank.org/data/download/GDP.pdf.
2. Ibid.
3. U.S. Central Intelligence Agency. "The World Factbook: The World: Economy." Accessed July 13, 2019. https://www.cia.gov/library/publications/the-world-factbook/geos/xx.html.
4. Carbon Pulse. "CP Daily: Monday October 15, 2018." October 15, 2018. Accessed January 14, 2020. https://carbon-pulse.com/60856/; Salmon, Felix. "The Cost of Climate Change." *Axios*. October 14, 2018. Accessed January 14, 2020. https://www.axios.com/climate-change-costs-wealth-carbon-tax-303d7cff-3085-49d9-accb-ec77689b9911.html.
5. Huwyler, Fabian et al. "Conservation Finance: From Niche to Mainstream: The Building of an Institutional Asset Class." 3.
6. Hoegh-Guldberg, O. et al. *Reviving the Ocean Economy.* 7.
7. Vertigo Lab. "Innovations for Coral Finance, ICRI Publication." 1.

8. UN Environment et al. "Analysis of International Funding for the Sustainable Management of Coral Reefs and Associated Coastal Ecosystems." 11.

9. Spergel, Barry and Melissa Moye. "Financing Marine Conservation: A Menu of Options." 9.

10. Huwyler, Fabian et al. "Conservation Finance: From Niche to Mainstream: The Building of an Institutional Asset Class." 3.

11. Hamrick, Kelley. "State of Private Investment in Conservation 2016." vii.

12. Milder, Brian. "Conservation Finance in the Galápagos Islands: Capitalizing the 'Missing Middle'." In *Conservation Capital in the Americas*, edited by James N. Levitt. 110.

13. American Pet Products Association. "Pet Industry Market Size & Ownership Statistics." Accessed December 13, 2019. http://www.americanpetproducts.org/press_industrytrends.asp.

14. Association of Zoos & Aquariums. "Zoo and Aquarium Statistics." Accessed December 13, 2019. https://www.aza.org/zoo-and-aquarium-statistics.

15. O'Connor et al., 2009, Whale Watching Worldwide: Tourism Numbers, Expenditures and Expanding Economic Benefits, a Special Report from the International Fund for Animal Welfare, Yarmouth, MA, USA, Prepared by Economists at Large. https://www.mmc.gov/wp-content/uploads/whale_watching_worldwide.pdf. 8.

16. The Numbers. "Finding Nemo (2003)." Accessed September 6, 2018. https://www.the-numbers.com/movie/Finding-Nemo#tab=international.

17. The Numbers. "Finding Dory (2016)." Accessed September 6, 2018. https://www.the-numbers.com/movie/Finding-Dory#tab=summary.

18. Clark, Story. *A Field Guide to Conservation Finance*. 184.

19. Ibid. 185.

20. Elliman, Kim and Peter Howell. "Anatomy of a Finance Program: Lessons from a Conservation Lender." In *Conservation Capital in the Americas*, edited by James N. Levitt. 148.

21. Robinson, Michele. "Debt Restructuring Initiatives Paper for the Commonwealth Secretariat." 7.

22. Ibid. 63.

23. XE. "Current and Historical Rate Tables." Accessed July 13, 2019. https://www.xe.com/currencytables/?from=USD&date=2016-03-23.

24. XE. "Current and Historical Rate Tables." Accessed July 13, 2019. https://www.xe.com/currencytables/?from=USD&date=2019-03-22.

25. XE. "Current and Historical Rate Tables." Accessed July 13, 2019. https://www.xe.com/currencytables/?from=EUR&date=2014-10-21.

26. XE. "Current and Historical Rate Tables." Accessed July 13, 2019. https://www.xe.com/currencytables/?from=EUR&date=2018-10-22.

27. XE. "Current and Historical Rate Tables." Accessed July 13, 2019. https://www.xe.com/currencytables/?from=GBP&date=2014-03-18.

28. XE. "Current and Historical Rate Tables." Accessed July 13, 2019. https://www.xe.com/currencytables/?from=GBP&date=2019-03-18.

29. XE. "Current and Historical Rate Tables." Accessed July 13, 2019. https://www.xe.com/currencytables/?from=AUD&date=2009-05-22.

30. XE. "Current and Historical Rate Tables." Accessed July 13, 2019. https://www.xe.com/currencytables/?from=AUD&date=2019-05-22.

31. Vanguard Asset Management. "Categorise and Evaluate Data: The RRTTLLU Framework." Accessed December 12, 2016. https://www.vanguard.co.uk/documents/adv/literature/advice_process/categorise-and-evaluate-data.pdf.

6

The Origins and History of Coral Reef Conservation Finance

Conservation finance is a subset field of finance that involves deploying funds for the conservation and restoration of ecosystems. The exact date when conservation finance for tropical coral reefs began is not known.

In 1634, the Boston Commons was established and is considered "the oldest public open space in the United States. Indeed, it is very likely the oldest public open space in the English-speaking world."[1] The Bogd Khan Uul National Park in Mongolia was established "by the Mongolian government in 1778, {but} it was originally chartered by Ming Dynasty officials in the 1500s as an area to be kept off limits to extractive uses, protected for its beauty and sacred nature."[2]

The first documented terrestrial protected area, according to the World Database on Protected Areas, appears to be a nature park in Russia known as the Donskoj rybny zapovednik (Donskoe zapretnoe rybnoe prostranstvo) which was established in 1819.[3] The first terrestrial state park appears to be Yosemite National Park in California, United States, established in 1864.[4] The first terrestrial national park – aside from the Bogd Khan Uul National Park – appears to be the U.S.' Yellowstone National Park, established in 1872.[5]

The first terrestrial and marine protected area, according to the World Database on Protected Areas, appears to be the nature reserve at Mistaken Island in 1874, followed by the Whakapirau River (1889) and Indian Island (1889). The first 100% marine protected area, according to the World Database on Protected Areas, appears to be the Garden of Tane (1883), followed by Letonnokan Merenrantaniitty (1900), Shell Keys (1907), and Key West (1908).[6] In addition, one of the world's oldest MPAs and the

© The Author(s) 2021

B. J. McFarland, *Conservation of Tropical Coral Reefs*,

https://doi.org/10.1007/978-3-030-57012-5_6

"oldest MPA along the West Coast {of the U.S.} is the Cabrillo National Monument, created in 1913 near San Diego, California."[7]

In 1891, the Trustees of Public Reservations (now the Trustees of Reservations, or TTOR) became the world's first regional land trust,[8] establishing the world's first conservation easement in Boston, Massachusetts.[9]

It is difficult to discern when the first government international budgetary expenditures for conservation were allocated. However, it is likely this mechanism was first established in the late nineteenth century or early twentieth century, when the world's first international conservation organizations were established. This includes the Society for the Preservation of the Wild Fauna of the Empire (now Fauna and Flora International, founded in 1903)[10] and the International Council for Bird Preservation (now BirdLife International, established in 1922).[11] Waterton Glacier International Peace Park, established in 1932 between the United States and Canada, is the world's first international, terrestrial peace park.[12]

In 1904, "Alfred Mayor found himself director of the Carnegie Institution's new department of marine biology at Loggerhead Key, the first tropical laboratory in the Western Hemisphere."[13]

In 1959, "the Exuma Cays Land and Sea Park {in the Bahamas} was the world's first protected area of its kind, when created in 1959 by the same legislation that established the Bahamas National Trust."[14]

The "first close approximation to a marine reserve was established in 1965 with the designation of the Scripps Coastal Reserve at the site of the existing San Diego Marine Life Refuge."[15]

In 1972, the "U.S. Congress passed legislation authorizing the first underwater equivalent of national parks, national marine sanctuaries, to provide a measure of security for natural, historic, and cultural values under the sea {...}."[16] In the 1970s, "Australia's Great Barrier Reef Marine Park Authority began by zoning broad areas of the 2,414-kilometer-long (1,500 mile) reef system for various uses, including about 3 percent of the area where fishing was restricted."[17]

The "{..} first national marine sanctuary {in the U.S.} was a small area designed in 1975 around the remains of the sunken Union warship *Monitor*, resting in more than 60 meters (200 feet) of water off the coast of North Carolina. In the same year, a small area of coral reefs at Key Largo became a national sanctuary, expanded in 1990 to encompass most of the 7,250 square kilometers (2,800 square miles) around the Florida Keys."[18]

Modern-day impact investing is comparable to socially responsible investing, which took "{...} off in the 1970s as part of grassroots campaigns to boycott firms in South Africa {as a result of Apartheid} as well as the nuclear weapons and tobacco industries."[19]

In the late 1970s, the concept of payments for ecosystem services (PES, referred to earlier as payments for environmental services) appears.[20]

The Bonaire National Marine Park, claims to be "the oldest marine reserve in the world. {...} The park, established in 1979, includes 2,600 ha (6,425 acres) of coral reef, sea grass and mangroves."[21]

Low-impact tourism likely began earlier, but in 1983, the term "eco-tourism" is first coined by Mexican architect Héctor Ceballos-Lascuráin.[22]

Debt-for-nature swaps came along shortly thereafter with the first such swap conducted in 1987.[23] One of the first debt-for-nature swaps related to the marine environment, known as the Seychelles' Climate Adaptation and Impact Investment Debt Swap, was completed in 2015.[24]

The sustainable, blue procurement of locally produced commodities—particularly by Indigenous Peoples and local communities—has existed for thousands of years. However, the origin of commercial, green procurement with independent certification standards likely started with the creation of the Forest Stewardship Council (FSC) in 1994[25] and the launch of the Sustainable Forestry Initiative (SFI) in 1995.[26] The Marine Stewardship Council (MSC) was launched shortly thereafter in 1997 after a statement of intent was signed between the WWF and Unilever in 1996. The MSC Fisheries Standard was then "born" in 1998,[27] and in 2010, the Aquaculture Stewardship Council (ASC) standard was created.[28]

The first widely recognized green bonds were issued in 2007 and 2008 by the World Bank[29] and European Investment Bank.[30] The first "blue bond" with funds ring-fenced to fund "sustainable fisheries and marine projects" was launched for the Seychelles in 2018.[31]

Looking forward, the blue economy is expected to grow. For instance, the UK Government forecasts the ocean economy to grow to USD$3 trillion by 2030.[32] The International Coral Reef Initiative-commissioned report, *Innovations for Coral Finance*, states four questions to model for coral reef conservation, which are:

1. Which stakeholders comprise your key beneficiary group?
2. What service/benefit do you offer to your targeted stakeholders (your proposition value)?
3. What are the main activities, partners and resources you need to mobilize to reach your coral ecosystem conservation objectives?
4. How much funding, and from which sources, do you need to reach your coral ecosystem conservation objectives?[33]

It is important to understand that "corals and mangroves provide a number of valuable services to local communities across the world but not all of them are within protected areas. For instance, globally, 28.6% of coral reef fisheries biomass, 20.4% of coral reef coastal protection (US$), and 44.3% of coral reef tourism value (US$) lies within protected areas."[34]

As summarized by Melissa Bos et al.:

> We identify five challenges: 1) funding for marine conservation is inadequate in terms of the size, duration, and diversity of revenue, 2) finance mechanisms are under-developed and under-utilised, 3) finance is often disconnected from conservation planning, 4) the environmental side-effects of economic activity increase the gap in global conservation funding, and 5) few individuals and programmes specialise in marine conservation finance and integrate its disparate lines of thinking. We then propose five solutions: 1) financial strategies for marine conservation, 2) increased research on and development of finance mechanisms, 3) integration of financial planning into conservation planning, 4) engagement of businesses in reducing the gap in conservation funding for marine ecosystems, and 5) definition, focus, and specialists for the emerging field of marine conservation finance. Multi-sector and interdisciplinary collaboration is essential to reduce the marine conservation-funding gap and sustain marine ecosystem services.[35]

Conservation finance developed alongside other trends, such as population growth, rural to urban migration, cultural changes, dietary changes, and the emergence of technology which made measuring and monitoring of fisheries and coral reefs more possible.

- 475 BCE: "The first manual of fish culture was written in 475 BCE by the Chinese author Fan Li {…}."[36]
- 1724, 1752: "Corals had long been a taxonomist's nightmare, a little-studied phenomenon that early theorists assumed to be some strange sort of plant. In 1724, the Frenchman Jean-André Peyssonnel overturned the work of a colleague in Montpellier with a letter to the Académie des Sciences, arguing for the first time that the so-called coral flower was in reality not a plant, but 'un insect' that could create bone. His idea was ridiculed until it was taken up some thirty years later by the Englishman John Ellis, who in 1752 told the Royal Society in London that these creatures were 'ramified [branchlike] animals,' after which his classification became increasingly accepted."[37]
- 1800 (approximately): The world population reaches 1 billion people.[38]

- 1844: "Professor Henri Milne-Edwards and a naturalist friend found the concept irresistible, and were the first known to 'dive for science' during their studies in the straits of Messina off the coast of Sicily in 1844."[39]
- 1851: *Moby-Dick* is published.[40]
- 1853: The "first aquarium designed for public display opened in Regents Park, London {…}."[41]
- 1856: "{…} P.T. Barnum opened an aquarium of 'curious creatures' from the sea in New York City."[42]
- 1859: Charles Darwin's *On the Origin of Species by Means of Natural Selection* is published.[43]
- 1865: "An underwater breathing system that enabled divers to swim freely in the sea was developed and used in 1865 by the Frenchmen Benoît Rouquayrol and Lt. Auguste Denayrouze, including use of a container filled with a reservoir of air carried on the diver's back."[44]
- 1868: Invention of an exploding harpoon gun.[45]
- 1870: *Twenty Thousand Leagues Under the Sea*, written by Jules Verne, is published.[46]
- 1872–1876: British HMS *Challenger* expedition, "the first global scientific exploration of the sea."[47]
- 1890s: "The first offshore oil wells were drilled off Santa Barbara in California in the 1890s {…}."[48]
- 1893: "{Louis} Boutan made the first undersea photos in the Bay of Banyuls-sur-Mer in 1893."[49]
- 1896: "The concept that some form of global warming might be triggered by the burning of fossil fuels was first proposed in 1896 by a Swedish chemist, Svante Arrhenius."[50]
- 1918: "Various versions of diving using tanks of compressed air were developed, including the first self-contained system, patented in 1918 by a Japanese inventor, Ohgushi, and a later diving 'lung' created by the Frenchman Yves Le Prieur."[51]
- 1920s (Late): "{…} one of the first scientists to use such gear for exploration was naturalist William Beebe, who adapted a copper helmet to explore coral reefs in Bermuda {…}."[52]
- 1926: "In 1926 Dr. W.H. Longley and Charles Martin of the National Geographic Society made the first underwater color photographs, in the Dry Tortugas {…}."[53]
- 1930: World population reaches 2 billion people.[54]
- 1934: William Beebe and Otis Barton in a bathysphere reach a depth of 3028 feet in the open ocean off Bermuda.[55]

- 1938: "In December 1938, off the southeast tip of Africa, an amazing fish {the coelacanth} was caught alive in a trawl – a fish that was supposed to have been dead for at least 60 million years!"[56]
- 1942: "In 1942, the brilliant engineer Émile Gagnan, and I (Jacques-Yves Cousteau) made the first completely automatic compressed-air apparatus, the aqualung."[57]
- 1947: "That year, the Oklahoma-based company Kerr-McGee completed the first drilling rig out of sight of land, in the Gulf of Mexico."[58]
- 1950s: "The introduction of echo sounders in the 1950s marked the onset of a new fishing revolution."[59]
- 1951: Rachel Carson's *The Sea Around Us* is published.[60]
- 1952: The last living Caribbean monk seal is seen alive.[61]
- 1953: Jacques Cousteau's *The Silent World* is published.[62]
- 1959: World population reaches 3 billion people.[63]
- 1960: "The *Trieste* {operated by Don Walsh and Jacques Piccard} reached the deepest known point on Earth, the Challenger Deep in the Mariana Trench, on January 23, 1960. Located 10,916 meters (35,813 feet) below the ocean's surface, the Challenger Deep is deeper than the height of Mount Everest!"[64]
- 1962: The U.S. research vessel "*Anton Bruun* began a four-year cruise that touched some of the same places visited by the Challenger scientists: first the Indian Ocean, as part of the International Indian Ocean Expedition, and later the southeastern Pacific and the Caribbean Sea."[65]
- 1966: International Whaling Commission bans the killing of blue whales.[66]
- 1967: "The first sign of real trouble was the 1967 wreck of the tanker *Torrey Canyon*."[67]
- 1968: The "term 'biological diversity' was coined by the conservation biologist Raymond F. Dasmann."[68]
- 1968: "While studying pollution control in the Great Lakes, University of Toronto economist John Dales hit on a way for the costs to be paid with minimal government intervention, by using tradable permits or allowances."[69]
- 1969: "while astronauts walked on the moon for the first time, four men spent two months living underwater in a subsea laboratory, Tektite I."[70]
- 1970s: "Shortages of valuable metals and greatly improved technologies for accessing the deep sea provoked a number of countries and companies to begin to take deep-sea mining seriously in the 1970s."[71]
- 1970: First Earth Day (April 22) is founded by U.S. Senator Gaylord Nelson.[72]

- 1971: Dr. Seuss' *The Lorax* is published.[73]
- 1972: The *Limits to Growth* is published.[74]
- 1972–1973: The first "imagery from NASA Landsat satellites had already become available in 1972-1973."[75]
- 1974: World population reaches 4 billion people.[76]
- 1975: "{...} the year that great white sharks were portrayed as evil, bloodthirsty people-eating villains in the blockbuster film *Jaws* {...}."[77]
- 1980s: "{...} since the 1980s, development of remotely operated vehicles (ROVs) and unmanned autonomous underwater vehicles (AUVs) has progressed from relatively simple 'flying eyeball' systems to a wide range of small, medium, and some very large systems devised for many jobs previously accomplished only by divers, and many that divers simply could not achieve."[78]
- 1984–1985: Year of the Ocean designated by the U.S.[79]
- 1987: *Our Common Future* is published.[80]
- 1987: World population reaches 5 billion people.[81]
- 1989 (March 24): *Exxon Valdez* runs aground in Prince William Sound in Alaska.[82]
- 1990s: "Soy expansion in South America started in the early 1990s with the rapid increase of meat consumption and the demand for protein-rich feed for chickens, pigs, and cattle. Fish meal, an important cattle feed, was no longer available after the collapse of the Peruvian anchovy fisheries."[83]
- 1997: The first International Year of the Reef (IYOR).[84]
- 1998: "When the *Washington Post* staff writer Joby Warrick published an article in 1998 under the title 'Mass extinction underway, majority of biologists say,' it triggered a huge wave of reactions. To date, the Mysterium.com website has accumulated over 300 links to authoritative reports and articles with evidence that a mass extinction is beginning."[85]
- 1999: World population reaches 6 billion people.[86]
- 2000: Census of Marine Life.[87]
- 2006: An Inconvenient Truth.[88]
- 2006: Sir Nicholas Stern's *Stern Review on the Economics of Climate Change*.[89]
- 2008: The second International Year of the Reef (IYOR).[90]
- 2009: Research on the nine planetary boundaries is first presented: (1) stratospheric ozone depletion, (2) loss of biosphere integrity (functional and genetic diversity), (3) chemical pollution and the release of novel entities, (4) climate change, (5) ocean acidification, (6) freshwater consumption and the global hydrological cycle, (7) land system change,

(8) biogeochemical flows including nitrogen and phosphorus, and (9) atmospheric aerosol loading.[91]

- 2009: David de Rothschild launches *Plastik*i, "an 18-meter (60-foot) catamaran built entirely of recycled plastic bottles, for an expedition from San Francisco, across the Great Pacific Garbage Patch, through islands of the South Pacific, to Sydney."[92]
- 2010, April 20: Deepwater Horizon oil spill occurs in the Gulf of Mexico.[93]
- 2010: Mission Blue I, "conceived of as a TED-at-sea," sails to the Galápagos Islands.[94]
- 2011: World population reaches 7 billion people.[95]
- 2012: Natural Capital Declaration launched at the Rio+20 conference.[96]
- 2012: "{…} the *Trieste* expedition remained the only manned dive to reach the Challenger Deep until the *DEEPSEA CHALLENGE* expedition in March 2012. In that expedition, Canadian inventor and filmmaker James Cameron became the first person to dive solo to the bottom of the Mariana Trench."[97]
- 2014: Years of Living Dangerously.[98]
- 2015: Pope Francis' Encyclical on the Environment is presented.[99]
- 2015: Mission Blue II sails from Papua New Guinea to the Solomon Islands.[100]
- 2017: An Inconvenient Sequel: Truth to Power.[101]
- 2018: The third International Year of the Reef (IYOR).[102]

Other Resources on the Origins and History of Coral Reef Conservation Finance

Conservation Finance Alliance

- http://www.conservationfinancealliance.org/; and
- A Framework White Paper: https://www.conservationfinancealliance.org/cfa-white-paper

Conservation Finance Network (CFN)

- http://www.conservationfinancenetwork.org/;
- Conservation Finance 101: http://www.conservationfinancenetwork.org/conservation-finance-101; and

- Conservation Finance Video Playlists: http://www.conservationfinancene twork.org/2016/01/12/explore-our-conservation-finance-video-playlists

Coral Reef Funding Landscape

- https://coralfunders.com/; and
- www.wcmc.io/corals

Funding the Ocean and Funding Map

- https://fundingtheocean.org/

OpenChannels

- https://www.openchannels.org/

Parley for the Oceans

- https://www.parley.tv/#fortheoceans

TED Talks' Mission Blue II

- https://www.ted.com/participate/ted-prize/mission-blue-2

UN Environment, International Coral Reef Initiative and UN Environment World Conservation Monitoring Centre's Analysis of International Funding for the Sustainable Management of Coral Reefs and Associated Coastal Ecosystems

- https://www.icriforum.org/sites/default/files/ICRI_Funding_landscape_coral_reef.pdf

Notes

1. Levitt. *From Walden to Wall Street*. 4.
2. TNC "The World's Oldest National Park: Ghosts of Monks and Red Deer." Last modified November 10, 2009. http://blog.nature.org/conservancy/2009/11/10/worlds-oldest-national-park-mongolia-nature-charles-bedford-bogdkhan/.

3. IUCN and UNEP-WCMC (2016). The World Database on Protected Areas (WDPA) [Online]. December 13, 2016. Cambridge, UK: UNEP-WCMC. Available at: www.protectedplanet.net.

4. California Department of Parks and Recreation. "Yosemite 'State Park'." Accessed November 23, 2016. http://www.150.parks.ca.gov/?page_id=27539.

5. National Park Service. "Yellowstone National Park: Frequently Asked Questions: History." Accessed November 23, 2016. https://www.nps.gov/yell/learn/historyculture/history-faqs.htm.

6. IUCN and UNEP-WCMC. "Discover the World's Protected Areas." Accessed December 13, 2019. https://www.protectedplanet.net/.

7. Sobel and Dahlgren. *Marine Reserves*. 238.

8. Foster, Dave. "Meeting the Conservation Challenge in New England." In *Conservation Capital in the Americas*, edited by James N. Levitt. 20–21.

9. TNC. "Easements 101." Accessed December 13, 2016. https://www.nature.org/about-us/private-lands-conservation/conservation-easements/conservation-easements-101.pdf.

10. FFI. "110 Years of Fauna & Flora International." Accessed November 23, 2016. http://www.fauna-flora.org/timeline/.

11. BirdLife International. "Our History." Accessed November 23, 2016. http://www.birdlife.org/worldwide/partnership/our-history.

12. UNESCO. "Waterton Glacier International Peace Park." Accessed December 13, 2016. http://whc.unesco.org/en/list/354.

13. McCalman. *The Reef*. 202.

14. Dahlgren. "Bahamian Marine Reserves—Past Experience and Future Plans." In: Sobel and Dahlgren. *Marine Reserves*. 268.

15. Nowlis. "California's Channel Islands and the U.S. West Coast." In: Sobel and Dahlgren. *Marine Reserves*. 240.

16. Earle. *The World Is Blue*. 244.

17. Ibid. 246.

18. Ibid. 246.

19. Daily, Gretchen C. and Katherine Ellison. *The New Economy of Nature*. 116.

20. Gómez-Baggethun et al. "The History of Ecosystem Services in Economic Theory and Practice: From Early Notions to Markets and Payment Schemes." https://doi.org/10.1016/j.ecolecon.2009.11.007.

21. Latin America & Caribbean Geographic. "Bonaire National Marine Park." October 15, 2018. Accessed December 13, 2019. https://lacgeo.com/bonaire-national-marine-park.

22. EcoClub. "International Ecotourism Monthly." Year 7. Issue 85. Accessed December 14, 2016. http://ecoclub.com/news/085.pdf. 2.

23. WWF. *Guide to Conservation Finance*. 41.

24. SeyCCAT. "Seychelles' Conservation and Climate Adaptation Trust (SeyCCAT): Achieving Conservation Through Innovative Finance and Creative Collaborations." Accessed August 7, 2019. https://seyccat.org/wp-content/uploads/2017/09/infographic_talking_points_bat_web-1.pdf.

25. FSC. "About Us." Accessed December 27, 2019. https://fsc.org/en/page/about-us.
26. Sampson, R. Neil. "The Sustainable Forestry Initiative Program: Seven Years of Sustainable Forestry." *FAO*. Accessed December 27, 2019. http://www.fao.org/3/XII/0700-A1.htm.
27. MSC. "Our History." Accessed August 7, 2019. https://www.msc.org/about-the-msc/our-history.
28. ASC. "Our History." Accessed December 31, 2019. https://www.asc-aqua.org/about-us/history/.
29. World Bank Group. "World Bank Green Bonds: Green Bond Issuances to Date." Accessed December 13, 2016. http://treasury.worldbank.org/cmd/htm/GreenBondIssuancesToDate.html.
30. Climate Bonds Initiative. "History." Accessed December 13, 2016. https://www.climatebonds.net/market/history.
31. Wahlén, Catherine Benson. "Seychelles Launches First Blue Bond." *International Institute for Sustainable Development*. 6 November 2018. Accessed 7 August 2019. https://sdg.iisd.org/news/seychelles-launches-first-blue-bond/.
32. UK Government Office of Science. "Foresight Future of the Sea." 2018. https://assets.publishing.service.gov.uk/government/uploads/system/uploads/attachment_data/file/706956/foresight-future-of-the-sea-report.pdf.
33. Vertigo Lab. "Innovations for Coral Finance, ICRI Publication." 3.
34. UNEP-WCMC, IUCN and NGS (2018). Protected Planet Report 2018. UNEP-WCMC, IUCN and NGS: Cambridge, UK; Gland, Switzerland; and Washington, DC, USA. https://livereport.protectedplanet.net/pdf/Protected_Planet_Report_2018.pdf. 13.
35. Bos, Melissa et al. "Marine Conservation Finance." 116.
36. Roberts. *The Ocean of Life*. 244.
37. McCalman. *The Reef*. 15.
38. Worldometers. "World Population." Accessed November 23, 2016.http://www.worldometers.info/world-population/#pastfuture.
39. Earle. *Sea Change*. 35.
40. HISTORY.com. "This Day in History: 1851: Moby-Dicks Is Published." Accessed September 18, 2018.https://www.history.com/this-day-in-history/moby-dick-published.
41. Helvarg. *50 Ways to Save the Ocean*. 121.
42. Ibid. 121.
43. HISTORY.com. "This Day in History: 1859: Origin of Species Is Published." Accessed March 27, 2017.http://www.history.com/this-day-in-history/origin-of-species-is-published.
44. Earle. *Sea Change*. 36.
45. Earle. *The World Is Blue*. 43.
46. Goodreads. "Twenty Thousand Leagues Under the Sea." Accessed December 28, 2018. https://www.goodreads.com/book/show/33507.Twenty_Thousand_Leagues_Under_the_Sea.

47. Earle. *The World Is Blue*. 126.
48. Roberts. *The Ocean of Life*. 178.
49. Cousteau. *The Silent World*. 234.
50. Roberts. *The Ocean of Life*. 62.
51. Earle. *Sea Change*. 37.
52. Earle. *The World Is Blue*. 187.
53. Cousteau. *The Silent World*. 236.
54. Worldometers. "World Population." Accessed November 23, 2016.http:// www.worldometers.info/world-population/#pastfuture.
55. Carson. *The Sea Around Us*. 38.
56. Ibid. 54.
57. Cousteau. *The Silent World*. 14.
58. Earle. *The World Is Blue*. 149.
59. Roberts. *The Ocean of Life*. 41.
60. RachelCarson.org. "The Sea Around Us." Accessed September 18, 2018. http://www.rachelcarson.org/SeaAroundUs.aspx.
61. The MarineBio Conservation Society. "Caribbean Monk Seals Declared Extinct." June 9, 2008. Accessed January 14, 2020. https://marinebio.org/car ibbean-monk-seals-declared-extinct/.
62. HISTORY.com. "This Day in History: 1953: Cousteau Publishes The Silent World." Accessed September 18, 2018. https://www.history.com/this-day-in-history/cousteau-publishes-the-silent-world.
63. Worldometers. "World Population." Accessed November 23, 2016.http:// www.worldometers.info/world-population/#pastfuture.
64. National Geographic Society. "Encyclopedia: Bathyscaphe." Accessed November 19, 2018. https://www.nationalgeographic.org/encyclopedia/ bathyscaphe/.
65. Earle. *Sea Change*. 25.
66. Greenpeace. "A Brief History of Whales and Commercial Whaling." 20 October 2016. Accessed December 28, 2018. https://www.greenpeace.org.uk/ brief-history-whales-and-commercial-whaling-20161020/.
67. Levinton, Jeffrey S. "Afterword." In Carson, *The Sea Around Us*. 241.
68. Martin. *On the Edge*. 135.
69. Smithsonian Magazine. "The Political History of Cap and Trade." Accessed November 23, 2016. http://www.smithsonianmag.com/air/the-political-his tory-of-cap-and-trade-34711212/?no-ist.
70. Earle. *The World Is Blue*. 189.
71. Ibid. 140.
72. Earth Day Network. "The History of Earth Day." Accessed November 23, 2016. http://www.earthday.org/about/the-history-of-earth-day/.
73. Random House. "Seuss Biography." Accessed November 23, 2016. http://m. seussville.com/biography.html.
74. Club of Rome. "The Limits to Growth." Accessed November 23, 2016.http:// www.clubofrome.org/report/the-limits-to-growth/.

75. Martin. *On the Edge*. 13.
76. Worldometers. "World Population." Accessed November 23, 2016.http://www.worldometers.info/world-population/#pastfuture.
77. Earle. *The World Is Blue*. 70–71.
78. Ibid. 199.
79. Ronald Regan Presidential Library & Museum. "Proclamation 5222—Year of the Ocean." July 13, 1984. Accessed January 9, 2020. https://www.reaganlibrary.gov/research/speeches/71384e.
80. Council on Foreign Relations. "Report of the World Commission on Environment and Development: Our Common Future (Brundtland Report)." Accessed November 23, 2016. http://www.cfr.org/economic-development/report-world-commission-environment-development-our-common-future-brundtland-report/p26349.
81. Worldometers. "World Population." Accessed November 23, 2016.http://www.worldometers.info/world-population/#pastfuture.
82. Levinton, Jeffrey S. "Afterword." In Carson, *The Sea Around Us*. 241.
83. Martin. *On the Edge*. 97.
84. International Coral Reef Initiative. "International Year of the Reef (IYOR)." Accessed September 6, 2018. https://www.icriforum.org/about-icri/iyor.
85. Martin. *On the Edge*. 263.
86. Worldometers. "World Population." Accessed November 23, 2016.http://www.worldometers.info/world-population/#pastfuture.
87. Earle. *The World Is Blue*. 132.
88. IMDb. "An Inconvenient Truth (2006)." Accessed March 29, 2017. http://www.imdb.com/title/tt0497116/.
89. Cambridge University Press. "The Economics of Climate Change: The Stern Review." Accessed March 30, 2017. http://www.cambridge.org/catalogue/catalogue.asp?isbn=9780521700801.
90. International Coral Reef Initiative. "International Year of the Reef (IYOR)." Accessed September 6, 2018. https://www.icriforum.org/about-icri/iyor.
91. Stockholm Resilience Centre. "About the Research." Accessed April 11, 2017.http://www.stockholmresilience.org/research/planetary-boundaries/planetary-boundaries/about-the-research.html.
92. Earle. *The World Is Blue*. 110.
93. Pallardy, Richard. "Deepwater Horizon Oil Spill." *Encyclopedia Britannica*. May 29, 2019. Accessed July 13, 2019. https://www.britannica.com/event/Deepwater-Horizon-oil-spill.
94. TED Talks. "Mission Blue II." Accessed August 12, 2019. https://www.ted.com/participate/ted-prize/mission-blue-2.
95. Worldometers. "World Population." Accessed November 23, 2016.http://www.worldometers.info/world-population/#pastfuture.
96. Natural Capital Finance Alliance. "The NCD and NCFA Story." Accessed December 13, 2016.http://www.naturalcapitalfinancealliance.org/about-the-natural-capital-declaration/.

97. National Geographic Society. "Encyclopedia: Bathyscaphe." Accessed November 19, 2018. https://www.nationalgeographic.org/encyclopedia/bathyscaphe/.

98. IMDb. "Years of Living Dangerously." Accessed March 29, 2017.http://www.imdb.com/title/tt2963070/?ref_=fn_al_tt_1.

99. Vatican. "Laudato si'." Accessed November 23, 2016. http://w2.vatican.va/content/dam/francesco/pdf/encyclicals/documents/papa-francesco_20150524_enciclica-laudato-si_en.pdf.

100. TED Talks. "Mission Blue II." Accessed August 12, 2019. https://www.ted.com/participate/ted-prize/mission-blue-2.

101. IMDb. "An Inconvenient Sequel: Truth to Power (2017). Accessed March 29, 2017. http://www.imdb.com/title/tt6322922/.

102. International Coral Reef Initiative. "International Year of the Reef (IYOR)." Accessed September 6, 2018. https://www.icriforum.org/about-icri/iyor.

7

Government Domestic Budgetary Expenditures

Introduction

Government domestic budgetary expenditures are an important financing mechanism for MPAs and particularly for the conservation of coral reefs.

Mechanisms of Instrument

Governments, whether subnational or federal, have a variety of mechanisms to raise funds for conservation. For example, "since enabling legislation was signed by President Lyndon Johnson in 1964, the Land and Water Conservation Fund (LWCF), largely funded from offshore oil and gas royalties paid to the U.S. government, has supported the acquisition of millions of acres of recreational landscapes and critical wildlife habitat."[1]

In 1975, the Great Barrier Reef was designated as the Great Barrier Reef Marine Park,[2] and the Great Barrier Reef Marine Park Authority was established.[3] In 2018, the Australian Government "announced the largest ever single investment in Reef protection. This includes a $443.3 million partnership with the Great Barrier Reef Foundation, aimed at sparking new and innovative responses from, and collaborations with, private investors and philanthropists."[4]

In general:

Three main forms of public funds are used: general revenue, bonding, and dedicated funding. Funding from general revenue is the least predictable

© The Author(s) 2021

B. J. McFarland, *Conservation of Tropical Coral Reefs*,

https://doi.org/10.1007/978-3-030-57012-5_7

because it is subject to fluctuating annual budgetary decisions. Bonding, once authorized, is earmarked for conservation. Because it typically needs to be appropriated each year by the legislature, however, it can also be undependable. The surest form of public spending comes from dedicated funds, which accrue money generated by a specific income source and therefore cannot be diverted by state agencies as priorities shift.[5]

Examples of conservation finance tools for jurisdictions with dedicated funding sources include but are not limited to: lotteries; sales and use taxes; general obligation bonds; property tax levies; real estate transfer taxes; impact fees; special assessments; business improvement districts; and revenue bonds. In addition, there are environmental penalty monies, such as fines for oil spills, illegal fishing, or poaching, that are collected, along with taxes and royalties from oil, gas, and/or mineral concessions.

Governments also have the ability to raise funds through the issuance of specialty license plates and stamps, such as semipostal stamps including the Save Vanishing Species Stamp. Although the U.S. Congressional authorization had expired as of December 2019, one of the targeted species of the US Fish & Wildlife Service's (USFWS) program was "stopping leatherback sea turtle decline in Côte d'Ivoire {Ivory Coast}."[6] In addition:

The 1998 {German Bundespost} stamp featuring coasts and seas financed local income-generating projects for cultivation of algae in South Africa, projects empowering women in the fisheries sector in the Philippines, and environmental education and ecotourism projects in Kaliningrad.[7]

Regarding specialty license plates:

Out of 56 specialty license plates issued by the state of Florida, three marine species – manatees, dolphins, and sea turtles – ranked in the top ten for sales in 2002 (the manatee was number one until a ferocious panther edged it out). Since its introduction in 1990, over 500,000 'Save the Manatee' license plates have been sold with revenues of over $30 million deposited in the Save the Manatees Trust Fund for Florida Fish and Wildlife Commission manatee protection programs. Marketing can encourage vehicle owners to buy a plate for a charismatic species.[8]

Recreational fishing can be another source for government financing of marine protected areas:

Recreational fishing can generate significant revenues for conservation through payment of fishing license fees and taxes and duties on fishing tackle and equipment and boat fuel {...} Every U.S. state requires recreational fishers to purchase a fishing license, and most of the revenues thereby collected are used for sustainable management of aquatic species and public access and education programs. {...} Based on the 'user pays' concept, the Federal Aid in Sport Fish Restoration Act, which was enacted by the U.S. Congress in 1950 and later amended, earmarks revenues collected by the U.S. Treasury for programs to improve fish habitat, public access and aquatic education, and for coastal wetlands conservation and restoration. The revenue sources include a 10 percent excise tax on fishing rods, reels, creels, artificial baits, lures, flies, tackle boxes and other types of recreational fishing equipment; a 3 percent excise tax on electric trolling motors and fish finders; import duties on sport fishing equipment, pleasure boats and yachts; and, a portion of taxes on motorboat fuel sales.[9]

As categorized by the International Coral Reef Initiative-commissioned report, *Innovations for Coral Finance*, there are:

- Tourism-related taxes such as "entry or departure fee, daily fee (based on overnight stays), or specific activity or entrance fee (scuba divers, to access an MPA, etc.);"
- Indirect impact taxes such as "products that can be harmful for coral ecosystems (plastic bags, etc.), or inland or offshore activities (agriculture, etc.);" and
- Resource-extraction taxes such as on "industries extracting directly in the marine environment, or industries with upstream extractive activities that may impact coral ecosystems."[10]

On a more local level, "ballot measures may be referred by the legislative body (termed a referendum) or placed on the ballot by citizen petition (termed an initiative). Some measures are advisory in nature, others create statutory obligations, and yet others may actually amend government charters."[11] This situation is not typical around the world. While state and national parks are often provided funds from domestic governments, some marine parks do not have concession stands due to their remoteness, while others do not collect visitor fees. Furthermore, in many developing countries with other pressing development needs such as access to quality education, healthcare, and electricity, along with relatively young park services, MPA managers struggle for identity and funding.

Size of Instrument

As reported in the Protected Planet Report 2018, there were 238,563 designated protected areas recorded in the World Database on Protected Areas.[12] This includes approximately 20 million square kilometers (km^2), equivalent to two billion hectares or 4.9 billion acres, of terrestrial protected areas.[13]

As of December 2019, there were 16,977 marine protected areas covering 27.63 million km^2 and which represented coverage of approximately 7.6% of the ocean. National waters, which represent 39% of the ocean, had 25.01 million km^2 of designated protection, which represents 17.7% of the total national waters. In contrast, the high seas represent 61% of the ocean and only 1.2% of the high seas are designated for protection.[14]

It is important to note:

Over the last several years the number and spatial extent of MPAs have increased rapidly. In 2000 the area covered by MPAs was approximately 2 million km^2 (or 0.7% of the Ocean), since then there has been over a ten-fold increase in MPA coverage with 27,631,644 km^2 (or 7.63%) of the ocean being covered by MPAs.

Since 2010, marine protected area coverage has increased by over 14 million km^2. The progress in growth results from a combination of sites being expanded e.g. US Papahānaumokuākea Marine National Monument in the USA which increased to just over 1.5 million km^2, and new sites being created, e.g. the Pitcairn Islands Marine Reserve which covers an area greater than 800,000 km^2, and the recent designation of Marae Moana Marine Park in the Cook Islands covering an area of 1.97 million km^2.

As of 2018, the USA, France and United Kingdom and their overseas countries and territories make up over 50% of the area covered by MPAs while Australia, Cook Islands, New Zealand and Mexico cover an additional 30%.[15]

Similarly:

{…} MPAs have increased more than 15-fold since 1993 when the CBD entered into force. A larger area of the ocean is now protected than on land, though proportionally the much larger ocean realm has lower percentage coverage than does the terrestrial realm. Since April 2016, more than 8 million km^2 of new marine protected areas have been added to the WDPA, strengthening protection of ecological regions and Key Biodiversity Areas in the marine realm. {…} This growth in marine protection is largely the result of several countries declaring very large reserves, e.g. Brazil, Mexico, and some protecting their entire EEZ, e.g. the designation of the approximately 2 million km^2 Marae Moana Marine Park in the Cook Islands in 2017. The four largest

Table 7.1 Top 10 largest MPAs

	Name of MPA	General region	Size (km^2)
1	Ross Sea Region Marine Protected Area	Antarctica	2,060,058
2	Marae Moana	Cook Islands	1,981,965
3	Réserve Naturelle Nationale des Terres australes françaises	French Overseas Territory	1,654,999
4	Papahānaumokuākea Marine National Monument	Hawaii, United States	1,516,557
5	Parc Naturel de la Mer de Corail	New Caledonia	1,291,643
6	Pacific Remote Islands	U.S. Minor Outlying Islands	1,277,784
7	South Georgia and South Sandwich Islands Marine Protected Area	South Georgia and the South Sandwich Islands	1,069,872
8	Coral Sea	Australia	995,251
9	Steller Sea Lion Protection Areas, Gulf	Alaska, United States	866,717
10	Pitcairn Islands Marine Reserve	Pitcairn Islands, UK	839,568

Source Protected Planet

marine protected areas were created or expanded in the last two years. (CBD Secretariat, 2018a)[16]

The total global coverage of the largest twenty MPAs is 17.57 million km^2, which is 63.6% of the total global coverage of all MPAs (i.e., 27.63 million km^2).[17] Below are the largest top ten largest MPAs[18] (Table 7.1).

There are 46 sites on the IUCN Green List in fourteen countries and sixteen of these sites are MPAs.[19] These sites "{…} on the IUCN Green List are certified as being effectively managed and fairly governed, with long-term positive impact on people and nature. Every five years, they are evaluated against a set of demanding criteria defined by the IUCN Green List Standard. These criteria include the quality of protection of natural values and the effectiveness of actions against threats."[20] The sixteen MPA sites are:

<table>
<tr><td>

1. Paracas;
2. Guadeloupe;
3. Cape Byron;
4. Cerbère - Banyuls;

5. Complexe lagunaire de Salses-Leucate;

6. Con Dao;

</td><td>

7. Gorgona;
8. Iroise;
9. Malpelo;
10. Marawah;

11. Zona marina del Archipiélago de Espíritu Santo;

</td><td>

12. Montague Island;
13. Ras Mohammed;
14. Isla San Pedro Mártir;
15. Terres Australes Françaises; and
16. Cote Bleue.[21]

</td></tr>
</table>

In addition, as of October 2019, there were 2372 sites on the Ramsar List of Wetlands of International Importance covering 253.6 million hectares.[22] As of December 2019, there are approximately 701 biosphere reserves in 124 countries.[23]

With respect to funding, there is funding coming from a variety of domestic governments. As previously noted by the International Coral Reef Initiative (ICRI), "{…} an average US$270 million commitment per year for marine conservation was recorded between 2010 and 2016."[24] Most of this is from governments' domestic expenditures. More recently, in 2018, the Australian Government "announced the largest ever single investment in Reef protection. This includes a $443.3 million partnership with the Great Barrier Reef Foundation, aimed at sparking new and innovative responses from, and collaborations with, private investors and philanthropists."[25] Also in 2018, the U.S. NOAA awarded "over $8.3 million to advance coral reef conservation science and management."[26] More recently in 2019, the National Fish and Wildlife Foundation "announced more than $1.4 million in grants to support community-based efforts to improve the health and resilience of coral reefs."[27]

Introduction to Case Studies

The following case studies will examine Australia's Great Barrier Reef Marine Park, the Papahānaumokuākea Marine National Monument, and the Chagos Archipelago. While the Great Barrier Reef Marine Park and the Papahānaumokuākea Marine National Monument are relatively well-financed via government domestic budgetary expenditures, the Chagos Archipelago is underfunded.

Case Study #1: Australia's Great Barrier Reef Marine Park

Introduction

Australia's Great Barrier Reef (GBR) is perhaps the quintessential tropical coral reef example. Located on the East Coast of Australia, the Great Barrier Reef system is the largest on Earth, running approximately 2300 kilometers (approximately 1400 miles), covering 344,400 square kilometers (34.44 million hectares; approximately half the size of Texas), and is literally visible from space with the naked eye. In fact, the Great Barrier Reef Marine Park (GBRMP) crosses 14 degrees of latitude and is approximately the same size as the entire country of Germany, Italy, Japan, or Malaysia[28] (Fig. 7.1).

The GBR starts at its most Southern coral cay, Lady Elliot Island, and runs along the coast of Queensland "north at distances from 15 to 200 km offshore to Torres Strait which separates Australia and Papua New Guinea."[29] The Great Barrier Reef World Heritage Area (GBRWHA) "occupies approximately the same area as the GBRMP, although some variations exist in their coastal boundaries: the GBRWHA also includes the islands of the Great

Fig. 7.1 Map of Great Barrier Reef and Queensland State (*Credit* Lady Elliot Island Eco Resort)

Barrier Reef, while the GBRMP consists of the marine environment alone."[30] The primary body responsible for the management of the Great Barrier Reef is the Australian Government's Great Barrier Reef Marine Park Authority (GBRMPA, "Authority").

In early 1970, the Australian Government "{...} agreed to dedicate three million Australia dollars for an institute of marine science at Townsville, to study 'ways of researching and protecting the Great Reef' {...}."[31] This financing has significantly increased over the years. Likewise, the Great Barrier Reef Marine Park Authority's "total operating revenue for 2018–19 was $79.707 million, compared with $78.880 million for 2017–18."[32] The top three sources of revenue for the Authority in 2018–2019 was from Commonwealth Appropriation ($41.77 million), Special Appropriation / Environmental Management Charge ($11.45 million) and from the Queensland Government ($13.28 million).[33] Further, the "operating expense for managing the Marine Park in 2018–19 was $78.268 million, compared with $68.241 million for 2017–18."[34] Australia also "launched a A$500 mln ($379 mln) plan to preserve the Great Barrier Reef mainly by tackling invasive species and reducing water pollution. {This said,} Environmentalists criticised the plan, saying the biggest threat to the reef is from unchecked climate change."[35]

Zoning is also an important management tool for the GBR, with approximately 33% of the GBR a no-take zone and other areas are eligible for multiple uses. Such multiple uses include commercial and recreational fishing, shipping, aquaculture, and recreation. In certain zones, explosives for defense training and port development are allowed.

Identify the Problem

The GBR has been impacted by a wide range of problems including: coral bleaching as a result of global climate change; water quality issues as a result of land use and land-use change; coastal developments and increasing recreation; unsustainable fishing; and increased impacts from pollution and shipping.[36] Similarly, "{...} it is important to recognise that changes in the ecosystem have occurred as a result of multiple, combined impacts whose effects have varied geographically and temporally."[37]

<u>Coral Bleaching</u>
Coral bleaching occurs due to increased water temperatures which stress corals to the degree where they expel their symbiotic zooxanthellae.

If temperatures rise more than a degree or two above the typical maximum, they {corals} bleach. Bleaching happens when there is a breakdown in the relationship between corals and the microscope plants that live within their tissues and provide them with much of their food. At higher temperatures these microscopic plants, or zooxanthellae, as they are called, begin to harm their hosts, at which point the corals kick them out. Since it is the zooxanthellae that give corals their color, the tissue becomes transparent and colonies turn deathly white, as the chalky skeleton beneath is revealed. Bleached corals are starving; they die within weeks unless temperatures revert to normal.[38]

This, in part, has contributed to a decline in coral reproduction within parts of the GBR.[39] Sadly, the GBR has experienced coral bleaching in 1998, 2002, 2016 and 2017,[40] and another coral bleaching is underway in 2020 which has been described as "like watching the Louvre 'burn to the ground'."[41]

Water Quality Issues from Runoff
Queensland State has a large sugarcane and livestock industry. In fact, sugarcane cultivation first expanded from approximately 1864–1884.[42] Today, the Murray-Darling Basin is Australia's breadbasket:

The Murray-Darling Basin covers one-seventh of the Australian land mass and ranks as the 21st largest basin in the world. It is very important for its diverse flora and fauna and contribution to the national economy. There are over 30,000 wetlands in this basin, and 15 are listed as Ramsar Wetlands. About 85 per cent of all irrigation in Australia takes place in this basin, which supports an agricultural industry worth around $9 billion annually. However, the Murray-Darling Basin is under threat due to climate change, deforestation, weeds and dryland salinity.[43]

Yet, such agricultural and livestock production has come at the expense of Queensland's forests:

Queensland, which accounts for over half of Australia's annual deforestation, could introduce a new law that will restrict clearing as soon as next week, the Guardian reports. A state commission released a report Monday night proposing only a few changes to the state Labor government's proposal. That news coincided with a new report by the Climate Council, which found that more than 1 mln hectares of woody vegetation were cleared in Queensland during 2012-16 after the former Liberal party government eased clearing rules.[44]

Deforestation leads to more runoff and erosion, which increases siltation of the offshore waters. In addition, the water quality—due to increased nitrogen, phosphorus, and fertilizers from fields or untreated animal waste from pastures—has declined over the years. Likewise, "in particular, the GBRWHA is strongly influenced by its sources of freshwater, sediments, and nutrients, especially the 35 drainage basins of eastern Queensland that form the Great Barrier Reef Catchment Area (GBRCA)."[45]

More specifically:

Many reports suggest that the condition of the Great Barrier Reef has declined since European settlement commenced in Queensland, as a result of direct exploitation and the development of adjacent coastal land. {… it has been …} argued that sediment discharges have increased by three to four times, nitrogen discharges have doubled and phosphorus quantities have increased by six to ten times since 1800.[46]

Coastal Developments and Increasing Recreation
Australia has a relatively low population density. As the sixth largest country on Earth with a total land area of 7.7 million square kilometers, Australia is close to the size of the United States.[47] Yet, Australia has a population of "just" 25.4 million people as of June 2019,[48] which equals two-thirds the population of California (i.e., 39.5 million people as of July 2019).[49]

Since 1981, a "very rapid expansion of the Queensland tourism industry has occurred, driven by increases in international tourism and domestic migration to Queensland."[50] Today:

Land modification associated with the increased human population, urban development and agricultural expansion in the Great Barrier Reef catchment area has reduced the quality of water flowing into the Great Barrier reef lagoon. This area comprises some 25 per cent of the land area of Queensland, and is home to more than one million people in addition to the two million tourists that visit the region each year.[51]

In addition to pollution from fertilizers, pesticides, and sugar-plant effluent, this increased coastal development and recreation can lead to dredging and urban sewage.

Unsustainable Fishing
While less of a concern today, historically there were unsustainable harvesting for the pearl-shell trade, along with unsustainable fishing for bêche-de-mer (sea cucumbers) and trochus, a type of sea snail. This said, "although reef organisms (including corals) have been removed from the Great Barrier Reef since the period of earliest European exploration, the first sustained European

commercial fisheries in the Great Barrier Reef were the *bêche-de-mer* and pearl-shell fisheries."[52]

In fact:

> pearl-shell became one of the most economically significant exports from Queensland, and was used in the manufacture of buttons and ornaments; the shell was exported to Europe and south-east Asia. While pearls were sometimes taken with the shells, those were not the commercial object of the trade and were usually kept by the divers. {...} synthetic plastics replaced pearl-shell in the manufacture of buttons and the pearl-shell market collapsed.[53]

In addition to the trade of pearl-shells, bêche-de-mer, and trochus, there were also unsustainable fishing for sea turtles, dugong, and whales:

> The dugong is listed as vulnerable to extinction due to various factors, including variations in seagrass availability, incidental drowning in shark nets set for bather protection, accidental by-catch in commercial gill nets, vessel strikes, habitat loss and over-fishing {commercial hunting for oil and hides took place from approximately 1847-1969}.[54]
>
> Since European settlement in Queensland, various human activities have exploited marine turtles in the Great Barrier Reef and adjacent areas: the production of tortoise-shell {1871-1940s}, the commercial marine turtle fisheries {1867-1962}, the 'sport' of turtle-riding at tourism resorts {1900s-1960s} and the traditional hunting of turtles by Indigenous people.[55]
>
> The GBRMPA (2000) identified numerous anthropogenic impacts on cetaceans, including commercial whaling, harassment, vessel strikes, entanglement in nets, ingestion of litter, underwater explosions, pollution (including noise pollution), disease, live capture and habitat degradation.[56]

Furthermore, coral and shell collecting "{...} were widespread, cumulative and probably severe for some species. Coral and shell collecting occurred in more places, and for longer periods, than has previously been documented. Four main types of coral and shell collecting occurred in the Great Barrier Reef: informal collecting, scientific collecting, unregulated commercial collecting and licensed collecting."[57]

<u>Mining</u>

Mining has taken several forms throughout the GBR over the years. This includes coral mining and guano mining, along with petroleum leases and mining.

Coral mining "{...} initially took place for the manufacture of agricultural and industrial lime; however, terrestrial sources of agricultural and industrial lime eventually replaced lime manufactured from coral, and coral collected since the 1950s increasingly supplied the curios and ornamental trades instead."[58] Similarly, guano mining (i.e., extraction of hardened bird droppings) led to the devastation of some islands, such as Lady Elliot Island (i.e., see case study in Chapter 12).

In the late 1960s, the Queensland Government "openly signaled its intention of allowing petroleum companies to explore for oil and gas anywhere within an area encompassing 80 percent (eighty thousand square miles) of the Great Barrier Reef."[59] While large scale oil spills near the GBR are not common, one such damaging event was the *Oceanic Grandeur* oil tanker spill in the Torres Strait in 1970.[60]

Increased Impacts from Pollution and Shipping

Another issue facing the Great Barrier Reef is the increased impacts from pollution and shipping:

> Adani's plans for the 60 Mt/year Carmichael coal mine in northern Queensland have been delayed for nearly a decade amid legal challenges and a lack of willing funders. But on Thursday {November 2018}, Adani announced it will fund the mine itself, and it hopes to begin construction before the end of the year. The project will be smaller than originally planned, initially producing 15 Mt coal annually, gradually rising to 27 mln. Green groups have attacked the proposal for the massive carbon emissions expected to come from the production itself and from burning the coal in major Asian nations, primarily India. Those groups have also fought the project tooth and nail over concerns that transporting so much coal over the struggling Great Barrier Reef will deteriorate the reef even more.[61]

This said, there are some large ports in Queensland relatively near the GBR, including Gladstone, which is the sixth largest port in all of Oceania.[62]

Crown-of-Thorns Starfish

Another major problem is the crown-of-thorns starfish (*Acanthaster planci*). Noel Monkman "was among those, too, who at Green Island in 1963 discovered the first coral damage wrought by crown-of-thorns starfish – an infestation of which was soon reported to be spreading all over the Reef."[63] The problem with crown-of-thorns starfish became so significant that "coral cover has declined by about 50 percent over the past thirty years, the report stated, deeming the crown-of-thorns starfish responsible for almost half of this decline."[64]

Charlie Vernon, the world-renowned coral reef scientist:

> became convinced that numbers of the latter {crown-of-thorns starfish} were soaring due to overfishing of the starfish's natural predators, and that survival of

the millions of larvae expelled annually into the ocean currents was enhanced by the growing levels of chemical pollution (Crown-of-thorns larvae thrive in polluted waters).[65]

Crown-of-thorns multiply after being cut, so one of the techniques now is to inject them with vinegar. There has even been the development of COTsbot, a robot designed to take out the crown-of-thorns starfish.[66]

Cyclones

One major threat to the GBR is cyclones, which can act like "a bulldozer clearing the rainforest."[67] This said, Cyclone Yasi in February 2011 and Cyclone Larry in March 2006 were particularly damaging to parts of the GBR.

Other Issues

Other issues facing the GB, whether historically or in contemporary times, includes the "{...} creation of coconut palm plantations on many islands; the destruction of island vegetation by introduced goats; the misuse of fire; and the introduction of exotic types of vegetation, such as *Lantana spp.* and prickly pear (*Opuntia spp.*)."[68]

In addition to *Lantana spp.* and prickly pear, other invasive species throughout Queensland and the offshore islands include cane toads, feral pigs, rats, and cats.

Historically, there have also been disputes between the Australian Government and Queensland State over sovereignty claims and who controls the GBR.

Why the Problem Is Important

The GBR is the largest on Earth, consisting of more than 2900 separate coral reefs, home to more than 450 species of hard coral, more than 1600 species of fish, and iconic species such as humpback whales (*Megaptera novaean-gliae*), minke whales (*Balaenoptera acutorostrata*), giant manta rays (*Manta birostris*), and orange clownfish (*Amphiprion percula*).[69] Furthermore, of the "seven extant species of marine turtles, six occur in Queensland waters: green (*Chelonia mydas*), hawksbill (*Eretmochelys imbricata*), loggerhead (*Caretta caretta*), flatback (*Natator depressus*), olive ridley (*Lepidochelys olivacea*) and leatherback (*Dermochelys coriacea*) turtles. Most are defined by the IUCN (2013) as critically endangered, endangered, or vulnerable"[70] (Fig. 7.2).

This said, Raine Island is "the largest rookery for nesting green sea turtles in the world, as well as being the most significant nesting site for sea birds, with over eighty-two species recorded there."[71]

Fig. 7.2 Picture of Green Sea Turtle at Lady Elliot Island (*Credit* Brigitta Jozan)

The GBR is also a sacred place for Australia's Aboriginal and Torres Strait Islander peoples.

The GBR is adjacent to the Wet Tropics World Heritage Area,[72] and the Coral Sea Reserves,[73] along with the Elizabeth and Middleton Reefs Marine National Nature Reserve which are on the Ramsar List of Wetlands of International Importance.[74]

Furthermore, two sites (Far Northern Great Barrier Reef and the Whitsundays) are listed among the fifty most important bioclimatic units and the GBR is listed as a World Heritage Marine Programme.

How Problem Was Identified

There is a long history associated with the GBR. It is noted that "Matthew Flinders, not James Cook, was the true European father of the Reef, because he was the first person to infer its unified existence and to conceive of it as a whole."[75] In time "Joseph Jukes, whom Griffin nicknamed 'the geologist,' was officially charged with investigating the geological character of the Great Barrier Reef and the structure, origins, and behavior of the reef-growing corals – the first scientist ever to be specifically assigned such a task."[76]

Since this time, numerous entities have worked to identify and address the problems facing the GBR. This includes: local and international NGOs, such as the Citizens of the Great Barrier Reef and the Great Barrier Reef Foundation; academic and scientific institutions, such as the Australia Research Council Centre of Excellence for Coral Reef Studies at James Cook University and the Global Change Institute at the University of Queensland; businesses, such as Lady Elliot Island Eco Resort and GreenCollar; and governmental agencies, such as the Australian Institute of Marine Sciences, Great Barrier Reef Marine Park Authority, and the Queensland State Government.

Effectiveness of Process for Identifying Problem

According to Dr. Petra Lundgren of the Great Barrier Reef Foundation, regarding the effectiveness of the process:

> for all intents and purposes it has been. The re-zoning plan that was adopted in 2003 was a massive undertaking, which involved a lot of stakeholder and community consultation. It would be madness to say that it was not opposed (it was hugely controversial), but with time, I would argue that most Australians now agree that it has been a net positive outcome. The RRAP {Reef Restoration and Adaptation Program} program includes a dedicated program area around stakeholder, community and Traditional owner engagement. As part of the funding mechanism for this program (the Reef Trust Partnership), 10% of the total allocated funding is set aside, especially to support activities led and co-designed by the Traditional Owner groups that have connection to the GBR (there are 70 Traditional owner groups along the GBR).[77]

Thus, the process for identifying the multitude of problems facing the GBR appears to be effective.

Steps Taken to Address the Problem

Over time, many steps have been taken to address the problems facing the GBR.

First, the "{…} earlier decision {in 1936}[78] to protect Green Island reef was a significant one: it created one of the earliest marine protected areas in existence."[79]

Then, "on April 16, 1970, Prime Minister Gorton introduced a bill into federal parliament claiming Commonwealth sovereignty over the Reef and its waters."[80] An important next step occurred when the Great Barrier Reef

Marine Park Act was "passed in 1975 but not fully promulgated until 1979. The act prescribed that most of the Reef province became a marine park governed by a committee of the Great Barrier Reef Marine Park Authority, which was in turn answerable to a federal government minister."[81]

Shortly afterwards, "{...} on October 26, 1981, the Great Barrier Reef received what two of its finest historians, James and Margarita Bowen, have called a 'conservation climax' – World Heritage listing 'as the most impressive marine area in the world.'"[82]

Several decades later, "the 2004 Zoning Plan for the Great Barrier Reef Marine Park increased the proportion of the Marine Park that was highly protected by 'no-take' zones from less than 5 per cent to more than 33 per cent."[83]

Results

There have been numerous results attributable to the Great Barrier Reef, including:

- Designation of the GBR as the Great Barrier Reef Marine Park in 1975,[84] along with the establishment of the Great Barrier Reef Marine Park Authority in 1975[85];
- Distinguished as a World Heritage Site in 1981, the first coral reef ecosystem to achieve such a status[86];
- The adjacent Wet Tropics World Heritage Area was also distinguished as a World Heritage Site in 1988[87];
- The Coral Sea Reserves,[88] along with the Elizabeth and Middleton Reefs Marine National Nature Reserve,[89] were designated as Ramsar List of Wetlands of International Importance in 2002[90]; and
- The "2004 Zoning Plan for the Great Barrier Reef Marine Park increased the proportion of the Marine Park that was highly protected by 'no-take' zones from less than 5 per cent to more than 33 per cent."[91]

More recent results include, in 2017, the production of the "Great Barrier Reef blueprint for resilience (the Blueprint) which prioritised 10 actions to build the Reef's resilience."[92] The Authority also "manages an Australian Government funded Crown-of-thorns Starfish Control Program to protect a network of high ecological and economic value coral reefs from outbreaks. Additional funding of $13.2 million during 2018–19 allowed the Authority to increase the program's capacity from two to six vessels in late 2018."[93] The Reef 2050 Long-Term Sustainability Plan (Reef 2050 Plan) and the

Reef 2050 Integrated Monitoring and Reporting Program (RIMReP) are being developed and "once established, RIMReP will be a game-changer — providing coordinated access to information about the Reef, its catchment and human use of the Region."[94]

A further result is the "Great Barrier Reef Ocean Observing System (GBROOS) {which} is the world's first reef-based IP data network, giving researchers an unprecedented range of data on reef conditions and enabling them to better track changes and impacts."[95]

Challenges and How They Were Met

There are numerous challenges facing the GBR.

For instance, there is a marine-terrestrial cycle that takes place between the lagoon of the Great Barrier Reef Marine Park and the Wet Tropics World Heritage Area, particularly near Port Douglas. This said, rain is generated from the coast and eventually precipitates on the Western slopes of the Daintree Rainforest. Over time, the rainwater trickles back to the coast, helps to cool off the shallower waters surrounding the coral reefs of the Lagoon, and this can help mitigate the effects of coral bleaching. However, as of September–October 2019, Queensland had been in a 3–4 year drought.[96]

Another challenge for the GBR is that there are a lot of historical sugarcane plantations throughout Queensland going back generations as a result of Italian immigrants. This said, how do you tell grandparents that they now need to stop the cultivation of sugarcane? What alternative livelihoods could they undertake? A somewhat similar challenge is that while cattle ranching is not allowed on, or near the coast, it still takes place at times.[97]

Clearing of forests in Queensland, particularly for sugarcane cultivation and cattle ranching, has come close to riverbanks, such as the Daintree River. Normally, there would be a buffer of natural forests between the sugarcane plantation (or cattle ranch) and the rivers. However, over time, the rivers change course and erode the riverbanks. This process reduces the extent of the buffer zone and as a result, more sediments and nutrients flow into the rivers. It is important to understand that most of what flows into the rivers, such as the Daintree River, will eventually make its way to the offshore reefs.[98]

Adani's new coal mine in Queensland, called the Carmichael Coal Mine, will see its coal exported across the GBR which will lead to dredging, increased shipping, and its combustion will contribute to climate change along with ocean acidification and warmer sea temperatures.

Another challenge is that the GBR management has transitioned from a complete focus on conservation, to a focus on both coral conservation and restoration.[99] According to Dr. Lundgren:

> Actively restoring coral and coral reefs is a new thing in Australia. The first permit to do coral restoration on the GBR is less than two years old. Following the two back-to-back mass bleaching events in 2016 and 2017, the Great Barrier Reef Marine Park Authority recognized that protection alone was no longer enough and commenced work around adjusting policies and regulation to allow more active forms of intervention and management.
>
> The primary challenge here, like everywhere else, is to find ways to restore ecosystem function and resilience. Restoring small patches of high value reefs (like a single tourism site) is still achievable, because there are still many healthy areas of coral remaining that can act as sources of brood stock, but going from restoring small patches (which you can keep doing over and over again) to restoring a resilient ecosystem is a huge step, which will require a large amount of resources and R&D. The Reef Restoration and Adaptation Program is built on an extensive concept feasibility study around tackling this challenge, and the program is scheduled to be rolled out over the next four years.[100]

Furthermore, there are some challenges associated with the fact that Australia's A$500 million (USD$379 million) plan to preserve the GBR had its funding allocated to a single organization, the Great Barrier Reef Foundation.[101]

> Just days after the public announcement, the full amount was transferred to the Foundation's bank account before any financial due diligence or plan for investing the money had begun. The Foundation had been in discussion with the government regarding a AU$5 million grant, and without requesting more money, on short notice, the government informed them that they would be receiving one hundred times the requested amount. Critics say that the government rushed through the investment due to political reasons – to balance the budget before the end of the financial year and ahead of elections – with no effort to complete a transparent and fiscally-responsible decision. There was no competitive tender process, no requirement for the Foundation to tell the public what the funds would be used for before the transfer of money, and no financial due diligence prior to the transfer of funds.[102]

Nevertheless, it appears as though the Foundation is doing some exceptional work.

Beyond Results

The management of the GBR, particularly its 2004 zoning plan, can serve as a model for other countries.

According to world-renowned coral scientist, Charlie Vernon:

It's Charlie's hope that some as yet unknown strains of symbiotic algae, better able to cope with a heat-stressed world, might eventually form new partnerships with the corals. Or that the adaptive energies of fast-growing corals like *Acropora* might somehow outpace the rate of bleaching. Or that pockets of coral lying in shadowed refuges on cool, deep reef slopes, or in recently discovered deep waters, might survive to become agents of future renewal.[103]

As of March 2018, Queensland State had "{...} begun work to establish a A$500-million ($390 mln) fund that will use carbon markets to cut greenhouse gas emissions in the land sector,"[104] which would have the dual benefits of helping to mitigate global climate change and to reduce topsoil runoff onto the GBR.

Furthermore, with respect to education and outreach, the "Aquarium welcomed more than 128,000 visitors during 2018–19, with more than 83,000 people participating in educational talks and tours. Through its formal programs and outreach education via the award-winning Reef Videoconferencing program, reef education was delivered to more than 8000 students across Australia and the world."[105]

Lessons Learned

Upon reflecting on our travels down the coast of Queensland, it came to the author and his wife's mind that if we cannot save the GBR, then it is challenging to see how we can save the world's remaining coral reefs. Furthermore, pressing questions—which are also facing many other places around the world—include:

- How do you balance tourism with local needs for income versus the health of the natural environmental and its capacity to absorb the impacts?
- How do you maintain the curiosity of tourists and locals, while at the same time, managing the level of tourism?
- How do you maintain the interest and financial support of past visitors, especially in light of the daunting challenges?
- How do you ultimately make tourism sustainable?

According to Dr. Lundgren, a few lessons learned are:

That there are no easy solutions. I came into this as an ecologist, and throughout my University and research days I kept thinking that if we are clever, and come up with ecological / genetic solutions, then we should be able to fix this. What you eventually realise is (and the issues around climate change has made this so abundantly clear), is that unless people / politicians / communities WANT change, you can have all the answers you need, and it won't make much difference. It is about behavior change, trust, and management of fear, more than about finding the right gene in the coral. I have also learned that I am not a social scientist or a very good communicator, so lesson number 2 is that every major program needs social scientists and good communicators, - it is critical.[106]

Other Resources on the Great Barrier Reef

Australian Bureau of Statistics' Experimental Environmental-Economic Accounts for the Great Barrier Reef, 2017

- http://www.abs.gov.au/ausstats/abs@.nsf/0/021873EB0FFA07A9CA258 1820077EE15?Opendocument

Australian Government's Reef 2050 Plan Investment Framework

- https://www.environment.gov.au/marine/gbr/publications/reef-2050-inv estment-framework

Australian Government's Reef Trust

- https://www.environment.gov.au/marine/gbr/reef-trust

Australian Institute of Marine Science

- https://www.aims.gov.au/

Australian Marine Environment Protection Association

- https://www.ausmepa.org.au/

Australian Research Council's Centre of Excellence for Coral Reef Studies at James Cook University

- https://www.coralcoe.org.au/

Charlie Vernon's A Reef in Time: The Great Barrier Reef from Beginning to End

- https://www.hup.harvard.edu/catalog.php?isbn=9780674034976

Citizens of the Great Barrier Reef

- https://citizensgbr.org/actions; and
- https://citizensgbr.org/toolkit

Fishes of Australia

- http://fishesofaustralia.net.au/

Global Change Institute at the University of Queensland

- https://gci.uq.edu.au/

Great Barrier Reef Foundation

- https://www.barrierreef.org/

Great Barrier Reef Outlook Report 2019

- http://www.gbrmpa.gov.au/our-work/outlook-report-2019

National Parks Association of Queensland

- http://www.npaq.org.au/

Queensland Department of Environment and Science

- https://parks.des.qld.gov.au/

Queensland Environmental Protection Agency

- https://environment.des.qld.gov.au/

Queensland Government Statistician's Office

- https://www.qgso.qld.gov.au/

Reef HQ Aquarium

- www.reefhq.com.au/

Wet Tropics World Heritage Area

- https://www.environment.gov.au/heritage/places/world/wet-tropics; and
- https://whc.unesco.org/en/list/486/

Case Study #2: Papahānaumokuākea Marine National Monument

Introduction

The Papahānaumokuākea Marine National Monument (Monument), established in 2006, is "the largest contiguous, fully protected conservation area under the U.S. flag, and one of the largest marine conservation areas in the world. It encompasses 582,578 square miles of the Pacific Ocean (1,508,870 square kilometers) - an area larger than all the country's national parks combined."[107] The Monument encompasses the Northwest Hawaiian Islands and is a World Heritage Marine Site (Fig. 7.3).

With respect to funding, in September 2019, the U.S. Senate Appropriations Committee "approved $4.1 million in federal funding to support management and research projects for the Papahānaumokuākea Marine National Monument."[108]

Identify the Problem

Historically:

In the late 1800s and early 1900s, the NWHI {Northwestern Hawaiian Islands} were exploited and ravished by seal hunters, whalers, feather hunters, pearl divers, and guano miners. Seals, sea turtles, seabirds, sharks, and whales

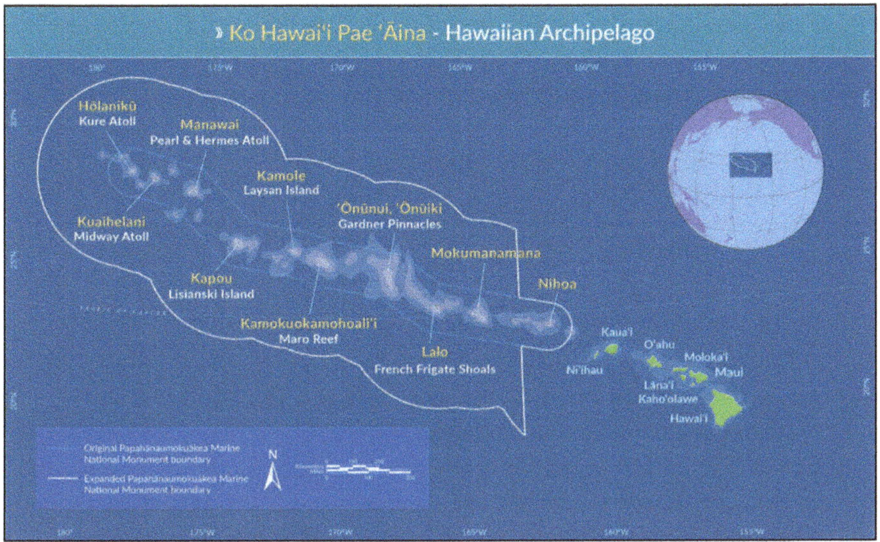

Fig. 7.3 Map of Papahānaumokuākea Marine National Monument (Open Source Map–*Credit* National Oceanic and Atmospheric Administration) (National Ocean Service, Office of National Marine Sanctuaries, National Oceanic and Atmospheric Administration. "Papahānaumokuākea Marine National Monument." December 20, 2019. Accessed January 14, 2020. https://www.papahanaumokuakea.gov/; Also see: Office of Marine & Aviation Operations. "Image Licensing & Usage Info." May 17, 2019. Accessed January 14, 2020. https://www.omao.noaa.gov/find/media/images/image-licensing-usage-info)

were slaughtered in mass. Alien and exotic plants and insects drastically changed the unique ecosystems by destroying or out-competing many of the endemic native species. In 1894, entrepreneurs from a rabbit canning industry released rabbits that devoured nearly all the vegetation on some islands. In the early part of the 20th century, Japanese feather hunters slaughtered millions of seabirds.[109]

Today, there are a few problems facing the Monument. In an expedition to the Monument in August 2019, researchers observed damage to coral reefs and to the atolls at French Frigate Shoals due to Hurricane Walaka of October 2018.[110]

In addition, an invasive species of red alga was also identified by the researchers at the Pearl and Hermes Atoll.

Lastly, are the problems associated with invasive mice and rats:

Introduced, damaging (invasive) mice and rats infested Midway Atoll more than 75 years ago during military occupancy. Invasive mice persisted after rats were removed from the island in 1996 and are now the sole non-native mammal present in the Northwestern Hawaiian Islands. The majority of seabird extinctions around the world have occurred on islands and were caused by invasive mammals, in particular invasive rodents.[111]

Such problems are common throughout the world.

Why the Problem Is Important

These problems are important because the Monument, as of December 2019, was the fourth largest marine protected area in the world,[112] and is host to an enormous amount of biodiversity. For instance, the Monument:

> supports a dynamic reef ecosystem with more than 7,000 marine species, of which approximately one quarter are unique to the Hawaiian Islands. This diverse ecosystem is home to many species of coral, fish, birds, marine mammals, and other flora and fauna, including the endangered Hawaiian monk seal, the threatened green sea turtle, and the endangered leatherback and hawksbill sea turtles. In addition, this area has great cultural significance to the Native Hawaiian community and a connection to early Polynesian culture worthy of protection and understanding.[113]

Similarly, the Hawaiian Island Chain, including the Monument, is home to more than 60% of the U.S.' coral reefs.[114]

With respect to Midway Atoll, "more than 3 million birds representing 29 species rely on Midway Atoll." Midway Atoll also provides "nesting habitat for the world's largest Laysan Albatross colony" and is an "important habitat for Pacific seabirds facing increasing threats."[115] This includes 73% of all the world's Laysan albatross (*Phoebastria immutabilis*),[116] and 36% of all black-footed albatross (*Phoebastria nigripes)* in the world.[117] The problems regarding invasive mice are particularly important because:

> Researchers first observed mice preying on albatross on Midway in 2015. An exponential increase in attacks the following year signals the potential for an alarming escalation in the number of birds being preyed upon and dying from their wounds. A combination of factors, including extensive and costly control efforts have reduced impacts in recent years, but the risk of population-scale impacts to albatross persist. Their lack of defense mechanisms and dedication to their eggs leaves albatross defenseless against invasive mice. Albatross lay only one egg every one or two years. Both parents invest their energy in hatching and raising the chick. This reproductive cycle means that losses to the colony each year continue to impact the population for decades to come.[118]

The Monument is a very important cultural site as well. For instance:

> The island of Mokumanamana is situated in the center of the Hawaiian Islands chain and it was believed to be in a unique position on the northern Tropic, acting as an axis point between the world of the supernatural and the living. This island is dominated by 34 individual heiau (temples), sites that were used for ritual purposes. {...}
>
> Nihoa Island, with over 89 cultural sites that range from habitation, religious, and agricultural, was developed in conjunction as a remote elite outpost for recurrent staging and use of Mokumanamana and its temples.[119]

The Monument is also listed as a World Heritage Marine Programme.

How Problem Was Identified

The problems were identified, specifically at Midway Atoll, by several NGOs and U.S. Government agencies:

> The U.S. Fish and Wildlife Service is coordinating with the Papahānaumokuākea Marine National Monument co-managers and working with Island Conservation, American Bird Conservancy, National Fish and Wildlife Foundation, National Oceanic and Atmospheric Administration, U.S. Geological Survey, and other members of the conservation community to protect and conserve Midway Atoll's seabird colony and are working to find solutions to this growing crisis.[120]

The co-trustees of the Monument are the NOAA and the USFWS, along with the State of Hawai'i's Department of Land and Natural Resources and the Office of Hawaiian Affairs.

Effectiveness of Process for Identifying Problem

There has been significant stakeholder engagement. For instance, "Native Hawaiians played key leadership roles in the public process to design, establish, and manage the monument."[121] In addition:

> The Expand Papahānaumokuākea coalition comprised a diverse community-driven effort that included kupuna (elders), fishermen, educators, cultural practitioners, non-profits, community groups, scientists, religious organizations, veterans, keiki (children), and many others across Hawai'i, and beyond

that requested the President to use the Antiquities Act to expand protection for PMNM.[122]

Thus, it appears as though the process for identifying problems has been effective.

Steps Taken to Address the Problem

Some of the steps taken to address the problems include:

- Establishment of the Monument;
- Declaration of World Heritage Marine Site;
- The Native Hawaiian Research Plan and Native Hawaiian Cultural Working Group were formed;
- The Monument's Management Plan was developed, commented upon, and approved. The Plan will guide work for fifteen years with reviews conducted every five years[123]; and
- Funding was appropriated by the U.S. Senate.

Results

Two major results attributed to the Monument are its declaration as a marine national monument and the appropriation of funding.

The Monument was initially established in 2006, and then later expanded in 2016:

George W. Bush made a huge splash with his Pacific protected areas and single-handedly added over 30 percent to the total area under protection in the world's oceans. That makes him one of the world's great marine conservationists! That expression might stick in some people's throats given his rather less glorious environmental record on land. Whatever your feelings about Bush, he left a wonderful legacy in the sea.[124]

{Then in 2016} US President Barack Obama expanded a national monument off Hawaii, creating the world's largest marine reserve, the White House says. His announcement on Friday quadruples in size a monument originally created by President George W Bush in 2006. The Papahānaumokuākea Marine National Monument will now span 1.5m sq km (582,578 sq miles), more than twice the size of Texas.[125]

In 2010, "Papahānaumokuākea was also inscribed as a mixed (natural and cultural) World Heritage Site by the United Nations Educational, Scientific, and Cultural Organization (UNESCO). It is the first mixed UNESCO World Heritage Site in the United States."[126] Similarly, it is important to note the Monument "was designated the first mixed conservation site in the United States due to its natural and cultural importance. It is also the world's first cultural seascape, being recognized for its continuing connections to indigenous people."[127]

With respect to funding, as previously mentioned, the U.S. Senate Appropriations Committee "approved $4.1 million in federal funding to support management and research projects for the Papahānaumokuākea Marine National Monument" in September 2019.[128]

Another important result is the steps taken to eradicate the invasive mouse:

The U.S. Fish and Wildlife Service has finalized the Midway Seabird Protection Plan to remove the predatory invasive house mouse from Midway Atoll. The Final Environmental Assessment and Finding of No Significant Impact associated with the project are now available to the public. The U.S. Fish and Wildlife Service will continue to work with partners and continually evaluate all aspects of the project as it progresses to ensure that the expectations outlined in the project plan are being met.[129]

Challenges and How They Were Met

The main contemporary challenges facing the Monument are dealing with marine debris, terrestrial pollution, and alien species. To address marine debris, there is "an ongoing multi-agency marine debris clean-up program {which} has removed more than 586 tons of debris from the property in the past ten years."[130] There are also cleanup efforts underway to address the legacy terrestrial pollution and quarantine measures in place to help prevent the further introduction of alien species.[131]

Beyond Results

The results of the Monument appear sustainable as the Monument will be managed by several state and federal government agencies:

The proclamation will also include a stipulation that the Office of Hawaiian Affairs be named a co-trustee of the monument, joining the National Oceanic and Atmospheric Administration, the U.S. Fish and Wildlife Service and the

State of Hawaii's Department of Land and Natural Resources in the management of the area. All commercial resource extraction activities, including commercial fishing and any future mineral extraction, are to be fully prohibited in the monument.[132]

The Mokupapapa Discovery Center in Hilo, Hawai'i Island was established in 2003 for ongoing education and outreach. There are also monitoring efforts:

Originally, biological monitors from federal management agencies were assigned to access trips to ensure that natural resources were protected. More recently, cultural monitors were assigned to biological access so that cultural values and dimensions of management were also addressed. It is now common for western scientists to join Native Hawaiian practitioners on expeditions aboard traditional voyaging canoes. {...} Education and public engagement is also a priority in the management of PMNM, and several steps have been taken toward integrating culture into educational and outreach efforts.[133]

Lessons Learned

An important lesson learned is that:

When Papahānaumokuākea was established, it was a groundbreaking idea that management of a large-scale protected area could recognize both natural resources and cultural resources. Papahānaumokuākea's successful co-management structure and incorporation of Native Hawaiian values in its creation and management serves as a model for other conservation areas around the world. Since the establishment of Papahānaumokuākea, a number of nations have established large-scale MPAs (LSMPAs > 100,000 km2), and many have integrated traditional knowledge and cultural values into these Processes. Cooperation, alliances, and sharing of information have been essential to the success of LSMPA networks, as many of the existing LSMPAs consist of remote islands in the Pacific, and thus are connected by common history, culture, and ancestry.

{Similarly,} PMNM provides concrete examples of how cultural knowledge is integrated across four dimensions: (1) governance and management structure, (2) policies, plans, and protections; (3) cultural practices and research; and (4) education.[134]

Other Resources for Papahānaumokuākea Marine National Monument
<u>Big Ocean Managers</u>

- https://bigoceanmanagers.org/

Hawai'i Institute of Marine Biology

- http://www.himb.hawaii.edu/

Hawai'i Wildlife Fund

- https://www.wildhawaii.org/index.html

Island Conservation

- https://www.islandconservation.org/midway-atoll/

Oceans Caucus Foundation

- http://www.ocfoundation.us/

Ocean Champions

- https://www.oceanchampions.org/

Office of Hawaiian Affairs

- https://www.oha.org/papahanaumokuakea/

Papahānaumokuākea Marine National Monument

- https://www.papahanaumokuakea.gov/; and
- https://www.papahanaumokuakea.gov/management/mp.html

State of Hawaii's Division of Aquatic Resource

- https://dlnr.hawaii.gov/dar/

Case Study #3: The Chagos Archipelago

Introduction

The Chagos Archipelago, also referred to as the Maldives-Chagos-Lakshadweep Atolls or as the Great Chagos Archipelago, is a United Kingdom (UK) marine reserve encompassing 545,000 square miles (1.4 million square kilometers) and 55 reefs.[135] In the past, the area was previously referred to as the British Indian Ocean Territory (BIOT) Marine Protected Area (Fig. 7.4).

In part due to its remoteness and the fact that it is not part of mainland UK, the Chagos Archipelago has not received substantial government funding. This said, one of the first things the Former BIOT Commissioner, Mr. Nigel Wenban-Smith, did around 1982, specifically for Chagos, was to write into the budget of the UK Foreign Office an annual visit by the commissioner to Chagos.[136]

Today, there is the Chagos Conservation Trust (CCT). The CCT is a:

not-for-profit, charitable incorporated organisation registered in the UK {, … that relies …} on membership fees, donations and funding through grants from the British government, trusts and foundations, and other charitable giving organisations to fund research and conservation work in the Chagos Archipelago. {Its founder…}, John Topp, generously provided an endowment that enables {the CCT} to pay for administration costs leaving any other funding received to go directly into achieving the organisation's vision and mission.[137]

Yet, funding remains a challenge for the Chagos Archipelago. There is no local lobbying efforts or local taxes. With no local government, there is no local government funding.[138]

Identify the Problem

In the early days, the Chagos Archipelago was not really an exclusion zone, except indirectly due to the military presence. There was some fishing, particularly for sea cucumbers and sharks, and at times, the military accidentally damaged the reefs.

In 1978, a heatwave struck the Chagos Archipelago, which killed 90% of the coral reefs. Over the next 10–15 years, the coral reefs recovered. However, in 2015 and 2016, the coral reefs were killed again due to another heatwave and now, in 2020, yet another heatwave in underway. If this present heatwave

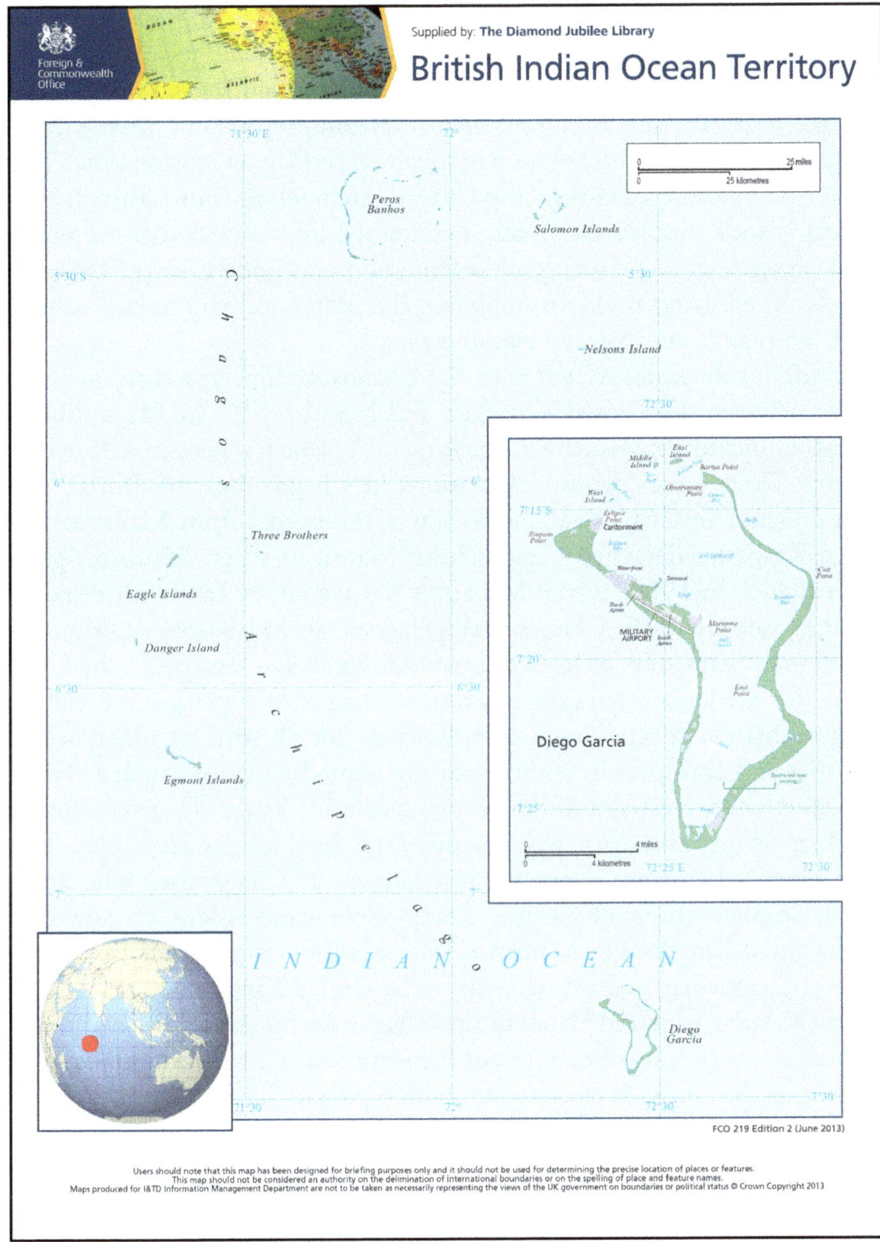

Fig. 7.4 Map of Chagos Archipelago (*Credit* Copyright Geoinnovations Ltd. This information map is licensed under the Open Government Licence v3.0. To view this licence, visit http://www.nationalarchives.gov.uk/doc/open-government-licence/)

is bad, it is "not wrong to say the corals are dead."[139] The dire situation is the heatwaves are getting more frequent and more severe. In fact, the heatwaves are hitting more frequently than the coral reefs can stabilize and breed. Essentially, the heatwaves are killing off the babies and they are not getting mature enough to breed. This leaves many of the coral reefs in an erosion state. There are no true volcanic islands in the Chagos Archipelago, but rather, they are all coral islands that are essentially maintained by the reefs. All the mechanisms which keep a reef going are dysfunction and not working. If there are no reefs, "there is no rubble to maintain the island and you have a situation where the islands are basically washing away."[140]

Another problem associated with the Chagos Archipelago is an ownership dispute between the United Kingdom and Mauritius.[141] In 1814, much of the region, including Mauritius, Chagos, and Seychelles, became UK overseas territory. There were coconut plantations in Chagos, but the British never had a presence in Chagos as Chagos was administered from Mauritius. The United Kingdom detached Chagos from Mauritius when Mauritius got its independence in 1968. While Mauritius has lost every case with respect to getting ownership of the Chagos Archipelago, a recent advisory decision from the International Court of Justice voted unanimously—except for the United States—for the United Kingdom to return Chagos Archipelago. If ownership changes, this could have huge complications for all sorts of other overseas territories and countries in addition to the United Kingdom such as France, the Netherlands, Russia, and the United States.[142] Yet, a UN resolution was clear that former colonists may not carve up their former territories. There is also a related concern about the resettlement of Chagossians who are the former, original settlers of Chagos. Many of the Chagossians are now either in the United Kingdom or Mauritius. There is also a multigenerational issue, as many Chagossians are now citizens of the United Kingdom and would not necessarily want to resettle back to the Chagos Archipelago.[143]

There is a very real concern about conservation. First, fishing pressure is a real issue as the region is close to one billion people. Fishing effort is demonstratively reduced, as it is a large area with large animals including sharks and marine mammals.[144] Poaching is not rampant, but there has been a good deal of shark poaching. However, there is pressure as there are more people in the region and fishing pressure is mainly coming from Sri Lanka and India.[145]

Lastly, rats and their impact on seabirds is a problem facing some of the islands in the Chagos Archipelago.[146]

Why the Problem Is Important

These problems, particularly related to the future of the Chagos Archipelago, are important because these are "fantastic coral reefs."[147] The MPA, which covers more than 500,000 km^2, is a fully protected, no-take zone. Although the MPA is not a World Heritage site, there are RAMSAR and nature areas zoned.[148]

In addition, Mauritius has an intent to develop a blue economy and Mauritius is neighbors to Seychelles, which forms a continuous geography.[149] Regardless whether the Chagos Archipelago remains part of the United Kingdom or becomes a part of Mauritius, there will be international pressure to maintain the area as an MPA because the area is unique, its right in the middle of the ocean, and there are very few remaining areas like the Chagos Archipelago. This said, the Pitcairn Islands and Phoenix Islands are kind of similar.[150] According to Dr. Sheppard, research needs to occur focusing on restoration, coral identification, and birds. Likewise, there are 8–9 important areas that are very rich in Indian Ocean birds.[151]

The problems are also important because many of us are experiencing a shifting baseline syndrome where people are not aware of what a reef could be like. The Chagos Archipelago was one of the only places, particularly in the Indian Ocean, of what a reef should look like. There are a lot of MPA managers throughout the region trying to stop their reefs from degrading and trying to restore their reefs, yet they have no idea what the ecosystem should look like. In fact, healthy reefs can support fifty times more biomass than what people thought. The Chagos Archipelago has avoided the decades of destruction that has occurred more recently and thus, one of the benefits of the Chagos region is that what a reef should be—in terms of species diversity, fish biomass, fish behavior, coral cover, etc.—has been shown. Likewise, the Chagos Archipelago is one of the few reefs that has "only" been impacted by climate change and not by overfishing and pollution runoff because the Chagos Archipelago is remote and uninhabited. Thus, the Chagos Archipelago presents a baseline for others to manage against.

Furthermore, the Chagos Archipelago is among the world's coral reef "bright spots."[152]

How Problem Was Identified

In the early 1980s, the Former BIOT Commissioner, Mr. Nigel Wenban-Smith, witnessed firsthand the general carelessness of environmental matters, such as the disposal of rusty military equipment.[153]

Due to the overall secrecy surrounding the U.S. military base Diego Garcia, the "Americans were coy about what exactly they could see (i.e., which was much less then as compared to what they can see now see from space)."[154] The United Kingdom was generally aware of poaching and illegal fishing as occasionally a ship, such as Taiwanese trawlers, would get into difficulty and they, for instance, would have a stock of shark fins onboard.[155]

Dr. Charles Sheppard became an advisor to the United Kingdom and undertook scientific expeditions to the Chagos Archipelago. These expeditions enabled Dr. Sheppard to see a lot of what was being done. The UK Government paid for the research ship (i.e., with capacity for 12–14 scientists), which also acts as a patrol boat.[156]

Effectiveness of Process for Identifying Problem

Research is conducted for approximately 4–6 weeks per year.[157] Prior to this, the UK Government wanted to keep the scientific input lowkey and science was a bit of a nuisance, but the United Kingdom wanted to maintain some control on the Diego Garcia military base.[158] There was generally an obsession with secrecy.[159] Thus, the effectiveness for identifying the problems was a bit mixed.

Steps Taken to Address the Problem

A few steps taken to address the problems facing the Chagos Archipelago include:

- In the mid-1970s, Dr. Sheppard took his first expedition to the Chagos Archipelago. There was a large gap in between voyages as there was "nothing because of secrecy and no funds for conservation." Then in the 1990s, Dr. Sheppard was able to return and such voyages helped to make the UK Foreign Service Office aware of the Chagos Archipelago's ecological value[160];
- A British Naval Officer, who was an amateur botanist, and Mr. Nigel Wenban-Smith reviewed the environment resulting in Mr. Wenban-Smith, in the early 1980s, negotiating a bilateral agreement with the United States that included a general environmental agreement. Soon thereafter, the United States and the United Kingdom held discussions every six months which included discussions about environmental matters. This, gradually, led to stricter control over environmentally destructive behaviors[161];

- In the early 1980s, the United Kingdom did not have any means to visit the outer islands. From time-to-time, Mr. Wenban-Smith visited the remote islands via naval vessels. In the late 1980s, the United Kingdom got its first, full-time patrol boat and this enabled the United Kingdom to gradually gain an understanding of the extent of poaching and fishing[162];
- In 1993, the Friends of the Chagos was established with a goal to get the UK Government to pay attention to the Chagos Archipelago's ecological value and its global importance from a conservation point of view. To achieve such a goal, members wrote letters to UK ministers, fielded questions from the UK parliament, and slowly persuaded key figures one by one[163]; and
- In 2008, the Chagos Conservation Trust, "led a network of conservation organisations that made up the Chagos Environment Network {CEN}. The CEN was formed to ensure that the globally-important natural environment of the Chagos Archipelago was conserved. It was a leading voice behind the campaign to establish the archipelago as a protected marine reserve in 2010."[164]

Results

Some of the results accomplished to date for the Chagos Archipelago are:

- In 1993, the Friends of the Chagos organization was established. The Friends of the Chagos would later be renamed the Chagos Conservation Trust (CCT)[165];
- Fishing was banned "on 1st April 2010 {when} the Chagos Archipelago was designated by the British government as a fully no-take marine reserve;" and
- Dr. Sheppard was able to get fishery observers onto French ships to see who was passing through the Chagos Archipelago. In the past, there was an old patrol boat called the *Pacific Marlin*, but it was slow and this made it hard to catch illegal fishing boats. Nowadays, the United Kingdom has a faster and more expensive patrol boat. However, this makes it more expensive and tougher to get a spot on the boat as researchers have to be self-funded in order to buy time on the patrol boat. Then again, researchers now have a much deeper understanding and scientific reports are more extensive. Further, by using drones, there is also much more coverage.[166]

Another result for the Chagos Archipelago has been the establishment of funding from the Bertarelli Foundation, including a USD$12 million investment for the years 2017–2021.[167] As explained by Helen Pitman of the

CCT, "results coming from Chagos are very slow and there have been long periods of time where no one could even get out there to conduct any research. It is also very, very expensive to work out there so you can't just pop along for a week's research in the outer islands. For example, CCT is committed to eradicating rats from 30 of the infested islands for a price tag of US$5 million!"[168]

Challenges and How They Were Met

Asides from the major problem of disputed ownership claims, there are several challenges facing the Chagos Archipelago. There are global stressors such as climate change, acidification, and heatwaves which resulted in a lot of corals dying during the 2015–2016 heat waves. However, it is important to note, in large part due to limited human pressures, the coral reefs are recovering very fast.[169]

There is some illegal fishing taking place in the Chagos Archipelago, particularly for sharks.[170] A related challenge to illegal fishing is that a perverse incentive is created when the illegal fisherfolk are retained, they are given a salary, food, and housing.[171]

Access to the Chagos Archipelago and access to funding are two other, related challenges. For example, Mr. Wenban-Smith would fly into Nairobi (Kenya) and take a military transport plane which would be waiting to take off to Chagos and nine hours later, would arrive to the Diego Garcia base.[172] More recently, one can fly from London to Bahrain to Diego Garcia via military transport. Dr. Sheppard has also boarded a patrol ship from Malé (the Maldives' capital) to the Chagos Archipelagos, but the voyage takes 2–3 days.[173]

Conservation work at the Chagos Archipelago is on more solid financial footing, due to support from the Bertarelli Foundation as well as from the UK Government.[174] The Pew Foundation was in the consortium and has also played a central role, including providing funding for meetings.[175]

Beyond Results

Unfortunately, long-term financing is not in place for the Chagos Archipelago. As explained by Helen Pitman, "Chagos has limited government funding and the conservation actions are largely taken on by NGOs and institutions with private or small amounts of government funding. There are BIOT conservation actions but no costed out management plan. {…} not a lot of {UK Government funding} goes to the MPA other than surveillance and enforcement."[176]

Lessons Learned

While not easy, when there are limited people, the reefs can recover much faster as there are fewer impacts. For instance, there are no tourist facilities in Chagos, tourism planes are not allowed, and while you might be able to get permission to sail a private yacht, it is not easy to get to Chagos. This said, the coral reefs in Chagos are noticeably more resilient.

There will be important developments, in the future, about what happens to the Chagossian people and how much force the United States exerts on the situation.

Another important lesson learned is the notion of ratty islands (i.e., with people) versus birdy islands (i.e., without people). As noted in Dr. Nicholas Graham et al.'s *Nature* publication, "Seabirds enhance coral reef productivity and functioning in the absence of invasive rats; there are two kinds of islands: there are islands with rats and there are islands without rats. Those islands without rats and with seabirds are usually much healthier."[177] As noted by Dr. Dunbar, "if I were a millionaire, I would spend it on rat eradication."[178]

Lessons learned by Dr. Charles Sheppard is that victory for conservation is just a victory for now and what has surprised Dr. Sheppard over the course of his experiences is the challenges to conservation. When the Chagos Archipelago banned the issuance of fishing licenses, the opposition was already stating that in three years, they would get the ban overturned.

Mr. Nigel Wenban-Smith has a few lessons learned distinct to the Chagos Archipelago. Because access to the Chagos Archipelago is limited and due to secrecy, conservationists had to start at ground zero. As explained, by Mr. Wenban-Smith, "we chose people to work with who had government connections in order to find the right people to help."[179] For instance, decision-makers who had worked for them in the past, could be approached in the future and thus, the conservationists could have some influence with current decision-makers. It was also important to know the people who carry influence in the nongovernmental world. Ultimately, it is important to see whose thoughts really matter.[180]

Other Resources for the Chagos Archipelago

Bertarelli Foundation

- https://www.fondation-bertarelli.org/tag/chagos/

Blue Marine Foundation

- https://www.bluemarinefoundation.com/film/chagos-marine-reserve/; and
- https://www.bluemarinefoundation.com/project/chagos-marine-reserve/

British Indian Ocean Territory

- https://biot.gov.io/; and
- https://biot.gov.io/environment/

Chagos: A History: Exploration, Exploitation, Expulsion by Nigel Wenban-Smith and Dr. Marina Carter

- https://www.ypdbooks.com/non-fiction/1532-chagos-a-history-exploration-exploitation-expulsion-YPD01728.html

Chagos Conservation Trust

- https://chagos-trust.org/

Chagos Information Portal

- https://chagosinformationportal.org/

iNaturalist

- https://www.inaturalist.org/projects/the-terrestrial-biodiversity-of-the-british-indian-ocean-territory-chagos-archipelago

Zoological Society of London

- https://www.zsl.org/regions/uk-overseas-territories/chagos-archipelago

Financial Analysis

The following financial analysis will look at return and risk.

Return

To generate revenue, or additional revenues, from government-owned MPAs, there are several options. As described in the International Coral Reef Initiative (ICRI)-commissioned report, *Innovations for Coral Finance*, blue taxes can "provide {a} regular and reliable source of revenue for conservation, {there is} no need to set up a new collection system or bureaucracy, {and blue taxes} can potentially create 'double dividends' by lowering existing taxes (e.g., labor taxes)."[181] Blue taxes may be "most appropriate in high-value tourism destinations; {one should consider the} importance of a feasibility study (interviews with tourism operators and tourists) to assess the willingness to pay; and study of local tourism market and direct regional competitors, to make sure implementation would not shift desire to travel."[182]

The MPA could start charging a visitation fee, if no such fee exists. For instance, in Costa Rica, one pays the dive shop and the reefs are "free"—there are no MPA access fees. However, this visitation fee might not be significant if the MPA is very remote with few visitors, as is the case with the Papahānaumokuākea Marine National Monument and the Chagos Archipelago.

The MPA could also offer differentiated pricing. This might include a higher price for foreign visitors and a lower, or no cost, option for local tourists. In addition, the MPA could offer season passes or look to lease concession stands.

Whether it is assessing visitor and researcher fees in addition to using the government's general funds, diversifying revenue sources can help contribute to financial sustainability. For instance, "{...} lotteries in the U.K., the Netherlands, and the U.S. state of Oregon have all funded marine or coastal conservation projects."[183] More specifically, the UK's lottery supported the Coral Cay Conservation Charitable Trust which developed "livelihood opportunities for coastal communities in the Philippines."[184]

Risk

As summarized in the ICRI-commissioned report, *Innovations for Coral Finance*, blue taxes can create "major challenges related to earmarking proceeds for conservation, and transparency, {there might be} social and economic challenges in introducing a new tax (e.g., acceptability), {a} clear definition of nature-damaging products and services is required, {and there is a} need for strong institutional and fiscal capacity."[185]

Business Risk

One significant business risk for an MPA that is primarily financed through government domestic expenditures is whether the same, or greater, amount of funds will be appropriated in the future. This scenario is especially true if the funds collected at the MPA go to the government's general fund before being redistributed to the MPA.

In addition, it is often difficult to predict the level of demand for the MPA's products and services. This could include the number of visitors such as divers or snorkelers paying an entrance fee, local tourism operator or local fisherfolk paying a user fee, or the number of future purchases at the concession stands. With respect to user fees, it is important that "fees are not too low so that rents are lost, but also should not exceed willingness to pay."[186]

Furthermore, there could be a lack of funding in the future, if there is a decline in budgets or a budget crisis.

Strategic Risk

There are numerous strategic risks associated with government domestic expenditures to local MPAs. Such strategic risks to establishing, maintaining, and expanding MPAs include:

- Where to locate the MPA. Ideally, the MPA would cover the coral reefs, but choosing what other areas are included (i.e., onshore mangroves and/or pelagic zones) can raise additional challenges. This also includes marine spatial planning and making sure the zoning of areas is done with proper stakeholder engagement.
- The size of the MPA. To some degree, the larger the better, but larger areas can be more difficult to monitor. In addition, a nonscientific design

and a lack of stakeholder engagement for the MPA can create additional challenges.

- Deciding on what recreational activities will be/are allowed. For example, is access to the MPA strictly forbidden to visitors, except qualified researchers? If tourists are allowed, how many are allowed, and are they allowed to stay overnight? Is recreational fishing allowed, or is fishing just for subsistence purposes allowed?
- Deciding on what strategic partners to work with, whether its university researchers, NGOs, and/or multinational companies.

Reputation Risk

Paper parks, where parks are merely designated areas without the proper management and funding to preserve the MPA, pose possibly the greatest reputational risk. This risk could be relevant to the host government, the responsible ministry or department, along with the park management. Illegal activities, whether it is being permitted or not addressed, pose reputational risks as well. Such activities could include illegal harvesting of coral, shark finning, and/or illegal fishing. The lack of engagement with local communities and other stakeholders can produce a reputation risk pertaining to the legitimacy of the MPA. Furthermore, the lack of proper enforcement, and especially if the MPA attracts poachers because the area is well protected while the other areas are experiencing a decline in fish biomass, can result in reputational risk.

Another reputational risk is known as ocean grabbing, which has been defined as "{…} the policies or initiatives that deprive local communities from traditional resources through reallocations by governments and the private sector."[187]

Liquidity Risk

Government domestic expenditures, particularly to MPAs, face unique liquidity risks. This is because presumably the MPA is illiquid and the manager—whether a national park service, the ministry of fisheries, and/or the ministry of environment—of the MPA is unable to sell their holdings (i.e., the MPA). Because they are unable to sell their holdings, they need to have a contingency plan if there are funding shortages. Such a shortage could impact local salaries, monitoring of the MPA, ongoing research studies, and/or additional actions needed to protect the reefs (e.g., eradication of invasive species). There can also be a liquidity risk presented if government budgeting comes after the MPA runs out of funds and thus, are temporarily unable to make payroll or upgrade its facilities.

Operational Risk

Operational risks facing MPAs are oftentimes similar to those operational risks experienced by other businesses. This includes management capabilities, staff morale, employee retention, adequate training, and promotional opportunities. Unique operational risks depend on the structure of the government domestic expenditure. For instance, are the concessions (e.g., gift shops, dive shops, visitor educational centers, etc.) subcontracted out to another entity?

Natural disasters, such as hurricanes, typhoons, tsunamis, the presence of coral disease, or infestations of invasive species (i.e., lionfish in the Caribbean Sea), can pose operational risks.

Other operational risks include technology risk (e.g., adequacy of monitoring technology, whether it is the use and classification of remote sensing technology or patrol boats) and the proper use of internal controls to minimize fraud (e.g., are the collections of visitor fees at a remote MPA entrance being properly recorded and submitted to management).

Legal and Regulatory Risk

There are lots of legal and regulatory risks, particularly surrounding traditional uses within and adjacent to the MPA. As noted by Dr. Earle, "hooking squirrels, spearing hawks, shooting owls, netting songbirds and butterflies, slaughtering the deer and bears – all would be permissible if 'marine sanctuary' rules applied to national parks."[188]

Additional complications arise if there are overlapping concessions granted (i.e., for the MPA and a deep-water gas and oil permit), or if a marine spatial planning exercise is poorly conducted. Similarly, there can be legal and regulatory risks associated with Free, Prior and Informed Consent (FPIC) if the MPA does not equitably work with local communities. Such risks could arise, for instance, if the MPA is expanding its boundaries at the expense of local communities or if the MPA restricts access (i.e., allows for little-to-no fishing by the local communities).

There can also be competing interests from local, state, or federal representatives who propose to divert funds from the MPA to other programs. As noted by WWF:

In many countries, fines for illegal logging, hunting, and fishing are paid into the national Treasury, and are not used for conservation purposes. This may also be the case for proceeds from sales of confiscated timber, fish, and wildlife that were illegally caught or harvested. In countries that require that money from fines and forfeiture must be paid into the national Treasury, it would be necessary to pass special new legislation in order to earmark these revenues exclusively for conservation.[189]

Somewhat similarly, legal and regulatory risk can occur if other agencies, departments, and/or adjacent government agencies are undertaking poor land use policy (i.e., leading to deforestation, erosion, and siltation) which impacts coral reefs.

Furthermore, newly elected governments could decide to abolish or defund the MPA.

Credit Risk

Credit risk includes default risk, bankruptcy risk, downgrade risk, and settlement risk.

When a government—whether it is a state or federal government—faces a credit risk such as a potential default or downgrade in its investment grade, then it becomes more expensive for the government to borrow money. With increased costs of borrowing, governments may have a smaller domestic budget, and thus, its MPA may experience a constricted budget.

Market Risk

Market risk includes interest rate risk, equity price risk, foreign exchange risk, and commodity price risk.

Interest rate risk could threaten a domestic government's ability to borrow funds if interest rates rise and make it prohibitively expensive. For example, as of July 2019, Papua New Guinea and Egypt both had a B rating from S&P and a B2 rating from Moody's, which would result in a higher interest rate, as opposed to the Bahamas, which had a BB+ rating from S&P and a Ba3 from Moody's.[190]

Equity price risk is not as relevant for a domestic government, but could threaten the MPA if significant private companies are devalued and no longer operate tours to the MPA or close down nearby hotels (i.e., thereby decreasing the user and visitor fees).

Foreign exchange risk may make it more (or less) difficult for international visitors (e.g., researchers, tourists, and/or students) to visit the MPA.

Commodity price risk can have a direct or indirect impact on government domestic expenditures to local MPAs. For instance, if oil prices increase, then it becomes more likely that drilling for oil will continue and potentially increase. This could occur in areas close to coral reefs, such as in the Gulf of Mexico or off the coast of Brazil. Similarly, an increase in the price of fish, such as tuna or shark fins, could increase the amount of illegal fishing taking place in and around MPAs.

Risk, Return, Time (Horizon), Taxes, Liquidity, Legal and Unique (RRTTLLU)

Risk and Return

Please see above for the risk and return associated with government domestic budgetary expenditures.

Time Horizon

The time horizon for government domestic budgetary expenditures, particularly investments into MPAs, is long term. In fact, the time horizon could be in perpetuity with ongoing investments into park infrastructure, monitoring, and staff salaries. Shorter time horizons may exist for smaller investments such as investments into the creation of mooring systems for divers or concession stands back onshore. As such, long-term budgeting needs to be driven by ways to frame out both financing in perpetuity alongside immediate short-term needs to strengthen the protection of the MPA (e.g., paying park rangers their salaries, expanding staff, offering training, and monitoring).

Taxes

There are likely limited, if any, taxes due to the government if the government is the sole provider of investments. However, there could be a sales tax levied on concessions, employment taxes, and/or some form of tourism tax for visitors (e.g., if visitors stay at onsite facilities). One incentive to attract concessioners would be to offer a reduced tax rate, even if for a short introductory time period, for their services rendered.

Liquidity

Liquidity might exist with concessions or other onsite facilities (e.g., tourist attraction or lodging) that could be transferred to another owner. However, the actual underlying MPA is likely to be illiquid unless the federal government can transfer the MPA to a state or province (i.e., or vice versa).

Legal

Legal considerations associated with government domestic budgetary expenditures will vary from country to country. Again, legal risks may arise as a result of improper Free, Prior and Informed Consent (FPIC) when deciding what activities can take place in the MPA and in whose name the access is granted. In addition, legal risks may arise when offering (or not offering) employment opportunities to local communities or when expanding the MPA's boundaries (i.e., particularly the no-take zones).

Unique

A unique aspect of financing MPAs is that their time horizon is in perpetuity. In addition, their revenue streams are often limited, and their activities must finance measures to prevent exploitation.

Another unique aspect of MPAs is that fauna migrates, and this migration may take place to an area outside of the MPA. Thus, an MPA may have been designed to protect whale sharks (*Rhincodon typus*) and attracts paying tourists to specifically see these whale sharks, but then the whale sharks migrate out of the MPA during a particular period of the year.

Policy Analysis

The following policy analysis will look at: defining the problem; establishing goals; selecting a policy; implementing a policy; and evaluating the policy.

Defining the Problem

One of the problems is that countries have not designated a sufficient amount of their seascape to legal protection, such seascape has not been properly zoned, there is/was a lack of public participation in the MPA's designation, and there is a lack of public funding.

For instance, "countries such as Australia, New Zealand, Columbia, Mexico, and Mozambique are far ahead of the United States in the percentage of their waters that they protect."[191]

In addition: some MPAs are considered paper parks; the onshore regional land use adjacent to the offshore coral reefs may not be well managed; there could be small domestic budgets due to other competing domestic interests (e.g., healthcare, energy access, and education); and resources might be squandered due to corruption.

Establishing Goals

The Aichi Biodiversity Targets of the Convention on Biological Diversity (CBD) are an important collection of twenty targets designed to outline and establish international conservation goals. More specifically:

> Target 10: By 2015, the multiple anthropogenic pressures on coral reefs, and other vulnerable ecosystems impacted by climate change or ocean acidification are minimized, so as to maintain their integrity and functioning.
>
> Target 11: By 2020, at least 17 per cent of terrestrial and inland water, and 10 per cent of coastal and marine areas, especially areas of particular importance for biodiversity and ecosystem services, are conserved through effectively and equitably managed, ecologically representative and well connected systems of protected areas and other effective area-based conservation measures, and integrated into the wider landscapes and seascapes.[192]

Thus, the goal is for countries to establish a significant (as defined, at a minimum, by the Aichi Targets), representative protected area for each biome and ensure that such a protected area is well-financed, conserves biodiversity, and benefits local communities.

Selecting a Policy

To select a policy, further analysis could be done in several manners.

First, one could look at which countries have established the highest percentage of marine protected areas relative to their territorial waters and examine the specific public policies and the specific financing mechanisms that enabled their establishment[193] (Table 7.2).

Table 7.2 Top 20 Countries/Regions with highest proportion of marine area covered by protected areas

	Country/Region	Proportion of marine area covered by protected areas (%)
1	Martinique	100
2	British Indian Ocean Territory	100
3	Pitcairn Islands	100
4	Guadeloupe	100
5	Slovenia	100
6	Palau	100
7	Monaco	100
8	Cook Islands	100
9	Mayotte	100
10	Saint Barthélemy	98
11	New Caledonia	96
12	Saint Martin	96
13	South Georgia and the South Sandwich Islands	85
14	U.S. Minor Outlying Islands	65
15	Saint Helena, Ascension and Tristan da Cunha	55
16	France	49
17	Germany	45
18	Norfolk Island	44
T-20	Chile	41
T-20	Australia	41
T-20	U.S.	41

Source Protected Planet

However, this analysis does not tell the whole story. For instance, some of the countries or regions on the list do not have tropical coral reefs such as Chile and Slovenia. Other areas, although a smaller percentage, such as the U.S. Minor Outlying Islands, are larger in total area than areas with a higher percentage. Meanwhile, other countries, such as Jordan at 36%, have made notable strides past the 10% Aichi target. In contrast, as of December 2019, and looking at the top five countries with the largest coral reefs, Indonesia had 3.06% of its marine area covered by protected areas, Australia had 40.56%, the Philippines had 1.16%, France had 48.93%, and Papua New Guinea had 0.14%.[194] Furthermore, a higher percentage of the marine area set aside for protection does not necessarily translate into quality protected areas as many of these areas might be considered "paper parks."

With a large percentage of its marine area protected and with the world's second largest coral reef area, Australia will be the primary focus of this policy analysis. This said, "the Great Barrier Reef Marine Park Zoning Plan 2003 {which came into effect July 2004} is the primary planning instrument for the conservation and management of the Marine Park. It also provides for a range of recreational, commercial and research opportunities and for the continuation of traditional activities."[195]

Implementing a Policy

The authority for implementing public policies and allocating government domestic expenditures—depending on the host country and the natural resource in question—can reside at the local, state/provincial, and/or federal levels.

Australia's policy officially came about:

> {…} in June 2003, announcing a new draft comprehensive zoning plan that would fully protect a representative 30 percent of the GBRMP in no-take reserves termed Marine National Park Zones. The new draft zoning plan covered the entire GBRMP, a departure from prior zoning plans that were done piecemeal, section-by-section; includes twenty-eight recently added coastal sections; and views the park as seventy interconnected bioregions.[196]

The following outlines Australia's zones[197] (Table 7.3).

Evaluating the Policy

Australia is widely seen as having the best managed, or at least one of the best managed, coral reef systems in the world. Prior to the current situation, "the administration of the Great Barrier Reef prior to the formation of the GBRMP was complex, involving six different Commonwealth and Queensland Government departments {…}."[198]

Some of the key lessons "{…} to draw from the Australian experience focus on the value of setting marine reserves in a broader context, the importance of a good public process, and the advantages of integrating site-specific development with national-level system planning."[199]

With this in mind:

> The GBRMP adopted a multi-use approach early on and developed strong cooperation with adjacent territorial governments and management plan and zoning plan processes to implement it. The ecosystem-oriented, public-driven planning processes developed for managing the GBRMP have often been well-received and viewed as state of the art by many.[200]

Table 7.3 Activities of the Great Barrier Reef Marine Park

Activity	General Use Zone	Habitat Protection Zone	Conservation Park Zone	Buffer Zone	Scientific Research Zone	Marine National Park Zone	Preservation Zone
Color	Light Blue	Dark Blue	Yellow	Olive Green	Green with Orange Outline	Green	Pink
Boating, Diving, Photography	Yes	Yes	Yes	Yes	Yes – Except for One Tree Island and AIMS which are closed to public access and shown as orange	Yes	No
Line Fishing (Max of 3 hand-held rods/lines per person and up to 6 hooks combined total)	Yes	Yes	Limited to 1 hand-held rod/line per person and 1 hook/lure per line	No	No	No	No
Trolling (Max of 3 hand-held rods/lines per person and up to 6 hooks combined total)	Yes	Yes	Yes	Limited to pelagic species	No	No	No
Limited spear fishing (not using a powerhead or underwater breathing apparatus other than a snorkel)	Yes	Yes	Unless identified as a Public Appreciation Area of State Closure	No	No	No	No
Bait Netting	Yes	Yes	Yes	No	No	No	No
Crabbing (trapping)	Yes	Yes	Limited to 4 crab pots or dillies per person	No	No	No	No

Image Credit Brian McFarland, Adapted from Anthony Walsh

Much groundwork was done prior to the release of the draft zoning plan, including an extensive biodiversity and habitat mapping initiative, a series of State of the Great Barrier Reef environmental status reports, and extensive consultation.[201]

Why was this ambitious proposal successful? The factors discussed above including diverse support, groundwork, sound science, mapping, status reports, and extensive consultation all contributed to its success. In addition, the GBRMPA's history and experience in dealing with complicated zoning and other issues, conducting extensive community outreach, and developing a good public process unquestionably helped. Finally, top-down national-level policy support and consistency with this plan developed over a number of years through extensive public consultation and outreach and a national commitment to safeguard Australia's ocean resources (e.g., Australia's National Ocean Policy and Strategic Plan for Establishing the National Representative System of Marine Protected Areas; see above) were essential to finalizing it.[202]

As described by Dr. Petra Lundgren:

The GBR is one of the best managed coral reefs in the world. The Great Barrier Reef Marine Park has an extensive zoning system that allows people to use, enjoy and make a livelihood from the park, while at the same time having a large proportion designated as national park (still open to enjoy, but closed to fishing or other extractive activities) or conservation zones (100% closed to the public). We have a well-functioning field management arrangement (joint undertaking by QLD and federal government) that ensures compliance. The new policies that are being developed to allow more active interventions are being developed in close collaboration with scientists, stakeholders and traditional owners. In addition, the Marine Park Authority is part of the RRAP {Reef Restoration and Adaptation Program}, thus ensuring that policy and regulation are considered and evolving in time with the development of new approaches and interventions.

However, impact from climate change and land-based activities remain the biggest threats to this system, and they are threats that cannot be managed through zoning plans, they require international and national efforts to ensure we reach the Paris targets and to reduce runoff from the land.[203]

Thus, the complete picture is complicated. While Australia's management of the GBR is widely seen as a model for others to replicate, Australia has permitted coal mines which contribute to global climate change and make the management of the GBR more difficult. In June 2019, Adani's Carmichael coal mine, which will be one of Australia's large coal mines, was approved.[204]

Future Outlook for Instrument

There are elements that are within the control of domestic governments and other elements that are out of their control. Governments in both developed and developing countries often own the vast majority of their country's marine areas, as opposed to privately owned coral reefs. This creates a potential opportunity for the expansion of existing national MPAs and the creation of additional MPAs, as has been the recent trend. For instance, "{...} a growing trend is evolving toward successful development of larger individual marine reserves and reserve networks utilizing improved design and public process principles tailored to local situations."[205]

A next potential large marine protected area is the Marine Arctic Peace Sanctuary.[206]

There will also be the need for more integrated coastal management plans. Likewise:

> In the long-term, marine reserves will need to be implemented together with improved water quality protection, fisheries management, climate change mitigation, larger scale and more comprehensive marine zoning, and a full array of effective marine protected areas to maximize their effectiveness and meet conservation goals.[207]

Although many governments are facing budget constraints, domestic budgetary allocations will remain a critically important source of financing for the conservation of coral reefs for the foreseeable future.

> From increased budget allocations to taxation, licensing and levying authority to the issuance of bonds in support of conservation actions to the setting of policy, the role of governments in financing conservation is broad and essential. {...} A conservation example that could provide a pathway in the future is land acquisition by governments. For example, the state of Florida in the United States has an annual land acquisition budget of $300 million. {...} In the case of coral reefs, targeting land buybacks adjacent to reefs could prove beneficial from a conservation and economic perspective in the long run.[208]

Thus, governments will likely play an increasingly large role in the design, implementation, and financing of MPAs.

Other Resources on Government Domestic Budgetary Expenditures

Atlas of Marine Protection; An Initiative of Marine Conservation Institute

- http://www.mpatlas.org/

Conservation International's Global Conservation Fund

- http://www.conservation.org/projects/Pages/global-conservation-fund.aspx

Global Greengrants Fund

- https://www.greengrants.org/

Inside Philanthropy's Compilation of Marine and Freshwater Funders

- https://www.insidephilanthropy.com/marine-conservation-grants

IUCN's Global Standard

- https://www.iucn.org/theme/protected-areas/our-work/iucn-green-list-pro
  tected-and-conserved-areas/global-standard; and
- https://iucn.my.salesforce.com/sfc/p/

National Fish and Wildlife Foundation, and the Coral Reef Conservation Fund

- https://www.nfwf.org/coralreef/Pages/home.aspx

NOAA's National Marine Sanctuaries

- https://sanctuaries.noaa.gov/

Parks Watch (Profiles on Select Parks in Central and South America)

- http://www.parkswatch.org/

<u>Protected Planet, Including World Database on Protected Areas and Protected Areas Management Effectiveness</u>

- https://www.protectedplanet.net/

<u>U.S. Coral Reef Task Force's Compilation of Federal Grants and Funding Opportunities</u>

- https://www.coralreef.gov/grants/

<u>USFWS's Cooperative Endangered Species Conservation Fund</u>

- https://www.fws.gov/endangered/grants/

<u>USFWS's Migratory Bird Conservation Commission and Fund</u>

- https://www.fws.gov/refuges/realty/mbcc.html

<u>U.S. National Marine Protected Areas Center</u>

- https://marineprotectedareas.noaa.gov/;
- https://marineprotectedareas.noaa.gov/aboutmpas/mpacenter/; and
- https://marineprotectedareas.noaa.gov/dataanalysis/mpainventory/mpaviewer/

Notes

1. More, Jeffrey T. "The Gray and the Green: The Built Infrastructure and Conservation Investment." In *From Walden to Wall Street*, edited by James Levitt. 179.
2. Australian Government. "The Great Barrier Reef Marine Park." Accessed February 28, 2020. https://www.environment.gov.au/epbc/what-is-protected/great-barrier-reef-marine-park.
3. Great Barrier Reef Marine Park Authority. "Our Story." Accessed February 28, 2020. http://www.gbrmpa.gov.au/about-us/about-us.
4. Great Barrier Reef Foundation. "Reef Trust Partnership Six year $443.3M—Largest Ever Investment in Reef Protection." Accessed December 19, 2019. https://www.barrierreef.org/science-with-impact/reef-partnership.

5. Elliman, Kim and Peter Howell. "Anatomy of a Finance Program: Lessons from a Conservation Lender." In *Conservation Capital in the Americas*, edited by James N. Levitt. 150–152.

6. U.S. Fish & Wildlife Service. "The Save Vanishing Species Stamp." Accessed December 13, 2019. https://www.fws.gov/international/save-vanishing-spe cies-stamp.html.

7. Spergel, Barry and Melissa Moye. "Financing Marine Conservation: A Menu of Options." 14.

8. Ibid. 13.

9. Ibid. 54.

10. Vertigo Lab. "Innovations for Coral Finance, ICRI Publication." 21.

11. Hopper and Cook, *Conservation Finance Handbook*. 15.

12. UNEP-WCMC, IUCN and NGS (2018). Protected Planet Report 2018. UNEP-WCMC, IUCN and NGS: Cambridge UK; Gland, Switzerland; and Washington, DC, USA. https://livereport.protectedplanet.net/pdf/Protected_ Planet_Report_2018.pdf. 6.

13. Ibid. 6.

14. Protected Planet. "Explore the World's Marine Protected Areas." Accessed December 12, 2019. https://www.protectedplanet.net/marine.

15. Ibid.

16. UNEP-WCMC, IUCN and NGS (2018). Protected Planet Report 2018. UNEP-WCMC, IUCN and NGS: Cambridge UK; Gland, Switzerland; and Washington, DC, USA. https://livereport.protectedplanet.net/pdf/Protected_ Planet_Report_2018.pdf. 9.

17. Protected Planet. "Explore the World's Marine Protected Areas." Accessed December 12, 2019. https://www.protectedplanet.net/marine.

18. Ibid.

19. IUCN. "IUCN Green List of Protected and Conserved Areas." Accessed December 12, 2019. https://www.iucn.org/theme/protected-areas/our-work/ iucn-green-list-protected-and-conserved-areas.

20. IUCN. "IUCN Green List Areas." Accessed December 12, 2019. https:// www.iucn.org/theme/protected-areas/our-work/iucn-green-list-protected-and-conserved-areas/iucn-green-list-areas.

21. Protected Planet. "Explore the World's Marine Protected Areas." Accessed December 12, 2019. https://www.protectedplanet.net/marine.

22. Ramsar. "The List of Wetlands of International Importance." 18 October 2019. Accessed December 12, 2019. https://www.ramsar.org/sites/default/ files/documents/library/sitelist.pdf.

23. UNESCO. "Biosphere Reserves—Learning Sites for Sustainable Development." Accessed December 13, 2019. http://www.unesco.org/new/en/natural-sciences/environment/ecological-sciences/biosphere-reserves/.

24. Vertigo Lab. "Innovations for Coral Finance, ICRI Publication." 2017. http:// vertigolab.eu/wp-content/uploads/2018/04/rapport-innovation-for-coral-fin ance.pdf. 1.

25. Great Barrier Reef Foundation. "Reef Trust Partnership Six year $443.3M—Largest Ever Investment in Reef Protection." Accessed December 19, 2019. https://www.barrierreef.org/science-with-impact/reef-partnership.

26. NOAA. "NOAA Awards Over $8.3 Million to Advance Coral Reef Conservation Science and Management." November 8, 2018. Accessed December 19, 2019. https://www.noaa.gov/media-release/noaa-awards-over-83-million-to-advance-coral-reef-conservation-science-and-management.

27. Globe Newswire. "$1.4 Million in Grants to Support Coral Reef Conservation Awarded by NFWF." December 3, 2019. Accessed December 19, 2019. https://www.globenewswire.com/news-release/2019/12/03/1955728/0/en/1-4-Million-in-Grants-to-Support-Coral-Reef-Conservation-Awarded-by-NFWF.html.

28. Day, Jon. "Two Perspectives on Evaluating MPA Management Effectiveness: Lessons Learned from Australia's Great Barrier Reef and India by Jon Day of the ARC Centre for Coral Reef Studies." Online webinar, ARC Centre for Coral Reef Studies, Townsville, Australia, March 13, 2018.

29. Walsh. *Lady Elliot Island*. 10.

30. Daley. *The Great Barrier Reef*. 3.

31. McCalman. *The Reef*. 246.

32. Great Barrier Reef Marine Park Authority 2019. Annual Report 2018–2019. GBRMPA, Townsville. http://www.gbrmpa.gov.au/about-us/corporate-information/annual-report. 21.

33. Ibid.

34. Ibid.

35. Carbon Pulse. "CP Daily: Monday April 30, 2018." April 30, 2018. Accessed December 30, 2019. https://carbon-pulse.com/51440/; Also see: Great Barrier Reef Marine Park Authority. "$500 Million Funding 'Game Changer' for the Great Barrier Reef." April 29, 2018. Accessed December 30, 2019. http://www.gbrmpa.gov.au/news-room/latest-news/latest-news/corporate/2018/$500-million-funding-game-changer-for-the-great-barrier-reef.

36. Day, Jon. "Two Perspectives on Evaluating MPA Management Effectiveness: Lessons Learned from Australia's Great Barrier Reef and India by Jon Day of the ARC Centre for Coral Reef Studies." Online webinar, ARC Centre for Coral Reef Studies, Townsville, Australia, March 13, 2018.

37. Daley, *The Great Barrier Reef*. 33.

38. Roberts. *The Ocean of Life*. 85.

39. Cooper, Rachel. "Great Barrier Reef Suffers 89% Decline in Coral Reproduction Rate After Bleaching." *Climate Action*. April 5, 2019. Accessed November 6, 2019. http://www.climateaction.org/news/great-barrier-reef-suffers-89-decline-in-coral-reproduction-rate-after-blea.

40. Schiffman, Richard. "A Close-Up Look at the Catastrophic Bleaching of the Great Barrier Reef." *Yale Environment 360*. April 10, 2017. Accessed April 3, 2020. https://e360.yale.edu/features/inside-look-at-catastrophic-bleaching-of-the-great-barrier-reef-2017-hughes.

41. Visser, Nick. "Mass Bleaching Hits Great Barrier Reef Again: Like Watching Louvre 'Burn To The Ground'." *HuffPost*. March 27, 2020. Accessed April 3, 2020. https://www.huffpost.com/entry/great-barrier-reef-mass-bleaching_n_5 e7d8e79c5b6256a7a280833.

42. Daley. *The Great Barrier Reef*. 48.

43. Maraseni, Tek N. and Munir A. Hanjra. "Payments to Landholders for Managing Water, Land and Ecosystems (WLE) in Coastal Agricultural Catchments for Protecting the Great Barrier Reef." In *Economic Incentives for Marine and Coastal Conservation*, edited by Essam Yassin Mohammed. 195.

44. Carbon Pulse. "CP Daily: Tuesday April 24, 2018." April 24, 2018. Accessed December 30, 2019. https://carbon-pulse.com/51127/.

45. Daley. *The Great Barrier Reef*. 4.

46. Ibid. 5.

47. WorldAtlas. "The Largest Countries in the World." Accessed October 5, 2018. https://www.worldatlas.com/articles/the-largest-countries-in-the-world-the-biggest-nations-as-determined-by-total-land-area.html.

48. Australian Bureau of Statistics. "Estimated Resident Population." December 12, 2019. Accessed February 19, 2020. https://www.abs.gov.au/AUSSTATS/abs@.nsf/mf/3101.0.

49. U.S. Census Bureau. "QuickFacts: California." 2019. Accessed February 19, 2020. https://www.census.gov/quickfacts/CA.

50. Daley. *The Great Barrier Reef*. 52.

51. Marsh, Helene's Foreword in Daley. *The Great Barrier Reef*. xi.

52. Daley. *The Great Barrier Reef*. 55.

53. Ibid. 62, 66.

54. Ibid. 95–96.

55. Ibid. 72, 73–92.

56. Ibid. 113.

57. Ibid. 128.

58. Ibid. 129.

59. McCalman. *The Reef*. 241.

60. Australia Maritime Safety Authority. "Oceanic Grandeur, 3 March 1970." Accessed January 14, 2020. https://www.amsa.gov.au/marine-environment/incidents-and-exercises/oceanic-grandeur-3-march-1970.

61. Carbon Pulse. "CP Daily: Thursday November 29, 2018." November 29, 2018. Accessed December 30, 2019. https://carbon-pulse.com/64239/.

62. WorldAtlas. "Busiest Cargo Ports in Oceania." Accessed December 30, 2019. https://www.worldatlas.com/articles/busiest-cargo-ports-in-oceania.html.

63. McCalman. *The Reef*. 233.

64. Braverman. *Coral Whisperers*. 178.

65. McCalman. *The Reef*. 269.

66. Platt, John R. "A Starfish-Killing, Artificially Intelligent Robot Is Set to Patrol the Great Barrier Reef." *Scientific American*. January 1, 2016. Accessed

December 30, 2019. https://www.scientificamerican.com/article/a-starfish-kil ling-artificially-intelligent-robot-is-set-to-patrol-the-great-barrier-reef/.

67. Wink, Jay. Interviewed by Brian McFarland. September 2019.
68. Daley. *The Great Barrier Reef*. 201.
69. Great Barrier Reef Marine Authority. "Biodiversity." Accessed February 28, 2020. http://www.gbrmpa.gov.au/the-reef/biodiversity.
70. Daley. *The Great Barrier Reef*. 72.
71. Fitzpatrick. *Shark Tracker*. 45.
72. UNESCO. "Wet Tropics of Queensland." Accessed December 30, 2019. https://whc.unesco.org/en/list/486/.
73. Australian Marine Parks. "Coral Sea Marine Park." Accessed December 30, 2019. https://parksaustralia.gov.au/marine/parks/coral-sea/.
74. Ramsar. "The List of Wetlands of International Importance." 18 October 2019. Accessed December 12, 2019. https://www.ramsar.org/sites/default/ files/documents/library/sitelist.pdf.
75. McCalman. *The Reef*. 50.
76. McCalman. *The Reef*. 82.
77. Lundgren, Petra. Email Message to Author. March 24, 2020.
78. Green Island Resort. "Green Island History." Accessed April 10, 2020. https:// www.greenislandresort.com.au/history.
79. Daley. *The Great Barrier Reef*. 175.
80. McCalman. *The Reef*. 246.
81. Ibid. 247.
82. Ibid. 247.
83. Marsh, Helene's Foreword in Daley. *The Great Barrier Reef*. x.
84. Australian Government. "The Great Barrier Reef Marine Park." Accessed February 28, 2020. https://www.environment.gov.au/epbc/what-is-protected/ great-barrier-reef-marine-park.
85. Great Barrier Reef Marine Park Authority. "Our Story." Accessed February 28, 2020. http://www.gbrmpa.gov.au/about-us/about-us.
86. Ibid. "Heritage." Accessed February 28, 2020. http://www.gbrmpa.gov.au/the- reef/heritage.
87. World Heritage Centre. "Wet Tropics of Queensland." Accessed February 28, 2020. https://whc.unesco.org/en/list/486/.
88. Ramsar Sites Information Service. "Coral Sea Reserves." Accessed February 28, 2020. https://rsis.ramsar.org/ris/1222.
89. Ibid. "Elizabeth and Middleton Reefs Marine National Nature Reserve." Accessed February 28, 2020. https://rsis.ramsar.org/ris/1223.
90. Ramsar. "The List of Wetlands of International Importance." 18 October 2019. Accessed December 12, 2019. https://www.ramsar.org/sites/default/ files/documents/library/sitelist.pdf.
91. Marsh, Helene's Foreword in Daley. *The Great Barrier Reef*. x.

92. Great Barrier Reef Marine Park Authority 2019. Annual Report 2018–2019. GBRMPA, Townsville. http://www.gbrmpa.gov.au/about-us/corporate-inform ation/annual-report. 12.

93. Ibid. 13.

94. Ibid. 14.

95. Lundgren, Petra. "Field Work." Accessed March 24, 2020. http://petrablun dgren.com/field-work.

96. Crew, Bill. Interviewed by Brian McFarland. September 2019.

97. Ibid.

98. White, David. Interviewed by Brian McFarland. September 2019.

99. Moore, Tom. Interviewed by Brian McFarland. December 2019.

100. Lundgren, Petra. Email Message to Author. March 24, 2020.

101. Chen, David and Laura Gartry. "Great Barrier Reef $444m Budget Funding Awarded to Small Foundation Without Tender Process." ABC News. May 22, 2018. Accessed December 30, 2019. https://www.abc.net.au/news/2018-05-22/great-barrier-reef-funding-labor-accuse-due-diligence/9785782.

102. Iyer, Venkat et al. Finance Tools for Coral Reef Conservation. 61.

103. McCalman. The Reef. 271.

104. Carbon Pulse. "Queensland to Set up A$500m Land-based Carbon Fund." October 25, 2018. Accessed December 30, 2019. https://carbon-pulse.com/48693/.

105. Great Barrier Reef Marine Park Authority 2019. Annual Report 2018–2019. GBRMPA, Townsville. http://www.gbrmpa.gov.au/about-us/corporate-inform ation/annual-report. 14.

106. Lundgren, Petra. Email Message to Author. March 24, 2020.

107. Papahānaumokuākea Marine National Monument. "About Papahānaumokuākea." Accessed February 7, 2019. https://www.papaha naumokuakea.gov/new-about/.

108. Brian Schatz Senate Office. "Appropriations Committee Approves More Than $4 Million For Papahānaumokuākea Marine National Monument." October 2, 2019. Accessed April 8, 2020. https://www.schatz.senate.gov/press-releases/appropriations-committee-approves-more-than-4-million-for-papahanaumok uakea-marine-national-monument.

109. Kikiloi, Kekuewa et al. "Papahānaumokuākea: Integrating Culture in the Design and Management of one of the World's Largest Marine Protected Areas." 440.

110. Maui Now. "Coral Reef Damage, Invasive Alga at Papahānaumokuākea." August 15, 2019. Accessed February 20, 2020. https://mauinow.com/2019/08/15/coral-reef-damage-invasive-alga-at-papahanaumokuakea/.

111. Island Conservation. "Midway Atoll National Wildlife Refuge, United States." Accessed February 21, 2020. https://www.islandconservation.org/midway-atoll/.

112. Protected Planet. "Explore the World's Marine Protected Areas." Accessed December 12, 2019. https://www.protectedplanet.net/marine.

113. The White House, Office of the Press Secretary. "Presidential Proclamation—Papahanaumokuakea Marine National Monument Expansion." August 26, 2016. Accessed February 20, 2020. https://obamawhitehouse.archives.gov/the-press-office/2016/08/26/presidential-proclamation-papahanaumokuakea-marine-national-monument.

114. USEPA. "America's Coral Reefs." May 4, 2018. Accessed February 26, 2020. https://www.epa.gov/coral-reefs/americas-coral-reefs.

115. Island Conservation. "Midway Atoll National Wildlife Refuge, United States." Accessed February 21, 2020. https://www.islandconservation.org/midway-atoll/.

116. Ibid.

117. Ibid.

118. Ibid.

119. Kikiloi, Kekuewa et al. "Papahānaumokuākea: Integrating Culture in the Design and Management of One of the World's Largest Marine Protected Areas." 439–440.

120. Island Conservation. "Midway Atoll National Wildlife Refuge, United States." Accessed February 21, 2020. https://www.islandconservation.org/midway-atoll/.

121. Kikiloi, Kekuewa et al. "Papahānaumokuākea: Integrating Culture in the Design and Management of One of the World's Largest Marine Protected Areas." 440.

122. Ibid. 441–442.

123. NOAA. "Management: Final Management Plan Completed." Accessed April 8, 2020. https://www.papahanaumokuakea.gov/management/mp.html.

124. Roberts. The Ocean of Life. 344.

125. BBC News. "World's Largest Marine Reserve Created off Hawaii." August 27, 2016. Accessed December 13, 2019. http://www.bbc.com/news/world-us-canada-37202045.

126. Sentman, Wayne. "Hawaii's Papahānaumokuākea to Become the World's Largest Marine Protected Area." Oceanic Society. Accessed February 20, 2020. https://www.oceanicsociety.org/blog/1759/hawaiis-papahnaumokukea-to-become-the-worlds-largest-marine-protected-area.

127. Kikiloi, Kekuewa et al. "Papahānaumokuākea: Integrating Culture in the Design and Management of One of the World's Largest Marine Protected Areas." 436.

128. Brian Schatz Senate Office. "Appropriations Committee Approves More Than $4 Million for Papahānaumokuākea Marine National Monument." October 2, 2019. Accessed April 8, 2020. https://www.schatz.senate.gov/press-releases/appropriations-committee-approves-more-than-4-million-for-papahanaumokuakea-marine-national-monument.

129. Island Conservation. "Midway Atoll National Wildlife Refuge, United States." Accessed February 21, 2020. https://www.islandconservation.org/midway-atoll/.

130. NOAA. "General Challenges to Conservation in Papahānaumokuākea Marine National Monument Development Pressures." Accessed April 8, 2020. https://www.papahanaumokuakea.gov/wheritage/challenges.html.
131. Ibid.
132. Sentman, Wayne. "Hawaii's Papahānaumokuākea to Become the World's Largest Marine Protected Area." Oceanic Society. Accessed February 20, 2020. https://www.oceanicsociety.org/blog/1759/hawaiis-papahnaumokukea-to-become-the-worlds-largest-marine-protected-area.
133. Kikiloi, Kekuewa et al. "Papahānaumokuākea: Integrating Culture in the Design and Management of One of the World's Largest Marine Protected Areas." 446.
134. Ibid. 447.
135. Roberts. The Ocean of Life. 340.
136. Wenban-Smith, Nigel. Interviewed by Brian McFarland. January 2020.
137. Chagos Conservation Trust. "FAQ." Accessed January 24, 2020. https://chagos-trust.org/faq.
138. Wenban-Smith, Nigel. Interviewed by Brian McFarland. January 2020.
139. Sheppard, Charles. Interviewed by Brian McFarland. January 2020.
140. Ibid.
141. Bowcott, Owen. "UN Court Rejects UK's Claim of Sovereignty Over Chagos Islands." The Guardian. February 25, 2019. Accessed April 10, 2020. https://www.theguardian.com/world/2019/feb/25/un-court-rejects-uk-claim-to-sovereignty-over-chagos-islands.
142. Sheppard, Charles. Interviewed by Brian McFarland. January 2020.
143. Dunbar, Rob. Interviewed by Brian McFarland. April 2019.
144. Ibid.
145. Sheppard, Charles. Interviewed by Brian McFarland. January 2020.
146. Gabbatiss, Josh. "Eradicating Rats from Tropical Islands Could Protect Coral Reefs from Effects of Climate Change, Scientists Find." The Independent. July 11, 2018. Accessed February 19, 2020. https://www.independent.co.uk/environment/rats-coral-reefs-eradicate-tropical-island-seabirds-fish-chagos-archipelago-indian-ocean-a8442371.html.
147. Dunbar, Rob. Interviewed by Brian McFarland. April 2019.
148. Pitman, Helen. Email Message to Author. January 27, 2020.
149. Dunbar, Rob. Interviewed by Brian McFarland. April 2019.
150. Ibid.
151. Sheppard, Charles. Interviewed by Brian McFarland. January 2020.
152. Cinner, Joshua E., et al. "Bright Spots Among the World's Coral Reefs." https://doi.org/10.1038/nature18607.
153. Wenban-Smith, Nigel. Interviewed by Brian McFarland. January 2020.
154. Ibid.
155. Ibid.
156. Sheppard, Charles. Interviewed by Brian McFarland. January 2020.
157. Ibid.

158. Ibid.
159. Wenban-Smith, Nigel. Interviewed by Brian McFarland. January 2020.
160. Ibid.
161. Ibid.
162. Ibid.
163. Ibid.
164. CCT. "Networks." Accessed January 24, 2020. https://chagos-trust.org/about/networks.
165. Wenban-Smith, Nigel. Interviewed by Brian McFarland. January 2020.
166. Ibid.
167. CCT. "Bertarelli Programme in Marine Science." May 10, 2018. Accessed April 10, 2020. https://chagos-trust.org/news/bertarelli-programme-in-marine-science.
168. Pitman, Helen. Email Message to Author. January 27, 2020.
169. Dunbar, Rob. Interviewed by Brian McFarland. April 2019.
170. Manson, Sophie. "Something Smells Fishy: Scientists Uncover Illegal Fishing Using Shark Tracking Devices." Mongabay. March 7, 2019. Accessed April 10, 2020. https://news.mongabay.com/2019/03/something-smells-fishy-scientists-discover-illegal-fishing-using-shark-tracking-devices/.
171. Dunbar, Rob. Interviewed by Brian McFarland. April 2019.
172. Wenban-Smith, Nigel. Interviewed by Brian McFarland. January 2020.
173. Sheppard, Charles. Interviewed by Brian McFarland. January 2020.
174. Ibid.
175. Ibid.
176. Pitman, Helen. Email Message to Author. January 27, 2020.
177. Graham, N.A.J., Wilson, S.K., Carr, P. et al. "Seabirds Enhance Coral Reef Productivity and Functioning in the Absence of Invasive Rats." Nature. 559, 250–253 (2018). https://doi.org/10.1038/s41586-018-0202-3.
178. Dunbar, Rob. Interviewed by Brian McFarland. April 2019.
179. Wenban-Smith, Nigel. Interviewed by Brian McFarland. January 2020.
180. Ibid.
181. Vertigo Lab. "Innovations for Coral Finance, ICRI Publication." 19.
182. Ibid. 21.
183. Spergel, Barry and Melissa Moye. "Financing Marine Conservation: A Menu of Options." 12.
184. Ibid. 12.
185. Vertigo Lab. "Innovations for Coral Finance, ICRI Publication." 19.
186. Iyer, Venkat et al. Finance Tools for Coral Reef Conservation. 22.
187. Mackelworth, PC et al. "Geopolitics and Marine Conservation: Synergies and Conflicts." 3.
188. Earle. Sea Change. 309.
189. Spergel, Barry and Melissa Moye. "Financing Marine Conservation: A Menu of Options." 54.

190. Trading Economics. "Credit Rating." Accessed July 23, 2019. http://www.tra dingeconomics.com/country-list/rating.

191. Helvarg. 50 Ways to Save the Ocean. 108.

192. CBD. "Aichi Biodiversity Targets." Accessed September 9, 2019. https://www. cbd.int/sp/targets/.

193. Protected Planet. "Aichi Target 11 Dashboard." Accessed December 12, 2019. https://www.protectedplanet.net/target-11-dashboard.

194. Ibid.

195. Walsh. Lady Elliot Island. 115.

196. Sobel and Dahlgren. Marine Reserves. 353.

197. Walsh. Lady Elliot Island. 117.

198. Daley. The Great Barrier Reef. 11–12.

199. Sobel and Dahlgren. Marine Reserves. 341.

200. Ibid. 341–342.

201. Ibid. 354.

202. Ibid. 354, 357.

203. Lundgren, Petra. Email Message to Author. March 24, 2020.

204. Carbon Pulse. "Queensland Gives Final Go-Ahead for Adani's Massive Carmichael Coal Mine." June 13, 2019. Accessed December 30, 2019. https:// carbon-pulse.com/76524/.

205. Sobel and Dahlgren. Marine Reserves. 358.

206. Parvati. "Marine Arctic Peace Sanctuary." Accessed December 16, 2019. https://parvati.org/maps/.

207. Sobel and Dahlgren. Marine Reserves. 361.

208. Iyer, Venkat et al. Finance Tools for Coral Reef Conservation. 16.

8

Conservation Easements

Introduction

As opposed to the earlier and more traditional approach where governments raise conservation finance through taxation (e.g., such as via sales taxes), there is also the role of conservation easements. With traditional land conservation easements, there is an ability for local entities such as governments and/or land trusts, depending on the jurisdiction, to utilize a tax deduction to raise conservation finance. This use of a tax deduction for conservation purposes is for the establishment of a conservation easement or to encourage donations to an eligible nonprofit organization which has a conservation-related mission. In the marine space, there are a few examples of nontraditional conservation easements.

Historical Overview

In 1891, the Trustees of Public Reservations (now the Trustees of Reservations) became the world's first regional land trust,[1] which established the world's first conservation easement in Boston, Massachusetts.[2]

Other notable moments in the history of conservation easements include:

- 1959: The term conservation easement is first "coined."[3]
- 1961: The Nature Conservancy (TNC) structures its first conservation easement.[4]

© The Author(s) 2021
B. J. McFarland, *Conservation of Tropical Coral Reefs*,
https://doi.org/10.1007/978-3-030-57012-5_8

- 1976: U.S. Federal Law "allows landowners to begin reporting the value of an easement as a tax deduction."[5]
- 1970s: "{…} the donation of conservation easements to a land trust, yielding a federal charitable tax deduction, a practice pioneered in the state of Maine in the 1970s,"[6] particularly by Tom Cabot along with Peggy and David Rockefeller.[7]
- 1989: "Conservation-minded developers were likely the first to use transfer fees and to use them successfully. A developer in South Carolina created what may be the earliest programs, adapting them from the model of early, publicly levied real estate transfer taxes. (Some government agencies tax land transfers to generate revenue for public programs). In 1989, Jim Light and his partner, James Chaffin, of Chaffin/Light Associates, instituted a private transfer 'assessment' at their Spring Island luxury development in South Carolina. This 3,000-acre development includes 1,200 acres of nature preserves and open space with live oak forests, marsh, and waterways {…}."[8]
- 1990s: In the late 1990s, "Colorado created one of the most generous conservation tax programs with an unusual twist. Landowners donating easements may sell the tax credits to individuals or corporations that can more readily use the tax credits."[9]
- 1990: TNC facilitates a conservation easement in Costa Rica, marking the first such easement in Latin America.[10]
- 2006: The U.S. West Coast Groundfish Fishery collapses and TNC "conducted a private buyout of some trawl fishing permits and vessels."[11]
- 2014: The Yela Forest Conservation Easement becomes one of the first conservation easements outside of the Americas.[12]

Mechanisms of Instrument

Conservation easements are essentially "a perpetual deed restriction placed on a property to protect its ecological and open space values."[13] These perpetual deed restrictions, which are far more common in terrestrial ecosystems, are permanent legal restrictions placed on a given property. For instance, in the U.S., the specific mechanism may involve the purchase and retirement of residential and/or commercial development rights for a given property. In addition, the property might be donated to a nonprofit organization such as a local land trust or a large conservation organization such as TNC, or possibly donated to a local government so the property can be maintained as conservation land. If the parcel is donated to an eligible entity, then the landowner—in the United States—is entitled to federal and/or state tax incentives. The eligible entity is often then responsible for regularly monitoring the property to ensure the deeded restrictions are being followed, and if not, the entity can legally enforce the restrictions by using litigation, if necessary.

In the United States:

> The conservation easement has become a very popular and successful tool for three reasons. First, the landowner does not have to relinquish ownership of the property and can still live on it, sell it, or pass it on to heirs. Second, the donation of a conservation easement entitles the donating landowner to tax savings for charitable contributions. Finally, conservation easements do not necessarily prohibit {all} development.[14]

A historical advantage of land trusts and nonprofit organizations, such as the Trust for Public Land (TPL), is that they were "nimbler and less burdened by the bureaucratic constraints than the government was and could offer tax incentives. TPL could often buy high-priority land or easements for less than their market value. Additionally, landowners usually preferred to negotiate land sales with private organizations rather than with governments."[15] Finally, in the United States, land trusts can insure their conservation easements via a variety of pooled insurance products.[16]

While conservation easements have been used in developed countries such as Australia (called "conservation covenants"), Canada, and the United States, their use in developing countries have occurred on a more limited basis such as in Costa Rica, Guatemala, and Mexico.[17] However, the difficult task is how to apply these conservation easement models to publicly owned, marine areas with tropical coral reefs?

Size of Instrument

According to the National Conservation Easement Database, there are more than 158,000 conservation easements in the United States which protected 27 million acres (11 million hectares) as of September 2019.[18] Although a bit dated, the 2015 National Land Trust Census Report stated that in the United States:

- 56 million acres (22.7 million hectares) were conserved by state, local, and national land trusts as of year-end 2015;
- 6.25 million people visited land trust properties in 2015;

- 77% of the total acres owned and under easement were held by an accredited land trust;
- USD$2.18 billion was the amount in endowments and dedicated funding managed by state, local, and national land trusts; and
- Land trusts had more than 8000 staff, nearly 16,000 board members, more than 200,000 volunteers, and nearly 5 million members and financial supporters.[19]

TNC, as of 2014, had worked in 23 countries on conservation easements including seven in Guatemala and one in Costa Rica.[20] However, the number of conservation easements in the marine environment, particularly for tropical coral reefs, is more difficult to obtain.

Introduction to Case Studies

The following case studies will look at the Palmyra Atoll purchase,the Phoenix Islands Protected Area Conservation Trust (PACT), the Pez Maya Land Purchase, and the Chumbe Island Coral Park Ltd (CHICOP). The Palmyra Atoll, which is now a part of the Pacific Remote Islands Marine National Monument, was a more traditional land conservation easement. The Palmyra Atoll Purchase is somewhat similar to the Pez Maya, which was a land purchase incorporated into Mexico's Sian Ka'an Biosphere Reserve. In contrast, the Phoenix Islands PACT involved the extinguishing of fishing licenses, which could be viewed as a type of easement. The CHICOP, located off the coast of Tanzania, was the first marine Privately Protected Area in the world and it is based on a Marine Conservation Agreement.

Case Study #1: The Palmyra Atoll Purchase

Introduction

One way the traditional, terrestrial conservation easement model can be applied to the marine space is via the purchase of atolls, as was done by The Nature Conservancy (TNC) and the Palmyra Atoll. The Palmyra Atoll is part of the Pacific Remote Islands Marine National Monument (PRIMNM), which was established in 2009 (Fig. 8.1).

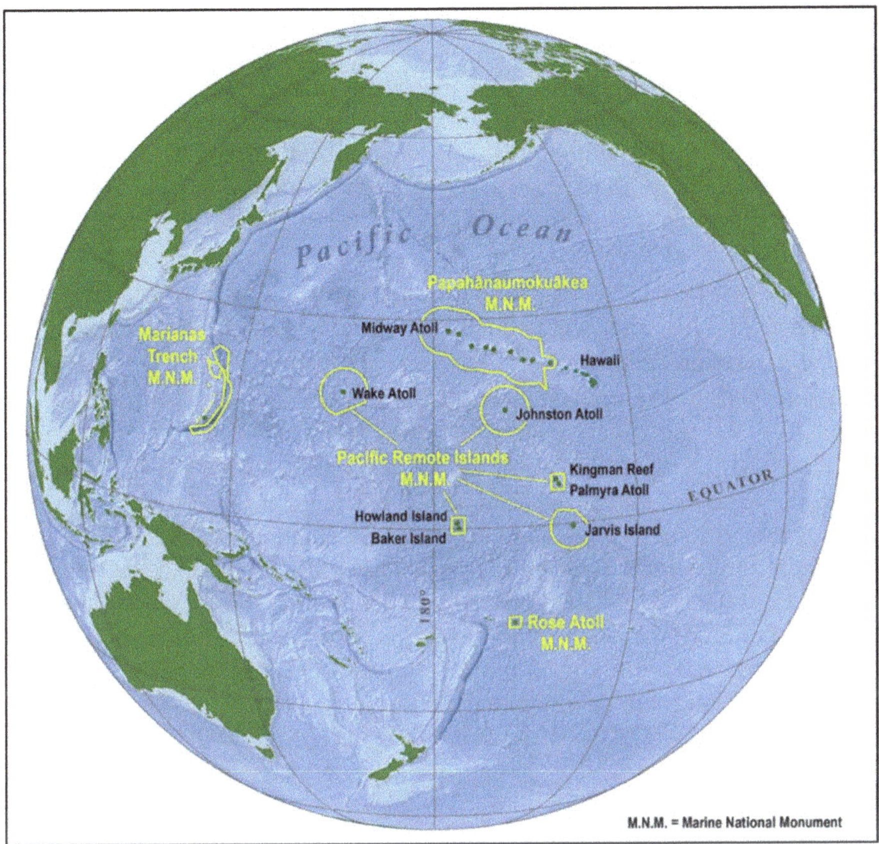

Fig. 8.1 Map of Palmyra Atoll and Pacific Remote Islands National Monument (*Credit* Open Source Map; U.S. Fish & Wildlife Service [USFWS. "Map of Pacific Islands Refuges and Monuments." September 28, 2016. Accessed January 14, 2020. https://www.fws.gov/refuges/refugeLocatorMaps/PacificIslands.html; Also see: USFWS. "Can I Download Images From USFWS Web Sites?" February 11, 2009. Accessed January 14, 2020. https://www.fws.gov/faq/imagefaq.html])

On January 6, 2009, President George W. Bush established the Pacific Remote Islands Marine National Monument under the authority of the Antiquities Act of 1906. The monument incorporates approximately 86,888 square miles within its boundaries, which extend 50 nautical miles from the mean low water lines of Howland, Baker, and Jarvis Islands; Johnston, Wake, and Palmyra Atolls; and Kingman Reef. {…} On September 25, 2014, the Obama Administration expanded the monument.[21]

With respect to the Palmyra Atoll:

In 2000, TNC purchased the Palmyra Atoll, the last intact marine wilderness in the U.S. tropics, for $37 million (including endowment and infrastructure costs) from the family that had been its owners and sole inhabitants for many years. The U.S. Fish and Wildlife Service subsequently approved the establishment of a National Wildlife Refuge for the waters off Palmyra Atoll. The atoll, located 1,693 km south of Hawaii, consists of 275 hectares of land and 6,277 hectares of pristine coral reefs (TNC 2000, TNC 2001).[22]

There were two phases to the purchase. Initially, the purchase was done from the reserves of TNC that were allocated to land acquisitions. There was then a capital campaign used.

With respect to the management of the PRIMNM:

At Howland Island, Baker Island, Jarvis Island, Palmyra Atoll, and Kingman Reef, the terrestrial areas, reefs, and waters out to 12 nautical miles (nmi) are part of the National Wildlife Refuge System. The land areas at Wake Atoll and Johnston Atoll remain under the jurisdiction of the U.S. Air Force, but the waters from 0 to 12 nmi are protected as units of the National Wildlife Refuge System. For all of the areas, fishery-related activities seaward from the 12-nmi refuge boundaries out to the 50-nmi monument boundary are managed by the National Oceanic and Atmospheric Administration.[23]

Identify the Problem

One problem facing the Palmyra Atoll is invasive, black rats:

Beginning in 1939, Palmyra Atoll was heavily used by the Navy as a base until the end of World War II. It is believed that black rats were introduced to the atoll by the military during this time. Approximately 6,000 military personnel were stationed on the atoll for several years. Rats are well known for having severe negative effects on the ecosystems they invade.[24]

The rats have been observed eating crabs and seedlings, along with having an impact on birds.

In addition, global climate change—particularly thermal stress and acidification—are problems facing PRIMNM.

Why the Problem Is Important

The problem of invasive black rats is important for several reasons. For instance:

Palmyra Atoll has one of the best remaining examples of Pisonia grandis forest found in the Pacific, and a large diversity of fish species (418 species). Many nationally and internationally threatened, endangered, and depleted species thrive at Palmyra and Kingman, including sea turtles, pearl oysters, giant clams, reef sharks, coconut crabs, fishes, and dolphins. Large schools of rare melon-headed whales reside off both atolls, and a potentially new species of beaked whale was recently described. Palmyra supports 11 nesting seabird species, including the third largest red-footed booby colony in the world, the largest black noddy colony in the Central Pacific, and large numbers of bristle-thighed curlews.[25]

The atolls, including the Palmyra Atoll, "are ideal 'laboratories' for assessing the effects of climate change without direct human impacts."[26]

The PRIMNM has a historical importance as well, as "Amelia Earhart and Fred J. Noonan planned to stop over at Howland on their famous world circumnavigation flight in 1937. Radio contact was lost, the airplane never arrived, and to this day their exact fate is unknown. The Amelia Earhart Day Beacon on Howland is named in her honor."[27]

It is important to note that all of the atolls, not just Palmyra Atoll, are important. For instance:

> Johnston is an important genetic stepping stone from the Line Islands to the Hawaiian Islands for invertebrates, fish, and corals. It supports 45 coral species, including a thriving table coral community and a dozen species found only in the Hawaiian and northern Line Islands. {…} Wake is the northernmost atoll in the Marshall Islands geological ridge and perhaps the oldest living atoll in the world. {…} Kingman is known to be the most undisturbed coral reef within the U.S., complete with a greater proportion of apex predators than at any other studied coral reef ecosystem in the world.[28]

Furthermore, the PRIMNM, particularly Baker and Jarvis Islands, are among the world's coral reef "bright spots."[29]

How Problem Was Identified

The problems facing the Palmyra Atoll were, in part, discovered by NGOs such as TNC and Island Conservation, as well as the U.S. Fish and Wildlife Service (USFWS). For instance:

> The Nature Conservancy of Hawai'i manages a small research camp at Palmyra Atoll for the Palmyra Atoll Research Consortium. Through this consortium of ten institutions from the United States and New Zealand, scientists conduct

research pertaining to biodiversity, conservation, natural history, ecosystem restoration, marine ecosystem dynamics, biogeochemistry, climate dynamics, and atmospheric processes.[30]

As explained by Chad Wiggins from TNC, in the early days, a conservation action plan developed. NOAA was a part of this exercise, along with TNC and the USFWS. This, in part, helped to identify the needs to better understand the ecosystem. What emerged was the Palmyra Atoll Research Consortium, which is a university–institutional partnership.[31]

Effectiveness of Process for Identifying Problem

The multi-stakeholder process for identifying the problems seems effective.

Steps Taken to Address the Problem

According to Island Conservation:

> This multistage project will restore the natural ecosystem, with each action building on the next. The first phase focused on the removal of black rats. Following removal will be the reduction of invasive coconut palm canopy cover and restoration of the native forest community. Next will come translocation of two endangered, insular bird species that are at risk in their current ranges. The final step will be active restoration of Palmyra's extirpated seabird colonies.[32]

An important step was development of a structure where members pay fees to TNC to do research at the Palmyra Atoll. This research is then fed back into the conservation action plan, while TNC is responsible for the costs associated with maintaining the research facility.[33]

Results

Fortunately, there have been some impressive results with respect to the invasive black rats. The USFWS, TNC, and Island Conservation:

> removed invasive rats from the atoll in 2011 to protect native species from extinction. In the absence of invasive species, scientists have documented a 5000% increase in native trees, an abundance of crabs, and an increase in seabirds. One species of mosquito was also eradicated as a result of the project.[34]

Challenges and How They Were Met

One challenge facing the PRIMNM, and to a lesser degree the Palmyra Atoll, is access:

> the seven atolls and islands included within the monument are farther from human population centers than any other U.S. area. {...} Only two refuges, Palmyra Atoll and Wake Atoll, are accessible by air; the rest require ship access. In some cases, it takes up to eight days to reach a refuge from its nearest port, and it may be visited only once every two years.[35]

Then again, this remoteness has helped conservation goals.

Beyond Results

It appears as though the results are sustainable. Likewise:

> The partners are committed to restoring Palmyra Atoll. The USFWS administers most of the atoll as part of its National Wildlife Refuge System; TNC, which owns Cooper-Menge Island and operates a small research field camp, is a co-manager. Island Conservation is a nonprofit organization dedicated to preventing extinctions by removing invasive species from islands.[36]

Lessons Learned

According to Mr. Chad Wiggins of TNC, a few lessons learned are:

- Without a partnership strategy and some level of agreement in place, the conservation activities are not going to go far;
- Identifying the long-term costs upfront and to determine whether it is viable to have these costs covered by a government, or whether the funds will need to come from elsewhere, is important;
- It is important to ensure good coordination of the research that is happening and to make sure such research is fed back into the management of the place;
- While the land acquisition did not do much for the reefs, what land acquisition did was bring in partners, and then the protection of the reefs came in when the marine park was established. The land acquisition was leveraged to advance policy and enabled TNC to become a stakeholder;

- There are lots of investments (i.e., such as investments in policy) which are not reflected in the land acquisition costs. In addition, there are lots of government investments, in some cases multiple governments, and Palmyra Atoll is not an exception. For instance, there are ongoing USFWS staff onsite; and
- The Palmyra Atoll is a great example of an intact ecosystem, with birds, marine animals, and the forest all connected. As an intact ecosystem with relatively few threats, the reefs have been resilient and this presents an interesting research angle from an ecology perspective. There are also researchers collaborating from different disciplines. Yet, it is important that such research is fed back into the conservation action plan and is not just "a pure research playground."[37]

Other Resources on the Palmyra Atoll Purchase

Island Conservation

- https://www.islandconservation.org/palmyra-atoll/

Palmyra Atoll Digital Archive

- http://www.palmyraarchive.org/

TNC, Hawaii

- https://www.nature.org/en-us/about-us/where-we-work/united-states/hawaii/

USFWS

- https://www.fws.gov/refuge/Pacific_Remote_Islands_Marine_National_Monument/;
- https://www.fws.gov/refuge/Pacific_Remote_Islands_Marine_National_Monument/wildlife_and_habitat/index.html; and
- https://www.fws.gov/refuge/Palmyra_Atoll/

Case Study #2: The Phoenix Islands Protected Area Conservation Trust

Introduction

Although not a traditional easement, the Phoenix Islands Protected Area Conservation Trust involved the extinguishing of fishing licenses, which could be viewed as a type of easement.

The country of Kiribati consists of three island groups: the Gilbert Islands, Line and Christmas Islands, and the Phoenix Islands. In addition, the Phoenix Islands consists of a number of submerged reefs, seamounts, deep sea, and tropical coral reefs. It is also important to note (Fig. 8.2):

Kiribati is classified as a microstate, a designation for countries with very small populations and landmasses, 280 square miles (726 km^2) in this case. But the enormous ocean area controlled by Kiribati – 1,370,300 square miles (2.2 million km^2) – makes it an important player on the world stage, especially as ocean resources and issues become more prominent in international affairs.[38]

As exclaimed by Anote Tong, the former president of Kiribati, exclaimed:

Fig. 8.2 Map of Kiribati (*Credit* Pomogayev)

The waters around us are some of the most productive fishing grounds for tuna and reef fish in the world. Kiribati has its own small-scale tuna fishery, but over the decades, with the rise of multinational fishing fleets, we came to rely on the sale of lucrative fishing rights to provide the most basic services to our people.

In 2006, Kiribati made news around the world by announcing that we were closing off some of our best coral reefs and tuna fishing grounds to create a marine protected area (MPA), the ocean equivalent of a Yellowstone National Park. Once the MPA was established, we began to phase out commercial fishing from 11 percent of our exclusive economic zone, or EEZ-some 400,000 square kilometers {approximately 154,441 square miles}.

At the center of the MPA were the eight islands of the Phoenix group, home to a few dozen people and millions of nesting seabirds. {…} Kiribati proposed to create a trust to offset the loss of fishing income and to permanently preserve the Phoenix Islands, with their birds and corals and fishes.[39]

Furthermore:

Our central task in creating the PIPA, then, was to channel a portion of global willingness to pay for conservation to Kiribati in a way that makes up for the potential loss of revenue from fishing licenses. Under this scheme, instead of letting commercial fishers pay for the right to harvest in the waters around the Phoenix Islands, Kiribati would allow the world to pay for the right to protect this unique ecosystem. In 2002, Minister Tetabo Nakara coined the phrase 'reverse fishing license' to describe this proposition – a phrase that perfectly captures an arrangement in which the world makes it financially viable for a developing country to establish one of the largest marine protected areas on Earth.[40]

As the right to fish was essentially purchased, the Phoenix Islands Protected Area can be viewed as a type of conservation easement or concession. This said, the "{…} model of the forest conservation concession was proposed: instead of leasing land to logging companies, landowners are paid not to log their forests."[41]

Identify the Problem

There were, and still are, several problems facing the Phoenix Islands. Such problems include impacts from climate change, overfishing, and the presence of invasive species.

With respect to global climate change:

{…} climate change is now becoming the most pressing threat to the nation. With three primary concerns: (1) eventual relocation from the islands as a result of sea level rise that will make them uninhabitable; (2) decreasing availability of freshwater and ability to grow food crops, due to changing rainfall patterns, drought and salinization of groundwater; and (3) changes in the marine environment, such as migration of fish stocks and bleaching of coral reefs, on which they have depended for centuries for protein and coastal protection.[42]

In fact, "freshwater availability is identified as one of the most important factors that drives abandonment of atoll settlements, as evidenced in the past by settlements in the Phoenix Islands which are in a dry belt, and from which all settlement schemes have retreated after some years."[43]

Furthermore, "the recent central Pacific El Niño events led to periods of abnormally high SSTs {sea surface temperature} which create heat stress on the PIPA coral reefs and may cause coral bleaching, as observed in 2002-03, 2010, and 2015."[44] Similarly, "in 2002-2003 the Phoenix Islands suffered the most intense sea temperature 'hotspot' yet recorded, which resulted in massive mortality, averaging 60% across the island group and increasing to near 100% in the most sensitive habitats."[45] It is important to note that "unlike bleaching events elsewhere in the world, this one was unique because it was not exacerbated by local human impacts."[46]

Prior to the creation of PIPA, there was extensive fishing in the Phoenix Islands:

Longline and purse seine fishery licenses constitute an important part of the Kiribati economy. Starting when under the United States and British control, what are today Kiribati waters were extensively fished for tuna and other large pelagics. Kiribati grants the majority of its licenses to foreign fishing vessels flying the flags of the United States, EU countries, South Korea, China, Japan and other nations. Purse seine tuna fishing is the major extraction method, with a purse seiner catching on average 32 tonnes of tuna per day, mostly skipjack.[47]

Furthermore, illegal fishing—particularly for sharks—has been a problem:

In 2001 a boat en route from American Samoa stopped in the islands to catch sharks for their most valuable product – their fins. Shark fins are used as an ingredient to thicken a soup traditionally served at weddings and other ceremonies throughout China and Southeast Asia, and the surging demand has pushed shark-fin fishing to the farthest corners of the globe. {…} The dried fins are sold for up to €500 or US$630 per kilo, many times the price of shark meat. Stripped of their valuable fins, the maimed sharks are thrown back into the water to die.

The shark-finning boat that visited the Phoenix Islands stopped at Kanton, Rawaki, Manra, and briefly at Orona to catch sharks using longlines. Longlining is a commercial fishing practice in which boats spool out up to 80 miles of fishing line baited with up to forty thousand hooks to reel in fish by the thousands. The shark-finning boat stayed in the islands for about three months. Luckily for the remaining sharks in the rest of the island group, engine trouble eventually forced the shark finners to head back to Samoa. But the damage was considerable: in a few short months, this single boat removed almost the entire adult population of sharks from half of the Phoenix Islands, including the three largest islands.[48]

In the past, whalers were also common and "at times, more than 600 whale ships plied these waters and whalers were so effective that even today the Phoenix Islands appear to be largely devoid of sperm whales."[49]

Invasive species, both marine and terrestrial species, have created additional problems:

The deliberate and accidental introductions of invasive species during all of these periods have been significant in the Phoenix Islands, and the faunal invaders include Pacific and Asian rats, rabbits, cats, ants, pigs and dogs; floral invaders include coconuts, lantana and other weeds. Some of the negative impacts of these biological invasions include the elimination of native seabirds through the destruction of eggs and young, and the elimination of native plants via competition for limited resources. Mammalian pests have been the most damaging invasives in PIPA, threatening the bird populations and modifying the entire natural island ecosystem. Rabbits, cats and different rat species have had notable impacts over time, with some birds (e.g. Phoenix petrels and White-Throated storm-petrels) and other biota being reduced to critically low numbers. The most sensitive fauna species survived on only the one island (Rawaki) that escaped rat infestation.[50]

This impact of invasive species, unforutnately, is a common theme throuhgout many islands.

Why the Problem Is Important

The Phoenix Islands are one of Earth's last intact oceanic coral archipelago ecosystems and the Islands include the largest remaining tuna fishery in the world, particularly for skipjack tuna (*Katsuwonus pelamis*).[51]

Another reason why the aforementioned problems—particularly climate change and overfishing—are important is due to the importance of subsistence fishing:

> The primary livelihoods in Kiribati are fishing, copra harvesting (a coconut cash crop for export only), government jobs, or small businesses. Kiribati has one of the highest rates of per capita fish consumption in the world, estimated at 72-207 kg/person/year. The overwhelming majority of this fish comes from local resources. Imported foods are expensive and unreliable and only a minority of people are cash employed (~25% overall, ~5% on outer islands). Almost all (97%) households engage in fishing to feed their families.[52]

PIPA is also important due to the significant biodiversity that is present. For instance, "PIPA is biogeographically situated in the centre of the equatorial Pacific, thus likely playing a pivotal role in the movements and dispersal of marine animals and larvae across the Pacific."[53] Likewise, PIPA is an important site for seabirds:

> PIPA hosts forested and arid seabird islands of high-global significance (>40 breeding colonies with several of the world's largest seabird breeding populations), including at least 19 breeding species, 51 species reported and a total population exceeding 1 million birds. In particular, the islands support globally important breeding colonies or concentrations of the following species: Audubon's Shearwater (*Puffinus l'herminieri*), Christmas Shearwater (*Puffinus nativitatis*), Phoenix Petrel (*Pterodroma alba*), White-throated Storm petrel (*Nesofregetta fuliginosa*), Great Frigatebird (*Fregata minor*), Lesser Frigatebird (*Fregata ariel*), Brown Booby (*Sula leucogaster*), Masked Booby (*Sula dactylatra*), Sooty Tern (*Sterna fuscata*), Greybacked Tern (*Sterna lunata*) and Blue Noddy (*Procelsterna cerulea*). Most of these species forage primarily within PIPA waters reflecting the local diversity and abundance of marine resources.[54]

Similarly,

> Compared with other remote island groups in the area, such as the Gilbert Islands to the west and the Line Islands to the east, we found a high diversity of both coral and fish species in the Phoenix Islands. Kanton proved to have the largest area of coral reef, with the highest number of species both within the Phoenix group and of the other islands surveyed so far.[55]

While many MPAs focus solely on the conservation of tropical coral reefs, it is important to take a more holistic, seascape approach. This said:

Most of the public's interest in the Phoenix Islands Protected Area is focused on the colorful coral reefs that surround each of the eight islands. But the coral reefs comprise only a tiny fraction of the PIPA. More than 99 percent of the ocean habitat in the marine protected area is in the pelagic zone. This zone is where a vast array of marine invertebrates live and drift; it's also home to schools of large oceangoing fish such as tuna and pods of whales and dolphins.[56]

PIPA is an important natural climate change research laboratory and protecting PIPA is also important due to its intrinsic value:

For instance, PIPA may have very high existence value arising from the benefits that people get from knowing that pristine marine ecosystems exist. The desire to protect this value may be motivated by the desire to share this knowledge or appreciation with friends or children or by sympathy for other living beings. PIPA also provides option value or the value of ensuring that some use value will be available in the future (however this depends on the duration or durability of the contract).[57]

Furthermore, Kiribati is listed among the world's coral reef "bright spots,"[58] and PIPA is listed as a World Heritage Marine Program.

How Problem Was Identified

The aforementioned problems were in part identified through a partnership between Conservation International (CI), the New England Aquarium (NEAq), and the Kiribati Government. This can be dated back to the {…} early 2000s, after exploratory expeditions by the NEAq to the Phoenix Islands under its 'Primal Oceans' project – a project led by Dr. Gregory S. Stone to find some of the last virtually untouched areas in the global ocean."[59]

Effectiveness of Process for Identifying Problem

The process for identifying problems appears to have been effective.

Steps Taken to Address the Problem

There have been many steps taken to address the aforementioned problems. This includes:

- In 2002, "Kiribati voluntary closed all commercial coral reef fisheries in the Phoenix Islands with no compensation {...}."[60]
- In 2003, the government launched the Kiribati Adaptation Program "to respond to climate change by capturing rainwater, planting mangroves, and taking other steps to reduce Kiribati's vulnerability to rising sea levels and other effects of climate change."[61]
- "{...} CI was brought in for technical expertise and funding support and a Memorandum of Understanding was signed between the three entities in 2005 and is still in use today."[62]
- In 2005, "a major revision of Kiribati laws was undertaken, beginning in 2005 and culminating in the passage of the Environmental (Amendment) Act of 2007, which took effect on September 4, 2007. Under this new authority, then MELAD {Ministry of Environment, Lands and Agricultural Development} Minister Nakara, acting on the advice of the Kiribati cabinet, formally enacted the new legal framework for the PIPA, known as the Phoenix Islands Protected Area Regulations 2008 (2/7/2008)."[63]
- PIPA Management plan was completed in 2009.[64]
- The "Boston-based law firm Ropes & Gray LLP drafted legislation establishing the Phoenix Islands Protected Area Conservation Trust {...}. After two years of development and negotiation, the PIPA Conservation Trust was established in May 2009."[65]
- Beginning in 2010 "all vessels visiting the PIPA waters must be inspected and verified as strictly pest-free, whether landing or anchored offshore. Landings obviously heighten biosecurity risks, and only those necessary for essential conservation and research work are permitted. All expeditions must follow stringent biosecurity protocols at their port of origin, en route to the PIPA, and before landing at the islands. Such measures include vessel fumigation, control of invasives at wharves at departure ports, checking of all equipment and provisions being brought on board the vessel, maintaining rodent bait stations on board, carrying out pest surveillance on board, and if landing is permitted, thorough checking of all material to be taken ashore."[66]

Furthermore, there was significant engagement with stakeholders, including fishing partners. For instance, "as part of the first conservation contract with Kiribati, a Tuna Working Group is being established to analyse the potential revenue impacts of the PIPA closure over the next 5 years."[67] In addition, "{...} the PIPA partners commissioned a natural resource valuation of the Phoenix Islands."[68]

For a detailed timeline of events, please see the book entitled, *Underwater Eden: Saving the Last Coral Wilderness on Earth*, and edited by Gregory S. Stone and David Obura.

Results

A few of the main results, thus far, include:

- The "PIPA Management Committee and Director have been operational since 2007, formulating and guiding the building of capacity for managing the protected area, and focused on the Management Plan."[69]
- In 2008, "3.1% of the MPA was no-take (12,714 km^2), which was expanded in 2015 to 99.4% (405,755 km^2) no-take. The remaining ~ 0.6% will remain a restricted use zone around Kanton Island to accommodate subsistence fishing for a small caretaker population."[70]
- The PIPA Trust Act was enacted in 2009 by Parliament.[71]
- In 2010, the Phoenix Islands were inscribed on the World Heritage list by UNESCO, "becoming Earth's largest and deepest World Heritage site."[72]
- In October 2013, the PIPA Trust was capitalized with USD$5 million, half of which was gifted by the Kiribati Government. This contribution was matched by CI through its Global Conservation Fund. The initial USD$5 million capitalization "of the PIPA Trust was a major step towards the goal of reaching a 2014 target of USD$13.5 million, which was originally intended to catalyze the second phase of protection, allowing an increase to the pelagic no-take zone by an additional 25%."[73]
- In 2014, "the Phoenix Islands Scientific Advisory Committee (SAC) was formed; this advising body can play a key role in evaluating plans for Kanton to minimize environmental impact and risk."[74]
- As of "January 1, 2015, the Phoenix Islands Protected Area was closed to commercial fishing."[75]

With respect to sharks, "since the implementation of no-take reef areas in 2008, an increase in shark populations have been noted, the most direct impact of protection seen to date."[76]

Furthermore, the natural resource valuation was completed. While the valuation did not include inshore fisheries, nor the scenario where Kiribati develops its own domestic fishing fleet, and nor the valuation of the coral reef, the PIPA partners "settled on methodology and a number of simplifying assumptions to derive an initial estimated valuation of roughly $1.7 million to $3.25 million a year. The team also had to consider the costs of managing the PIPA. {…} The PIPA partners determined that a management budget of $300,000 per year would be a bare minimum."[77] With respect to the natural resource valuation, "the resulting analysis required a tremendous effort and resulted in a groundbreaking study – the first at this scale for a marine protected area in a developing Pacific island nation."[78]

Management actions that have occurred to date in PIPA include the following:

- research and monitoring focused on coral reefs, bird populations, and invasive species, but with preliminary data collected on other natural assets of the protected area, including deep reef slopes, deep water and oceanic invertebrates, marine megafauna and island vegetation;
- invasive alien species eradications (rabbits, rats, cats, mice) on three islands with additional islands planned for the near future;
- consultations on Kanton with existing government staff and the resident community on the management of Kanton Atoll, including a draft zoning of the island and adjacent waters;
- consultations on enforcement with partner countries, including New Zealand, the US, and Australia;
- improved legislation and regulations concerning fisheries observers, management, and enforcement;
- outreach programs implemented in Tarawa and broader nationally (e.g. radio) to raise awareness, support and pride in PIPA;
- establishment of a PIPA office and a PIPA Trust office in Tarawa;
- apprehension for maritime offenses in other parts of Kiribati, demonstrating competence to do the same for PIPA;
- establishment of the PIPA Trust Fund;
- a Signed Conservation Contract between the Government of Kiribati and the PIPA Trust;
- implementation of satellite surveillance technology and geofencing to enforce fishing regulations and no-take status; and
- installation of MPA signs on 7 of the 8 Phoenix Islands informing any potential visitors of MPA status and rules.[79]

The Trust, which was established, is governed by a Board of Directors appointed by the Kiribati Government, the New England Aquarium, and CI.[80] The Trust was seeking an endowment of USD$25 million:

Based on our collective experience, we figured that an ambitious, yet realistic, initial target for the endowment would be $25 million. Working backward from this endowment amount (assuming an annual yield of 5 percent and taking into consideration management costs and the cost of operating the trust fund), we determined that this amount would be commensurate with the creation of a no-take zone (NTZ) equal to 25 percent of the total area of the PIPA. This would bring the no-take and restricted use zones within the MPA to a total of 28 percent. This decision might seem like a drastic compromise on the part of conservation, as only 28 percent of the PIPA would be completely off-limits to commercial fishing. However, this 28 percent would completely protect seven of the eight Phoenix Islands and associated coral reef systems (Kanton would still have a minor artisanal fishery), two submerged reef systems, and several key seamounts.[81]

With respect to the Trust:

The PIPA Conservation Trust has multiple activities, including provision of financial support for the management of PIPA and paying any fee that might be required to compensate the government for demonstrated declines in national fishing revenues as a result of the PIPA closures. These activities are managed and accomplished through the mechanism of a conservation contract, which was signed by the government of Kiribati and the PIPA Conservation Trust in 2014.[82]

Another important point to highlight regarding the Trust is that:

Fundamental to the project's success was the clear understanding that the trust would have no oversight role in developing or implementing the PIPA Management Plan. While the trustees are entitled to comment on the plan and any amendments to it, their role is limited to negotiating a conservation contract with the Kiribati government, accepting or rejecting the government's management plan as part of those negotiations, and ensuring that the reverse fishing license fee and the costs of managing the PIPA are paid to the government.[83]

This all said, it is important to note that PIPA is:

the largest marine protected area in the Pacific Islands region today by covering 408,250 square kilometers, which is about the size of California; it is the largest marine UNESCO World Heritage Site; it completely protects seven of the eight islands, along with lagoons and coral reefs; it protects over 80% of PIPA's high priority habitats including shark nursery areas and globally important seabird populations; it successfully removed all invasive species from three islands; it was the first Pacific Island Marine Protected Area to contain significant deep-sea habitat; and Kiribati is a large ocean state that is setting a powerful precedent for other island nations.[84]

Challenges and How They Were Met

There were many challenges experienced. This includes complex questions that needed to be addressed, the islands' remoteness, possible ongoing illegal fishing and engagement with the fishing industry, the need for new legislation, and enforcement.

In summary, "the Kiribati economy is constrained by isolation, limited land resources, frequent droughts, geographic fragmentation and a shortage of skilled workers. In addition, population growth, increasing consumerism, and the real and imminent threats from climate change are putting increasing pressure on Kiribati's most important natural capital assets."[85]

Take for example:

Before we could make such an arrangement possible, we first needed to answer several complex questions. How much money would be needed to make up for lost fishing license fees? What were the legal mechanisms needed in Kiribati to ensure that the PIPA was duly authorized under Kiribati law and protected by a comprehensive and enforceable regulatory structure? What mechanism could guarantee that this money would be available to the government of Kiribati on a timely, predictable basis? How could those who provided the financing for the reverse fishing license – the potential conservation investors – be assured that the funds would achieve their intended purpose?[86]

Due in part to the islands' remoteness, few detailed studies had been taken:

It was not until 1972–1973 that detailed marine surveys were conducted. A comprehensive study of Kanton atoll was undertaken, including work on lagoon circulation and biogeochemistry, coral taxonomy and biogeography and lagoon and leeward reef coral distributions and assemblages. Since 2000, regular terrestrial and marine expeditions have documented the biodiversity of algae, invertebrates, fishes and corals, as well as their changes in abundance over time (including over two known episodes of high-temperature-induced coral bleaching).[87]

Additional challenges related to the remoteness includes difficult diving conditions, the need for a medical doctor onboard research or tourist vessels, and seasickness among participants.

As previously mentioned, some of the problems facing the Phoenix Islands was illegal fishing. The possible continuation of illegal fishing presents management challenges going forward:

At this time, fisheries extraction on the shallow reef environments ceased, though some illegal activity still occurred. For example, a bunker vessel from Singapore, *Hai Soon* 28, was caught and fined $4.73 million USD for conducting bunkering 20 miles south of Nikumaroro without a license. Some illegal shark finning has also likely occurred, as indirectly suggested by smaller-than-expected shark abundance and biomass post-reef closure to fishing. Overall, however, the benefits of reef closure to fishing has already been observed on monitoring trips conducted by the NEAq after the 2008 closure; a seeming resurgence of shark populations (especially small juveniles) was noted on some of the islands, especially at Kanton, which had previously been heavily fished for shark fins. Continued evidence for resurgence of juvenile sharks continued in 2012, but the marked overabundance of juveniles compared to adults could be an indication of an illegal fishing event between 2009 and 2012.[88]

Furthermore, consider:

Given their relative health prior to legal protection, it is reasonable to ask whether and why conservation measures were necessary for the Phoenix Islands. Despite their remoteness, PIPA waters are threatened by climate change, by increasing fishing pressure on declining stocks of large pelagics like tuna and by potential increase in human usage given the acute overcrowding on several islands in Kiribati, plus the general increases in global human population, affluence, and mobility, all of which have the potential to enable increased traffic and use.[89]

Another challenge concerns the fishing industry. This said, "our fisheries partners vigorously contested the loss of some of their most fertile fishing grounds. Our own citizens protested over the potential loss of much-needed revenue. Naysayers maintained that the plan simply couldn't succeed, that such a tiny nation with so few resources couldn't police and enforce a huge marine protected area."[90]

Regarding the fishing industry:

The hardest task was calculating the appropriate compensation to the Kiribati government for lost fishing revenues. Kiribati receives fishing license fees through two types of arrangements. First, the fleets of other countries {...} pay annual access fees for the right to harvest in Kiribati's territorial waters as well as in its extensive EEZ {...}. Second, Kiribati is among a group of Pacific island countries that receive payments under a 1987 fishing treaty with the US that is renegotiated every ten years {...}.

To determine an appropriate fee for the reserve fishing license, we first needed to estimate the income that fishing licenses brought in from the PIPA.

But {...} schools of fish are in constant motion. Tuna, for example, follow fluctuating food sources, currents, and ocean temperature gradients, paying no attention to territorial boundaries. As a result, the catch from any one area {...} fluctuates from year to year {...}.

A further difficulty arose because the PIPA comprises only a portion of the Phoenix Island's EEZ, and existing catch data provided no way to tell how much of the reported catch from the Phoenix Islands' EEZ came from within the smaller PIPA zone. {...} Moreover, it was impossible to determine how, or even whether, the total Kiribati catch might decline if all or part of the PIPA were placed off-limits to fishing; the commercial fleet operating outside the PIPA would likely still catch a portion of what otherwise would have been caught inside the restricted area. {...}.[91]

In addition,

The second major question faced by the PIPA partners was how to ensure the payment of annual fees covering the reserve fishing license and the new management costs to the government of Kiribati. {...} The best way to guarantee these annual payments was through an endowed trust fund dedicated to the PIPA. {...} The PIPA endowment had to be large enough to ensure that the annual investment income could cover the reserve fishing license fees and the costs of managing the protected area as well as supporting the operations of the trust fund itself.[92]

It was recognized that the existing legislation was not adequate to underpin the creation of PIPA and the Trust, and likewise, new legislation was needed. As described by Peter Shelley:

When I arrived in Tarawa, however, there was no organized system of lawbooks or regulations either in the agency I was working with or in the National Library. The only complete set of laws seemed to be in the Office of the Attorney General, where it was in heavy use. In addition, Kiribati law was a patchwork of ordinances dating from British colonial days and new laws dating from Kiribati independence in 1979. {...} Just getting a copy of a single ordinance or regulation could turn into a full-day exercise.[93]

Similarly:

PIPA was the first and is still the only formal MPA in Kiribati. Because there was no precedent for creation or management of MPAs, Kiribati legislation could not accommodate or legally recognize special status for PIPA. Consequently, MELAD {Ministry of Environment, Lands, and Agriculture Development}, in coordination with the Office of the Attorney General and the Kiribati Cabinet, undertook a major revision of Kiribati law, beginning in 2005 and culminating in the passage of the Environmental (Amendment) Act 2007, which took effect on 4 September 2007.[94]

As is often the case, it is one thing to have adequate legislation in place and it is another matter to have effective enforcement of such legislation.

Kiribati requires 100% observer coverage on all purse seine vessels legally within Kiribati waters, and observers have responsibility of ensuring legality and enforcement of in-country rules, including the PIPA closures and the reporting of illegal activities onboard. Mandatory 100% observer coverage on DWFN vessels is required for all eight signatory nations to the Third Arrangement to the Nauru Agreement, together with seven other Pacific Island states. For PIPA enforcement, this means that all legal vessels will have observer coverage and be able to be tracked via video monitoring systems in real time; these legal fishing boats can also act as an additional surveillance tool to detect illegal/non-licensed vessels. In order to assist in the detection of illegal activities, visitors and residents are also required to assist with monitoring, control and surveillance and to report any suspicious activities.

The Kiribati Government already undertakes annual patrol visits within PIPA boundaries, and currently Australia, New Zealand and France provide aerial surveillance of the Phoenix Islands area. Kiribati also has a 'ship-riders' agreement with the US whereby Kiribati enforcement personnel are present on U.S. Navy and Coast Guard vessels operating in the area and the US fleet participates with Kiribati in investigating, reporting and arresting ships for illegal activities. {…} Additionally, Kiribati has signed a Declaration on Deep-Sea Bottom Trawling to Protect Biodiversity in the High Seas, which commits the members of the Pacific Islands Forum to urgently take actions consistent with international law to prevent destructive fishing practices on seamounts in the Western Tropical Pacific Islands Area.[95]

Furthermore, it is important to note that:

Additional plans for enforcement are also underway. For example, the recently signed conservation contract between Kiribati and the PIPA Conservation Trust (April 2014) states that the Ministry of Fisheries and Marine Resources Development will submit a request to the Pacific Islands Forum Fisheries Agency to create a virtual electronic perimeter using geofence technology that conforms to the boundaries of the PIPA set forth in the PIPA Regulations 2008. Vessel monitoring and reporting of any vessel movements within PIPA are expected to continue and will be enhanced by this virtual fencing technology. Furthermore, a new Kiribati Fisheries and Wildlife Conservation Unit is planned to be stationed on Kanton Island, which is hoped to significantly contribute to surveillance and enforcement of management rules. In May 2013, a PIPA Community Agreement was signed with Kanton Island caretaker residents whereby suspicious or illegal activities must now be reported. In this agreement, all members of the Kanton community indicated their commitment to safeguarding and protecting the universal values of Kanton Island. Finally, new technologies are being explored to further aid in surveillance and enforcement of PIPA boundaries. These include satellite surveillance and unmanned drone technology, and possibilities for use are currently being scoped to determine costs, feasibility and effectiveness.[96]

Beyond Results

On one hand, the results of PIPA appear sustainable. For instance, PIPA is "now a legally established, multiple-use, large Marine Protected Area (MPA)."[97] Furthermore:

PIPA has a unique conservation strategy as the first marine MPA to use a conservation contract mechanism with a corresponding Conservation Trust established to be both a sustainable financing mechanism and a check-and-balance to the oversight and maintenance of the MPA. {…} It was {also} the first MPA to contain substantial deep-water, pelagic, and seamount habitat in addition to shallow reefs and critical terrestrial habitat for nesting seabirds.[98]

There has been an effective public education and marketing campaign to raise awareness and garner support for PIPA:

News of PIPA is now regularly broadcast on the government-owned all-Kiribati radio station, Kiribati Broadcasting and Publications Authority (AM 1440), and national pride is apparent. Signs celebrating PIPA are displayed in Kiribati International Airports on both Tarawa and Kiritimati Island, and songs, including "PIPA You Are My Gift To Humanity" have been written and are regularly sung to celebrate major PIPA milestones and events. Outreach initiatives also include messaging in schools that extend beyond PIPA issues alone, with a focus on encouraging a conservation mindset, building in-country capacity and expertise in areas relevant to ocean conservation and research, and promoting knowledge of all-of-Kiribati geography, since very few Kiribati citizens have ever been to the Phoenix Islands. The I-Kiribati also are invoking the local word 'okai' with regards to PIPA, which means a traditional storehouse where reserved foods and treasures are kept for future use – especially in times of prolonged droughts and bad times. Considering PIPA as an okai for potential food security as well as a bank of Central Pacific biodiversity has been an important part of the outreach program to enable Kiribati residents to think about the multiple local and global benefits of ocean stewardship. With all of this education, outreach, and press coverage, coupled with social media, the domestic and international public are increasingly valuing PIPA's importance both as a national treasure and as an exploratory platform for climate change.[99]

Yet, on the other hand, climate change is having a significant impact:

The reefs of the Phoenix Islands experienced an extreme El Niño event in 2002-3, the worst so far recorded, which killed over 60 percent of the coral. Though tragic, this event provided scientists the opportunity to measure reef recovery. Most reefs suffering from catastrophic bleaching show few signs of major recovery a decade later, but thus far, the reefs of the Phoenix Islands appear to possess remarkable potential to recover. As of 2009, they appeared well on their way to recover from the catastrophic warming event.[100]

There are also the potential future impacts of tourism:

PIPA also has ecotourism potential that could benefit the Kiribati economy, though reservations have been expressed about the long-term economic viability and sustainability of tourism development in the Phoenix Islands given its isolation and remoteness. A growing niche tourism market might generate enough revenue to support the necessary infrastructure, as there are several potential niche markets. {...} Sadly, atolls in such an untouched state are likely to become increasingly scarce elsewhere notwithstanding the growing efforts to protect them, simply because of the unrelenting challenges of population growth and coastal development.[101]

The "elimination of commercial fishing may eliminate some risk of vessel grounding due to decreased vessel traffic, though increased tourism traffic may pose even higher threats as they will likely spend more time near islands (instead of in blue water habitats)."[102]

Yet, the leadership of Kiribati could have a strong and lasting impact on others:

In 2008 President Anote Tong of Kiribati presented a concept to fellow member nations of the Pacific Islands forum, promising a 'Pacific Oceanscape' initiative to protect the world's largest ocean. In 2010 the fifteen-member nations' leaders, including President Tong, met in Vanuatu and endorsed the Pacific Oceanscape Framework. Covering nearly 25 million square miles (40 million km^2) of ocean and island ecosystems – the size of Canada, the United States, and Mexico combined – the Pacific Oceanscape unites Pacific island states in a survival campaign for ocean conservation and management in the twenty-first century.[103]

Likewise, "following the inception of PIPA, a number of other large MPAs have been created, including several in the Pacific. This 'race to protect the oceans' is launching a new era of ocean protection and management."[104] It should be noted that "a recent meta analysis of 87 MPAs found that conservation benefits increased exponentially with the accumulation of five attributes: no take, well enforced, old (>10 years), large (<100 km^2) and isolated by deep water or sand. PIPA will soon meet three of these criteria (no-take, large and isolated). As of 2018, PIPA will formally be a decade old. However, enforcement remains a challenge due to the remoteness of the area {...}."[105]

Lessons Learned

Some of the lessons learned will be how some of the following questions get answered:

> Globally, however, success will likely be measurable only with time: do stocks of commercially important fishes within PIPA improve? Will PIPA be effectively enforced and maintained? Will the sustainable financing model be successful in the long term? Will PIPA be heralded as a model of marine conservation? And finally, even with all of the protections afforded to PIPA to mitigate habitat loss and overfishing, will the ecosystems succumb to the inevitable impacts of climate change, and even if so, will PIPA ecosystems have at least fared better than their unprotected neighbours as a result of its MPA status? These and many other questions remain to be answered in the coming decades.[106]

Other Resources on the Phoenix Islands Protected Area Conservation Trust

Anote Tong's TED Talks Page

- https://www.ted.com/speakers/anote_tong

Big Ocean

- https://bigoceanmanagers.org/

CI's Page on the Phoenix Islands Protected Area

- https://www.conservation.org/projects/Pages/phoenix-islands-protected-area.aspx

Kiribati Environment and Conservation Division of the Ministry of Environment, Lands and Agricultural Development

- http://www.environment.gov.ki/

NAI'A Cruises

- https://www.naia.com.fj/conservation-discovery/phoenix-islands/

New England Aquarium

- https://www.neaq.org/;
- https://www.neaq.org/tag/phoenix-islands-protected-area/; and
- http://pipa.neaq.org/

Obura, D., et al. Living Document. Phoenix Islands Protected Area Climate Change Vulnerability Assessment and Management, Report to the New England Aquarium, Boston, USA. 35 pp. Updated January 18, 2016.

- http://www.phoenixislands.org/pdf/PIPA-CC-scoping-study-Jan-18-2016.pdf

Phoenix Islands Protected Area

- http://www.phoenixislands.org/index.php

Presidential Web Portal for the Republic of Kiribati

- http://www.president.gov.ki/

Rotjan et al. "Establishment, Management, and Maintenance of the Phoenix Islands Protected Area." *Advances in Marine Biology*. Vol. 69. 2014. ISSN 0065-2881.

- http://dx.doi.org/10.1016.B978-0-12-800214-8.00008-6

Secretariat for the Pacific Community (SPC)

- https://www.spc.int/

Secretariat for the Pacific Regional Environmental Programme (SPREP)

- https://www.sprep.org/

Stone, Gregory S. and David Obura (Eds.). *Underwater Eden: Saving the Last Coral Wilderness on Earth.* Chicago: The University of Chicago Press, 2013.

Case Study #3: The Pez Maya Land Purchase

Introduction

Similar to the Palmyra Atoll Purchase, the Pez May Land Purchase demonstrates there is a role for onshore land conservation easements to assist with the protection of offshore coral reefs.

In 2001, the Mexican nonprofit organization Amigos de Sian Ka'an and The Nature Conservancy (TNC) helped structure the purchase of Pez Maya, a 26-hectare (64 acres) property with 2.4 km (1.5 miles) of beachfront and adjacent to the Sian Ka'an Biosphere Reserve ("Reserve"), a World Heritage Site (Figs. 8.3 and 8.4).

Prior to the purchase, the Pez Maya property was a fly-fishing lodge. The lodge ran into financial problems, Banco Mexicano de Comercio Exterior acquired the property, and allowed Amigos de Sian Ka'an to take tours through the location. Eventually, the idea of buying the property was proposed, the bank was interested in the sale, and the parties came to an agreement.[107]

In the end, the Pez Maya property was purchased for USD$1.8 million to USD$2.7 million, with a contribution of USD$325,000 to USD$400,000 from the Maine chapter of TNC,[108] along with approximately USD$100,000 from the Gillette Company and approximately USD$250,000 from the UN Foundation.[109]

Today, the Pez Maya property is privately owned by Amigos de Sian Ka'an and TNC holds the conservation easement on the property. A Trust was established, with both TNC and Amigos de Sian Ka'an acting as trustees, and the Trust is the legal mechanism to carry out the commitment of conserving the land in perpetuity.[110]

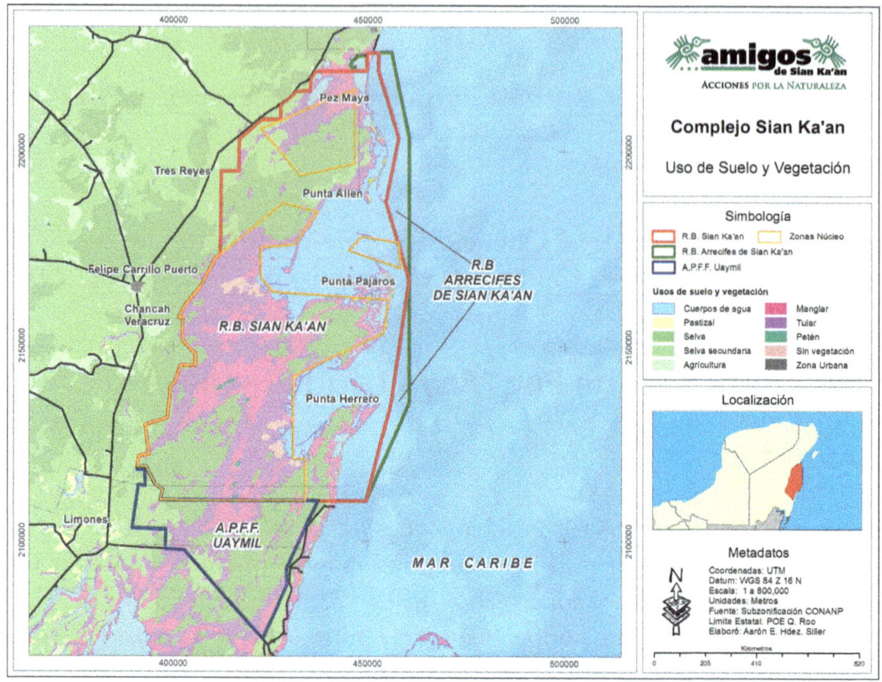

Fig. 8.3 Map of Sian Ka'an Biosphere Reserve and Pez Maya (*Credit* Amigos de Sian Ka'an A.C.)

The Sian Ka'an Biosphere Reserve is located in the Southern Yucatan Peninsula in Quintana Roo, Mexico. The Reserve is 528,000 hectares (1.3 million acres). This overall size includes approximately 120,000 hectares (296,400 acres) of marine area, including a portion of the Mesoamerican Barrier Reef—the world's second largest barrier reef—and 408,000 hectares (approximately one million acres) of tropical rainforests and palm savannas.[111] However, based on studies from Amigos de Sian Ka'an, the Sian Ka'an Biosphere Reserve was expanded in 1994 with the declaration of the Uaymil Flora and Fauna Protection Area, and in 1998, with the Sian Ka'an Reefs Biosphere Reserve. The three protected area complexes now have 652,000 hectares (1.6 million acres).

The Reserve was established with a Mexican Presidential Decree on January 20, 1986 and subsequently, Amigos de Sian Ka'an was founded as a private organization in 1986 by a group of citizens that had promoted the creation of Sian Ka'an. While the Reserve was declared a public protected area operated by the Mexican Government, Amigos de Sian Ka'an originally started working in the Reserve to assist the federal government with management and administration of the Reserve.[112] Through the years,

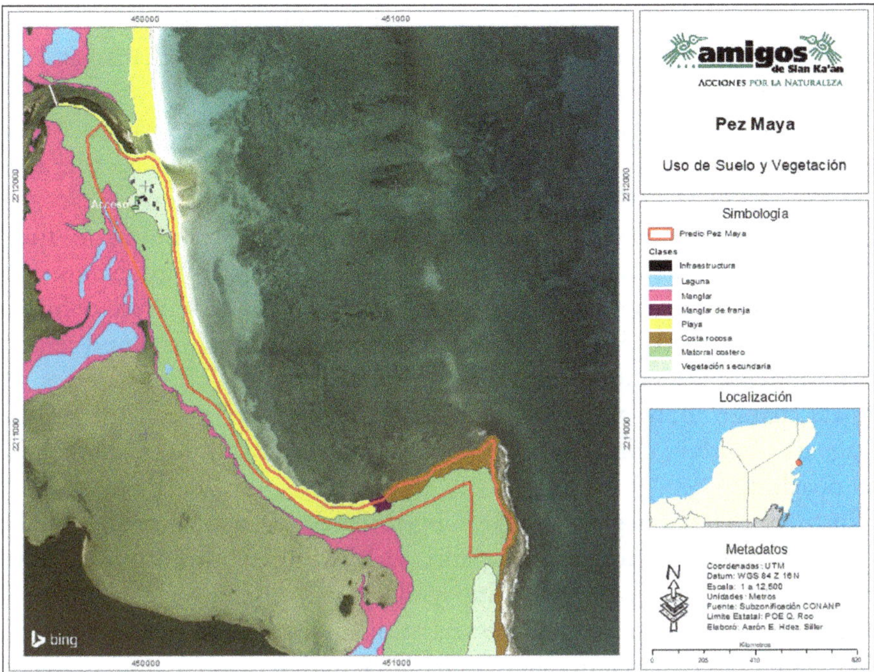

Fig. 8.4 Close Up Map of Pez Maya (*Credit* Amigos de Sian Ka'an A.C.)

the Mexican Government consolidated its institutions and established the Comisión Nacional de Áreas Naturales Protegidas (CONANP; The National Commission of Protected Natural Areas), which now manages the Sian Ka'an Biosphere Reserve.[113]

Identify the Problem

Prior to the purchase of the Pez Maya property, the area was well protected. There was a small impact from fishing, along with some minor illegal fishing, but the fisherfolk in Sian Ka'an are well organized, and there were no problems associated with forest fires.[114]

At the time, one of the major potential problems was that most of the coast in the Reserve is privately owned and some construction is allowed. This said, "Pez Maya lies 15 miles south of Tulum, and several developers {had} tried to acquire the property. With zoning designations as small as 50 meters, Pez Maya's shoreline had been threatened to be subdivided into as many as 60 parcels. The purchase will allow the conservationists to establish a barrier between the bioreserve and hotel development."[115] It is important to recognize that "{…} while there are strict, low-density zoning rules in place, officials acknowledge that such codes in Mexico often change and that perpetual protection was needed."[116]

As explained by Amigos de Sian Ka'an, when the Reserve was first established, there were not that many problems in the region. There were more potential, hypothetical problems, rather than actual problems, because the Reserve was more of a visionary project that was responding to future problems, rather than reactionary and responding to current problems.

Throughout Quintana Roo State, the main problems today are deforestation as a result of urban sprawl, water pollution, forest fires, and agriculture.[117] Such deforestation leads to a loss of freshwater, GHG emissions, and climate change, along with the threatened extinction of several species such as jaguars (*Panthera onca*) and ocelots (*Leopardus pardalis*) due to habitat loss and poaching.[118]

Furthermore, the region is beginning to face more global problems. The two main problems are disease, particularly white syndrome disease, and *Sargassum*. Some believe the white syndrome disease, which is impacting corals, is infectious and some believe the disease is related to the *Sargassum* issue. The *Sargassum* issue started in 2015 and is now a huge problem. Whether due to increased sea temperatures that are leading to increased algae growth or increased nutrients from agricultural and sewage runoff, there are literally millions of tons of *Sargassum*, a type of seaweed, from the Sargasso Sea arriving on Caribbean shores, including Mexico.[119] As explained by Mr. Kenneth Johnson, founder of the Mexico-based ecotourism company EcoColors, in 2018, approximately 170,000 tons of *Sargassum*—mostly in the summer, not so much in the winter—hit the shores of the Yucatan Peninsula. In 2019, authorities started monitoring the situation and were estimating 850,000 tons to one million tons would arrive.[120]

Additional big problems within the Reserve are pollution, particularly water pollution[121] and the trash that is washed up on beaches.[122]

Why the Problem Is Important

The problems of potential development in the past, along with the current problems of *Sargassum* and white syndrome disease, are important for several reasons.

First, the overall Reserve is important because it provides a refuge for a tremendous amount of terrestrial and marine biodiversity, and the Reserve provides a lot of connectivity. The Pez Maya coastal property "serves as critical wintering habitat for Maine birds,"[123] and hence the involvement of the Maine Chapter of TNC.

In general, the hydrological system throughout the region, including the Reserve, connects the tropical forests, wetlands, and the coral reefs together via the aquifer. The coral reefs offshore of Pez Maya and the Reserve are a part of the Mesoamerican Barrier Reef, the second largest barrier reef in the world, which stretches across Honduras, Guatemala, Belize, and Mexico. More specifically, Pez Maya provides an important ecological function as it is on the mouth of a river (Boca Paila) that connects the northern wetlands, the Boca Paila Lagoon, and the coral reefs offshore.[124]

The white syndrome disease could be infectious, and likewise, could spread to other parts of the Mesoamerican Barrier Reef.

With respect to *Sargassum* in the Atlantic Ocean, there are two main pelagic species known as *Sargassum fluitans* and *Sargassum natans*.[125] While *Sargassum* provides some ecosystem services such as an important habitat with shade and food for small animals (such as baby sea turtles), and *Sargassum* can also help support sand dunes and shoreline stabilization, "too much of a good thing is a bad thing."[126] Likewise, the *Sargassum* problem is important for several reasons. First, whether due to ocean temperatures being too high, upwelling off West Africa, dust disposition, or nutrient outflows from the Amazon River, the *Sargassum* blooms and inundations have increased. A recurrent Great Atlantic Sargassum Belt was discovered and the *Sargassum* inundation in the Caribbean and Florida are connected to this Belt. These *Sargassum* blooms and inundations can smother turtle nesting sites, causing turtle and fish mortality, attract insects, and can have an overall negative impact on tourism and the economy.[127] Similarly, *Sargassum* stinks terribly bad and people cannot, or refuse, to swim in areas with lots of *Sargassum*. *Sargassum* is also contributing to the creation of dead zones and is having a negative impact on the health of coral reefs.[128] The *Sargassum* problem is so bad at times that hotels in Tulum (i.e., mainly smaller, more expensive boutique hotels) that were normally charging USD$800 a night, were offering rooms at USD$40 a night because there was so much *Sargassum*. When the *Sargassum* is in the ocean, there is no problem, as it is alive and does not smell. However, when the *Sargassum* hits the shore, it decomposes fast and smells bad. Around Cancun, the problem is not too bad; however, the *Sargassum* situation is very bad around Tulum and Puerto Morelos.[129]

The problems are important because the Reserve also provides some important income to local communities. Furthermore, the Reserve is listed as a World Heritage Marine Programme.

How Problem Was Identified

In 2015, everyone started noticing the accumulation on the beach of *Sargassum*. The current levels of *Sargassum*, starting around approximately 2018, have reached unprecedented levels.[130]

The white syndrome disease, as it is known in Mexico, is also known as the stony coral tissue loss disease in the United States and throughout the Caribbean. This disease appears to have started in Biscayne Bay off Miami, Florida and was first identified shortly after the port was dredged. In addition, sewage is pumped offshore in this region and it was found, around this same time, that the sewage lines were leaking. Furthermore, the reefs had bleached that year so it could be one, or a combination of, these factors that created the white syndrome disease. Researchers believe it is a bacterium, as opposed to a virus, as the disease responds to antibiotics. Researchers also know that the disease is waterborne and can be transmitted through contact (i.e., or through water), but the actual pathogens have not been identified.[131]

Effectiveness of Process for Identifying Problem

In general, according to Amigos de Sian Ka'an, the strategies targeting the Reserve have been effective. The strategies are part of the institutional framework and allow for integrating new approaches and help identify threats to the region. In addition, mapping is a very useful tool as you can map just about everything.[132]

Steps Taken to Address the Problem

The problem with the *Sargassum* is the same all over the Caribbean and everyone is shocked by this relatively new problem. There was an international meeting in Cancun to discuss the *Sargassum* problem.[133]

As explained by Amigos de Sian Ka'an, it is always difficult to manage land. However, Amigos de Sian Ka'an has a small endowment with TNC. In addition, for approximately twelve years, there was an alliance with Global Vision International (GVI) and a lot of coral reef data was collected by GVI volunteers. As of June 2019, there is not much activity at Pez Maya and there is no longer a permanent presence. Despite this and although the data had some limitations, there were some interesting patterns identified. This said, over the last few years—especially because of the white syndrome disease—the patterns have been discouraging.[134]

As explained by Mr. Kenneth Johnson, where there are problems, there are opportunities. There is a local company starting to make bricks from *Sargassum* and some people have started to make animal feed, such as for chickens, from the *Sargassum*. However, you really need to get the *Sargassum* fresh from the ocean and it is hard to raise financing for the associated boats and equipment. Then again, if you can reach scale, you could generate a lot of animal feed and construction materials, and possibly bioenergy, from the *Sargassum*.[135]

Results

There are numerous results attributable to the Pez Maya land purchase. This includes:

- The funds of approximately USD$1.8 million to USD$2.7 million were raised by Amigos de Sian Ka'an and TNC and the land was acquired.[136,137]
- A perpetual conservation easement was placed on the Pez Maya property and to date, the property has been preserved since 2001 and there have been no major changes in the ecosystems.[138]
- Pez Maya has one of the largest databases on coral reef conditions for the region. There were about 12 to 13 years of ongoing coral reef data collected with assistance from volunteers with GVI, originating from the early 1990s. More recently, Amigos de Sian Ka'an has ended its partnership with GVI and partnered with the National Technical Institute of Mexico to install a tower in the Pez Maya property to monitor the weather. Such monitoring includes temperatures, wind speed, wind direction, precipitation, etc. This information is constantly received and will provide interesting and useful data for climate change research. In the future, Amigos de Sian Ka'an would like to install tide monitors and measure changes in sea level. The idea is to start gathering information on climate change, how climate is changing in this particular area, what is the effect on the ecosystems, understand what is going on with sea level, and explore how sea level change relates to temperature changes.[139]
- Another result is the modest infrastructure, which includes a small old lodge, is still in place. Amigos de Sian Ka'an are getting a little funding to improve these areas, and in the future, the long-term goal is to replace the current infrastructure with a new laboratory and with small buildings.[140]

There are numerous results achieved at the Sian Ka'an Biosphere Reserve. This includes the fact that the Reserve was designated a World Heritage Site in 1987[141] and 652,193 hectares (1.61 million acres) have been conserved between the Sian Ka'an Biosphere Reserve, the Uaymil Flora and Fauna Protection Area, and the Sian Ka'an Reefs Biosphere Reserve.[142]

The first Permanent Beach Cleaning Program was established at the Reserve and within the first two years, 29 tons (nearly 65,000 pounds) of waste was collected and removed from the Reserve.[143]

More recently, by studying the region's aquifer, Amigos de Sian Ka'an is helping to change paradigms. In addition to changing the hydrology models of the peninsula and changing how people see the aquifer, the results are also helping to promote development policies, planning tools, and policy actions.[144]

Challenges and How They Were Met

Although the Trust allows Amigos de Sian Ka'an to build a facility on the same building footprint, one challenge has been to raise the required funding to preserve the facilities at Pez Maya and to raise the required funding to build new facilities.[145]

The management plan for the coastline of the Reserve limits the average density of construction to 0.5 hotel rooms per hectares, which means approximately 800 hotel rooms along the 120 km of coastline. Although a challenge, all of the private landowners have agreed to this regulation.[146]

Another big challenge is water pollution. Water pollution is a growing problem because many cities throughout the region are expanding without proper wastewater treatment facilities. To address this challenge, many organizations—including Amigos de Sian Ka'an—are investing a lot of time and funds into studying water and the aquifer.[147]

Beyond Results

The results of the Pez Maya purchase are sustainable because the purchase allows the participants to guarantee the conservation of approximately 1.5 miles (2.4 km) of coastline and prevent its subdivision into smaller plots, and then subsequent development. Similarly, since 2001, the property has garnered legal protection. Furthermore, a similar purchase was also undertaken by Amigos de Sian Ka'an in 2011 near Cancun. However, with respect to Pez Maya, the ongoing management and financial sustainability have been a challenge. Amigos de Sian Ka'an, along with TNC, has a small endowment for ongoing management, but it has been difficult to raise funds in order to increase the endowment. This increase in the endowment would be to specifically cover the daily, operational expenses. Yet, the Pez Maya property has provided an operational base to facilitate the long-term monitoring of coral reef health and climate change and these activities have been sustained over the years.[148]

Lessons Learned

With more than 25 years of experience, including 12+ years as the Executive Director, Mr. Gonzalo Merediz Alonso of Amigos de Sian Ka'an offers a few important lessons. Amigos de Sian Ka'an has made sure the Pez Maya experience is shared through a variety of means including reports and workshops. It is important to understand that timing and valuation is critical. At the time, in 2001, it was possible to purchase the Pez Maya because it was affordable. Nowadays, it would not be realistic to purchase a similar, 26-hectare (64 acres) property with 2.4 km (1.5 miles) of beachfront in the Mexican Caribbean, due to the land being very expensive. It was a strategic purchase, but in general, Amigos de Sian Ka'an has a long-term strategy of promoting public protected areas. In addition, Amigos de Sian Ka'an has a current strategy that involves instead of purchasing private land, the organization directly works with private landowners to do things on their own land.[149]

In addition, having an NGO involved with all the projects, especially because NGOs can provide continuity, is invaluable. Establishing protected areas and all the associated conservation activities are long-term processes because it takes a lot of time to wait for the proper political moment, the right social mood, and sufficient financing opportunities—to have all of these three elements aligned is even more difficult. For instance, in Mexico, the municipal government changes every three years and the federal government changes every six years. NGOs provide continuity through (or despite) government change. If an NGO is unable to establish a protected area under a federal government, the NGO is able to adapt and change until the next administration when the possibility of establishing a protected area may arise. Some people complain about having to start all over with a new administration, but NGOs can work behind the scenes on other opportunities.

Another interesting observation made by Mr. Kenneth Johnson, owner of EcoColors, an ecotourism company based in Mexico, is that the cenotes and underground waters of Mexico were enough to cool off the coral reefs to some degree, where only 40% bleaching took place, as opposed to 90% bleaching that occurred elsewhere.[150]

Other Resources on Pez Maya Land Purchase

Amigos de Sian Ka'an

- http://www.amigosdesiankaan.org/en/

EcoColors' Tours to Sian Ka'an

- https://www.ecotravelmexico.com/

Global Vision International (GVI)

- https://www.gviusa.com/volunteer-in-mexico/; and
- https://gvi.org/turtle-monitoring-pez-maya/

Project Aware

- https://www.projectaware.org/diver/pez-maya

The Sargassum Watch System (SaWS)

- https://optics.marine.usf.edu/projects/saws.html

TNC's Valuing Nature in Mexico

- https://www.tncmx.org/en/home/; and
- http://www.nature.org/ourinitiatives/regions/northamerica/mexico/explore/valuing-nature-english.pdf

World Heritage Convention's Site on Sian Ka'an

- http://whc.unesco.org/en/list/410

Case Study #4: Chumbe Island Coral Park Ltd.

Introduction

The Chumbe Island Coral Park Ltd. (CHICOP) is located off the coast of Tanzania in East Africa. While CHICOP could be considered ecotourism, the Park involved a unique structure known as a "Privately Protected Area" (PPA), the first such marine PPA in the world, and is based on a "Marine Conservation Agreement" (MCA) (Fig. 8.5).[151]
   Furthermore:

CHICOP is a privately funded and managed reef conservation project on Chumbe Island, a small island 13 km off Zanzibar's west coast. The company was established specifically to create and manage the Chumbe Coral Reef Sanctuary. The site was initially chosen by investors for private conservation based on a number of favorable factors: formally owned by the military, the island was uninhabited and relatively undisturbed; fishing on the western side of the island was already closed from proximity to a shipping channel; and, few fishers could afford the cost of the outboard motor required to reach the island, thus ensuring that traditional fishers would not be displaced by conservation efforts. The Zanzibar Government approved the CHICOP project in 1992 and designated the Chumbe Reef Sanctuary in 1994 {…}.

   To fund its conservation activities, CHICOP established an eco-tourism concession on the island, including a visitor's center and 10 guest bungalows. Financing for park operating costs was to be generated through diving and snorkeling fees, glass-bottomed boat trips, nature trails, accommodation, and restaurant services. The estimated internal rate of return was 27 percent, with a capital payback after three years. Although originally estimated to cost about $200,000 to establish the main facilities, bureaucratic delays and high government fees increased the project's cost to over $1 million. To cover the higher costs, CHICOP shifted its target tourist clientele to an upscale market. A revised feasibility study indicated an internal rate of return of nine percent and a capital payback period of seven years. Currently, income from the ecotourism concession covers nearly all park running costs {…}.[152]

Identify the Problem

One problem facing the reefs surrounding CHICOP is destructive fishing, such as the use of trawling nets, the use of poisons (such as cyanide), and dynamite fishing. This said, the use of dynamite "destroys everything," including eggs, seagrass, and coral.

Fig. 8.5   Map of Zanzibar (*Credit* Rainer Lesniewski)

Another problem is the need for education. As explained by Sibylle Riedmiller, people, including people from Zanzibar, need to understand that corals are living things and not just rocks.[153]

Furthermore, "you get a coral trade, and of course like the rest of the world, over-development, pollution and coral bleaching. But the biggest challenges were dynamite fishing and over-fishing. As a result, Tanzania coral reefs were in dire straits."[154]

Why the Problem Is Important

The problems are important, in large part, because CHICOP serves as a valuable education center. As stated by Mr. Khamis Khafan Juma, environmental educator and assistant head ranger at CHICOP, the "main goal of environmental education is to create environmental awareness to the community and students, and to have a sustainable living, to value the nature, and to have a sustainable development in the future – in the classroom, we teach about the general concept of the environment, coral reef ecology, and the concept of ecotourism."[155]

This said, CHICOP was started in 1991 as an ecoresort with an education and conservation focus. This included, among many things, intertidal work, helping people practice snorkeling, guiding forest walks, and utilizing an ecological design to reduce its ecological impact.

Educational talks focused on:

- How the island was formed (i.e., fossilized corals on the land indicate the island was underwater in the past);
- How plants adapted to island life (i.e., cactus leaves are thick to store water);
- The different species found in the forest (balboa tree, cactus, mangrove forests, hermit and coconut crabs, antelope, and African fish eagle). For instance, the Balboa fruit can be used to make snacks which have 6x more vitamin C than oranges; and
- Students get to see coral, fish, lobster, sea urchins, sea cucumber, sponges, etc. and the interactions between the corals and these other organisms.[156]

There are many marine reserves throughout Kenya and Tanzania including:

- Kiunga Marine National Reserve;
- Malindi-Watamu Marine National Park;
- Mombasa Marine Park and Reserve;
- Diani/Chale Marine National Park and Reserve;
- Kisite-Mpunguti Marine National Park and Reserve; and
- Chumbe Reef Sanctuary Marine Protected Area.

Furthermore, Tanzania has several sites listed among the top fifty bioclimatic units, and Tanzania is one of seven key countries identified by the Coral Reef Rescue Initiative.

How Problem Was Identified

The problems facing CHICOP, and the surrounding reefs, were in part identified by the founder Sibylle Riedmiller, through government negotiations, and through extensive village meetings.[157]

Effectiveness of Process for Identifying Problem

The process for identifying and addressing the aforementioned problems appears to be effective.[158]

Steps Taken to Address the Problem

There were numerous steps taken to address the problem, which includes:

- There were approximately four years of negotiations with the Tanzania Government, which was one of the hardest parts according to Sibylle Riedmiller;
- Ongoing village meetings. This included meetings to get fishers onboard as park rangers and on-the-job training by volunteers, along with getting rangers to educate fishers. It is important to note that rangers could, because they were/are fishers themselves; and
- There were environmental education programs and advisory committees established.[159]

Results

There are many notable results from the CHICOP. This includes:

- The Park was the first MPA in Tanzania and is the first private marine park in the world. Likewise, the Park was privately established and privately funded with no government subsidies and with only ~14 tourists per day;
- The Park is collectively managed as both a marine park and as a forest reserve;

- Management plans were developed from 1995 to 2016;
- After twenty years of protection, there are more than 400 fish species, resident sea turtles, and more than 90% of all coral species represented in East Africa are found in the waters of the Park; note, this is just on a relatively small, one kilometer by 300-meter-wide stretch of corals; and
- The seven bungalows present at the Park, and which finance the Park, are utilizing sustainable technologies such as vegetative grey water filtration and no flush water toilets.[160]

The Park "has the only large scale marine education program in East Africa. The program is beneficial for girls, who often have fewer chances to access the sea. Chumbe provides a safe and culturally appropriate way for all students to snorkel. The education program has reached over 5,000 students and 850 teachers by 2010."[161]

Another important result is that the concepts of both Privately Protected Areas (PPAs) and Marine Conservation Agreements (MCAs) "have been acknowledged, analysed and documented by IUCN and the international conservation community."[162]

PPAs as "are now increasingly recognized by IUCN following the landmark IUCN-Resolution 36 on 'Supporting privately protected areas' In this Resolution, PPAs are defined as 'a protected area as defined by IUCN, under private governance, i.e. individuals and groups of individuals; non-governmental organisations; corporations – both existing commercial companies and sometimes corporations set up by groups of private owners to manage groups of Privately Protected Areas; for-profit owners; research entities (e.g. universities, field stations) or religious entities.' The Resolution 36 makes a strong case for giving PPAs much more attention and recognition, for their actual and potential contribution to biodiversity conservation on the continent."[163]

The experiences of CHICOP suggest that private "Marine Protected Areas may be attractive for private investment as they have a great potential for win-win partnerships with local fishers, for the well-known spillover effect of No-take zones, and the absence of human-wildlife conflicts, which makes them less risky than terrestrial PPAs."[164]

MCAs for private marine PPAs are "commonly created by contractual agreements of private investors with governments and/or local communities, based on so-called Marine Conservation Agreements (MCAs) for the purpose of conservation. Pioneering conceptual work on such Marine PPAs has been done by TNC and CI."[165]

MCAs are defined as "Any formal or informal contractual arrangement that aims to achieve ocean or coastal conservation goals in which one or more parties (usually right-holders) voluntarily commit to taking certain actions, refraining from certain actions, or transferring certain rights and responsibilities in exchange for one or more other parties (usually conservation-oriented entities) voluntarily committing to deliver explicit (direct or indirect) economic incentives."

The CHICOP was specifically mentioned in the UN Secretary Report for General Assembly for Rio+ 20.[166] Furthermore, Sibylle Riedmiller was "awarded the Order of Merit Cross from the Federal Republic of Germany for her tireless commitment to conserving Tanzania's natural heritage."[167]

Challenges and How They Were Met

According to Sibylle Riedmiller's TEDxSeaPoint on Chumbe Island, there were many challenges:

- 1. One of the biggest challenges is dynamite fishing. "It is big, big problem with big groups (i.e., organized crime) getting local fishermen behind it;"
- 2. Another big, related challenge is overfishing. For instance, a manta ray is one meal for a few people only, while a living manta ray can generate income from tourism for years;
- 3. Most MPAs are paper parks. You have governance issues, there are central bureaucracies and vested interests, and MPAs may crowd out local initiatives (i.e., outsiders are given ranger positions, who are not local fisherfolk and who may not speak local language, etc.);
- 4. Invasive species such as the Indian house crow (*Corvus splendens*), crown-of-thorns starfish (*Acanthaster planci*), the black rat (*Rattus rattus*) and sea urchins are a threat. Chumbe Island, to counter this threat, had to physically take out literally thousands of the crown-of-thorns starfish; and
- 5. The costs were three times more than planned, which led to CHICOP having to increase rates (which was also a challenge). This said, if visitors are spending a lot of money, do they really want to have school children running around the island? In fact, "yes, they love it – education actually goes along very well with upmarket (i.e., high end tourism)."[168]

Beyond Results

Terrestrial parks are seen as "anti-people," they sometimes exclude local people (i.e., local people are taken out of parks), there is a "spillover" of wild animals (i.e., which can be life-threatening as wild lions may attack children or wild elephants may trample crops), there are lots of human–wildlife conflicts (e.g., local people not allowed to kill wildlife), and there is little compensation for local people in the event of such conflicts. Marine parks have an advantage over terrestrial parks. While "no-take" marine parks exclude local people, they do create fish nurseries and this restocks neighboring, adjacent reefs which have a positive "spillover" effect and directly benefits fishers with more fish.[169]

There are a lot of community benefits incorporated, which shall help with the future sustainability of CHICOP:

> In addition to conservation benefits, CHICOP was designed to provide significant local community benefits. CHICOP used a capacity building approach to train fishers from neighboring villages as park rangers to monitor and patrol the reef. The park rangers proved instrumental in raising awareness of marine conservation in their communities and in generating a positive local response to the sanctuary. CHICOP also runs an education program for children in which schools throughout Zanzibar run trips to the island to visit the park and learn about marine conservation. Chumbe hosts volunteer biologists, educators, and students who perform research in the Island's waters and provide continued local training in marine conservation techniques.[170]

Similarly, women and girls are given the opportunity to learn how to swim and snorkel, which otherwise would not be provided in the traditional, government-funded educational system.[171]

Other aspects of CHICOP's sustainability include the ecological design of the bungalows which includes: composting toilets; rainwater harvesting with gravel and sand filtration; solar heating for hot water in the bathrooms; and solar photovoltaic panels for electricity generation.[172]

Furthermore, TNC and CI have developed comprehensive Guidelines for MCAs and a Toolkit that presents case studies from around the world (i.e., including CHICOP). It should also be noted that "{…} East Africa possesses a remarkably well-developed system of such {marine} reserves that is among the best studied in the world."[173]

Lessons Learned

Although more focused on Kenya, the following lessons learned appear to also be applicable to nearby Tanzania:

> Among the lessons learned from this research to date are: (1) fishing impacts appear to be the primary driver behind both major declines in East African nearshore fish communities and serious degradation of coral reef ecosystems; (2) certain large fish, including keystone predators and herbivores, are especially vulnerable to fishing pressure, and their removal can profoundly alter urchin populations and degrade reefs; (3) Kenya's no-take marine national parks have been remarkably successful in protecting, restoring, and reversing declines with respect to both fish and coral reefs; (4) no-take marine reserves are likely the best and may be the only way to protect Kenya's rare and vulnerable fish species and precious coral reef systems; (5) protection from fishing impacts also provides resistance and resilience to other stresses; and (6) marine reserves can provide real fisheries, tourism, and overall economic benefits, but ensuring or optimizing these may be complicated and design-dependent.[174]

According to Sibylle Riedmiller's TEDxSeaPoint on Chumbe Island, some of the many lessons learned include:

- 1. Red tape and corruption can be big issues;
- 2. Eco-technology was more expensive than thought and even to keep it running (i.e., like solar photovoltaic panels) is a major challenge;
- 3. CHICOP had to go to the upmarket tourism sector; and
- 4. It is important to note that tourism can be volatile; if you are frugal in your management, you can survive (i.e., Chumbe can be at about 40% occupancy). There were slumps along the way, but Chumbe is the #1 hotel in Zanzibar according to TripAdvisor.[175]

Other Resources on Chumbe Island Coral Park Ltd.

Chumbe Island Coral Park Ltd.

- http://www.chumbeisland.com/;
- https://chumbeisland.com/about-chumbe/downloads-links/; and
- http://www.facebook.com/pages/Chumbe-Island-Coral-Park/119009334 818409

Earth Changers

- https://www.earth-changers.com/sustainable-development/sibylle-riedmiller-chumbe-island

Hotel Association of Tanzania (HAT)

- www.hat-tz.org

IUCN World Commission on Protected Areas' (WCPA) Guidelines for Privately Protected Areas

- https://portals.iucn.org/library/node/47916

Reef Resilience Network's Marine Conservation Agreements

- http://reefresilience.org/management-strategies/marine-conservation-agreements/

Sibylle Riedmiller's TEDx Talk

- https://www.youtube.com/watch?time_continue=42&v=yWldTssTRBA

TNC and CI: Practitioner's Field Guide for Marine Conservation Agreements: Best Practices for Integrating Rights-based Incentive Agreements into Ocean and Coastal Conservation Efforts

- http://www.reefresilience.org/wp-content/uploads/Practitioners_Field_Guide_for_Marine_Conservation_Agreements_2012.pdf

Financial Analysis

The following financial analysis will look at return and risk.

Return

There are numerous returns—both financial and non-financial—from conservation easements. With traditional land conservation easements, the landowner, at least in the United States, can generate state and/or federal tax savings by donating their land to an eligible nonprofit (e.g., local land trust, the Trust for Public Land, and/or TNC) or by donating the conservation easement. Alternatively, the landowner may generate a financial return by undertaking a limited development project.

There are also numerous non-financial returns. For example, with the onetime or staggered cost of purchasing an easement, the ecosystem services provided may include the protection of a stretch of coral reef and its storm protection attributes and protect an important fish nursery. Conservation easements can also provide an educational opportunity for schools and recreation opportunities for local communities. Furthermore, food security can be enhanced through preserving fish nurseries and preventing its conversion to an alternative use such as a port for cruise liners.

There are many costs associated with purchasing a conservation easement. This includes, but is not limited to, the costs of:

- 1. Obtaining a credit report and purchasing insurance;
- 2. Appraisal of the ecosystem service values;
- 3. Legal expenses, including contract review and environmental quality assessment (e.g., presence of hazardous materials or invasive species, and/or water quality issues); and
- 4. Site visits.

Risk

There are a variety of risks associated with identifying areas for possible acquisition via a conservation easement, along with structuring the purchase and maintaining the easement in perpetuity.

Business Risk

There are numerous business risks associated with purchasing, maintaining, and monitoring a conservation easement. These business risks include:

- Raising sufficient funds in time at a competitive rate; and
- Presence, or later identification, of infestations of invasive species and/or coral disease.

Conservation easement insurance, if available, is a mechanism to mitigate business risk.

Strategic Risk

When selecting a particular seascape for a possible conservation easement, some of the strategic risks that should be assessed include:

- The extent of restoration work to be done. This could include little-to-no restoration work to more extensive work with respect to removing piers, cleaning up marine waste, and/or eradicating invasive species;
- Availability of additional, adjacent parcels to scale up the impact;
- Intergenerational equity, which takes place when the next generation assumes ownership and they may possibly seek to unwind or ignore the conservation easement; and
- Managing the relatively few cash inflows (e.g., a one time, upfront payment for the easement) with the relatively more frequent cash outflows (e.g., monitoring conservation easement).

Reputation Risk

Although related to land, the following may also be applicable to the marine space:

> Purchasing private land can sometimes be an expensive or politically controversial option, particularly if current residents or businesses need to be relocated and compensated. Yet, often, it can be relatively cost-effective, particularly in areas where land prices are low, where funding is available from donors, and where there is strong local support for protecting the resource by restricting its use or access.[176]

Likewise, the Kiribati Government faced a possible reputation risk with fishers.

Liquidity Risk

Conservation easements, due to their perpetual existence, can pose a unique liquidity risk to the easement holder because the owner of the easement is unable to remove the easement and sell the seascape at a presumably higher value.

Operational Risk

Operational risks for conservation easements include insufficient internal controls (e.g., poor recordkeeping of contractual obligations associated with an easement), incompetent management, and fraud (e.g., falsifying information). Furthermore, there is significant operational risk that is associated with easements not being properly monitored. (i.e., due to a lack of will from the easement holder to undertake monitoring).

Legal and Regulatory Risk

Several types of legal and regulatory risks exist for conservation easements. This includes:

- The slight chance of eminent domain taking the land (or sea), even after the land has been encumbered with a perpetual conservation easement;
- Political opposition to conservation in favor of more international fishing access and/or commercial development;
- Situations where the current, or future, owner violates the legal and regulatory requirements embedded in the conservation easement. This could result in an expensive and time-consuming litigation between the owner and the trust. Such violations may be intentional, or could be unintentional as a result of a confusing, or poorly translated, contract; and
- The conservation easements not being monitored (i.e., due to a lack of will from the easement holder to undertake monitoring).

Perhaps the greatest legal and regulatory risk is adapting the conservation easement model from the United States and applying the concept to countries with tropical coral reefs. For instance, "{…} TNC and other conservation organizations have worked to adapt the legal framework that allows for conservation easements to work within the Napoleonic legal system that prevails in Chile and most of Latin America. Easement templates have been drafted and tested in several Latin American countries, including Chile. Adapting English common law legal instruments to Napoleonic law poses a significant challenge, however."[177]

Furthermore, there is a risk associated with what is, or is not, included in the easement such as who assumes the responsibilities for force majeure as outlined in the conservation easement.

Credit Risk

Credit risk includes default risk, bankruptcy risk, downgrade risk, and settlement risk.

To properly value a conservation easement and to mitigate the risk of a default or bankruptcy, Story Clark suggests that "{...} land trusts should encourage their donors to retain state-certified, respected appraisers who are knowledgeable in this area of valuation and to follow Uniform Standards of Professional Appraisal Practice. However, in many parts of the country, such appraisers are hard to come by."[178] This said, a potential credit risk is that if it is tough to appraise a property in some parts of the United States, it is presumably much more difficult to appraise the appropriate easement price for a conservation easement on a seascape, for example, in the remote Phoenix Islands.

An example of settlement risk is if the property owner transfers the tax credit and/or the conservation easement to another party, but the other party does not pay for the tax credit and/or conservation easement.

Market Risk

Market risk includes interest rate risk, equity price risk, foreign exchange risk, and commodity price risk.

Interest rate risk and foreign exchange risk could make the easement purchase prohibitively expensive and simultaneously make the tax deduction, if any, insignificant. Similarly, commodity price risks could increase the value of fishing licenses beyond the affordability for a conservation easement. Equity price risk is less relevant to the purchase of a conservation easement for tax deductions.

Risk, Return, Time (Horizon), Taxes, Liquidity, Legal and Unique (RRTTLLU)

Risk and Return

Please see above for the risk and return associated with tax deductions and conservation easements.

Time Horizon

The time horizon for tax deductions and conservation easements will likely vary. Thus, the requirement of maintaining a conservation easement (e.g., maintaining snorkeling trails, monitoring for illegal fishing, and/or removing invasive species) is a long-term commitment that may exist in perpetuity. In contrast, the tax deductions, if any, may exist over a shorter time frame than the length of the conservation easement.

Taxes

With land-based conservation easements, there are likely reduced taxes (e.g., inheritance taxes, sales taxes, income taxes, and particularly reduced property taxes) in exchange for establishing a conservation easement. Such taxes are less applicable to seascapes, which are almost always owned by a national government. However, there could be various fines, penalties, and possibly backdue taxes reinstated as a result of failing to maintain the conservation easement.

Liquidity

Seascapes with conservation easements may be less liquid than comparable, non-encumbered seascapes.

Legal

Legal considerations associated with tax deductions, if any, and conservation easements will undoubtedly vary from country to country. In the United States, for instance, these legal considerations for conservation easements and tax deductions will even vary from state to state. The key legal issues will be spelt out in the conservation easement contract and the legal risk is that those legal obligations are not abided by, or not transferred, with the seascape. In addition, neighboring jurisdictions or other stakeholders may violate the terms of the conservation easement (e.g., by harvesting fish, building telecommunication pipelines, or drilling for deepwater oil).

Unique

Conservation easements may be challenged when legally reviewed by succeeding generations. Thus, as a seascape encumbered with a conservation easement is passed from generation to generation, future generations may challenge the legal viability of a previous generation's conservation easement. Similarly, there can be a lack of understanding of what exactly perpetuity means in easement terms. It is very important to spend a lot of time on this topic, especially in remote areas, because the easement owner may offer a relatively large sum of money and people do not really care about the details when faced with an upfront, lump sum of money. To mitigate this risk, the decision must be the communities' decision and not what a government or NGO wants.

Conservation easements may also face challenges when others seek to use the seascape that is encumbered with the easement. In some cases, the conservation easement monitoring funds—usually assigned to a third-party NGO—become a liability in perpetuity. This means that funds are safe kept, or should be kept, at a financial institution to fund monitoring in perpetuity. Yet these funds may not be sufficient when a well-funded adversary—such as a new government—seeks to use eminent domain or other legal tools to negatively impact the legal strength of the conservation easement.

Policy Analysis

The following policy analysis will look at: defining the problem; establishing goals; selecting a policy; implementing a policy; and evaluating the policy.

Defining the Problem

Two problems associated with tax deductions and conservation easements are the fact that there is a low level of private ownership in many of the countries that host tropical coral reefs, particularly of marine ecosystems, and the fact that few, if any, conservation easements have been structured in these countries.

For example:

- How does one identify privately owned land parcels (i.e., like Pez Maya) adjacent to offshore coral reefs? Are they affordable?

- How would countries such as Indonesia, which follow the civil law system, grant conservation easements?
- To achieve a tax deduction, will the country's tax code need to be revised?
- Would there be some type of clawback provision if the terms of the conservation easement were violated?
- How would the terms of the conservation easement be enforced, especially if the proposed conservation easement is on a remote atoll?
- Can local fishing communities and/or Indigenous Peoples participate in conservation easements?

Establishing Goals

From a government's perspective with a goal of promoting marine conservation, one option—among the diversity of options—is to develop public policies that enable private landowners to receive a financial incentive (i.e., possibly through a tax deduction for a conservation easement) to conserve their privately owned lands which are adjacent to offshore coral reefs.

Selecting a Policy

As noted,

> Under the 1982 U.N. Convention of the Law of the Sea, coastal countries may seek compensation from distant water fishing fleets in return for granting access to their waters. Compensation may take the form of financial payments, development projects, technical assistance, and research assistance. "Compensation for access" agreements typically include payments made by the government of a distant water fishing fleet to the government of a coastal country (state-to-state payments) or payments made by individual fishers or a fishers' association to the government of the coastal country (enterprise-to-state payments), often through licensing arrangements. State-to-state payments offer a stable source of revenue for developing countries (which is often not used for fisheries management), but represent a form of subsidization for distant water fishing fleets that may lead to unsustainable fishing. With the adoption of enterprise-to-state payments, payments may fluctuate widely depending on the state of the fishery.[179]

Similar to the Phoenix Islands case study, one could review the Small Island Developing States (SIDS) to see which countries have entered into such "compensation for access" agreements and assess whether a conservation easement could be structured. For example:

An estimated $60.3 million was paid to Pacific island countries for access to tuna fisheries. Most of the license fees were generated through fees paid by fishers from countries such as Japan, Korea, China (Taipei), and the U.S.A., most often through bilateral arrangements. For many countries in the Pacific region, such as the Marshall Islands, Kiribati, Tuvalu, Palau, and Vanuatu, tuna access fees represent a large share of government revenues.[180]

Other tools to select a policy include the Ocean Health Index,[181] and the Environmental Performance Index.[182]

Implementing a Policy

The "compensation for access" payments, as previously mentioned, can be structured as state-to-state or enterprise-to-state. Thus, the entity responsible for implementing the policy could be the Office of the President or Prime Minister, or at a cabinet-level such as the ministry of fisheries or the ministry responsible for international affairs.

Evaluating the Policy

Evaluation of the easement policy could include assessing:

- How much funding was raised and what were funds used for;
- Has monitoring and enforcement resulted in the reduction of fishing fleets in the designed area and buffer zones (i.e., is leakage occurring);
- Is the ecosystem (e.g., fish biomass, coral cover, etc.) rebounding; and
- Is the policy sustainable in the long term?

Time will tell with respect to Kiribati. However, as of April 2020, the Ocean Health Index ranks Kiribati #163 out of 221 countries,[183] and the Environmental Performance Index[184] ranks Kiribati at #95.

Future Outlook for Instrument

A potentially useful tool for conservation easements would be to establish a comparable website to the Database of State Incentives for Renewables & Efficiency (DSIRE),[185] which could compile all the local, federal, and/or international incentives to purchase and manage conservation easements, and this could spur the future development of more conservation easements.

The future of conservation easements could start to look at alternative sites that are not privately owned in the traditional sense. For example, even if you do not own the land, but you have a long-term lease such as a fishing concession, there could be an opportunity for conservation easements.

Although more of a conservation concession:

Conservation concessions have so far been introduced primarily in forest areas, but in Indonesia, conservation organizations such as the CCIF and TNC are testing the concept in the marine environment. For example, in partnership with CI, a pioneer in developing conservation concessions, CCIF has analyzed the potential for a conservation concession in the Togean Islands in Central Sulawesi, whereby communities would acquire the rights from the government to manage an area for conservation. Conservation organizations would provide compensation to the villagers for enforcement costs and for reduction in income experienced by villages substituting conservation for exploitation of marine species (e.g., live-fish harvesting). TNC is developing an ecotourism concession through a joint venture with an Indonesian company, Putri Naga Komodo, which will operate in and around Komodo National Park. Although the park authority will retain its mandate over management and enforcement in the park, the joint venture will be responsible for marketing the park, setting entry fees, collecting revenues, investing in park infrastructure, licensing dive operations, and investing in local business development. (CCIF 2003, Komodo National Park Collaborative Management Initiative).[186]

It remains unclear whether these concessions were ultimately approved. As noted by Robert Deacon and Dominic Parker:

Our analysis also raises a number of questions that could be the subject of future research. It is a practical necessity to ask which legal constraints exist that might limit the use of marine easements in different fisheries. Are there limitations on who can 'participate' in a fishery and is an NGO 'participating' by owning easements? Could legal rules prohibit easements from encumbering a permit or ITQ when the identity of the permit or ITQ owner changes? Considering the possibility of institutional or regulatory change, can marine easements be framed so that they will respond to such change and remain effective? For example, how could the terms of an easement on a limited entry permit be modified to appropriately encumber an ITQ?[187]

Other Resources on Conservation Easements

Agricultural Conservation Easement Program of the US Department of Agriculture

- https://www.nrcs.usda.gov/wps/portal/nrcs/main/national/programs/easements/acep/

Conservation Almanac

- http://www.conservationalmanac.org/secure/

Conservation Resource Center's Tax Credit Exchange

- http://taxcreditexchange.com/

Foundation Center's GrantCraft

- http://www.grantcraft.org/

Land Trust Standards and Practices from the Land Trust Alliance

- http://www.landtrustalliance.org/topics/land-trust-standards-and-practices

National Conservation Easement Database

- http://conservationeasement.us/

The Conservation Registry

- http://www.conservationregistry.org/

The Learning Center from the Land Trust Alliance

- http://learningcenter.lta.org/login

TNC

- http://www.nature.org/about-us/private-lands-conservation/conservation-buyer/properties/index.htm;
- http://www.nature.org/about-us/private-lands-conservation/conservation-easements/all-about-conservation-easements.xml; and
- http://www.tnclands.tnc.org/

Notes

1. Foster, Dave. "Meeting the Conservation Challenge in New England." In *Conservation Capital in the Americas*, edited by James N. Levitt. 20–21.
2. TNC. "Easements 101." Accessed December 13, 2016. https://www.nature.org/about-us/private-lands-conservation/conservation-easements/conservation-easements-101.pdf.
3. Deacon, Robert T. and Dominic P. Parker. "Encumbering Harvest Rights to Protect Marine Environments: A Model of Marine Conservation Easements." 40.
4. TNC. "Easements 101." Accessed December 13, 2016. https://www.nature.org/about-us/private-lands-conservation/conservation-easements/conservation-easements-101.pdf.
5. Ibid.
6. Levitt, James N. *Conservation Capital in the Americas*. 3.
7. Ibid. 27.
8. Clark, Story. *A Field Guide to Conservation Finance*. 164.
9. Ginn. *Investing in Nature*. 124.
10. TNC. "Easements 101." Accessed December 13, 2016. https://www.nature.org/about-us/private-lands-conservation/conservation-easements/conservation-easements-101.pdf.
11. TNC. "California Groundfish Project." 2020. Accessed March 14, 2020. https://www.nature.org/en-us/about-us/where-we-work/united-states/california/stories-in-california/california-groundfish-project/.
12. Conner, Mike. Interviewed by Brian McFarland. April 2017.
13. Meyer, Shannon. "Markets for Ecosystem Services and Reclaiming the Great Dismal Swamp." In *Conservation Capital in the Americas*, edited by James N. Levitt. 199.
14. Tepper, Henry and Victoria Alonso. "The Private Lands Conservation Initiative in Chile." In *Conservation Capital in the* Americas, edited by James Levitt. 57.
15. Clark, Story. *A Field Guide to Conservation Finance*. 184.
16. Land Trust Alliance. "Insurance: Insurance for Land Trusts." Accessed March 29, 2017. http://www.landtrustalliance.org/topics/insurance.
17. Tabas, Philip. "New Developments in Conservation Easements." *American College of Environmental Lawyers*. Last modified June 16, 2014. http://www.acoel.org/post/2014/06/16/New-Developments-in-Conservation-Easements.aspx.
18. National Conservation Easement Database. "Easement Holder Profiles." Accessed September 9, 2019. https://www.conservationeasement.us/eholderprofile/.
19. Land Trust Alliance. "2015 National Land Trust Census Report: Our Common Ground and Collective Impact." Accessed February

25, 2020. http://s3.amazonaws.com/landtrustalliance.org/2015NationalLandTrustCensusReport.pdf. 3 and 10.

20. TNC. "Easements 101." Accessed April 4, 2017. https://www.nature.org/about-us/private-lands-conservation/conservation-easements/conservation-easements-101.pdf.

21. USFWS. "Pacific Remote Islands: About the Monument." Accessed August 12, 2019. https://www.fws.gov/refuge/Pacific_Remote_Islands_Marine_National_Monument/about.html.

22. Spergel, Barry and Melissa Moye. "Financing Marine Conservation: A Menu of Options." 42.

23. USFWS. "Pacific Remote Islands Marine National Monument." Accessed February 21, 2020. https://www.fws.gov/uploadedFiles/Region_1/NWRS/Zone_1/Pacific_Remote_Islands_Marine_National_Monument/Documents/PRIMNM%20brief(2).pdf.

24. Island Conservation. "Palmyra Atoll: Restoration Project, USA." Accessed February 21, 2020. https://www.islandconservation.org/palmyra-atoll/.

25. USFWS. "Pacific Remote Islands Marine National Monument." Accessed February 21, 2020. https://www.fws.gov/uploadedFiles/Region_1/NWRS/Zone_1/Pacific_Remote_Islands_Marine_National_Monument/Documents/PRIMNM%20brief(2).pdf.

26. Ibid.

27. Ibid.

28. Ibid.

29. Cinner, Joshua E., et al. "Bright Spots Among the World's Coral Reefs." https://doi.org/10.1038/nature18607.

30. USFWS. "Pacific Remote Islands Marine National Monument." Accessed February 21, 2020. https://www.fws.gov/uploadedFiles/Region_1/NWRS/Zone_1/Pacific_Remote_Islands_Marine_National_Monument/Documents/PRIMNM%20brief(2).pdf.

31. Wiggins, Chad. Interviewed by Brian McFarland. March 2020.

32. Island Conservation. "Palmyra Atoll: Restoration Project, USA." Accessed February 21, 2020. https://www.islandconservation.org/palmyra-atoll/.

33. Wiggins, Chad. Interviewed by Brian McFarland. March 2020.

34. Island Conservation. "Palmyra Atoll: Restoration Project, USA." Accessed February 21, 2020. https://www.islandconservation.org/palmyra-atoll/.

35. USFWS. "Pacific Remote Islands Marine National Monument." Accessed February 21, 2020. https://www.fws.gov/uploadedFiles/Region_1/NWRS/Zone_1/Pacific_Remote_Islands_Marine_National_Monument/Documents/PRIMNM%20brief(2).pdf.

36. Island Conservation. "Palmyra Atoll: Restoration Project, USA." Accessed February 21, 2020. https://www.islandconservation.org/palmyra-atoll/.

37. Wiggins, Chad. Interviewed by Brian McFarland. March 2020.

38. Stone, Gregory S. "In Search of Paradise." In *Underwater Eden*, edited by Gregory S. Stone and David Obura. 6.

39. Tong, Anote. "Foreword." In *Underwater Eden*, edited by Gregory S. Stone and David Obura. VII–IX.
40. Niesten, Eduard and Peter Shelley. "Protecting Paradise." In *Underwater Eden*, edited by Gregory S. Stone and David Obura. 107.
41. Ibid. 296.
42. Obura, David, et al. Living Document. Phoenix Islands Protected Area Climate Change Vulnerability Assessment and Management, Report to the New England Aquarium. 5.
43. Ibid. 25.
44. Ibid. 12.
45. Ibid. 18.
46. Obura, David and Randi D. Rotjan. "Coral Reefs and Climate Change." In *Underwater Eden*, edited by Gregory S. Stone and David Obura. 84–86.
47. Rotjan et al. "Establishment, Management, and Maintenance of the Phoenix Islands Protected Area." 308.
48. Stone, Gregory S. "In Search of Paradise." In *Underwater Eden*, edited by Gregory S. Stone and David Obura. 9–12.
49. Rotjan et al. "Establishment, Management, and Maintenance of the Phoenix Islands Protected Area." 304.
50. Ibid. 319.
51. CI. "Establishing the Phoenix Islands Protected Area." Accessed October 9, 2018. https://www.conservation.org/publications/Documents/CI_Asia-Pac ific_PIPA-Factsheet.pdf. 1.
52. Obura, David, et al. Living Document. Phoenix Islands Protected Area Climate Change Vulnerability Assessment and Management, Report to the New England Aquarium. 22.
53. Rotjan et al. "Establishment, Management, and Maintenance of the Phoenix Islands Protected Area." 300.
54. Ibid. 307.
55. Obura, David and Randi D. Rotjan. "Coral Reefs and Climate Change." In *Underwater Eden*, edited by Gregory S. Stone and David Obura. 82.
56. Stone, Gregory S. "Expedition Diary: Diving the Pelagic Zone, Nikumaroro, 2009." In *Underwater Eden*, edited by Gregory S. Stone and David Obura. 74.
57. Obura, David, et al. Living Document. Phoenix Islands Protected Area Climate Change Vulnerability Assessment and Management, Report to the New England Aquarium. 24.
58. Cinner, Joshua E., et al. "Bright Spots Among the World's Coral Reefs." https://doi.org/10.1038/nature18607.
59. Stone, Gregory S. "Expedition Diary: Diving the Pelagic Zone, Nikumaroro, 2009." In *Underwater Eden*, edited by Gregory S. Stone and David Obura. 294.
60. Niesten, Eduard and Peter Shelley. "Protecting Paradise." In *Underwater Eden*, edited by Gregory S. Stone and David Obura. 113.

61. Tong, Anote. "Foreword." In *Underwater Eden*, edited by Gregory S. Stone and David Obura. VII–IX.
62. Rotjan et al. "Establishment, Management, and Maintenance of the Phoenix Islands Protected Area." 294.
63. Niesten, Eduard and Peter Shelley. "Protecting Paradise." In *Underwater Eden*, edited by Gregory S. Stone and David Obura. 117.
64. Rotjan et al. "Establishment, Management, and Maintenance of the Phoenix Islands Protected Area." 296.
65. Niesten, Eduard and Peter Shelley. "Protecting Paradise." In *Underwater Eden*, edited by Gregory S. Stone and David Obura. 118.
66. Pierce, Raymond J. "Birds and Invaders." In *Underwater Eden*, edited by Gregory S. Stone and David Obura. 52–54.
67. Rotjan et al. "Establishment, Management, and Maintenance of the Phoenix Islands Protected Area." 297.
68. Niesten, Eduard and Peter Shelley. "Protecting Paradise." In *Underwater Eden*, edited by Gregory S. Stone and David Obura. 108.
69. Obura, David, et al. Living Document. Phoenix Islands Protected Area Climate Change Vulnerability Assessment and Management, Report to the New England Aquarium. 8.
70. Ibid. 8.
71. Rotjan et al. "Establishment, Management, and Maintenance of the Phoenix Islands Protected Area." 296.
72. Tong, Anote. "Foreword." In *Underwater Eden*, edited by Gregory S. Stone and David Obura. VII–IX.
73. Rotjan et al. "Establishment, Management, and Maintenance of the Phoenix Islands Protected Area." 297.
74. Obura, David, et al. Living Document. Phoenix Islands Protected Area Climate Change Vulnerability Assessment and Management, Report to the New England Aquarium. 28–29.
75. Ibid. 5.
76. Ibid. 314.
77. Niesten, Eduard and Peter Shelley. "Protecting Paradise." In *Underwater Eden*, edited by Gregory S. Stone and David Obura. 114.
78. Ibid. 109.
79. Obura, David, et al. Living Document. Phoenix Islands Protected Area Climate Change Vulnerability Assessment and Management, Report to the New England Aquarium. 8–9.
80. CI. "Establishing the Phoenix Islands Protected Area." Accessed October 9, 2018. https://www.conservation.org/publications/Documents/CI_Asia-Pacific_PIPA-Factsheet.pdf. 2.
81. Niesten, Eduard and Peter Shelley. "Protecting Paradise." In *Underwater Eden*, edited by Gregory S. Stone and David Obura. 115.

82. Obura, David, et al. Living Document. Phoenix Islands Protected Area Climate Change Vulnerability Assessment and Management, Report to the New England Aquarium. 23.
83. Niesten, Eduard and Peter Shelley. "Protecting Paradise." In *Underwater Eden*, edited by Gregory S. Stone and David Obura. 119.
84. CI. "Establishing the Phoenix Islands Protected Area." Accessed October 9, 2018. https://www.conservation.org/publications/Documents/CI_Asia-Pacific_PIPA-Factsheet.pdf. 2.
85. Obura, David, et al. Living Document. Phoenix Islands Protected Area Climate Change Vulnerability Assessment and Management, Report to the New England Aquarium. 22.
86. Niesten, Eduard and Peter Shelley. "Protecting Paradise." In *Underwater Eden*, edited by Gregory S. Stone and David Obura. 107–108.
87. Rotjan et al. "Establishment, Management, and Maintenance of the Phoenix Islands Protected Area." 314.
88. Ibid. 316–317.
89. Ibid. 294, 299.
90. Tong, Anote. "Foreword." In *Underwater Eden*, edited by Gregory S. Stone and David Obura. VII–IX.
91. Niesten, Eduard and Peter Shelley. "Protecting Paradise." In *Underwater Eden*, edited by Gregory S. Stone and David Obura. 109.
92. Ibid. 114.
93. Shelley, Peter. "Expedition Diary: A Legal Education, Kiribati Style, Tarawa, 2004." In *Underwater Eden*, edited by Gregory S. Stone and David Obura. 120.
94. Rotjan et al. "Establishment, Management, and Maintenance of the Phoenix Islands Protected Area." 294, 296.
95. Ibid. 317–318.
96. Ibid.
97. CI. "Establishing the Phoenix Islands Protected Area." Accessed October 9, 2018. https://www.conservation.org/publications/Documents/CI_Asia-Pacific_PIPA-Factsheet.pdf. 1.
98. Rotjan et al. "Establishment, Management, and Maintenance of the Phoenix Islands Protected Area." 290–291.
99. Obura, David, et al. Living Document. Phoenix Islands Protected Area Climate Change Vulnerability Assessment and Management, Report to the New England Aquarium. 24–25.
100. Obura, David and Randi D. Rotjan. "Coral Reefs and Climate Change." In *Underwater Eden*, edited by Gregory S. Stone and David Obura. 89.
101. Rotjan et al. "Establishment, Management, and Maintenance of the Phoenix Islands Protected Area." 316.
102. Ibid. 320.
103. Stone, Gregory S., et al. "The Future of the Phoenix Islands." In *Underwater Eden*, edited by Gregory S. Stone and David Obura. 127–129.

104. Rotjan et al. "Establishment, Management, and Maintenance of the Phoenix Islands Protected Area." 298.

105. Ibid. 298.

106. Ibid. 320–321.

107. Merediz Alonso, Gonzalo. Interviewed by Brian McFarland. June 2019.

108. Spergel, Barry and Melissa Moye. "Financing Marine Conservation: A Menu of Options." 43.

109. Environmental News Network. "Corporate and Environmentalist Alliance Buys Sensitive Coast Land in Mexican Nature Reserve." October 28, 2004. Accessed June 17, 2019. https://www.enn.com/articles/250-corporate-and-env ironmentalist-alliance-buys-sensitive-coast-land-in-mexican-nature-reserve.

110. Merediz Alonso, Gonzalo. Interviewed by Brian McFarland. June 2019.

111. World Heritage Centre. "Sian Ka'an." Accessed February 13, 2017. http:// whc.unesco.org/en/list/410.

112. Merediz Alonso, Gonzalo. Interviewed by Brian McFarland. January 2017.

113. National Commission of Natural Protected Areas. "Who We Are / About Us." Accessed February 24, 2017. http://www.conanp.gob.mx/english.php.

114. Merediz Alonso, Gonzalo. Interviewed by Brian McFarland. June 2019.

115. Newswire. "Gateway to Yucatan Biosphere Reserve Protected." February 6, 2002. Accessed June 17, 2019. ens-newswire.com/ens/feb2002/2002-02-06-02.asp.

116. Environmental News Network. "Corporate and Environmentalist Alliance Buys Sensitive Coast Land in Mexican Nature Reserve." October 28, 2004. Accessed June 17, 2019. https://www.enn.com/articles/250-corporate-and-env ironmentalist-alliance-buys-sensitive-coast-land-in-mexican-nature-reserve.

117. Amigos de Sian Ka'an. "Deforestation." Accessed December 29, 2016. http:// www.amigosdesiankaan.org/en/threats/deforestation.

118. Ibid. "Extinction." Accessed December 29, 2016. http://www.amigosdesian kaan.org/en/threats/extintion.

119. Merediz Alonso, Gonzalo. Interviewed by Brian McFarland. June 2019.

120. Johnson, Kenneth. Interviewed by Brian McFarland. October 2019.

121. Amigos de Sian Ka'an. "Water Pollution." Accessed February 13, 2017. http:// www.amigosdesiankaan.org/en/threats/water-pollution.

122. Ibid. "Beach Pollution." Accessed February 13, 2017. http://www.amigosdes iankaan.org/en/threats/beach-pollution.

123. Spergel, Barry and Melissa Moye. "Financing Marine Conservation: A Menu of Options." 43.

124. Merediz Alonso, Gonzalo. Interviewed by Brian McFarland. June 2019.

125. Hu, Chuanmin. "OCTO and EBM Tools Network Co-Sponsored Webinar: Sargassum Watch System warns of incoming seaweed by Chuanmin Hu of University of South Florida." August 27, 2019. https://www.openchannels. org/webinars/2019/sargassum-watch-space.

126. Ibid.

127. Ibid.

128. Merediz Alonso, Gonzalo. Interviewed by Brian McFarland. June 2019.
129. Johnson, Kenneth. Interviewed by Brian McFarland. October 2019.
130. Merediz Alonso, Gonzalo. Interviewed by Brian McFarland. June 2019.
131. McField, Melanie. Interviewed by Brian McFarland. January 2020.
132. Merediz Alonso, Gonzalo. Interviewed by Brian McFarland. January 2017.
133. Merediz Alonso, Gonzalo. Interviewed by Brian McFarland. June 2019.
134. Ibid.
135. Johnson, Kenneth. Interviewed by Brian McFarland. October 2019.
136. Spergel, Barry and Melissa Moye. "Financing Marine Conservation: A Menu of Options." 43.
137. Environmental News Network. "Corporate and Environmentalist Alliance Buys Sensitive Coast Land in Mexican Nature Reserve." October 28, 2004. Accessed June 17, 2019. https://www.enn.com/articles/250-corporate-and-env ironmentalist-alliance-buys-sensitive-coast-land-in-mexican-nature-reserve.
138. Merediz Alonso, Gonzalo. Interviewed by Brian McFarland. June 2019.
139. Ibid.
140. Ibid.
141. World Heritage Centre. "Sian Ka'an." Accessed February 13, 2017. http:// whc.unesco.org/en/list/410.
142. Merediz Alonso, Gonzalo. Interviewed by Brian McFarland. January 2017.
143. Amigos de Sian Ka'an. "Sian Ka'an: Permanent Beach Cleaning Program." Accessed February 24, 2017. http://www.amigosdesiankaan.org/en/projects/sia nka-an/143-limpieza-permanente-de-playas.
144. Merediz Alonso, Gonzalo. Interviewed by Brian McFarland. January 2017.
145. Merediz Alonso, Gonzalo. Interviewed by Brian McFarland. June 2019.
146. Ibid.
147. Merediz Alonso, Gonzalo. Interviewed by Brian McFarland. January 2017.
148. Merediz Alonso, Gonzalo. Email message to author. July 26, 2019.
149. Merediz Alonso, Gonzalo. Interviewed by Brian McFarland. June 2019.
150. Johnson, Kenneth. Interviewed by Brian McFarland. October 2019.
151. Riedmiller, Sibylle. Email Message to author. March 4, 2019.
152. Spergel, Barry and Melissa Moye. "Financing Marine Conservation: A Menu of Options." 64–65.
153. Green Renaissance. "Zanzibar—Chumbe Island Coral Park." November 2, 2009. Accessed March 12, 2020. https://www.youtube.com/watch?v=TelrgQ yvEeg.
154. Earth Changers. "Sibylle Riedmiller Is an Earth Changer." 2019. Accessed April 11, 2020. https://www.earth-changers.com/sustainable-development/sib ylle-riedmiller-chumbe-island.
155. Marcus, Lucy. "Chumbe Island Coral Park." December 2, 2010. Accessed March 12, 2020. https://vimeo.com/17396919.
156. Riedmiller, Sibylle. "TEDxSeaPoint-Sibylle Riedmiller-Chumbe Island." YouTube. May 9, 2012. Accessed June 23, 2019. https://www.youtube.com/ watch?v=yWldTssTRBA.

157. Ibid.
158. Ibid.
159. Ibid.
160. Ibid.
161. Marcus, Lucy. "Chumbe Island Coral Park." December 2, 2010. Accessed March 12, 2020. https://vimeo.com/17396919.
162. Riedmiller, Sibylle. Email message to author. March 4, 2019.
163. Ibid.
164. Ibid.
165. Ibid.
166. UN General Assembly. "Protection of Coral Reefs for Sustainable Livelihoods and Development." 2011. Accessed April 10, 2020. https://www.un.org/esa/dsd/resources/res_pdfs/ga-66/SG%20report_Coral%20Reefs.pdf. 21.
167. CHICOP. "Chumbe Press Release from 01.07.16." Accessed April 11, 2020. https://chumbeisland.com/news/newsdetails/article/chumbe-founder-sibylle-riedmiller-awarded-with-the-cross-of-the-order-of-merit-of-the-federal-repu-1/.
168. Riedmiller, Sibylle. "TEDxSeaPoint-Sibylle Riedmiller-Chumbe Island." YouTube. May 9, 2012. Accessed June 23, 2019. https://www.youtube.com/watch?v=yWldTssTRBA.
169. Ibid.
170. Spergel, Barry and Melissa Moye. "Financing Marine Conservation: A Menu of Options." 64–65.
171. Riedmiller, Sibylle. "TEDxSeaPoint-Sibylle Riedmiller-Chumbe Island." YouTube. May 9, 2012. Accessed June 23, 2019. https://www.youtube.com/watch?v=yWldTssTRBA.
172. Ibid.
173. Sobel and Dahlgren. *Marine Reserves.* 332–333.
174. Ibid. 333.
175. Riedmiller, Sibylle. "TEDxSeaPoint-Sibylle Riedmiller-Chumbe Island." YouTube. May 9, 2012. Accessed June 23, 2019. https://www.youtube.com/watch?v=yWldTssTRBA.
176. Spergel, Barry and Melissa Moye. "Financing Marine Conservation: A Menu of Options." 42.
177. Tepper, Henry and Victoria Alonso. "The Private Lands Conservation Initiative in Chile." In *Conservation Capital in the* Americas, edited by James Levitt. 57.
178. Clark, Story. *A Field Guide to Conservation Finance.* 81.
179. Spergel, Barry and Melissa Moye. "Financing Marine Conservation: A Menu of Options." 53.
180. Ibid. 53.
181. Ocean Health Index. "Home." Accessed April 11, 2020. http://www.oceanhealthindex.org/.

182. Yale University. "Environmental Performance Index." Accessed April 11, 2020. https://epi.envirocenter.yale.edu/.

183. Ocean Health Index. "Annual Scores and Rankings." Accessed April 11, 2020. http://www.oceanhealthindex.org/region-scores/annual-scores-and-rankings.

184. Yale University. "2018 EPI Results." Accessed April 11, 2020. https://epi.env irocenter.yale.edu/epi-topline?country=Kiribati.

185. DSIRE. "Home." Accessed February 25, 2020. http://www.dsireusa.org/.

186. Spergel, Barry and Melissa Moye. "Financing Marine Conservation: A Menu of Options." 46.

187. Deacon, Robert T. and Dominic P. Parker. "Encumbering Harvest Rights to Protect Marine Environments: A Model of Marine Conservation Easements." 56.

9

Government International Budgetary Allocations

Introduction

While governments can utilize the budgetary process to finance domestic conservation of tropical coral reefs, governments can also utilize a part of their budget to finance tropical coral reef conservation in other countries or to finance their own domestic MPAs that link to international peace parks.

International peace parks, also known as transfrontier conservation areas (TFCAs) or transboundary protected areas (TBPAs), are protected areas that span more than one country. Such international peace parks can help establish positive relationships between countries, while expanding protected areas to increase wildlife habitat and preserve the territories of Indigenous Peoples and local communities.

Historical Overview

Notable dates for government international budgetary allocations and international peace parks include:

- 1932: Waterton Glacier International Peace Park, established between the U.S. and Canada, becomes the world's first international peace park.[1]
- 1991: The "Bhutan Trust Fund for Environmental Conservation (BTF) – the first environmental fund in the developing world – established a solid foundation for biodiversity conservation through enduring legal, institutional, and technical frameworks."[2]

© The Author(s) 2021
B. J. McFarland, *Conservation of Tropical Coral Reefs*,
https://doi.org/10.1007/978-3-030-57012-5_9

- "Established in 1991, the GEF has been one of the leading sources of project funding for marine conservation through the biodiversity and international waters focus areas."[3]
- 2016 (December): The Blue Action Fund is established by the German Ministry for Economic Cooperation and Development (BMZ), through the KfW Development Bank (KfW).[4]

Mechanisms of Instrument

The source of funds for a government's international budgetary allocations is likely similar to, or the same as, the source of funds for a government's domestic budgetary allocations. These sources may include taxes (e.g., income taxes), environmental fines or penalties (e.g., due to oil spills), or via lottery sales. However, structuring government international budgetary allocations comes in a variety of distinct forms. Such financing mechanisms can be done via bilateral or multilateral channels. Funds can also go toward the establishment of a protected area in a different country, as well as toward the establishment of a joint, international peace park. Often, a designated entity such as a trust fund, like the Brazilian Biodiversity Fund (FUNBIO), is established to receive and manage the funds.

Another aspect of government assistance to a foreign country is via sister parks, which might involve financial and/or non-financial, technical assistance. For instance, protected areas may enter into a partnership to exchange ideas through in-person training, share experiences with new technology, and pay for food and lodging for visiting rangers. Sister park agreements exist between the Phoenix Islands Protected Area and the Papahānaumokuākea Marine National Monument,[5] along with between New Caledonia's Coral Sea Marine Protected Area and the Cook Islands Marine Park.

Size of Instrument

The overall size of net Official Development Assistance (ODA) from the Organization for Economic Cooperation and Development's (OECD) Development Assistance Committee (DAC), "a unique international forum of many of the largest providers of aid, including 30 members,"[6] was estimated at USD$147 billion in 2017.[7] Below are the top ten largest recipients,[8] and the top ten largest providers of net ODA (Table 9.1).[9]

Table 9.1 Top 10 recipients and providers of ODA

	Country	Net ODA (US dollar, millions, 2017)		Country	Aid disbursements (US dollar, millions, 2017)
1	United States	34,732	1	Afghanistan	2831
2	Germany	25,005	2	India	2570
3	United Kingdom	18,103	3	Syria	2566
4	Japan	11,463	4	Iraq	2279
5	France	11,331	5	Bangladesh	2225
6	Turkey	8121	6	Ethiopia	2207
7	Italy	5858	7	Jordan	1878
8	Sweden	5563	8	Nigeria	1743
9	Netherlands	4958	9	South Sudan	1695
10	Canada	4305	10	Vietnam	1540

Source OECD

Specific to the marine environment, a "total of 314 projects were identified from 60 funders between 2010 and 2016, with a total value of USD$1.9 billion in total project costs, including USD$343.2 million in primary funding (i.e., in-cash donor funding). {…} Multilateral and intergovernmental agencies and funds and philanthropic foundations awarded the largest amount of funding, with the top five funders collectively allocating nearly USD$1.6 billion of the total project cost commitments identified to date."[10]

Although outdated, as of 2007, the World Database on Protected Areas identified "227 TBPA complexes incorporating 3,043 individual protected areas or internationally designated sites."[11]

With respect to bilateral assistance, the OECD reports that the:

> Bilateral biodiversity-related ODA by members of the DAC reached USD$8.3 billion per year in 2015-16, accounting for 6% of total bilateral ODA.
>
> Africa accounted for the highest share of biodiversity-related ODA (34%), followed by Asia (23%), America (15%), Europe (13%), and Oceania (1%). 14% was unallocated by country and region. {…}
>
> Over the period 2012-16, the majority of biodiversity-related ODA was committed by a small number of providers, with the top three accounting for 46%. Over the same period, biodiversity-related ODA was distributed across a number of countries, with some receiving large commitments for individual activities that significantly impacted the average commitments received.[12]

With respect to multilateral assistance for "sustainable management of coral reefs and associated coastal ecosystems," there are several large initiatives:

- The Global Environment Facility (GEF) Trust Fund (24 projects funded between 2010 and 2016 with USD$168 million in primary funding and USD$1.43 billion in total project cost);
- Green Climate Fund (GCF) (1 project funded, in Tuvalu, between 2010 and 2016 with USD$36.0 million in primary funding and USD$38.9 million in total project cost);
- European Union (1 project funded between 2010 and 2016 with USD$19.13 million in primary funding and USD$19.13 in total project cost);
- The European Commission Framework Programme 7 (8 projects funded between 2010 and 2016 with USD$15.82 million in primary funding and USD$18.62 million in total project cost)[13]; and
- The Asian Development Bank provided "technical and financial support to the CTI {Coral Triangle Initiative}, with an assistance package of $242.4 million, including two regional technical assistance and two national projects."[14]

In 2014, "the launch of the sixth replenishment of the GEF Trust Fund included a particular focus on expanding the area of coral reefs within marine protected areas, as per Program 6 ('Ridge to Reef Program') {…}."[15]

The Oak Foundation funded 45 projects between 2010 and 2016 (including USD$20.77 million in primary funding and USD$93.61 million in total project cost), has been a major donor.[16]

There are several additional, bilateral and multilateral initiatives established. This includes:

- Commonwealth Clean Oceans Alliance with the World Bank[17];
- Municipal Waste Recycling Program through USAID's Development Innovations Group, a four-year, USD$3.5 million grant program[18];
- Plastic Solutions Fund, which as of January 2020, had provided funding to 75 organizations working in 13 countries and "$7 million in grants {were provided} from our fund's partners and members"[19];
- PROBLUE Trust Fund on Sustainable Oceans, "is part of the World Bank's overall blue economy program, which is worth around US$5 billion, with a further $1.65 billion in the pipeline"[20];
- The GEF Marine Plastics program[21];
- The European Commission and European Investment Fund recently launched the €75 million BlueInvest Fund in February 2020[22]; and
- USFWS's Species Programs, including the Marine Turtle Conservation Fund.[23]

Other bilateral and multilateral initiatives include:

- Western Indian Ocean Marine Science Association;
- World Heritage Fund;
- Norway's Development Cooperation for Marine Conservation;
- U.S. Government's White Water to Blue Water Initiative in the Wider Caribbean;
- International Coral Reef Action Network (ICRAN); and
- USAID's support for Mesoamerican Caribbean Reef and the Sulu Sulawesi Marine Ecoregion. USAID has also endowed large conservation trust funds such as the Indonesian Biodiversity Foundation-KEHATI.

The GCF has the potential to dwarf all these multilateral sources of financing as:

> {…} advanced economies have formally agreed to jointly mobilize USD 100 billion per year by 2020, from a variety of sources, to address the pressing mitigation and adaptation needs of developing countries. Governments also agreed that a share of new multilateral funding should be channeled through the newly established Green Climate Fund.[24]

Further, with respect to international funding from 2010 to 2016:

> The top five most funded countries in total project costs were Indonesia, Mexico, Sri Lanka, United States (including overseas territories), and Tuvalu {…}. Large, individual coastal management and climate resilience projects drive the pattern for Tuvalu and Sri Lanka. {…} Eastern Africa, the Wider Caribbean and the East Asian Seas regions received comparably more funding than the South-East Pacific, West and Central Africa, and the Red Sea and Gulf of Aden. {…} Project addressing 'conservation and sustainability' of coral reefs (98 projects) and 'marine protected area management' (46 projects) have received the greatest proportion of funding, equating to USD 1.2 billion and USD 239.4 million respectively.[25]

Introduction to Case Studies

The following case studies look at the Marine and Coastal Protected Areas Project of Brazil, the Binational Red Sea Marine Peace Park between Jordan and Israel, and the Mesoamerican Reef Fund (MAR Fund). Funding for the Marine and Coastal Protected Areas Project of Brazil helped support 17 protected areas which cover 1.6 million hectares along with seven research centers. In contrast to the other case studies presented, the Binational Red Sea Marine Peace Park between Jordan and Israel is not considered a success story. While joint management of the area and some collaborative projects was the optimistic vision, this vision was not achieved. The MAR Fund focuses on the largest transboundary reef and the largest barrier reef in the Atlantic Ocean.

Case Study #1: The Marine and Coastal Protected Areas Project of Brazil

Introduction

Approximately ten years ago, in 2009, FUNBIO (Fundo Brasileiro para a Biodiversidade or the Brazilian Biodiversity Fund) and Brazil's Ministry of Environment convened a workshop that brought together stakeholders including academia, NGOs, and government officials to look at Brazil's marine issues and what would be relevant conservation initiatives. These discussions were anchored on a lot of public policies such as Brazil being a signatory of the Convention on Biological Diversity (CBD), along with Brazil's National Policy on Protected Areas and the 2006 National Strategic Plan for Protected Areas. Ultimately, what followed was that one aspect which needed focus was the establishment of a representative system of marine protected areas (MPAs), including the prioritization of sustainable use and benefit sharing. Secondly, the discussion focused on what would be a proposal that would help accomplish this objective? This, eventually, became the Marine and Coastal Protected Areas Project, better known as the GEF MAR Project (Fig. 9.1).[26]

The World Bank, soon thereafter, got involved and the Global Environment Facility (GEF) provided financing via the World Bank. The key part was looking at long-term financing. FUNBIO used its experience from a previous program focused on the Amazon Rainforest called the Amazon Region Protected Areas (ARPA) Program which: created new terrestrial protected areas and undertook consolidation in the first phase; during the second phase, there was a refinement of the strategy about long-term financing; and the third phase focused on phase transition financing.

ARPA then led to the GEF MAR Project, which had a focus on the marine environment and which was more comprehensive and more holistic as compared to the ARPA Program. In the GEF MAR Project, in a way, the proponents were following the same process (i.e., looking into a strategy for long-term financing, prioritization exercises for areas of global significance,

Fig. 9.1 Map of areas covered by the Marine and Coastal Protected Areas Project (*Credit* FUNBIO)

etc.) even though the World Bank was the only financial supporter at the time and the key supporter. It is important to note, there are no terrestrial areas supported by GEF MAR; if the areas are not fully marine areas, they have at least a component.[27]

Identify the Problem

As explained by FUNBIO, you cannot think about marine conservation without thinking about coastal development, especially for a country as large as Brazil. Back around 2009, the focus was to look more at marine conservation, but not to look at coastal development. While Brazil has initiatives dealing with coastal development, there are a number of challenges in addition to conservation and the management of MPAs. In addition, some places had an increasing impact from tourism, while other places were more threatened by offshore oil exploration. Regarding industrial fishing fleets, Brazil does not have a major problem with the fleets in the MPAs. Likewise, there are too few MPAs to have issues from foreign fleets and those few MPAs that are established are close to shore. Yet, there are different levels of industrial fishing and fisheries management, in general, is a challenge in Brazil.[28]

Why the Problem Is Important

One of the reasons why problems facing Brazil's coastline are important is because the areas covered by the GEF MAR Project are home to "over 1,600 species of fish, 100 species of bird, 2,300 species of invertebrates and 54 species of mammal {...}."[29]

Furthermore, the Abrolhos Archipelago is listed among the fifty most important bioclimatic units and the Brazilian Atlantic Islands (Fernando de Noronha and Atol das Rocas Reserves) are listed as a World Heritage Marine Program.

How Problem Was Identified

These problems were identified via the initial workshop with attendees including FUNBIO and Brazil's Ministry of Environment.

Effectiveness of Process for Identifying Problem

The process seems to be effective. As explained by FUNBIO, it was a learning process from all sides. Currently, there are representatives from artisanal fisheries and they participate via the protected areas councils and in the Project's councils.

Steps Taken to Address the Problem

FUNBIO has been involved with the GEF MAR Project from the very beginning, as the Project was initiated by both FUNBIO and Brazil's Ministry of Environment. Some of the steps included:

- Initial workshop[30];
- Working with stakeholders; and
- The GEF approved the Project and the contract was signed in 2014.[31]

Results

There have been numerous results attributable to the GEF MAR Project. The Project reportedly received a USD$18.2 million grant from the GEF and USD$99,656,000 in co-financing.[32] Such funding helped the Project support 17 protected areas which cover 1.6 million hectares (3.95 million acres or 16,000 km^2) along with seven research centers.[33] Furthermore,

> Brazil surpassed its target of extending protection to 10% of its seas. In 2018, with support from GEF MAR, four new federal MPAs were created, taking the percentage of marine areas under protection from 1.5% to 26.5% overall. Together, these parks, sanctuaries and no-take zones cover 940 thousand km^2 of ocean, an area almost the size of Iceland.[34]

According to FUNBIO, the biggest result was achieved in 2018, when the Project very successfully created two mosaics (i.e., extremely large MPAs) with support from the Brazilian Navy. This resulted in the creation of 92 million hectares of new MPAs. The GEF MAR Project had a goal to go from 1.5 to 5% of Brazil's EEZ, while the CBD target is 10%. Yet, the Project results in 26.5%.[35]

It is important to note that the GEF MAR Project supports both fully protected and sustainable use areas, along with several extractive reserves and research centers. One related result of the Project has been implementing monitoring strategies in conjunction with the different research centers involved. For instance, TAMA is focused on sea turtles and the Aquatic Mammal Research Center in the state of Alagoas is focused on the rehabilitation and release of manatees, which are a threatened species. In addition, FUNBIO is undertaking the first call for proposals directly aimed at strengthening the associations in charge of managing extractive reserves.[36]

A further important result to mention is that the GEF MAR Project has taken a more integrated approach and is not just focused on strengthening marine protected areas, but rather focused on more cross-cutting work.[37]

Challenges and How They Were Met

There were numerous challenges over the past decade associated with the GEF MAR Project. For instance, there are clear mandates for fisheries regulation in Brazil's MPAs. In the past, there were clear command and control policies regarding who was responsible for MPA management, but outside of the MPAs was sort of "nowhere."[38]

In addition, there were challenges associated with trying to secure private financing and there was a rather long process (i.e., took approximately three years) to formalize the submission process to the GEF.[39]

As a three-dimensional place, it was very challenging to work in the region, particularly due to the high costs and difficult logistics. However, the partnership with the Brazilian Navy was key and enabled the creation of these additional MPAs.[40]

Lastly, understanding the needs of artisanal fisheries was a challenge. To address such challenges, it was useful to have representatives from the artisanal fisheries participate via the protected areas councils and in the Project's councils.[41]

Beyond Results

The results seem to be sustainable, as long-term financing has been secured and the MPAs have been designated. In addition, there has been a lot of stakeholder engagement before, during, and after the GEF funding was received.

Lessons Learned

There are several lessons learned and which were shared by FUNBIO. First, one lesson learned from the GEF MAR Project, and which also came from the ARPA Program, is the importance of the planning process. Likewise, having a clear planning process that is directly linked to targets is very important. For instance, what do you want to reach? Can the targets meet the objectives? The planning process is so important that you can get lost in the operational process. For instance, how do you organize and structure the needs to most effectively use your funding?

The second lesson learned is that involvement of local communities is key, regardless if the MPA is full protection, for sustainable use, or for extractive usage. Through local traditional populations' representation, it is important to undertake direct dialogue and to have communities directly involved with project governance.[42]

Other Resources on the Marine and Coastal Protected Areas Project of Brazil

Federal Government of Brazil's GEF MAR Page

- http://www.mma.gov.br/areas-protegidas/programas-e-projetos/projeto-gef-mar

Federal Government of Brazil's Programa ARPA

- http://programaarpa.gov.br/en/

FUNBIO's GEF MAR Project Page

- https://www.funbio.org.br/en/programas_e_projetos/gef-mar/

GEF

- https://www.thegef.org/project/marine-and-coastal-protected-areas

Case Study #2: Binational Red Sea Marine Peace Park Between Jordan and Israel

Introduction

In contrast to the other case studies presented in this book, the Binational Red Sea Marine Peace Park between Jordan and Israel (Red Sea Marine Peace Park) is not considered a success story. While joint management of the area and some collaborative projects was the optimistic vision, this vision was not achieved (Fig. 9.2).

The Red Sea Marine Peace Park was "{…} supported and financed, at least in its initial stages, largely by the U.S. government with the U.S. State Department and the U.S. National Oceanographic and Atmospheric Agency (NOAA) heavily involved."[43]

Identify the Problem

There were numerous problems, both historically and currently, facing the Red Sea. As reported back in 2001:

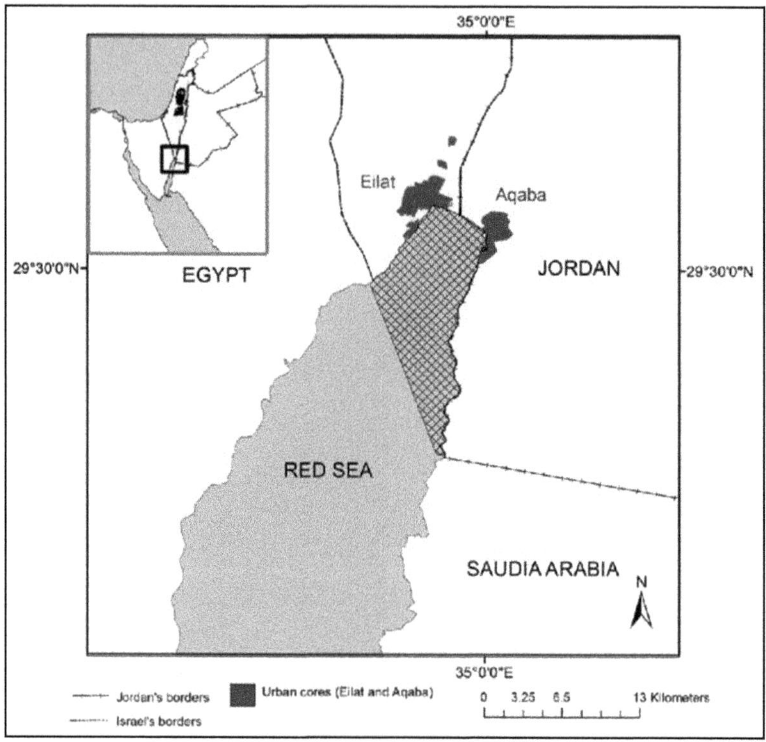

Fig. 9.2 Map of the proposed Red Sea Marine Peace Park Area in the Northernmost part of the Red Sea between Israel and Jordan (*Credit* Portman, Michelle E. and Yael Teff-Seker. "Factors of success and failure for transboundary environmental cooperation: projects in the Gulf of Aqaba." Journal of Environmental Policy and Planning. 31 January 2017)

The five-member panel, including a German and an American as well as Prof. Yehudith Brick from Israel, reached the conclusion that pollution in the bay is destroying the coral reefs and said that sources of pollution include sewage from the city, phosphate dust that escapes into the air and water during loading at Eilat port, and fish farming pens in the depths of the bay.[44]

There are constant threats from pollution and planned coastal development.[45]

A unique feature of the region is that "{…} the Gulf of Aqaba is isolated from the rest of the Red Sea by the narrow Straits of Tiran, {and thus,} its water circulation is sluggish. That leaves the gulf exquisitely vulnerable to agricultural runoff and sewage, and to silt dumped offshore by dredging and landfill operations on both sides of the border. The reefs are also under direct assault from a thriving coral trade and from hulls and keels that graze the reefs."[46]

Similarly, as highlighted by Dr. Yael Teff-Seker, coastal development and poor water quality are the two greatest threats to biodiversity in the Red Sea. In Jordan in 1995, and more recently, there are periodic oil spills. Land-based pollution, such as sewage and other types of waste from Eilat, particularly from the hotels, and some from Jordan, poses problems to the offshore coral reefs. There were fish cages for aquaculture, which have since been taken out, and there is some irresponsible diving and cutting out of corals. Furthermore, there is a lack of management enforcement and there is a need for better public awareness.[47]

Why the Problem Is Important

The Red Sea is home to some tremendous coral reefs off of Aqaba, Jordan and Eilat, Israel, along with Thistlegorm, Shark, and Yolanda Reefs of the Egyptian Red Sea. With this in mind:

> A final known factor that requires emphasis, relates to the threat of environmental crisis. The degraded environmental status of the northern Gulf's extraordinary marine ecosystem loomed threatening enough to maintain cooperation for improved (or at least stable) environmental quality and thus supporting one of the goals of the RSMPP for some time.[48]

In addition, the problems facing the Red Sea can also impact the region's tourism and recreational uses:

> The RSMPP program included three major elements: (1) joint management efforts; (2) outreach to the adjacent communities about the park; and (3) scientific cooperation for joint monitoring of marine ecosystem health. Economic development was an important goal of the RSMPP, since tourism and recreational uses in the northern Gulf of Aqaba are important to both countries.[49]

The region is home to approximately 140 species of stony coral and nearly 1000 species of fish. As raised by Carl Zimmer in *Science*, "particularly intriguing is why they exist at all: nowhere else in the Indian or Pacific oceans do reef-building corals grow so far north of the equator."[50]

Furthermore, the Northern Red Sea, which is also home to Egypt's Sharm-el-Sheikh and Ras Mohammed reefs, is considered one of the "Seven Wonders of the Underwater World."[51]

How Problem Was Identified

There were numerous stakeholders involved with the initial design and with the early attempts of implementing the Red Sea Marine Peace Park. Such stakeholders included: the Israel National Parks Authority, the Israel Ministry of Environmental Protection, the Interuniversity Institute (IUI) for Marine Sciences in Eilat, Aqaba Special Economic Zone Authority (AESZA), and the Jordanian Marine Science Station, along with USAID and the U.S. NOAA.[52]

Effectiveness of Process for Identifying Problem

Although the Red Sea Marine Peace Park did not materialize, the process for identifying the problems was effective. In fact, scientific cooperation and the joint monitoring of coral health were perhaps the most successful aspects of the initiative.[53]

Steps Taken to Address the Problem

Some of the earlier steps taken to address the problem include:

- Peace treaty that Jordan and Israel signed in 1994;
- Funding from USAID, which involved a "$2 million, 3-year effort, launched in 1999{...}[54];" and
- Joint monitoring of coral health.

Results

Despite not being an overall success story, there were a few results attributable to the Red Sea Marine Peace Park. In the early stages, there was a relatively successful, joint scientific monitoring program and scientific collaboration. Thus, "the bi-annual joint research boat tours that were started in 2003 stopped in September 2014."[55] Furthermore, from 2005 to 2014, "live coral cover at the Eilat reefs had gradually increased."[56]

Challenges and How They Were Met

There were numerous challenges associated with the proposed Red Sea Marine Peace Park and most were not met. In sum:

The Red Sea marine peace park shared by Israel, Jordan, and Egypt presents an interesting example as well where the term 'peace park' has been used to describe a joint conservation management regime for ecosystem services. However, here the challenge is that the establishment of the peace park has not really led to a reduction of tensions among the players, and the subsequent implementation has been limited.[57]

Specifically, some of the many challenges facing the Red Sea Marine Peace Park included: low profitability; minimal funding; minimally involved third-parties (particularly after the initial engagement from the United States and World Bank); low transparency and low public awareness; low levels of motivation and commitment; low stakeholder interest; unbalanced relationship between the two host countries; and the lack of a promoting party.[58]

A major reason why the Red Sea Marine Peace Park failed is because the U.S. Department of State ultimately pulled out as a result of its internal champion leaving. When the peace agreement was signed between Jordan and Israel, there was a real desire to find collaborative projects. However, the relationships between Israel and the Palestinians have progressively gotten worse as Israel has more of a right-of-center government and with its actions in the West Bank, it has hurt relations with Jordan. The relationship was much better, almost great, when the peace agreement was signed, but now the relationship has deteriorated and there is little official contact. While there are still some diplomatic relationships, the Jordanians are constantly expressing their displeasure of what is going on with Israel. As of now, Jordan and Israel are not collaborating on the Red Sea Marine Peace Park and in addition, the money that was allotted has, over the years, been used for other things or just not allocated.[59]

Beyond Results

Due to the many challenges, the results of the Red Sea Marine Peace Park are not sustainable. One slight possibility is that there are many programs between Germany and Israel and Germany is in touch with both Jordanian and Israeli officials.[60]

Approximately 25 years ago, Dr. Michelle Portman recalls a conversation where she was told that—in part due to marine pollution—there would be no corals in the region within ten years. Fortunately, there are still reefs and they are in better shape in Jordan due to less coastal development, as compared to Israel.[61]

For a historical perspective, "if it does live up to its promise, the Peace Park may even serve as a model for other parts of the world, where coastal ecosystems are shared by countries that want to move from conflict to cooperation–from the river deltas between Pakistan and India to the Adriatic coastline of the former Yugoslavia."[62]

Lessons Learned

One lesson learned is that cross-boundary reefs can complicate things and it is easier if the reefs are completely within a single country. Then again, the Great Barrier Reef faces water quality and pollution runoff risks while located entirely in one country.

In the region, there are ongoing conflicts between Israel and Palestine. Peace in Jordan was dependent on the beginning of a peace process with Israel. While the royal family in Jordan has a stabilizing effect in the region, the Jordan Monarchy cannot publicly be seen as cooperating with Israel—especially in the eyes of people in Jordan who closely identify with Palestine.

If the Peace Park were to be revisited, it is likely to be an incremental success. There is a desire to cooperate on a range of issues, such as clean drinking water and electricity. With respect to the Peace Park, incremental success could start with something like zoning or by restarting joint research dives. In addition, USAID was very helpful in the beginning and USAID, or perhaps the World Bank or another country such as the United Kingdom or Canada, could help restart the process. Furthermore, establishing the Peace Park's headquarters in a different location such as Cyprus could help. Lastly, it would be great if the Peace Park could incorporate neighboring Egypt.[63]

Other Resources on the Binational Red Sea Marine Peace Park Between Jordan and Israel

Israel Ministry of Environmental Protection

- http://www.sviva.gov.il/English/Pages/HomePage.aspx;
- http://www.sviva.gov.il/English/env_topics/MajorBodiesOfWater/Pages/GulfOfEilat.aspx; and
- http://www.sviva.gov.il/English/env_topics/InternationalCooperation/IntlConventions/Documents/CBD-report-5-2016.pdf

Case Study #3: Mesoamerican Reef Fund

Introduction

The Mesoamerican Reef Fund (MAR Fund) was established to support conservation in the Mesoamerican Reef ecoregion, as a U.S. private 501(c)(3) nonprofit whose mission is to:

> {…} drive regional funding and partnerships for the conservation, restoration, and sustainable use of the Mesoamerican Reef.
>
> To do so, the MAR Fund operates as an ecoregional planning and coordinating body which prioritizes projects and allocates funding. The Mesoamerican Reef Fund aspires to be known and respected as a trustworthy and transparent fundraising mechanism able to sustain and finance effective transnational alliances, policies, and practices that conserve the Mesoamerican Reef and advance the health and well-being of the region's people.

The MAR Fund raises funds for the following initial program areas:

- 1. Saving our sanctuaries: a legacy of caring. Establishment and protection of an interconnected network of priority coastal and marine protected areas in the region.
- 2. Fishing for the future: sustainable fisheries for a thriving reef. Community participation in co-management of their fisheries, establishment of fish replenishment zones and monitoring and protection of fish spawning aggregations.
- 3. Climate Change: the need is now. Supporting adaptation to climate change and its impacts on the reef.
- 4. Belize Marine Fund: enhancing marine conservation in Belize with the participation of local stakeholders, for wider regional benefit.
- 5. Clean water for the reef: improving sewage and solid waste management for a healthier ecoregion and its people.[64]

The MAR Fund focuses on the largest transboundary reef and the largest barrier reef in the Atlantic Ocean (Figs. 9.3 and 9.4 and Table 9.2).

WWF supported its initiation via its Mesoamerican program. Eventually, WWF approached the four preexisting national trust funds to co-develop a long-term regional fund. These four national trust funds are known as the Protected Areas Conservation Trust (PACT; of Belize), Fundación para la Conservación de los Recursos Naturales y Ambiente en Guatemala (FCG; Foundation for the Conservation of Natural Resources and Environment in

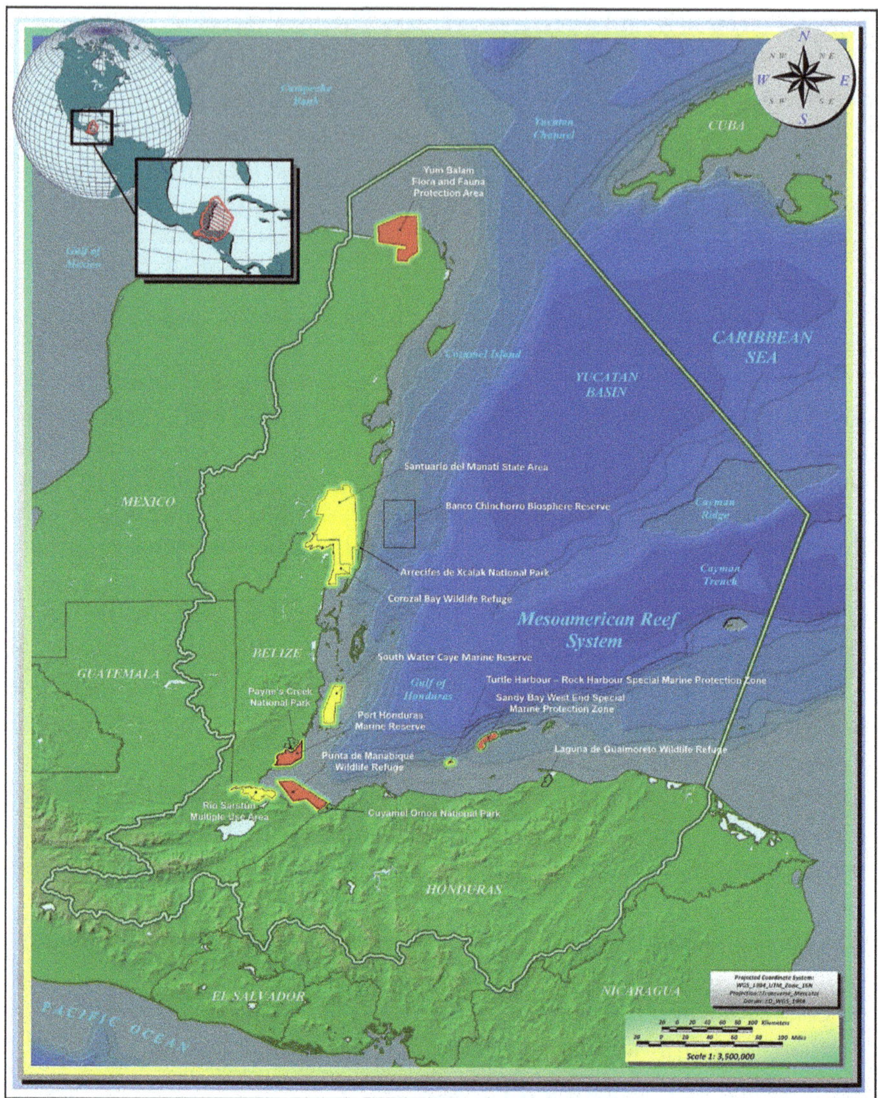

Fig. 9.3 Map of priority coastal and marine protected areas targeted by Phases I and II of the Project "Conservation of Marine Resources in Central America," Funded by KfW (*Credit* MAR Fund)

Guatemala), Fundación Biósfera (Biosphere Foundation; of Honduras), and Fondo Mexicano para la Conservación de la Naturaleza (FMCN; Mexican Fund for the Conservation of Nature). TNC eventually joined WWF in the effort to create the regional fund. Collectively, in 2004, the MAR Fund was launched.[65]

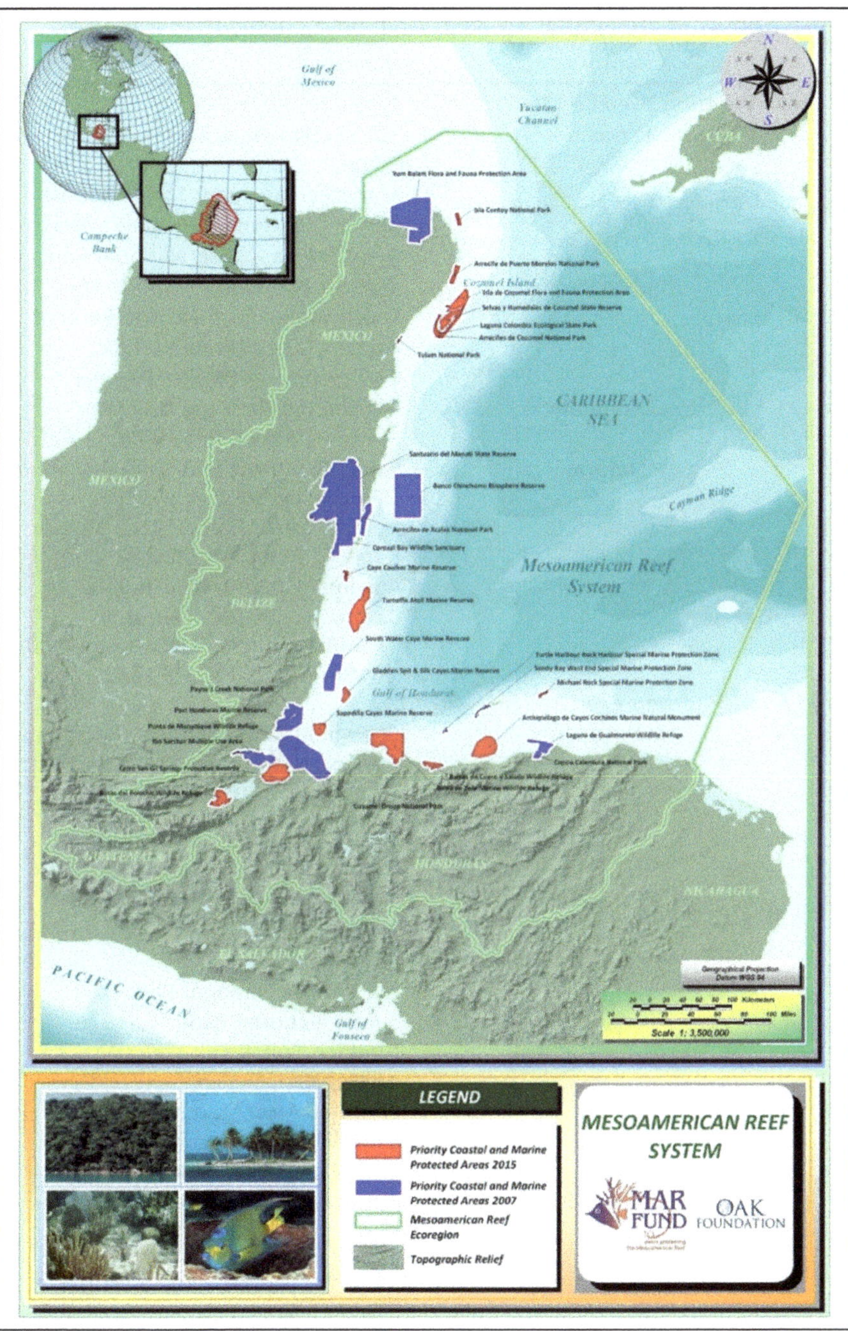

Fig. 9.4 Map of the first and second priority coastal and marine protected areas in the MAR fund region (*Credit* MAR Fund)

Table 9.2 List of priority protected areas from both priority-setting exercises

	Country	Name
First priority setting exercise (2007)		
1	Mexico	Yum Balam Flora and Fauna Protection Area
2		Manatee Sanctuary State Reserve
3		Banco Chinchorro Biosphere Reserve
4		Arrecifes de Xcalak National Park
5	Belize	Corozal Bay Wildlife Sanctuary
6		South Water Caye Marine Reserve
7		Payne's Creek National Park
8		Port Honduras Marine Reserve
9	Guatemala	Río Sarstún Multiple Use Area
10		Punta de Manabique Wildlife Refuge
11	Honduras	Sandy Bay West End Special Marine Protection Zone
12		Marina Turtle Harbor Rock Harbor Special Marine Protection Zone
13		Cuyamel Omoa National Park
14		Capiro y Calentura National Park/Laguna de Guaimoreto Wildlife Refuge
Second priority-setting exercise (2015–2016)		
1	Mexico	Cozumel Island: includes the following areas, Arrecifes de Cozumel National Park/Área Arrecifes de Cozumel Flora and Fauna Protection Area/Forests and Wetlands of Cozumel State Reserve/Laguna Colombia State Ecological Park
2		Arrecife de Puerto Morelos National Park
3		Isla Contoy National Park
4		Tulum National Park
5	Belize	Gladden Spit & Silk Cayes Marine Reserve
6		Turneffe Atoll Marine Reserve
7		Sapodilla Cayes Marine Reserve
8		Caye Caulker Marine Reserve
9	Guatemala	Bocas del Polochic Wildlife Refuge
10		Cerro San Gil Reserve for Natural Springs

(continued)

Table 9.2 (continued)

	Country	Name
11	Honduras	Archipiélago de Cayos Cochinos Natural Marine Monument
12		Barras de Cuero y Salado Wildlife Refuge
13		Michael Rock Special Marine Protection Zone
14		Bahía de Tela Wildlife Marine Refuge

Credit MAR Fund

The MAR Leadership Program (MAR-L Program), hosted in the Mexican Fund for the Conservation of Nature (FMCN), was designed by the Summit Foundation, FMCN and with the participation of MAR Fund. The MAR-L Program fills an important gap by focusing on people from Mexico, Belize, Guatemala and Honduras who demonstrate leadership potential and by providing them with training, mentoring, and networking opportunities that may otherwise be inaccessible to them. The Program targets early-to-mid career professionals working on marine conservation. The mission of the MAR-L Program is to protect the Mesoamerican Reef by developing the capacities of people from different backgrounds (i.e., communication experts, architects, fisherfolk, biologists, etc.) working in academia, community leaders, government officials, representatives of fisher cooperatives, and civil society. Each year, the MAR-L Program targets an emerging threat and then the Program opens up the application process for candidates from the four countries to apply. These fellows receive training on topics such as project design and technical training. Fellows will also learn leadership skills (i.e., strategic communication, storytelling, negotiation, conflict resolution, fundraising, etc.) and ultimately are provided the tools that will help them throughout their professional career and to become change agents. In the beginning, the Summit Foundation and the FMCN were the main donors to the MAR-L Program, and then in 2018, the Program received its first grant of USD$50,000 from the MAR Fund.[66] Throughout the years, at least 25 of the leaders that have been through the Program have received, or are receiving, funding from the MAR Fund for their projects in the four countries or have been contracted as consultants.

Identify the Problem

As per the 2015 Report Card on the health of the Mesoamerican Reef, prepared by the Healthy Reefs for Healthy People Initiative, the four main threats to the reef are:

1. The discharge of effluents and contaminants from human activities in the watersheds and marine areas;
2. Unsustainable coastal development;
3. Chronic fishing pressure which endangers ecosystem function in many places across the MAR; and
4. Climate change impacts on the reef.[67]

MAR Fund's programs respond to most of these threats.

As indicated above, one of the big threats for the region is sewage and solid waste. The MAR Fund began looking at the sewage problem initially through the Conservation of Marine Resources in Central America Project and then through its small grants program. A really great wastewater treatment facility was financed in Roatan, Honduras. It is a great example because it is small and hopefully can be replicated. Small is good because if the wastewater facility breaks down, the funds can be raised to fix it.[68]

The Stony Coral Tissue Loss Disease (SCTLD), also known as the white disease, was initially discovered in Florida, then was registered in Mexico in mid-2018 and in Belize in mid-2019. This is another threat facing the Mesoamerican Reef. In October 2019, the MAR Fund organized a meeting in Belize through two of its programs, the Reef Restoration Initiative and the Belize Marine Fund, with several experts and coral conservation practitioners from the MAR, the Dominican Republic and the United States, to discuss applicable recommendations for: monitoring; treatment; ex situ conservation; policy and regulation; and communication and outreach for the MAR region. Mexico already has a national action plan and Belize is working on a plan.[69]

Sargassum is another main problem, but not a current focus of the MAR Fund. Researchers think the bloom is because of all the nutrients flowing into the Amazon River, that are then carried out into the Atlantic Ocean. The *Sargassum* is coming from Africa and then back to Quintana Roo.[70]

Why the Problem Is Important

The Mesoamerican Reef (MAR) is the largest barrier reef in the Atlantic Ocean. In addition to the aesthetic beauty, the "MAR coral ecosystems provide revenue and food services to at least half of the 1 to 2 million people living on the coastline (FFEM, 2017). MAR's contribution to the local economy has been estimated at US$2.7 billion in terms of tourism, US$395 million in terms of fisheries, and between US$944 million and US$2.8 billion in terms of coastal protection."[71]

The region is home to some iconic reefs such as Utila and Roatan (i.e., especially West Bay of Roatan) in Honduras and Cozumel (i.e., particularly Palancar Reef) in Mexico.

How Problem Was Identified

There were several steps taken to identify the problems.

When the MAR Fund was established, it had no funding, and began to fundraise from zero. The Board agreed to focus its efforts on creating and strengthening a functional network of coastal and marine protected areas. Although the region encompasses the watersheds of the four countries that drain into the Caribbean, the coasts and the sea, the MAR Fund agreed that the coastal and marine protected areas were the backbone of the MAR and supporting them would contribute to the permanence of ecological processes, environmental services, connectivity, etc. In 2006, because there were over 65 coastal and marine protected areas in the region, the MAR Fund facilitated the first MPA priority-setting process. A very participative methodology was developed, which although the same for each country, allowed every country to determine the factors they considered most important.[72] An initial set of 14 coastal and marine protected areas resulted from this process (i.e., four in Mexico, Belize and Honduras, and two in Guatemala). The MAR Fund focused its fundraising efforts on this initial set of areas. In 2016, the priority-setting methodology was updated and a second regional exercise was carried out, with a resulting new set of 14 coastal and marine priority areas.[73] Although the MAR Fund has been providing support to several of these areas through its Small Grants Program, it is searching for additional funds for the areas.

As the MAR Fund supported coastal and marine protected areas, it began to also respond to other issues/threats affecting the protected areas, such as overfishing. As a result, the MAR Fund established the program, Fishing for the Future: Sustainable Fisheries for a Thriving Reef. This program began with a focus on community participation in co-management of their fisheries through the establishment, management, and monitoring of fish replenishment zones (i.e., no-take zones). Now, this program also incorporates monitoring and protection of fish spawning aggregations. As with the threat of overfishing, the MAR Fund began to focus on other threats. The Reef Rescue Initiative, for example, responds to the threat of climate change to reefs.[74]

Effectiveness of Process for Identifying Problem

The process of identifying problems, from utilizing a methodology to priority-setting exercises to establishing regional networks—such as the Reef Restoration Network and the Sustainable Fisheries Network—has been very effective.[75]

Steps Taken to Address the Problem

A few of the many steps taken, include:

- The MAR Fund was established;
- Coordination was, and is continuing, with national funds, such as PACT; and
- Fundraising is an ongoing effort.

Results

While it has been complicated to raise funds, the MAR Fund has been successful and a major result is that KfW is the largest funder to the MAR Fund. Similarly, KfW has also provided funding for an endowment fund for the MAR Fund, whose revenues may be used for the Small Grants program, Fishing for the Future program, and funding for the MAR Fund's operations. In addition, the Fonds Français pour l'Environnement Mondial (FFEM; The French Facility for Global Environment) also provided endowment funding to the MAR Fund for the Small Grants program. In addition to the use of interest from both endowments, the MAR Fund also does annual fundraising.[76]

Recently, the MAR Fund got some big initial funding from FFEM for a new program focused on fish spawning aggregation. For many species, during certain times of the year, the species will aggregate to spawn and can be quickly overfished. The MAR Fund is supporting partners who are monitoring these fish spawning aggregations. While Mexico has done great work monitoring and protecting several fish spawning aggregations and Belize has protected thirteen aggregations, in some areas these fish numbers are still declining (i.e., something is going on). The MAR Fish Project is contributing to the establishment of a regional monitoring network that can provide information to more effectively protect this natural phenomenon that sustains local economies and provides food security.

The Oak Foundation is a large Switzerland-based foundation and they were one of the largest funders in the Mesoamerican Reef region. The Oak Foundation revised their strategy and moved out of the region in 2016. The Belize Marine Fund (BMF), a program of the MAR Fund and which launched in November 2016 at the 18th RedLAC Annual Assembly in Brasilia, Brazil, was created through the USD$10 million endowment challenge grant from the Oak Foundation. The MAR Fund must raise USD$15 million as a match to mobilize this Oak Foundation commitment. The vision of the BMF is to provide long-term financial sustainability for addressing high-priority marine resources management and conservation issues in Belize for greater impact throughout the Mesoamerican Reef ecoregion. While the USD$10 M gift is contingent on the MAR Fund raising the 1.5–1.0 match, the Oak Foundation has made USD$500,000 available for marine conservation annually, for five years, as the MAR Fund raises the matching funds. To date, the BMF has awarded over USD$1.1 million to sixteen grantees, through thirteen small, sixteen targeted and six discretionary grants. These grants have leveraged over three times that amount in matching funds from the partners on the ground. Similarly, the MAR Fund has advanced discussions with KfW for Phase 3 of the Conservation of Marine Resources in Central America Project, and some of that funding could potentially go to an endowment and would trigger a match.[77]

The MAR Leadership Program has also achieved many results. In 2020, the Program celebrated its 10th anniversary and as of September 2019, a total of 105 fellows have went through the Program. In addition, the MAR Leadership Program helps fellows to launch their projects and over fifty projects are being implemented. Such projects include: sustainable fisheries; valuing mangroves; tourism and coastal development; solid waste management; blue economy; marine protected areas management; and sustainable use projects. Furthermore, many fellows have grown in their professional careers, are growing in their responsibilities (i.e., such as becoming CEOs of civil society organizations), and many have been very active in advocacy in their countries. There are also a lot of synergies between cohorts (different fellows over the years connecting to new or old cohorts) and within their cohorts. One of the most important products of this program is the network of leaders it has created. Lastly, many universities in Latin America do not touch leadership and the MAR Leadership Program is focused on developing leadership skills such as conflict resolution.[78]

Challenges and How They Were Met

Some major challenges facing the Mesoamerican Reef and where the MAR Fund operates are:

- Limited institutional capacity;
- Lack of financial sustainability and self-reliance in projects and institutions;
- Poor enforcement of protection laws and regulations;
- Limited private sector engagement;
- Lack of training opportunities;
- Incomplete scientific knowledge (i.e., information gaps); and
- Gender and social/economic inequities.[79]

The MAR Fund is a U.S. based nonprofit. There was a debate about having the organization based in one of the host countries, but being based in the United States makes it easier to raise financing. In addition, being established in one of the four countries could have placed an undue, additional amount of work or pressure on the member fund from that country. A major related challenge for the MAR Fund was when it started, they had zero funding— nothing. The MAR Fund asked themselves, what is the best way to raise funds and what is the best way to position themselves? The MAR Fund board made an important decision that the marine and coastal areas are the back-bone of the reef and that these areas should be operating and exchanging information among themselves. The MAR Fund started to fundraise for this and developed a regional process to identify key areas. Back in 2006, there were over 65 coastal and marine areas between the four countries and this was too much to cover. The MAR Fund facilitated the process using a very interesting and participative methodology where each country could focus on what aspects to define their priority areas. A total of fourteen priority areas (i.e., four each in Mexico, Belize, and Honduras, and two in Guatemala) were identified and the MAR Fund got KfW funding to target nine specific marine protected areas. The fourteen areas were eligible for small grants program. The MAR Fund supported their management plans (i.e., training their staff, helping with monitoring, working with local communities who were living in and around the protected areas on their economic activities and training communities, etc.). A related challenge is that with the four countries, each had different legislation and different needs. Yet, herein was an opportunity, as one country might be very good at surveillance or training and then they could exchange information with another country and operate as a network. In addition, councils were created in each of the protected areas in the countries to participate in the decision-making. These councils incorporate, in addition to the managers and co-managers, community representatives.

The first priority setting was completed in 2007 and KfW gave €5 million for Phase 1 and an additional €5 million for Phase II of the Conservation of Marine Resources in Central America Project. This Project focused on nine of the fourteen sites (see above). In 2015 and 2016, a second priority-setting activity took place. An opportunity for Phase 3 of the Project came up, this time for €13 million. The second priority fourteen sites will be targeted by Phase 3. The MAR Fund now has to fundraise for the new priority areas.[80]

For the MAR Leadership Program, the first challenge has been the difficulty of raising funds. A second related challenge is, how do you link capacity building to conservation impacts—particularly when the impacts will occur five years after the project is implemented (i.e., grouper populations do not increase in one year, but it takes three–five years). More specifically, in 2011, a fellow named Gabriella Navas entered the program about establishing nurseries for elkhorn coral in the Southern part of Quintana Roo, Mexico. At the time, the MAR Leadership Program had more money and the Program sponsored her trips to conferences so she could present her work and get feedback from experts. She was very successful in the community, started another nursery in Puerto Morelos, and then had nurseries all over Quintana Roo with thousands of transplanted colonies. Gabriella is an expert and says she owes all the credit to the program (i.e., for all the tools, access to the experts, etc.), but it took eight years. Another challenge for the MAR Leadership Program is to keep fellows engaged with the network. Maria Eugenia Arreola, the Director of the Program, is always sending out information (i.e., new data that is interesting, funding opportunities, etc.). María Eugenia Arreola sometimes learns that people are not collaborating, yet some are active and some are not. Some fellows in the program are becoming closed off to outsiders, like a fraternity. Before they were competing for funds, but now they are sharing opportunities with other fellows, but closing off others.[81]

Beyond Results

The MAR Fund has achieved a lot to date and its positioned well for the future. For instance, there are two other, related regional initiatives. There is the MAR Leadership Program and there is the Healthy Reefs for Healthy People Initiative (i.e., the Healthy Reef Initiative).

The Healthy Reef Initiative measures the health of the reefs with the help of seventy partners and then publicly releases its findings every two years in the form of a report card. Since 2008, the Healthy Reef Initiative has produced a report card every two years that assesses the four indicators of live coral cover, biomass of coral reef fish, biomass of commercial reef fish, and macroalgae cover. Marks are given for each country and an overall regional mark is also given. Each report card also focuses on a particular aspect. This report card is a unique thing as no one else is doing a constant report card. These report cards help to tell the rest of the people working in the region what is working and what is not working. For instance, despite bleaching and overfishing, there is still a slight trend of increased coral growth. While commercial fish stocks are still decreasing, there is 10x more fish biomass in no-take zones than other areas. Whether it is wildlife reserves, marine parks or marine reserves, there is a big amount of the ocean surface that is being protected in the region. Yet, only about 3% is no-take zones and this is one of the aspects targeted by the MAR Fund Sustainable Fisheries Program: to support the creation, management and monitoring of more no-take zones throughout the region. The Healthy Reef Initiative has also put out a few Eco-Audits whereas the Eco-Audits see if the recommendations are moving forward in the countries and the region as a whole. With respect to beyond results, the Healthy Reef Initiative has framed their methodology for insular Caribbean to help set up similar monitoring programs.[82]

The other regional initiative is the Mesoamerican Leadership Program, which as previously mentioned, builds human capital, provides capacity training to upcoming leaders, and helps develop their leadership skills so they can go out to the region and make a change. Many of these leaders are now in higher decision-making positions and are making a change.[83]

The MAR Fund works very closely with both initiatives. The MAR Fund started focusing on marine protected areas, then started focusing on key areas outside these marine protected areas (i.e., like fisheries), and then started to focus on other activities and the associated fundraising (i.e., it is very rare to get unrestricted funding). Now one of the things that the MAR Fund is looking at is climate change. The Reef Rescue Initiative, funded by KfW, is looking at contributing to the conservation of the Mesoamerican Reef by increasing the resilience and recovery ability of the MAR and thereby the environmental and cultural services it provides.

Through the initiative, the regional Reef Restoration Network was revived and two regional biennial meetings of the network have been supported. The objective of the network is to be the meeting point between groups dedicated to restoration, conservation and protection of coral reefs in Mexico,

Belize, Guatemala, and Honduras, to exchange experiences and create closer links between practitioners. Mexico and Belize are leading in reef restoration experience and Honduras is learning.[84]

The MAR Fund is working to fine-tune a parametric insurance to cover the cost of reef restoration after hurricanes, piloted by TNC for Puerto Morelos, Mexico. The MAR Fund started discussing the insurance product ten years ago and is now working with Willis Towers Watson to design the parametric insurance. The studies that have been carried out to design the insurance are:

- 1. A risk and vulnerability assessment, which includes analysis to determine the characteristics of hurricanes which are a particular risk to reefs, based on historical analysis of hurricane damage to reefs (i.e., based on meta-analysis of data from the Caribbean Basin). Results of this analysis provide correlation between hurricane characteristics and damages. A digital database with information related to the hurricane impacts and the potential damages to coral reefs (pre- and post-hurricane), has been prepared as an additional result of this study;
- 2. Cost estimation for restoration and emergency response actions. Information obtained will be the basis for estimation of insurance coverage requirements, based on specific reef sites, as well as the cost for emergency response and short and long-term restoration. This provides the basis for the risk modelling required to underpin parametric index design; and
- 3. Preliminary requirements for a simple parametric insurance structure, for example pay-out estimation at pure hazard thresholds, have been completed.

The MAR Fund has helped the countries in the MAR select coral sites to pilot the parametric insurance. Seven pilot sites across the four countries were selected. Willis Towers Watson will help design the insurance and after the first iterations of the parametric insurance with TNC, the Proponents are looking to see how to make it more sensible.

The MAR Fund has adopted and replicated the Early Warning Emergency Response Protocol, developed by TNC initially for the coast of Quintana Roo, in the rest of the MAR countries. The Protocol includes the profile of the emergency response brigades and the actions required to secure rapid attention to reefs damaged by hurricane impacts. The MAR Fund, in collaboration with TNC and local partners, has carried out the dissemination of the Emergency Response Protocol at each site proposed for the insurance.

Further, the MAR Fund developed a GIS database and 278 maps with overlaid scenarios to identify reef exposure to various threats, including hurricanes and shipping lanes. The MAR Fund also facilitates the exchange and implementation of reef restoration methodologies through a Small Grants Program to support innovation in restoration techniques, ongoing fieldwork of reef-restoration practitioners, and now also response to the Stony Coral Tissue Loss Disease.[85]

Another collaboration is the Caribbean Pacific Alliance for Marine Conservation Finance:

> a partnership of the Mesoamerican Reef (MAR) Fund, the Caribbean Biodiversity Fund, and Pacifico (Environmental Funds Platform of the Tropical Eastern Central Pacific) has plans underway to launch Blue Challenge, an incubator that will focus on fisheries, ecotourism and waste management businesses, three industries that have significant impact on marine ecosystems and in which promoting good conservation actors can have beneficial impact on system health.[86]

This Alliance will also seek to attract private finance to the region.[87]

Lessons Learned

According to María José González, Executive Director of the MAR Fund, one of the key lessons learned is that you need to be very flexible as a mechanism and be ready to adapt and to take opportunities as they come up that respond to the threats to the MAR. It is important to know how to take on new initiatives. For instance, the MAR Fund is in the position to bring actors together on different topics—now its mangroves, reef restoration, lionfish invasion, and reef fisheries. With respect to mangroves, the MAR Fund held a workshop in September 2019 on regional mangroves to identify what are the gaps, who is doing what, and for what is funding needed. As a result, a regional mangrove strategy was developed and the MAR Fund has begun to look for funding for some of the actions established in the strategy.

Similarly, the MAR Fund does not do activities on the ground, but is flexible regarding what is needed on the ground. For instance, there needs to be an easy mechanism for locals to acquire funds, but simultaneously, the MAR Fund is able to adopt procedures for grantees on the ground and still have adequate control of the funding so that donors know what the results were, how the money was spent, and that the money was properly spent. Ultimately making it as easy as possible, but being accountable to donors is key.

Another lesson learned is that each country has different legislation and as a regional fund, you need to take this into account. For mangroves, the legislation is very clear for the three countries of Mexico, Guatemala, and Honduras, however, in Belize, there is new mangrove legislation, mainly resulting from the effort to better manage the Belize Barrier Reef System World Heritage Site.[88]

From María Eugenia Arreola of the MAR Leadership Program, one lesson learned was that the program was very expensive because there were four in-person workshops (i.e., one week workshop in each country). The costs were high, in part due to flights being expensive and due to hiring the consultants to teach the modules. Because of the challenge of raising funds and to make it more cost-effective, in 2017, the Program did not open cohort applications, and instead, did a review of the Program. Starting in 2018, the Program is now undertaking just 1 in-person workshop for 15 days in Mexico. During those two weeks, the Program offered the most useful trainings that it thought should be prioritized based off those trainings that were the most well-received in the past. There were also four webinars offered and a short, four-day site visit to Baja California, Mexico. This said, it was easier for fellows to take two weeks off (i.e., for the in-person training in Mexico) and then four days off (i.e., for the field trip) approximately six months later; in contrast, it was very hard to take four weeks off. Essentially, you do not need an expensive program to achieve a successful program which provides opportunities to learn about local projects and to create bonds between fellows.[89]

Another lesson learned is that it is about the attitude of the fellows and about maintaining the excellency of the trainings. Likewise, the more the Program links fellows to outside experts and to have exchanges with people outside, the more the fellows feel satisfied about the benefits they receive from the Program. In addition, having a cohort with representatives from different sectors and different backgrounds is very useful as many are dealing with similar challenges. By working together and telling their stories, the fellows can learn from one another. For instance, some fellows in the Program are against poor coastal development, but you have fellows from the government who are approving the development plans.[90]

An additional lesson learned is that it is tough to raise funding for capacity building when more donors are interested in funding projects. The MAR Fund was one of the founders of the MAR Leadership Program and has supported 25 fellows and their projects. The Summit Foundation is also supporting some of the fellows' projects. This said, if you have a great project, but not a great leader implementing the project, the project can fail. Similarly, you can have a great idea and a great leader, but if the project is not well thought through (i.e., lack of funding plan, etc.), it will fail.[91]

Other Resources on Mesoamerican Reef Fund

Comisión Centroamericana de Ambiente y Desarrollo (CCAD; Central American Commission for Environment and Development)

- https://www.sica.int/ccad/

Fondo Mexicano para la Conservación de la Naturaleza, A.C. (FMCN; Mexican Fund for the Conservation of Nature, A.C.) of México

- https://fmcn.org/?lang=en

Fundación Biosfera (Biosphere Foundation) of Honduras

- https://www.biosfera.org/en/

Fundación para la Conservación de los Recursos Naturales y de Ambiente en Guatemala (FCG; Foundation for the Conservation of Natural and Environmental Resources in Guatemala)

- http://fcg.org.gt/?lang=en

Healthy Reefs for Healthy People

- www.healthyreefs.org/cms/

Integrated Transboundary Ridges-to-Reef Management of the Mesoamerican Reef

- https://www.thegef.org/project/integrated-transboundary-ridges-reef-man agement-mesoamerican-reef

Kanan Kay Alliance

- www.alianzakanankay.org/en/

MAR Fund

- http://marfund.org/en/;
- https://marfund.org/en/annual-reports/; and
- https://marfund.org/en/reef-rescue-initiative/

Mesoamerican Reef Leadership Program

- http://liderazgosam.org/en/

Protected Areas Conservation Trust (PACT) of Belize

- https://www.pactbelize.org/

Sistema de la Integración Centroamericana (SICA; Central American Integration System)

- https://www.sica.int/

Financial Analysis

The following financial analysis will look at return and risk.

Return

The returns generated from a government's international budgetary allocations can come in a few different forms. For instance, this transfer of payment could generate a financial return or a non-financial return, such as a change in policy. If the funding from one government to another government is in the form of a loan, then the lender may earn interest. In addition, the host government could earn a financial return if the MPA incorporates revenue streams (e.g., visitor fees for tourists, user fees for fisherfolk, etc.) as outlined in the domestic budgetary allocations chapter.

Risk

The following is a summary of some of the challenges and possible solutions to the formation of an international peace park:

{1.} Endogeneity, or the perception that conservation is a consequence rather than a constituent of peace-building. *Solution*: Dialectical policy – consider conservation as a trust-building activity in a feedback loop.

{2.} Preexisting local conflicts undermine peace-building by labeling it as co-optation and dispossession. *Solution*: Resolve micro-conflicts beforehand, acknowledge past grievances and make process transparent to local residents.

{3.} Conservation agencies are external to security decision apparatus. *Solution*: Make conservation a strategic asset in foreign policy matters with participation of scientists and environmental agency staff in deliberations.

{4.} International NGOs that are hesitant to interfere in border issues. They may fear denied access or political retribution – or that a confrontational approach will lead to their marginalization. *Solution*: NGOs should play an epistemic role – exchanging knowledge between parties and mediating for community members on all sides.[92]

Business Risk

The specific business risk of international budgetary allocations depends on the underlying program or project that is being invested in—such as large-scale ecotourism operations, undertaking a marine spatial planning exercise, and/or establishing a Marine Stewardship Council (MSC) certified-fishery. Additional business risks are that funds may not be renewed in future years and the underlying activities that are being funded may not be designed to become a commercially viable business in the future.

Strategic Risk

There are many strategic risks for international budgetary allocations to MPAs. One such strategic risk includes choosing which country to support and choosing the specific partners among local NGOs, local community associations, the private sector, government agencies, and/or international organizations.

Another strategic risk is choosing what approach to take. This may include: decisions over whether to set up an in-country trust fund or a trust fund registered outside of the host country; whether to establish a collective, international peace park or whether to establish a single country MPA; and whether to concentrate financing efforts offshore and/or to finance watershed conservation.

There is also "donor drift" when donor countries may shift their funding to other countries, other conservation financing instruments, to different challenges (for example, away from coral restoration efforts to marine pollution cleanup) or to different biomes.

Reputation Risk

Reputation risks may arise for international budgetary allocations if funds are squandered due to mismanagement, corruption, or improper budgeting. Another reputational risk may arise if the underlying conservation activities or the conservation financing instrument does not succeed (i.e., illegal fishing continues). Reputational risks can also arise if local stakeholders are not properly consulted and feel their fishing grounds are being misappropriated. Another concern "{…} with the establishment of peace parks is the increase in state presence in these rural areas. In the past state actors, most often military, were unwelcome and dangerous, as they often brought corruption and human rights violations, such as looting and violence."[93] Additional reputational risks include, "in some instances, states have used MPAs to extend rights over disputed marine resources, restrict the freedom of others, and establish sovereignty over maritime space."[94]

Liquidity Risk

Similar to a government's domestic budgetary allocations, a government's international budgetary allocations are often illiquid. Thus, a government's contribution to a multilateral agency such as the UN Green Climate Fund, the creation of transfrontier conservation area, or funding an international trust fund (i.e., like the Phoenix Islands Protected Area Conservation Trust), is likely to have few options for liquidation.

From the recipients' perspective, the disbursements from a government's international budgetary allocation may come at a later time then when their expense (i.e., paycheck for marine patrols and park administration) is due. In turn, this may create another liquidity risk, for instance, if a park ranger gets paid yearly, but the park ranger needs to pay his or her child's school fees on a monthly basis.

Operational Risk

There are many operational risks related to a government's international budgetary allocation to MPAs. For instance, the management structure of an international peace park could present a risk if there is not equal representation and equal benefit sharing. There could be additional operational risks for an international peace park due to ineffective coordination such as not harmonizing legislation, not harmonizing master plans, and not sharing scientific research across national borders. Similarly, "there are the practical questions of how the bureaucracies of both countries could jointly manage differences in organization, language, culture, and budgetary politics. The nature and scope of marine conservation could also lead to serious disagreement and political backlash."[95]

The role of third parties can impact the operations of a joint, marine protected area. For instance, the Mammellone Bank, located in the Strait of Sicily between Italy and Tunisia, has been managed through bilateral fishing agreements. However, "the suspicion is that fishing vessels from Libya and Egypt, which face lower fuel and crew costs, are depleting the target species in the area."[96] Likewise, there could be operational risks to MPAs due to other third-party initiatives such as fisheries or exploration for undersea fossil fuel resources.

One operational risk for international budgetary allocations is that there might be delays in disbursements (i.e., similar to the previous liquidity risk) and the lack of a sufficient project pipeline. Additional operational risks can be the challenges associated with getting funding "to the beach" (i.e., lack of internet, few passable roads to the shoreline, informal banking sector, etc.) and the lack of audited financial statements for the MPA.

Legal and Regulatory Risk

One of the greatest legal and regulatory risks associated with international peace parks concerns border security. For instance, "cross-border cooperation efforts must be supported and/or guarded by corresponding laws and regulations of participating parties."[97] If not, there is a legal risk. If such international budgetary allocations involve a multilateral agency, such as the World Bank, which often have near-universal national government support, there is likely to be fewer legal and regulatory risks.

Significant legal and regulatory risks can occur with transfrontier conservation areas when one party attempts to exit the partnership. Similarly, the MPA may entirely dissolve, may slowly fall apart, or there could be internal disputes between which level of government (i.e., state versus federal) has authority over the MPA.

There could be legal risks if one nation is a party to a convention, while the other nation is not. There can also be legal and regulatory risk associated with bilateral policy and funding priority changes with the donor country. There can be overlapping legal title claims, such as between an offshore oil exploration concession and an adjacent MPA with designated areas for tourism and use by local fisherfolk. Furthermore, counties may find themselves in between a conflict among two other countries, such as China and Australia's competition in the South Pacific.[98]

Credit Risk

Credit risk includes default risk, bankruptcy risk, downgrade risk, and settlement risk.

One example of settlement risk is that the local entity's bank or credit union (i.e., such as a remote outpost for an international peace park) may not be able to receive a very large international transfer of funds. In contrast, multilateral agencies often have near-universal national government support and, thus, are less likely to have credit risks.

Market Risk

Market risk includes interest rate risk, equity price risk, foreign exchange risk, and commodity price risk.

A specific market risk for a government's international budgetary allocation is that, due to foreign exchange fluctuations, the amount committed to another country may rise or fall overtime. A similar situation could occur due to commodity price risk, where a donor country's level of support is tied to the revenue they earn from a globally traded commodity.

Risk, Return, Time (Horizon), Taxes, Liquidity, Legal, and Unique (RRTTLLU)

Risk and Return

Please see above for the risk and return associated with government international budgetary allocations. It is important to note that returns, should be viewed not just as financial returns but also returns from net positive climate, community, and biodiversity outcomes as "regional seas agreements and regional fisheries management organizations appear to provide a better forum for neighboring states to air grievances, arrange face-to-face negotiations and developer cooperation, outside the global spotlight."[99]

Time Horizon

The time horizon for government international budgetary expenditures will depend on the nature of the commitment. For instance, many countries have contributed to the World Bank and the UN for several decades. In contrast, other government international budgetary allocations can be more short-term in nature such as the USAID multiyear commitment to the Red Sea Marine Peace Park. Furthermore, the ARPA Trust Fund is a sinking 25-year fund, while the Mesoamerican Reef Fund is set up for perpetual financing.

Taxes

There are likely limited, if any, taxes due to the donor government or host government if the governments are the sole providers and recipients of the international budgetary allocations. However, there could be local taxes such as sales tax levied on concessions, employment taxes, and/or some form of tourism tax for visitors. Depending on what exactly is being financed, there could be a loss of tax revenue such as fewer excise taxes, royalties, or income taxes—especially when compared to an offshore oil concession or the granting of international fishing access.

Liquidity

Similar to government domestic budgetary allocations, liquidity might exist with concessions or other onsite facilities (e.g., tourist attraction or lodging) that could be transferred to another owner. However, the actual underlying MPA is likely to be illiquid unless the federal government can transfer the protected area to a state or province (i.e., or vice versa).

If a protected area has multiple cash flows via different conservation financing instruments (e.g., ecotourism, international budgetary allocations, and impact investments), there could be a liquidity risk if one of the more significant sources of financing slows down.

Legal

Legal considerations associated with government international budgetary expenditures will vary from country to country. When considering the establishment of a trust fund, it is important to consider the jurisdiction of the legal entity that will manage the trust fund.

Unique

Unique aspects include the establishment of multi-country, joint marine protected areas. Most of the other conservation financing instruments—such as debt-for-nature conversions, payments for ecosystem services, or ecotourism—focus on one country.

Policy Analysis

The following policy analysis will look at: defining the problem; establishing goals; selecting a policy; implementing a policy; and evaluating the policy.

Defining the Problem

Governments, whether its domestic or international budgetary expenditures, are facing competing interests such as healthcare, energy access, and education. Tropical coral reef degradation and loss continues today and many host countries do not have adequate funds, hence, the need for international support. Furthermore, many tropical coral reefs, such as the Mesoamerican Reef and the Coral Triangle, stretch across national borders and there is a need for countries to establish cross border, international peace parks.

Establishing Goals

The goal of developed countries earmarking a certain percentage of their revenue to ODA dates back to 1958, when the World Council of Churches proposed a target of at least 0.7%. In 1969,

> {..} the Pearson Commission – in its report Partners in Development – proposed a target of 0.7% of donor GNP. {…} With the revised System of National Accounts in 1993, gross national product was replaced by gross

national income (GNI), an equivalent concept. DAC members' performance against the 0.7% target is therefore now shown in terms of ODA/GNI ratios.[100]

Thus, a subset goal is to have a significant portion of this ODA be allocated to marine conservation and toward effective policies and projects to reduce the pressure on tropical coral reefs.

Selecting a Policy

The largest providers of ODA in 2017,[101] and the largest recipients of ODA in 2016 or 2017,[102] in terms of a percentage of GNI, were as follows (Table 9.3).

In contrast, the largest providers of ODA in 2017,[103] and the largest recipients of ODA in 2017,[104] in US dollars were as follows (Table 9.4).

This said, "the United States, Germany, France, and Japan provided the largest amounts of bilateral biodiversity-related ODA. Norway, Iceland, France and Belgium dedicated the highest shares of their ODA portfolios to biodiversity-related activities."[105]

Specifically related to financial support for marine conservation:

A large portion of funds over the period from 2010-2016 comes from the Global Environment Facility (GEF), Green Climate Fund, Oak Foundation,

Table 9.3 Largest ODA providers and recipients as a percentage

	Country	Net ODA provided in 2017 (% of GNI)		Country	Net ODA received in 2016 or 2017 (% of GNI)
1	United Arab Emirates	1.027	1	South Sudan	60.5 (2016)
2	Sweden	1.019	2	Tuvalu	45.6 (2017)
3	Luxembourg	0.996	3	Marshall Islands	26.9 (2017)
4	Norway	0.993	4	Micronesia	25.1 (2017)
5	Turkey	0.954	5	Somalia	24.8 (2017)
6	Denmark	0.737	6	Malawi	24.1 (2017)
7	United Kingdom	0.699	7	Central African Republic	23.4 (2017)
8	Germany	0.667	8	Kiribati	21.2 (2017)
9	Netherlands	0.604	9	Liberia	20.8 (2017)
10	Switzerland	0.457	10	Afghanistan	18.7 (2017)

Source OECD and World Bank

Table 9.4 Largest ODA providers and recipients in US dollars

	Country	Net ODA in 2017 (billions of USD$)		Country	Net ODA received in 2017 (current USD$, billions of USD$)
1	United States	USD$34.732	1	Syria	USD$10.361
2	Germany	USD$25.005	2	Ethiopia	USD$4.118
3	United Kingdom	USD$18.103	3	Afghanistan	USD$3.804
4	Japan	USD$11.463	4	Bangladesh	USD$3.740
5	France	USD$11.331	5	Nigeria	USD$3.359
6	Turkey	USD$8.121	6	Yemen	USD$3.234
7	Italy	USD$5.858	7	Turkey	USD$3.142
8	Sweden	USD$5.563	8	India	USD$3.094
9	Netherlands	USD$4.958	9	Jordan	USD$2.921
10	Canada	USD$4.305	10	Iraq	USD$2.907

Source OECD and World Bank

European Union, and the European Commission Framework Programme 7. Together, these five organizations have contributed around $300 million in the given six-year period (UN, 2018). The top five funded countries in terms of total project receipts have been Indonesia, Mexico, Sri Lanka, United States, and Tuvalu which have collectively received more than a quarter of all project funding, around $488 million (UN/ICRI, 2018). The majority of the funding ($1.8 billion, 95% of total funds) has been focused on four main areas: conservation and sustainability, marine protected area management, promoting sustainable living and alternative livelihoods, and fisheries management.[106]

This said, Germany and France are among the leaders when it comes to providing international finance for marine conservation.

Implementing a Policy

Germany has several windows of marine financing, including KfW. In addition, Germany's Federal Ministry for Economic Cooperation and Development has a ten-point Plan of Action for Marine Conservation and Sustainable Fisheries.[107]

France utilizes its Agence Française de Développement (AFD; French Development Agency) along with the French Facility for Global Environment (FFEM) for marine conservation.

Evaluating the Policy

Germany and France have both established policies and committed significant funding. For instance:

> To respond to the need for new financial resources, in December 2016 the German Ministry for Economic Cooperation and Development (BMZ), through the KfW Development Bank (KfW), established the Blue Action Fund. In June 2017, the Swedish Ministry for Foreign Affairs joined as a donor. The Agence Française de Développement (AFD) became a donor in October 2018.[108]

Further:

> KfW Group on behalf of the German Federal Government, the European Investment Bank (EIB) and the Agence Française de Developpement (AFD) launched the Clean Oceans Initiative to support the development and implementation of sustainable projects that will reduce pollution in the world's oceans over the next five years. This partnership will provide EUR 2 billion long-term financing for projects aiming at reducing marine litter, especially plastics, as well as untreated wastewater discharge, with a view to crowding-in private sector investment.[109]

KfW is the largest funder to the MAR Fund. KfW and FFEM have provided funding toward the MAR Fund's endowment fund.[110] KfW also provided financing for Mirova's Althelia Sustainable Ocean Fund.[111]

As described by Dr. Petra Lundgren:

> One of the major issues that many of the developing countries experience is the loss of academics to developing countries and thus relying on short term aid project and donor expertise to provide solutions. There is a parallel to restoring coral reefs here, ensure you build the capacity within, so that it eventually does not have to rely on inputs from the outside.[112]

Thus, it is important for countries and philanthropies to also financially support higher education and research such as "establishing research grants for local scientists to work with issues affecting their country and their coral reefs."[113]

Future Outlook for Instrument

Many governments face budget constraints, and if structured properly, international budgetary allocations to MPAs will provide opportunities for both the host country's government and the private sector. One difficulty in this context of limited public funds has been the attempt to move the funding out of the aid budgets and over to the finance ministries because this funding is more budgetary support than traditional aid.

With respect to international peace parks, additional transfrontier protected areas should also be established. While there have been discussions of a Tri-National Park planned for the Gulf of Honduras between Belize, Guatemala, and Honduras, the actual implementation appears far away.[114] There are some potential areas for future peace parks, such as the South China Sea as well as between Kenya and Tanzania.

The following is some well heeded guidance on establishing international peace parks:

- Establish interagency steering committees and subsidiary project management systems;
- Prepare a stakeholder analysis and include stakeholder consultation mechanisms and processes;
- Use prior studies and 'lessons-learned' from earlier project work in the area and in neighboring countries;
- Identify the capacity-building and staff-training needs;
- Draw up an inventory of financial and human resources needs;
- Make a plan for long-term capacity building and institutional strengthening in the project's implementation;
- Analyze all recent land-use, natural resource management, and conservation planning activities and development proposals from former cases of protected areas to determine their current status; and
- Support community-driven economic development and alternative livelihood proposal activities {...}.[115]

Furthermore, it is noteworthy to pay attention to the emerging Global Fund for Coral Reefs, which saw the Prince Albert II of Monaco Foundation and Vulcan commit USD$250,000 each toward establishing the Global Fund for Coral Reefs.[116]

Other Resources on Government International Budgetary Allocations

Global Transboundary Conservation Network

- http://www.tbpa.net/page.php?ndx=12

IUCN's "Transboundary Conservation: A systematic and integrated approach" Report

- https://portals.iucn.org/library/node/45173

OECD Statistics on External Development Finance Targeting Environmental Objectives Including the Rio Conventions

- http://www.oecd.org/dac/stats/rioconventions.htm

Transboundary Conservation: A New Vision for Protected Areas

- https://www.press.uchicago.edu/ucp/books/book/distributed/T/bo3771 136.html

Notes

1. UNESCO. "Waterton Glacier International Peace Park." Accessed December 13, 2016. http://whc.unesco.org/en/list/354.
2. Gonzales, José and Hermilio Rosas. "The Caral Archeological Site and Supe Valley Restoration." In *Conservation Capital in the Americas*, edited by James N. Levitt. 91.
3. Spergel, Barry and Melissa Moye. "Financing Marine Conservation: A Menu of Options." 22.
4. Blue Action Fund. "About." Accessed December 18, 2019. https://www.blu eactionfund.org/about-2/.
5. NOAA. "'Sister Site' Partnership Established Between Papahānaumokuākea and Kiribati's Phoenix Islands MPA." July 31, 2017. Accessed December 30, 2019. https://sanctuaries.noaa.gov/news/features/0909_kiribati.html.
6. OECD. "Development Assistance Committee (DAC)." Accessed January 3, 2020. http://www.oecd.org/dac/development-assistance-committee/.
7. OECD. "Net ODA: Total, Million US Dollars, 2000–2017." Accessed January 3, 2020. https://data.oecd.org/oda/net-oda.htm.

8. OECD's OECD.Stat. "Aid (ODA) Disbursements to Countries and Regions [DAC2a]." Accessed January 3, 2020. https://stats.oecd.org/Index.aspx?datasetcode=TABLE2A.
9. OECD. "Net ODA: Total, Million US Dollars, 2000–2017." Accessed January 3, 2020. https://data.oecd.org/oda/net-oda.htm.
10. UN Environment et al. "Analysis of International Funding for the Sustainable Management of Coral Reefs and Associated Coastal Ecosystems." 4.
11. Global Transboundary Conservation Network. "UNEP-WCMC Transboundary Protected Areas Inventory-2007." 2007. Accessed February 25, 2020. http://www.tbpa.net/page.php?ndx=78.
12. OECD. "Biodiversity-Related Official Development Assistance 2016." Accessed April 11, 2020. https://www.slideshare.net/OECDdev/biodiversityrelated-official-development-assistance-2016. 3.
13. UN Environment et al. "Analysis of International Funding for the Sustainable Management of Coral Reefs and Associated Coastal Ecosystems." 5.
14. Asia Development Bank. "Saving the Coral Triangle." Accessed April 13, 2020. https://development.asia/case-study/saving-coral-triangle.
15. UN Environment et al. "Analysis of International Funding for the Sustainable Management of Coral Reefs and Associated Coastal Ecosystems." 6.
16. Ibid. 5.
17. Commonwealth Secretariat. "Commonwealth Clean Ocean Alliance." Accessed January 15, 2020. https://bluecharter.thecommonwealth.org/action-groups/marine-plastic-pollution/.
18. Development Innovations Group. "MWRP APS." Accessed January 15, 2020. http://www.developinnovations.com/MWRP-APS.aspx.
19. Plastic Solutions Fund. "Partners." Accessed January 15, 2020. https://plasticsolution.org/partners/.
20. World Bank Group. "PROBLUE: Healthy Oceans, Healthy Economies, Healthy Communities." Accessed January 15, 2020. https://www.worldbank.org/en/programs/problue/overview.
21. GEF Marine Plastics. "Home." Accessed January 15, 2020. https://gefmarineplastics.org/.
22. European Commission. "Press Release: European Commission and European Investment Fund Launch €75 Million BlueInvest Fund." February 2, 2020. Accessed February 10, 2020. https://ec.europa.eu/commission/presscorner/detail/en/IP_20_167.
23. USFWS International Affairs. "Species Programs." Accessed April 11, 2020. https://www.fws.gov/international/wildlife-without-borders/species-programs/.
24. Green Climate Fund. "Global Context: Response Description." Accessed April 4, 2017.http://www.greenclimate.fund/about-gcf/global-context#the-big-picture.
25. UN Environment et al. "Analysis of International Funding for the Sustainable Management of Coral Reefs and Associated Coastal Ecosystems." 7.

26. Marques, Fernanda F.C. Interviewed by Brian McFarland. March 2019.

27. Ibid.

28. Ibid.

29. FUNBIO. "GEF MAR: What Is." Accessed September 10, 2019. https://www.funbio.org.br/en/programas_e_projetos/gef-mar/.

30. Marques, Fernanda F.C. Interviewed by Brian McFarland. March 2019.

31. GEF. "Marine and Coastal Protected Areas." Accessed September 10, 2019. https://www.thegef.org/project/marine-and-coastal-protected-areas.

32. Ibid.

33. FUNBIO. "GEF MAR: What Is." Accessed September 10, 2019. https://www.funbio.org.br/en/programas_e_projetos/gef-mar/.

34. Ibid.

35. Marques, Fernanda F.C. Interviewed by Brian McFarland. March 2019.

36. Ibid.

37. Ibid.

38. Ibid.

39. Ibid.

40. Ibid.

41. Ibid.

42. Ibid.

43. Portman, Michelle E. and Yael Teff-Seker. "Factors of Success and Failure for transboundary Environmental Cooperation: Projects in the Gulf of Aqaba." 6.

44. Alon, Gideon. "Ministries Accept Int'l Report on Eilat Bay Pollution." *Haaretz.com.* August 21, 2001. Accessed September 10, 2019. https://www.haaretz.com/1.5397377.

45. Portman, Michelle E. Interviewed by Brian McFarland. August 2019.

46. Zimmer, Carl. "The Partitioning of the Red Sea." 627–628.

47. Teff-Seker, Yael. Interviewed by Brian McFarland. January 2020.

48. Portman, Michelle E. and Yael Teff-Seker. "Factors of Success and Failure for transboundary Environmental Cooperation: Projects in the Gulf of Aqaba." 14.

49. Ibid. 6.

50. Zimmer, Carl. "The Partitioning of the Red Sea." 627–628.

51. Denny, Megan. "Seven Wonders of the Underwater World." *PADI*. May 27, 2013. Accessed September 19, 2019. https://www2.padi.com/blog/2013/05/27/seven-wonders-of-the-underwater-world/.

52. Portman, Michelle E. and Yael Teff-Seker. "Factors of Success and Failure for Transboundary Environmental Cooperation: Projects in the Gulf of Aqaba." 6.

53. Teff-Seker, Yael. Interviewed by Brian McFarland. January 2020.

54. Zimmer, Carl. "The Partitioning of the Red Sea." 627–628.

55. Portman, Michelle E. and Yael Teff-Seker. "Factors of Success and Failure for transboundary Environmental Cooperation: Projects in the Gulf of Aqaba." 7.

56. Shaked, Dr. Yonathan and Prof. Amatzia Genin. "The Israel National Monitoring Program in the Northern Gulf of Aqaba Funded by Israel's Ministry of Environmental Protection." March 2014. Accessed September 10, 2019. www.sviva.gov.il/English/env_topics/marineandcoastalenvironment/Docume nts/Gulf-of-Eilat-Monitoring-Program-2013-Scientific-Report-English-Abs tract.pdf. 11.

57. Ali, Saleem H. "Conclusion: Implementing the Vision of Peace Parks." In *Peace Parks*, edited by Saleem H. Ali. 338.

58. Portman, Michelle E. and Yael Teff-Seker. "Factors of Success and Failure for Transboundary Environmental Cooperation: Projects in the Gulf of Aqaba." 8.

59. Portman, Michelle E. Interviewed by Brian McFarland. August 2019.

60. Ibid.

61. Ibid.

62. Zimmer, Carl. "The Partitioning of the Red Sea." 627–628.

63. Teff-Seker, Yael. Interviewed by Brian McFarland. January 2020.

64. MAR Fund. "What Is MAR Fund?" Accessed September 10, 2019. https:// marfund.org/en/what-is-marfund/.

65. Gonzalez, María José. Interviewed by Brian McFarland. August 2019.

66. Arreola, María Eugenia. Interviewed by Brian McFarland. August 2019.

67. Healthy Reefs for Healthy People Initiative. "2015 Report Card." 2016. http://www.healthyreefs.org/cms/wp-content/uploads/2015/05/MAR-EN-small.pdf.

68. Gonzalez, María José. Interviewed by Brian McFarland. August 2019.

69. Ibid.

70. Ibid.

71. Vertigo Lab. "Innovations for Coral Finance, ICRI Publication." 66.

72. López-Gálvez, Ileana Catalina. "Prioritization of Coastal and Marine Protected Areas in the Mesoamerican Reef Region." May 2007. https://www.marfund. org/wp-content/uploads/2014/10/Priorization-Coastal-and-Marine-Protected-Areas.pdf.

73. Rojas, Oscar E. "Review and Update of the Prioritization of Coastal-Marine Protected Areas in the Mesoamerican Reef Ecoregion." April 2016. https://www.marfund.org/wp-content/uploads/2014/10/Final-report-priority-CMPA-OR-20160716.pdf.

74. Gonzalez, María José. Email message to author. April 12, 2020.

75. Gonzalez, María José. Interviewed by Brian McFarland. August 2019.

76. Ibid.

77. Gonzalez, María José. Email message to author. April 12, 2020.

78. Arreola, María Eugenia. Interviewed by Brian McFarland. August 2019.

79. MAR Fund. "MAR Fund Prospectus." 2014. Accessed August 12, 2019. https://ingles.marfund.org/wp-content/uploads/2014/10/1.-MAR-FundProspectus.pdf. 6.
80. Gonzalez, María José. Interviewed by Brian McFarland. August 2019.
81. Arreola, María Eugenia. Interviewed by Brian McFarland. August 2019.
82. Gonzalez, María José. Interviewed by Brian McFarland. August 2019.
83. Ibid.
84. Gonzalez, María José. Email message to author. April 12, 2020; Also see Mesoamerican Reef Restoration Network. "Home." Accessed April 13, 2020. https://coralmar.org.
85. Gonzalez, María José. Email message to author. April 12, 2020.
86. Iyer, Venkat et al. *Finance Tools for Coral Reef Conservation*. 43.
87. Gonzalez, María José. Email message to author. April 12, 2020.
88. González, María José from MAR Fund. Interviewed by Brian McFarland. August 2019.
89. Arreola, María Eugenia. Interviewed by Brian McFarland. August 2019.
90. Ibid.
91. Ibid.
92. Ali, Saleem H. "Conclusion: Implementing the Vision of Peace Parks." In *Peace Parks*, edited by Saleem H. Ali. 340.
93. Blundell, Arthur G. and Tyler Christie. "Liberia: Securing the Peace Through Parks." In *Peace Parks*, edited by Saleem H. Ali. 236.
94. Mackelworth, PC et al. "Geopolitics and Marine Conservation: Synergies and Conflicts." 1.
95. Lambacher, Jason. "Nesting Cranes: Envisioning a Russo-Japanese Peace Park in the Kuril Islands." In *Peace Parks*, edited by Saleem H. Ali. 269.
96. Mackelworth, PC et al. "Geopolitics and Marine Conservation: Synergies and Conflicts." 6.
97. Portman, Michelle E. and Yael Teff-Seker. "Factors of Success and Failure for Transboundary Environmental Cooperation: Projects in the Gulf of Aqaba." 3.
98. Hollingsworth, Julia. "Why China Is Challenging Australia for Influence over the Pacific Islands." *CNN*. July 22, 2019. Accessed December 30, 2019. https://www.cnn.com/2019/07/22/asia/china-australia-pacific-investment-intl-hnk/index.html.
99. Mackelworth, PC et al. "Geopolitics and Marine Conservation: Synergies and Conflicts." 14.
100. OECD. "The 0.7% ODA/GNI Target—A History." Accessed April 6, 2020.http://www.oecd.org/dac/stats/the07odagnitarget-ahistory.htm.
101. OECD. "Net ODATotal, % of Gross National Income, 2017." Accessed July 23, 2019.https://data.oecd.org/oda/net-oda.htm.
102. World Bank. "Net ODA Received (% of GNI)." Accessed July 23, 2019.http://data.worldbank.org/indicator/DT.ODA.ODAT.GN.ZS?year_high_desc=true.

103. OECD. "Net ODA: Total, Million US Dollars, 2000–2017." Accessed July 23, 2019.https://data.oecd.org/oda/net-oda.htm.

104. World Bank. "Net Official Development Assistance Received (Current US$)." Accessed July 23, 2019.http://data.worldbank.org/indicator/DT.ODA.ODAT. GN.ZS?year_high_desc=true.

105. OECD. "Biodiversity-Related Official Development Assistance 2016." Accessed April 11, 2020. https://www.slideshare.net/OECDdev/biodiversityrel ated-official-development-assistance-2016. 7.

106. Iyer, Venkat et al. *Finance Tools for Coral Reef Conservation.* 12.

107. German Federal Ministry for Economic Cooperation and Development. "Marine Conservation and Sustainable Fisheries: Ten-Point Plan of Action." Accessed April 6, 2020. www.bmz.de/en/publications/type_of_publication/inf ormation_flyer/information_brochures/Materialie262_marine_conservation. pdf.

108. Blue Action Fund. "About." Accessed April 6, 2020. https://www.blueactio nfund.org/about-2/.

109. KfW. "The World's Major Climate Financiers EIB, KfW and AFD Launch a EUR 2 Billion Initiative." October 12, 2018. Accessed April 6, 2020. https://www.kfw.de/KfW-Group/Newsroom/Latest-News/Pressemit teilungen-Details_491008.html.

110. Gonzalez, María José. Interviewed by Brian McFarland. August 2019.

111. Marchant, Christopher. "Althelia Sustainable Ocean Fund Raises Additional $50m." *Environmental Finance.* November 19, 2019. Accessed December 31, 2019. https://www.environmental-finance.com/content/news/althelia-sus tainable-ocean-fund-raises-additional-$50m.html.

112. Lundgren, Petra. Email message to author. March 24, 2020.

113. Ibid.

114. Betoulle, Jean-Luc and Astrid Alvarado (Editors). "Site Conservation Planning: Gulf of Honduras: Belize, Guatemala and Honduras." 2005. Accessed January 28, 2020. https://www.cbd.int/doc/meetings/mar/rwebsa-wcar-01/ other/rwebsa-wcar-01-guatemala-03-en.pdf.

115. Fuller, Stephan. "Linking Afghanistan with Its Neighbors Through Peace Parks: Challenges and Prospects." In *Peace Parks*, edited by Saleem H. Ali. 308–309.

116. Prince Albert II of Monaco Foundation and Vulcan. "Global Fund for Coral Reefs." Accessed February 21, 2020. https://www.icriforum.org/sites/default/ files/Flyer%20Coral%20Reef%20Fund%20final.pdf.

10

Impact Investing

Introduction

As defined by the Forum for Sustainable and Responsible Investment, a member of the Global Sustainable Investment Alliance, "sustainable, responsible and impact investing (SRI) is an investment discipline that considers environmental, social and corporate governance (ESG) criteria to generate long-term competitive financial returns and positive societal impact."[1]

Impact investing, particularly related to tropical coral reef conservation, is a relatively new class of emerging conservation financing instruments. Such instruments include impact investments structured by nonprofit and for-profit organizations, along with through dedicated funds such as Mirova's Althelia Sustainable Ocean Fund.

Historical Overview

The modern socially responsible investment movement, which is related to impact investing, has its origins in the "1970s as part of grassroots campaigns to boycott firms in South Africa as well as the nuclear weapons and tobacco industries."[2]

Other notable dates associated with impact investing include:

- 1970s and 1980s: The "concept that the economy and ecology can be mutually beneficial has its roots in the 1970s and evolved in the 1980s to include discussions about sustainable development."[3]

© The Author(s) 2021
B. J. McFarland, *Conservation of Tropical Coral Reefs*,
https://doi.org/10.1007/978-3-030-57012-5_10

- 1990s: In "the 1990s, the term 'triple bottom line' to denote economic, ecological, and social performance became a popular catchphrase among businesses that aimed for more than just finance profits."[4]
- 2009: The Global Impact Investing Network (GIIN) is founded.[5]
- 2010: World Bank's Wealth Accounting and the Valuation of Ecosystem Services (WAVES) partnership is launched.[6]
- 2010: TNC developed an impact investment strategy and formed a dedicated team with support from the Robertson Foundation. TNC would later form NatureVest in 2014 with assistance from JPMorgan Chase & Co.[7]
- 2011: Althelia Ecosphere is launched.[8]
- 2012: Natural Capital Declaration is launched.[9]
- 2012: The Economics of Ecosystems and Biodiversity (TEEB) for Business Coalition is launched. The Coalition would later become the Natural Capital Coalition in 2014.[10]
- 2012: TNC's Conservation Notes are launched.[11]
- 2014: Credit Suisse's Nature Conservation Notes are issued.[12]
- 2018: The Sustainable Blue Economy Finance Principles are publicly announced.[13] The initiative "is currently being spearheaded by the European Commission, WWF and the World Resources Institute; and is supported by the European Investment Bank."[14]
- 2018 (December)–2019 (January): Although the initial public offering was postponed,[15] "a first-of-its kind investment trust{called Global Sustainability Trust Plc, managed by Aberdeen Standard Investments} targeting private 'impact investments' was trying to raise £200 million ($250 million) via a London listing," and was claimed to "be the first board-led sustainable investment trust."[16]
- 2018 (December): The world's "first benchmark index to price financial water risk to equities launched on Bloomberg."[17]
- 2019 (December): Circulate Capital, "the investment management firm dedicated to incubating and financing companies and infrastructure that prevent ocean plastic in South and Southeast Asia (SSEA), today announced the first close of the US$106 million Circulate Capital Ocean Fund (CCOF). It is the world's first investment fund dedicated to address Asia's plastic crisis and is also one of the ten largest ASEAN-based Venture Capital Funds in the market."[18]

Mechanisms of Instrument

On a general level, impact investing involves either an equity investment or the extension of an interest-bearing loan to a particular entity, project, or program that is generating a return while also having a positive impact on the environment. As explained by WWF:

> Sustainable capital is an area of public and private financing that utilizes access to favorable equity, credit, and microfinance as an incentive to promote environmental sustainability and responsible business practices. {…} Examples of sustainable capital {…} include environmental investment or biodiversity enterprise funds; forest securitization, or more broadly, eco securitization; favorable credit or loan terms tied to performance incentives or standards required by public or private lenders; and emerging microfinance programs to promote community conservation and social development.[19]

Impact investors "can be classified into three categories: Impact First (primarily seeking to maximize impact while secondarily expecting financial returns); Investment First (fiduciaries primarily seeking market-rate or premium returns and secondarily seeking a positive social or environmental impact); and Catalyst First (primarily seeking to give or invest to help build the impact investing industry and infrastructure)."[20]

As broad examples of impact investing, the Cambridge, Massachusetts-based organization Root Capital offers "{…} trade credit, which is available for up to one year and oriented around a production cycle such as a harvest."[21] The Caral-Supe Trust Fund is structured in such a way that "investors will receive an average market return for participating in the investment fund, and all residual capital gains over the average return will be donated to the trust fund."[22] The Meloy Fund for Sustainable Community Fisheries is an impact investment fund that "incentivizes the development and adoption of sustainable fisheries by making debt and equity investments in fishing-related enterprises that support the recovery of coastal fisheries in Indonesia and the Philippines."[23] The Overseas Private Investment Corporation, which transformed into the U.S. International Development Finance Corporation (DFC), offers political risk insurance.[24] In addition, Conservation International (CI) launched CI Ventures, LLC which is "an investment fund that provides loans to small- and medium-sized enterprises that operate in the forests, oceans and grasslands where Conservation International works."[25]

Further, there are different smaller scale, "green banking" models such as credit cards and checks with a portion of proceeds donated to conservation activities

Size of Instrument

There are several different ways to assess the size of the impact investing sector and its ability to assist with the financing of coral reef conservation.

For instance, the U.S. SIF Foundation found that as of 2018, "investors in the United States consider environmental, social and governance (ESG) factors across $12 trillion of professionally managed assets {...}."[26]

According to the Global Impact Investing Network (GIIN) 2019 Annual Impact Investor Survey, there was an estimated USD$33 billion committed in 2018 to 13,000+ impact investments. These respondents planned to invest more than USD$37 billion in 2019 to 15,000+ impact investments.[27] Furthermore, more than 2300 signatories with approximately USD$80 trillion in assets under management have agreed to the six voluntary principles of the United Nations' Principles for Responsible Investment.[28]

However, conservation activities, in order to attract impact investments and especially funding from institutional investors, will need to demonstrate their activities, among many things, are scalable, have a proven track record, and can absorb large sums of funding.

There are several large funds such as:

- Alliance to End Plastic Waste, an alliance with 40+ global organizations and a 5-year commitment to provide USD$1.5 billion[29];
- Circulate Capital Fund, "the investment management firm dedicated to incubating and financing companies and infrastructure that prevent ocean plastic in South and Southeast Asia (SSEA), today {December 2019} announced the first close of the US$106 million Circulate Capital Ocean Fund (CCOF). It is the world's first investment fund dedicated to address Asia's plastic crisis and is also one of the ten largest ASEAN-based Venture Capital Funds in the market[30];"
- March Asset Management's Mediterranean Fund[31];
- Sky Ocean Rescue and Sky Ocean Ventures; and
- Sustainable Oceans Fund, which "over five years, we're investing £25 million in companies that can help us all to give up plastic for good."[32]

Introduction to Case Studies

The following case studies will examine Mirova's Althelia Sustainable Ocean Fund, the Parametric Insurance for the Mesoamerican Reef, and the Arrecifes del Sureste Marine Sanctuary of the Dominican Republic. The Althelia Sustainable Ocean Fund is a first of its kind fund that aims to scale up sustainable fishing and other marine protection activities while making investors a commercial return. The Parametric Insurance for the Mesoamerican Reef is the first case of parametric insurance applying to a natural structure. The Arrecifes del Sureste Marine Sanctuary, which is the second largest MPA of the Dominican Republic and the third largest MPA for the whole Caribbean region, entered into a first such Public–Private Partnership in 2018.

Case Study #1: Mirova's Althelia Sustainable Ocean Fund

Introduction

The Althelia Ecosphere was launched by Sylvain Goupille and Christian del Valle[33] in 2011 and closed its first round of financing in June 2013 with €60 million (USD$80 million at the prevailing exchange rate) for the Althelia Climate Fund. In February 2017, the Althelia Sustainable Ocean Fund hit the market as "a first-of-its-kind fund that aims to scale up sustainable fishing and other marine protection activities while making a commercial return for investors."[34] In October 2017, Althelia Ecosphere was acquired by Mirova, the affiliate of Natixis Global Asset Management.[35]

The initial aim of the Althelia Sustainable Ocean Fund was to "{…} reach a first close of up to $60 million by the summer {of 2017), reaching up to $100 million next year {2018.}."[36] The Fund's initial focus was to:

{…} invest in three main areas:

1. Sustainable fisheries and aquaculture – such as investing in stock and biodiversity recovery, or businesses that reduce wastage.
2. Responsible supply chains and infrastructure – such as investing in processing or improved traceability.
3. Broader seascapes and the blue economy – such as a share of revenues from ecotourism or blue carbon credits. [37]

Furthermore, the Fund is:

{…} structured as an eight-year, Luxembourg SICAV. {…} As well as directly courting institutions, the fund will also issue loan notes via Credit Suisse. These Sustainable Ocean Notes will be issued in ticket sizes starting at $150,000. {…} The notes also help provide a secondary market for investors, explained Fabian Huwyler, a Zurich-based director of sustainability affairs at Credit Suisse, which is structuring the notes. {…} PwC will be the auditor and Allen & Overy the legal advisor to the fund. Huwyler added that a Sustainable Fisheries Finance Toolkit that will set best-in-class environmental, social and governance investment standards for the sector was being jointly developed by Conservation International, EDF, Althelia and Credit Suisse.[38]

A SICAV stands for "Société d'investissement à Capital Variable, or SICAV, {which} is a publicly-traded open-end investment fund."[39]

More specifically, the Althelia Sustainable Ocean Fund:

{…} invests in scalable, impact-aligned businesses with the objective of generating real assets, build resilience in coastal ecosystems and create sustainable economic growth and livelihoods in the blue economy. {…} It aims to generate a market-based rate of return by making between 10 and 20 investments in marine and coastal enterprises that can deliver marine conservation and improved livelihoods. Target impacts include improved food and climate security, stabilisation or increased fish stocks, improved livelihoods and economic value for local stakeholders, prevention of marine waste and plastic pollution, as well as ecological biodiversity and conservation gains in ocean and coastal ecosystems.[40]

Identify the Problem

One of the main problems is that there is a lot of money available from the private sector, and specifically from institutional investors, that want to invest in the conservation finance asset class. Yet, this is a challenging asset class and investment proposition for many reasons including the fact that many local projects are too small in size and many of their business models are unproven when viewed from a traditional, commercial standpoint.[41]

Why the Problem Is Important

As discussed in a white paper produced by Althelia Ecosphere, UNEP:

estimates that over 60 percent of the world's total Gross National Product comes from areas within 100 km of our coastlines. Worldwide revenue from seafood amounts to more than USD 190 billion, while marine and coastal tourism generate approximately USD 161 billion annually. Fishing, aquaculture and tourism combined provide over 300 million jobs worldwide. In short, oceans help underpin global prosperity and food security and are fundamental to the long-term existence of society.[42]

Yet, there is not enough funding going into oceans generally speaking and getting private finance into the space is one of the biggest problems.

How Problem Was Identified

The problems of climate change and the lack of investable opportunities for institutional investors were identified by Althelia Ecosphere's visionary founders, a team of world-class experts, and their extensive network including Conservation International (CI) and the Environmental Defense Fund (EDF).

Effectiveness of Process for Identifying Problem

With visionary founders, a team of world-class experts, and their extensive network, the Althelia Ecosphere has been very effective at identifying the problems of climate change and the lack of investable opportunities for institutional investors.[43]

Steps Taken to Address the Problem

The first step taken was to establish Althelia Ecosphere. Althelia Ecosphere then launched the Althelia Climate Fund. Althelia Ecosphere took steps to identify strategic partners, identify investors, and assess projects alongside the development of their Environmental, Social, and Governance (ESG) Policy[44] and an Impacts Monitoring Framework.[45]

Additional steps with respect to the Sustainable Ocean Fund include:

- Strategic partners, such as EDF, were identified;
- The Sustainable Fisheries Finance Toolkit was developed; and
- The Sustainable Ocean Fund was launched.

Results

The Althelia Ecosphere was launched by Sylvain Goupille and Christian del Valle in 2011 and closed its first round of financing in June 2013 with €60 million (USD$80 million at the prevailing exchange rate) for the Althelia Climate Fund. The Althelia Climate Fund reportedly raised approximately USD$117 million by the end of 2014 with a goal to reach USD$150 million.[46]

With respect to the Althelia Sustainable Ocean Fund, some notable results include:

- The Fund "reached a second close for its Althelia Sustainable Ocean Fund, raising an additional $50 million and taking the fund's total commitments to $92 million. The additional commitments come from European institutional investors, including German development bank KFW and insurance providers Garance and BNP Paribas Cardif {...} The fund will remain open to investment until the end of this year, and the target final close is expected to be more than $100 million. The fund opened in 2017 and had initially planned to reach the $100 million mark by the end of 2018."[47]
- As of August 2020, Mirova's Sustainable Ocean Fund reached a final close at USD$132 million in committments, surpassing its original target.[48]
- The Fund reported already made "{...} investments in aquaculture in Mexico and in a company developing technology to reduce bycatch in global fisheries."[49]
- The Fund established a partnership with USAID "and has executed a {USD$50 million} risk sharing guarantee through USAID's Development Credit Authority (DCA) that will provide loan guarantees directly to portfolio investments made by the fund[50];" and
- Strategic partnerships have been developed between the Fund and leading, international nonprofit organizations CI and EDF.[51]

This said, the Sustainable Ocean Fund (SOF) and CI "signed a framework agreement in 2016 whereby the Sustainable Ocean Fund gets access to technical assistance funding, pipeline and technical capabilities."[52] Similarly, "Althelia and EDF signed a framework agreement in 2016 for EDF to act as scientific and technical advisor to the SOF."[53]

The Arrecifes del Sureste Marine Sanctuary of the Dominican Republic, spearheaded by Blue Finance, was expected to receive its initial capital expenditures from the Sustainable Ocean Fund.[54]

It should also be noted that just the fact the Sustainable Ocean Fund was created at this scale (i.e., got a team together with strong partners and secured financing), despite all the challenges and despite its first-mover status, is a major result.[55]

Challenges and How They Were Met

As discussed in a white paper produced by Althelia Ecosphere, some of the challenges facing investments in sustainable landscapes, some of which are unique to oceans, includes:

- lack of perception and understanding of the scale and urgency of the ocean challenge and opportunity;
- limited investment-ready marine projects and a lack of knowledge regarding bankable opportunities, as well as few credible investment case studies to point to in the space;
- several NGOs are doing credible work around policy and capacity-building, however few organizations are developing a pipeline of investment opportunities;
- investment ticket sizes are typically smaller than larger institutional investors' minimum investment size but larger than many individual impact investors' desired allocation;
- lack of understanding of the composition and nature of marine/ ocean assets and how they can be organized to receive and repay investment;
- ocean industries and in particular aquaculture are perceived as the 'Wild West' with little regulation and few formal and reputable market structures to provide investors with comfort;
- projects can be perceived as difficult and risky with many operational and political challenges to overcome, and issues such as leakage and enforcement are particularly acute in developing country contexts;
- few common legal frameworks and market structures across the 'Ocean Economy', particularly in the developing world where the need for investment in the transition to sustainability is the greatest; and
- individual projects can be quite different and require context-specific interventions and financing packages, for example, fishery and aquaculture species have widely varying biological growth and recovery rates.[56]

To overcome these challenges, the Sustainable Ocean Fund's strategy is to:

- Have a "diversified project pipeline with multiple revenue streams {…}";

- Select "mature and semi-mature projects {…}"; and
- Use "rigorous screening criteria {…}."[57]

Further, the financial structure of the Sustainable Ocean Fund is to:

- Use "impact-focused capital {…}";
- Incorporate a "first loss facility {…}"; and
- Provide "debt financing {…}."[58]

An additional challenge for fisheries is there is a high technical hurdle that a lot of people cannot get over. While there are lots of investors interested in fisheries and blended financing opportunities, questions such as what does sustainable fisheries look like, what should investors be asking project developers, and what should investors be looking at in general, had to be resolved.[59]

To address this challenge, *the Principles for Investment in Sustainable Wild-Caught Fisheries*, outlining the following nine principles was developed:

- 1. Compliance with local, national and international fisheries laws and regulations;
- 2. Current environmental status;
- 3. Future environmental status;
- 4. Monitoring and enforcement;
- 5. Traceability and transparency;
- 6. Human rights;
- 7. Stakeholder engagement;
- 8. Stakeholder access; and
- 9. Food, nutrition and livelihood security.[60]

Similarly, determining how much money are we talking about, where will the money come from, how can things be standardized, what kind of personnel is needed, figuring out who needs to talk to who, and seeing what tools have worked elsewhere, can be a challenge.[61]

Beyond Results

It is difficult to discern the long-term sustainability of the Sustainable Ocean Fund as the Fund is still relatively new. Ideally, the pace and size of the Fund will pick up, all the money will be deployed, investors will be happy with the demonstrated returns, and additional Sustainable Ocean Funds—perhaps with slightly shifted parameters or focus—will be developed.[62] This said, in addition to the Sustainable Ocean Fund, Althelia Ecosphere also established several other initiatives including the:

- Althelia Biodiversity Fund Brazil[63];
- Althelia Climate Fund[64];
- Madagascar Climate and Conservation Fund[65]; and
- Nature Conservation Notes.[66]

As discussed in a white paper produced by Althelia Ecosphere:

> The investment thesis is simple – better-managed fisheries and coastal habitats are more profitable in the longer-term than poorly managed ones. They can support higher levels of harvest, require lower levels of effort (are more operationally efficient) and can access premium and value-added markets.[67]

In addition:

> Further value can be built both on and off the water by improving fish handling, processing and routes to market to increase the value of landed catch and avoid environmental and economic waste. In addition, improved integrated management programs can work to turn marine waste and pollution – which can themselves have significant negative impacts on fish stocks, biodiversity and quality of life for coastal communities – into cost-neutral or profitable recycling and waste-to-energy systems.[68]

Furthermore, some of the recent projects financed by the Althelia Climate Fund include the:

- Cordillera Azul National Park REDD+ Project[69];
- Guatemalan Conservation Coast Project[70];
- Sumatra Merang Peatland Project[71];
- Taita Hills Conservation and Sustainable Land Use Project[72]; and
- Tambopata-Bahuaja REDD+ and Agroforestry Project.[73]

Lessons Learned

A key lesson learned by Tim Fitzgerald of EDF is that it has become abundantly clear that the money is the easy part. Likewise, finding the money is not nearly as hard as coming up with a good, investable project that meets all the specific criteria laid out. There are a lot of great ideas and potential projects, but in some cases, they might not be big enough, or are too big, be in the wrong sector, lack a gender or food security lens, be in the wrong geography, or any number of other things. To receive funding, "you need about a hundred things to go right, whereas you just need one thing to go wrong to not proceed."[74]

In the past, financing had always been a challenge in the ocean space. Up until now, such investments had been characterized as unique, boutique investments. However, we are starting to talk about funds or investments in the nine or ten-figure range and this led to larger investors and a different class of players to start paying attention. The general ballpark to fix fisheries around the world is about USD$250 billion. This said, a couple of USD$1 million fisheries deals is not going to get it done. However, stitching together nine or ten-figure deals and commitments may get us to some inflection point at which point financing really turns up.[75]

Another lesson learned, although not related to the Sustainable Ocean Fund, is that there is a lot of money available for climate change mitigation work, but not so much for adaption and particularly for oceans. EDF is working with a number of countries to look at financing from the Green Climate Fund for fisheries because there is a lot of climate adaptation and resilience work needed with fisheries.[76]

Other Resources on Althelia Sustainable Ocean Fund

Althelia Fund

- News and Press: https://althelia.com/news-press/; and
- Sustainable Ocean Fund: https://althelia.com/sustainable-ocean-fund/

Bloomberg

- http://www.bloomberg.com/research/stocks/private/snapshot.asp?privcapId=264550904

EDF's Sustainable Fisheries Toolkit

- http://fisherysolutionscenter.edf.org/fisheries-toolkit/project-phases

European Investment Bank's Althelia Climate Fund Page

- http://www.eib.org/projects/pipelines/pipeline/20100720

<u>Principles for Investment in Sustainable Wild-Caught Fisheries</u>

- http://www.fisheriesprinciples.org/files/2019/05/updated-PrinciplesInves tmentWEB_final.pdf

Case Study #2: Parametric Insurance for the Mesoamerican Reef

Introduction

The Mesoamerican Reef encompasses coral reef systems of Mexico, Belize, Guatemala, and Honduras. Mexico "is home to the northernmost coral reef in the eastern Pacific and home of the Mesoamerican Reef, the second largest barrier reef in the world, that begins off the tip of the Mexican Yucatan {…}."[77]

Several stakeholders worked together to develop a parametric insurance product for a portion of the Mesoamerican Reef off the coast of Puerto Morelos, Mexico (Figs. 10.1 and 10.2).

Traditionally,

> {…} parametric insurance has mainly been used to compensate farmers for agricultural losses in developing areas, even though other economic sectors at risk (transportation, etc.) are also beginning to use it. There is only one case of parametric insurance applying to a 'natural structure': the Coastal Zone Management Trust in the resort towns of Cancun and Puerto Morelos in Mexico. This parametric insurance program is a Public-Private Partnership between hotel owners (insured party), the state of Quintana Roo, Mexico, an insurance company (Swiss Re), and an environmental NGO (The Nature Conservancy), and presents several advantages for coral ecosystem conservation.[78]

As previously mentioned, there are several stakeholders involved:

- Hotel owners will "pay Swiss Re an amount between US$1 million and US$7.5 million for parametric insurance premiums on the policy[79];"
- The international insurance company Swiss Re's role is "if any destructive storms damage the reefs, {Swiss Re} pays out between US$25 million and US$70 million in any given year, mainly for restoration of the reef[80];"
- Quintana Roo State "acts as a self-insurance when beaches and reefs are damaged by a storm but the policy threshold value is not met[81];" and

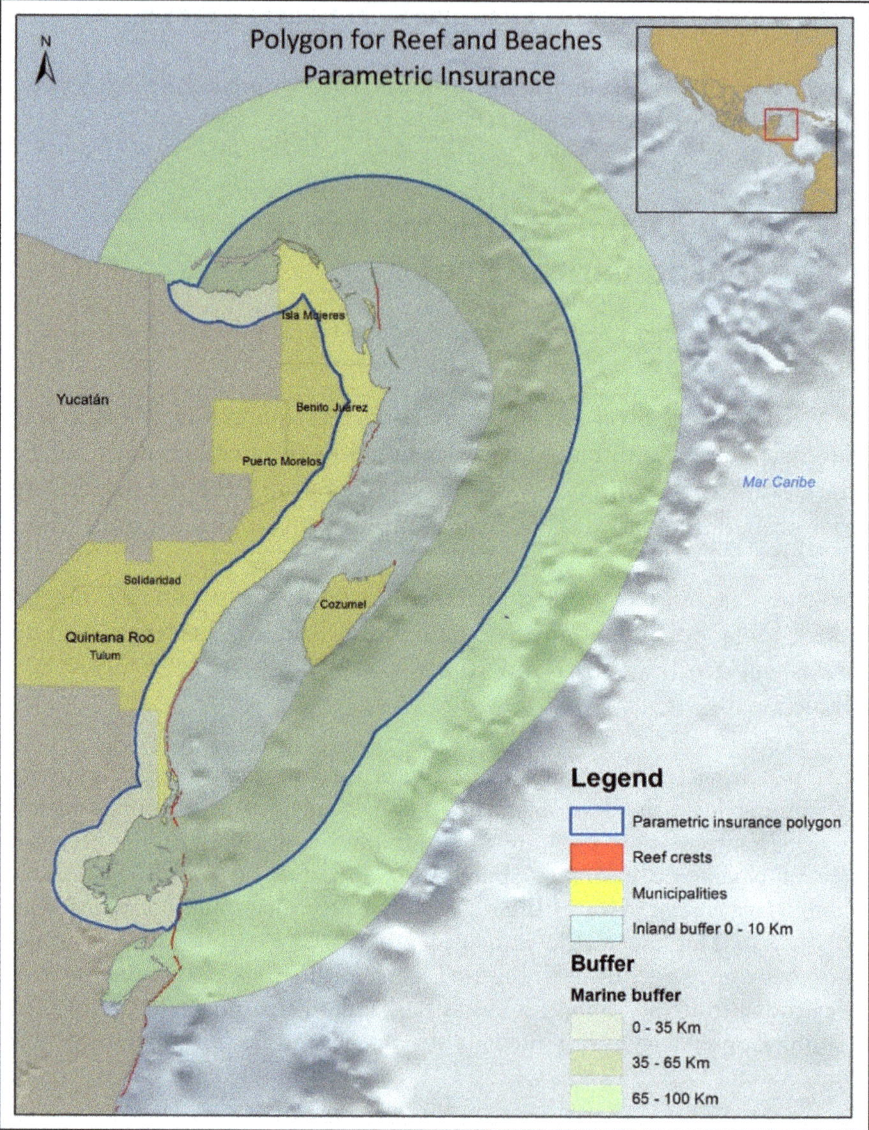

Fig. 10.1 Map of Mesoamerican reef covered by parametric insurance (*Credit* The Nature Conservancy, Reef Insurance Primer, 2019)

- TNC "pays for the science-based restoration and maintenance of the reefs."[82]

The advantages of general parametric insurance, although not specific to coral ecosystems, include: "decrease in administrative costs (insurance purchasing

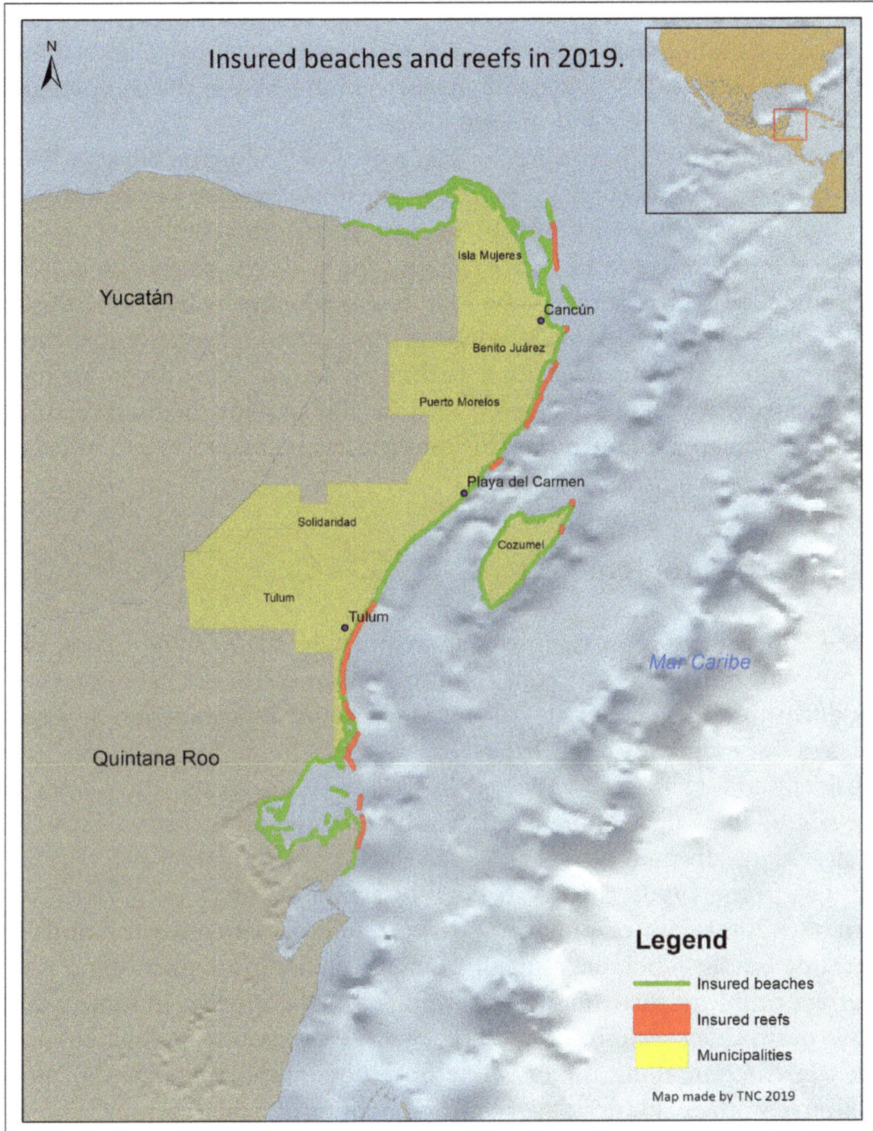

Fig. 10.2 Map of insured beaches and reefs covered by parametric insurance (*Credit* The Nature Conservancy, Reef Insurance Primer, 2019)

and closure); limitation of moral risk (possibility that the insured party overestimates their compensation); transparency; and compensation available more quickly, right after the climatic events."[83]

Inconveniences of general parametric insurance, again not necessarily specific to coral ecosystems, include: "only works for risks that can be modeled with a high level of certainty; from one individual to another, calculated compensation can be lower than real losses; potential market limitations; and risk that there is actually no correlation between the compensation established in the policy and the real losses."[84]

If a relatively small hurricane, say category 1 or 2, hits the Yucatan Peninsula, there are volunteers (e.g., tour operators may provide boats and hotels may donate gasoline) who step in to help restore the coral reefs offshore. In addition, there could be some funding for such restoration provided by environmental NGOs such as TNC, or local governments such as the Quintana Roo Government. However, having a parametric insurance instrument in place allows the impacted local entities to transfer the risk to an insurance company which in turn would then quickly provide the funding, in the event of a catastrophic hurricane, for the significant coral reef restoration that would then be required.

Identify the Problem

In the 1970s and 1980s, there was approximately 40–50% light coral cover off the coast of Mexico's Yucatan Peninsula. As of 2018, the light coral cover was down to approximately 8–10% due to extensive degradation as a result of a series of events since the 1980s.

Initially, the reefs were impacted by yellow band disease, then there was an overgrowth of algae in the late 1980s because the sea urchins, which graze on algae, virtually disappeared. The population of herbivorous fish decreased and then in the 1990s, there were coral bleaching events.

In the 2000s, water pollution—as a result of agricultural runoff and untreated sewage—became a major problem. In addition, overfishing is also a current problem; particularly, overfishing of the herbivores. In some specific areas, overtourism has impacted the reefs. However, it should be noted that construction along the coastline does not impact the reefs too much. Collectively, these local stressors are compounded when a hurricane hits the Mesoamerican Reef and similar scenarios are happening throughout the Caribbean. This is where the parametric insurance instrument comes in.[85] For instance, "{…} Hurricanes Wilma and Emily caused a combined damage of over \$17 billion to Mexican infrastructure in 2015."[86]

Why the Problem Is Important

These problems are important because hurricanes can put a lot of stress on the local ecosystem, which in turn impacts coastal protection, can reduce revenue and the availability of jobs attributed to tourism, and decrease the productivity of the local fisheries. For instance, the reefs off the coast of the Yucatan Peninsula are degrading approximately 2% per year where no hurricanes have hit, but in contrast, the reefs are degrading more than 6% (i.e., three times greater degradation) where a category four or five hurricane has hit. Hurricanes can have such an impact on reefs that up to 20% of an area's coral cover can be lost in the matter of a few hours from one hurricane. Typically, when coral colonies break, a new colony will be produced. However, because of the other stresses (e.g., water pollution, bleaching, etc.), when the hurricanes hit, there is no capacity for the reefs to rebound. In essence, all of the restoration work will be lost unless a mechanism, such as the parametric insurance product, is implemented. While some may take the reefs work for granted (e.g., coastal protection of reducing wave height and wave intensity, tourism revenue and fisheries production), the reefs are a valuable asset that need to be protected.[87]

How Problem Was Identified

The problems associated with overfishing and water pollution are the result of extensive ongoing studies throughout the Mesoamerican Reef. This includes research from international environmental NGOs such as TNC and local organizations such as the Healthy Reefs for Healthy People Initiative, the MAR Fund, and Amigos de Sian Ka'an. Furthermore, the problem associated with a lack of funding for coral reef restoration after a large hurricane was well known throughout the region.

Effectiveness of Process for Identifying Problem

The process for identifying the problems appears effective as a wide range of stakeholders, from local communities, TNC, and government officials, to the private sector (e.g., Swiss Re and hotel owners) have been active participants.

Steps Taken to Address the Problem

The main, general steps taken to establish the parametric insurance instrument were:

- 1. Assess if asset needs insurance;
- 2. Identify potential buyers;
- 3. Design the insurance;
- 4. Define the institutional arrangement;
- 5. Make the transaction; and
- 6. Build the capacities to repair the reef.[88]

With respect to designing the insurance, one can consider compensatory insurance or parametric insurance. Compensatory insurance is "based on damage or losses. Compensation costs are hard to assess and takes time, and it is subject to ligation."[89] In contrast, parametric insurance "is a pre-agreed amount when certain conditions are met. It is designed for quick payment which allows prompt response to the reef."[90]

In addition, three important elements of parametric insurance are:

- 1. A parameter and a threshold, or a trigger;
- 2. A polygon "that covers how far the event can cause damages;" and
- 3. The payout.[91]

Another design element is to design a damages' curve which helps to establish the threshold, or trigger event. For the Mesoamerican Reef Parametric Insurance, there was a lot of research with clear data that demonstrated below 110 knots of wind speeds, the reefs may or may not have minimal damage. Such minimal damage, which could have impacted 5–10% of the coral cover, could be restored with one's own financial resources. However, events above 110 knots of wind speeds, the impact is significant and often beyond one's own financial resources.[92]

Regarding the polygon, the Mesoamerican Reef Parametric Insurance covers a 28-kilometer section of the marine reserve from Cancun south to approximately two–three kilometers past Puerto Morelos. In addition, the polygon also includes a 65-kilometer buffer zone. This said, if wind speeds reach the established threshold within any section of the polygon (i.e., in the marine reserve or within the 65-kilometer buffer), the insurance policy is triggered.[93]

The payout scenario is as follows when a trigger event occurs in the polygon:

- Wind speeds above 110 knots to 129 knots = 50% of money paid out;
- 130–159 knots = 60% of money paid out; and
- Over 160 knots = 100% of money paid out.[94]

Another way to envision this is with four scenarios:

- Scenario 1: Without restoration;
- Scenario 2: Limited ecological restoration;
- Scenario 3: Extended ecological restoration and limited structural restoration; and
- Scenario 4: Extended ecological restoration and structural restoration.[95]

When defining the instuitional arrangement, it is important to:

- 1. Involve the stakeholders to decide the institutional arrangement (i.e., those responsible for the asset, the beneficiaries of the services, and the experts who know how to repair the asset);
- 2. Identify or agree on the buyer (i.e., based on the legal framework and institutional capacity, who is entitled to buy an insurance);
- 3. Identify who receives the payout (i.e., identify or create a reliable mechanism that can receive the payout and ensure the funds are spent according to the intent); and
- 4. Define the rules to manage the payout (i.e., develop the governance and rules to facilitate the decision making).[96]

When making the transaction, one should:

- 1. Develop the terms of coverage (i.e., such as a technical document that describes the asset, the risk, the parameter, the payout, and the buyer);
- 2. Find a willing seller (i.e., open a procurement process and request business proposals); and
- 3. Negotiate the terms (i.e., the buyer and the seller could adjust the premium by changing the payout, the polygon, and/or the threshold of the parameter).[97]

Finally, when it comes to building the capacities to repair the reef, TNC is leading the process, but TNC works very closely with the national parks service, with universities (i.e.,who are developing the techniques of how to do coral reef repairs), and with the local communities (i.e., local fisherfolk, tour operators, and communities).[98]

To build capacities, one should:

- 1. Develop a post-storm protocol (i.e., agree on the activities to responds to after a disaster affects the asset);
- 2. Form and train a post-storm brigade such as volunteers for a rapid response;
- 3. Develop a reef repair guide (i.e., a technical document that describes the actions to restore the damages); and
- 4. Train restoration specialists (i.e., such as reef mangers and coral ecologists).[99]

Results

To date, some of the major results include:

- Initially engaged Swiss RE, which provided significant assistance with designing the parametric insurance product;
- TNC hired an expert to help further assist with the parametric insurance instrument;
- A protocol was developed for how to do coral reef restoration;
- Guidelines were developed for how to insure nature. The report, "*A Guide on How to Insure a Natural Asset*," was released in 2019[100];
- Important questions were answered such as, how much money would be needed in the case of a trigger event (e.g., expenses for a certain amount of boats, divers, snorkelers, fuel, and supplies)?
- Participants selected a specific area, which consists of a 28-kilometer section of the marine reserve. This section extends from Cancun south to approximately 2–3 kilometers past Puerto Morelos. In addition, a 65-kilometer buffer zone around this specific area was also insured;
- Three parameters of atmospheric pressure, storm surge, and wind speeds were reviewed. Wind speeds was ultimately chosen as the sole variable, since it was the best variable that correlated with damage to the reefs. In addition, NOAA's Hurricane Center was chosen as the independent, third-party provider of this data on wind speeds; and
- The Coastal Zone Management Trust, which will collect fees from local hotels and use these funds to both maintain the coral reef and to pay for the parametric insurance policy, was established in November 2018 by TNC and the Quintana Roo Government.[101]

Challenges and How They Were Met

The main challenges with the parametric insurance can be categorized into four topics:

- 1. Design features;
- 2. Engaging stakeholders, particularly the government;
- 3. Institutional arrangements; and
- 4. Having the capacity to do the restoration.

Design features that posed a challenge included:

- Determining how much money is required;
- Figuring out who can legally buy the insurance, and from whom?
- Determining what response(s) you need to do after an event;
- Choosing the exact polygon; and
- Deciding whether there really is a buyer (i.e., whether someone is really willing to pay for the insurance product).

To address this concern, there was a lot of research conducted throughout the Caribbean. Such research included how reefs were degraded as a result of hurricanes and techniques for proper restore degraded reefs.

Engaging stakeholders, particularly the government was a challenge. More specifically, identifying the right people in the Quintana Roo Government who are interested in exploring the use of parametric insurance was the challenge. In addition, while some of the design was already there (i.e., regional research and established local groups) and the funding (i.e., hotels paying the fee to the government) was already there, it took a lot of education and outreach to tour operators and hotel operators about the science and the impacts of hurricanes on coral reefs and how degraded coral reefs impact their businesses. The tour operators were more eager to participate, but have less money. In contrast, the hotel operators have more money, but were not as eager (i.e., in part because the reefs are literally a bit farther from their establishments). Yet, TNC showed the hotels how much they will lose if reefs are damaged as a result of hurricanes because the hotels would lose their beaches if the reefs are damaged and would surely lose tourists.[102]

Structuring the appropriate institutional arrangement was very complicated. This includes:

- Engaging all the appropriate stakeholders from the private sector actors (i.e., hotel industry, tour operators, etc.) to government officials;
- Agreeing on the bylaws;
- Determining how things will be managed to ensure the money intended for the beach goes there; and
- Reviewing the laws about insurance, such as who actually can buy insurance.

Establishing the Coastal Zone Management Trust helped addressed the concern about getting the appropriate institutional arrangement to help ensure transparency and proper financial management. However, it is important to understand that there are many different potential institutional arrangements. For instance, the policyholder could—instead of being the Coastal Zone Management Trust—could be a government entity or a particular hotel.

In the case of Quintana Roo, the government wanted to set up a trust fund to be financed by fees collected from hotels and a portion of these fees will specifically go to the Coastal Zone Management Trust to, in part, pay for the parametric insurance. As of now, the Coastal Zone Management Trust would eventually receive the insurance payout, but this matter is still being decided upon. As of February 2019, the consortium is now doing all of the financial arrangements to determine who would receive the funds and by June 2019, before the hurricane season, this should be in place. There is already an advisory committee and a board for the trust fund.

Furthermore, there will be an open, bidding process to purchase the insurance and legally, the insurance must come from a Mexican insurance company. Generally, this insurance company would then reinsure with a larger, multinational company.

With respect to the capacity to do the restoration, if you do not have the restoration expertise in place, why do you need the money? The analogy used by Fernando Secaira, is that it would be like having health insurance, but not having a doctor or hospital to visit. To address this challenge, TNC built a restoration brigade and a protocol for coral reef restoration.[103] Likewise, there is a lot of work that can be done to reduce the impacts on coral reefs from a hurricane. For instance, scattered debris can be collected, coral fragments can be glued back to reduce mortality, and coral colonies can be consolidated.

Other challenges include the fact that the Mesoamerican Reef Parametric Insurance is the first of its kind and the "use of parametric insurance for coral reefs is a far more specific application."[104] In general, other challenges associated with parametric insurance is its cost and the required modeling:

With a parametric insurance plan, a pre-set payment is triggered automatically, so it can come within days. It's usually more expensive than traditional insurance and requires extensive modeling from the insurer to get right, but for governments that want to respond to a disaster immediately, it's become a good option to supplement traditional insurance and foreign aid.[105]

In addition, the parametric insurance has been described as:

A 'Band-Aid' for coral reefs that won't treat the bleaching that's chronically ailing them. She also took issue with the way the approach values nature for its human applications, rather its own inherent value.[106]

Furthermore, there is the possible risk or challenge of free-riding, where some businesses (i.e., non-participating hotels, restaurants, etc.) do not pay the fees but who nevertheless benefit from the parametric insurance.

Beyond Results

Many elements of the Mesoamerican Parametric Insurance instrument could be applied in other jurisdictions. However, it is important to recognize that "this project has 'the perfect storm' of backers in a major charity like TNC, a focused local government and willing private sector, so it's not guaranteed to work in other cases. But if it does work in Mexico, it will have 'a huge demonstration effect'."[107] However, "Secaira, suggested parametric insurance could help countries begin to factor their natural wealth into their national accounting, the same way they do highways and hospitals."[108] Furthermore, TNC received a USD$1 million grant from Bank of America to assess how a parametric insurance product could be applied in Florida and Hawaii.[109]

In one analysis, "Kirkpatrick (2018) thinks premiums paid by the hotel owners should be between $1 million and $7.5 million in 2017, with around $25 million to $70 million in payouts a year in the event of an extreme storm."[110] This could provide a long-term, sustainable financing source for coral restoration. Furthermore, it has been reported that new underwriters are interested in the Mesoamerican Parametric Insurance product.[111]

Lessons Learned

According to Mr. Fernando Secaira of TNC, a few lessons learned include:

- 1. First, you need to make sure you need the parametric insurance for your particular situation and you need to answer a lot of questions. For instance, how do I insure dunes or mangroves? What are the risks you are facing? Is someone willing and able to repair these ecosystems? How do you define that you have a willing buyer who is legally entitled to buy it? At times, you may find a situation where someone or some entity is interested in the coral reef or whatever natural asset, but are not as interested in the repair. It is then easy to jump into the design of the insurance, but contributors to the Mesoamerican Reef Parametric Insurance struggled a lot trying to answer these very questions—but you have to answer the questions;
- 2. Next, upon determining the parametric insurance is appropriate, you really need to determine how much the cost of the repairs will truly be. You need the estimated costs for all of it and you need to determine, what is the breaking point? When do you transfer the risk, and when do you hold the risk? This, in part, depends on one's financial capacity;
- 3. You should try to keep the design of the parametric insurance as relatively simple as possible;
- 4. In the next design, there could be other parameters incorporated. For instance, the length of time the hurricane stays in the polygon could be a trigger, as opposed to solely the wind speed;
- 5. Transparency is very important; and
- 6. We should not oversell the insurance. Likewise, it is important to understand that the parametric insurance product is just one piece in an overall comprehensive policy.[112]

Other Resources on Parametric Insurance for the Mesoamerican Reef

Artemis

- https://www.artemis.bm/

Bloomberg

- https://www.bloomberg.com/news/articles/2017-07-20/a-coral-reef-gets-an-insurance-policy-of-its-own

Environmental Finance

- https://www.environmental-finance.com/content/news/swiss-re-develops-first-ever-parametric-insurance-policy-for-a-coral-reef.html; and
- https://www.environmental-finance.com/content/news/swiss-re-dives-into-coral-reef-insurance-in-mexico.html

Global Ecosystem Resilience Facility

- https://www.willistowerswatson.com/en-US/Insights/trending-topics/csp-global-ecosystem-resilience-facility

Ogden, P., Bovarnick B. and Hoshijima, Y. (2015). Key Principles for Climate-Related Risk Insurance. Center for American Progress.

- https://cdn.americanprogress.org/wp-content/uploads/2015/08/261 31302/ClimateRiskInsurance-report.pdf

Reef Resilience Network

- http://reefresilience.org/case-studies/mesoamerican-reef-mpa-design-2/

TNC

- https://blog.nature.org/conservancy/2017/07/21/insuring-nature-ens uring-resilience/

Case Study #3: Arrecifes Del Sureste Marine Sanctuary of the Dominican Republic

Introduction

At nearly 8,000 km^2 (800,000 hectares), the Arrecifes del Sureste Marine Sanctuary of the Dominican Republic ("Sanctuary") is the second largest MPA of the Dominican Republic and the third largest MPA for the whole Caribbean region (Figs. 10.3 and 10.4).[113]

The Sanctuary was established approximately ten years ago in 2009.[114] However, the Sanctuary was historically considered a paper park until the Sanctuary became the first such Public–Private Partnership (PPP) on February 23, 2018.[115]

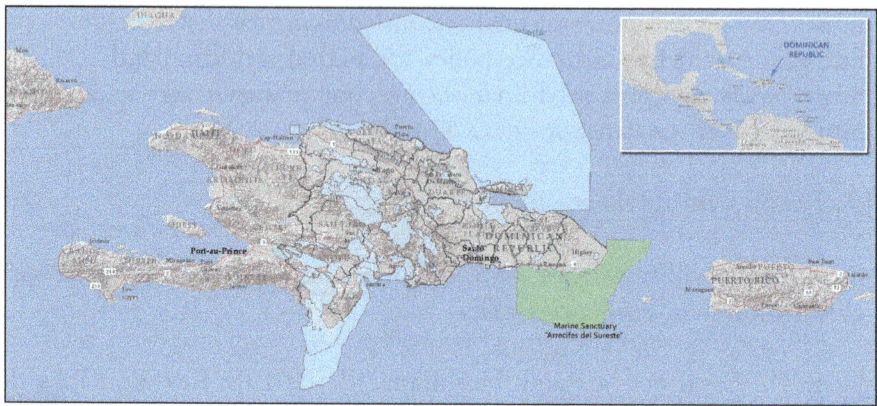

Fig. 10.3 Map of protected areas of the Dominican Republic (*Credit* Blue Finance)

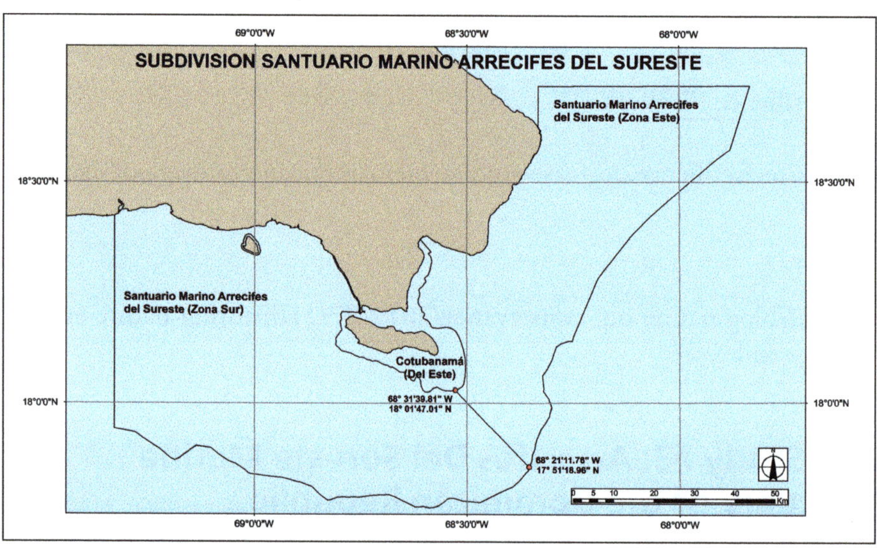

Fig. 10.4 The Arrecifes del Sureste Marine Sanctuary of the Dominican Republic
(*Credit* Blue Finance)

Blue finance identified PPP agreements as an effective means of co-managing MPAs with the private sector. The private sector is expected to provide the majority of required funds to improve and manage the area, and receive a return on investment mainly through user fees and innovative tourism products. The main advantages of PPPs include their flexibility to set fees and charges; their ability to establish funding mechanisms such as concessions, respond to customer needs, and retain the money they earn (which provides an incentive for greater entrepreneurship); and their freedom to implement staffing policies based on efficiency and market salaries. {…} The government will maintain its core functions and be responsible for regulation and enforcement of uses and zonation, the set-up or user fees and maintenance of specific on-shore facilities.[116]

For instance, user fees may come from tourists or fisherfolk, while innovative tourist products may include a "world-class visitor centre."[117]

With respect to financing:

The idea is that a collateral-free loan will be issued to the NGOs acting as a special purpose entity through the Sustainable Ocean Fund, an impact investment fund run by conservation investment manager Althelia. Investment will be sourced chiefly from development finance institutions – hopefully with private investors joining. The structure is still in negotiation but is expected to have an eight- to 10-year term with an annual coupon.[118]

Identify the Problem

Problems facing the Sanctuary, which are common threats to other marine reserves, are overfishing and unregulated coastal development. With respect to unregulated coastal development, there are lots of hotels which cater to the large volume of tourists (i.e., which is possibly overcapacity) and compliance is not properly enforced. This, in turn, results in marine pollution and deteriorating water quality.[119]

Why the Problem Is Important

The Sanctuary is very unique for several reasons and this is why addressing the problems of overfishing and unregulated coastal development is important. One of the main attributes of the Sanctuary is its size.

A second unique attribute of the Sanctuary is that there are very different habitats throughout the areas of the Sanctuary that are in the Atlantic Ocean and those areas that are in the Caribbean Sea (i.e., different morphology). Similarly, the Sanctuary includes both pelagic and coastal areas, whereas most conservation projects focus on either pelagic or coastal areas, but rarely focus on both.[120]

Another unique attribute of the Sanctuary is that it abuts one of the largest human developments and tourist hubs, particularly Punta Cana, throughout the Caribbean. This context of tourism and the number of visitors is very unique, because many MPAs, such as the Papahānaumokuākea Marine National Monument, are located in remote places. In contrast, Punta

Cana hosts up to four million visitors a year. In addition, there are approximately 500,000 people living in the perimeters of the Sanctuary and another 1+ million living within the Sanctuary, many of whom are dependent on fishing and tourism.[121] Thus, the Sanctuary is both an opportunity for tourism revenue and as a source of seafood.

How Problem Was Identified

The problem associated with coastal development and tourism is rather obvious as you can see the coastal development and the tourism statistics are readily available. For instance, between 2010 and 2019, monthly tourist arrivals to the Dominican Republic have fluctuated between 200,000 and 700,000 visitors.[122]

With respect to overfishing, there are some precise points for long-term monitoring of live coral cover, water quality, and fishing effort. In addition, there are two conservations NGOs in the consortium—one in the East and one in the South—that have assisted.[123]

Effectiveness of Process for Identifying Problem

The process for identifying the problems appears effective.

Steps Taken to Address the Problem

The major steps to establish the Sanctuary and to work toward addressing the problems of overfishing and unregulated coastal development were to form the Public–Private Partnership, engage stakeholders, develop the management plan, and secure the financing.

The Public–Private Partnership (PPP) was formally established on February 23, 2018.[124] This PPP can be understood as an "Entrepreneurial Marine Protected Area (EMPA). An EMPA is a management area that is primarily supported by a profit-bearing business model, typically associated with nature tourism."[125]

Since mid-2017, "Blue Finance has been working closely with the Minister of Environment and Natural Resources of the Dominican Republic, who spearheaded the project."[126] In March 2018, technical staff from Blue Finance visited the Dominican Republic and:

met the major NGOs, funders and marine service providers for both the north and south zones. Discussions were held with Foundation Grupo Punta Cana, Dressel Divers, Iberostar, Fundemar, The Nature Conservancy (TNC) and the Hotel Association & Tourism Cluster among others. These discussions have laid the foundation for the formation of the management plan and informed both technical and administrative procedures.[127]

This said, Grupo Puntacana Foundation supports one of the largest coral nursery programs in the Caribbean, and Fundación Dominicana de Estudios Marinos (FUNDEMAR) monitors reef users such as tourists and fishers.[128]

The management plan should incorporate all the relevant regulations and zoning, along with a plan to incorporate all the stakeholders.[129]

Blue Finance, which was instrumental in the establishment of the Sanctuary, does not do the day-to-day operations, but rather provides a project management office, does stakeholder meetings and engagement, and works on communication. In addition, Blue Finance provides support to the marine stakeholders such as dive shops, hotels, and other service providers.[130]

With respect to financing:

The company will receive major financing for its initial capital expenditures from international impact investors. In time, it is expected to be financially sustainable and generate its own incomes from user fees and innovative tourism models. The non-profit company is comprised of 2 local conservation NGOs, 2 local foundations of the major tourism holdings in the country and other associations. The company will be responsible for hiring and managing the staff to carry out the activities agreed to by the Government, within the co-management agreement for a period of 10 years. This will be carried out with no increase in public debt, as the body will purchase required equipment and pay personnel for the Sanctuary. The company itself will be guided by a multidisciplinary Advisory Council, of public and private citizens. An independent, internationally recognised institution will audit the performance of the consortium annually.

Initial capital expenditures are expected to be provided by the Sustainable Ocean Fund, an impact investment fund managed by Althelia, dedicated to creating, accelerating, and executing sustainable fishery, aquaculture and coastal conservation projects globally, whilst applying best-in-class social and environmental governance.[131]

The Blue Finance "technical team valued the financing needs for the site: an upfront investment cost of US$3 million required to achieve sustainable reef management and develop visitor attractions, along with an annual operational cost of US$800,000."[132]

Results

Despite being a relatively new Sanctuary, there have been many results to date. This includes:

- The PPP was signed February 23, 2018[133];
- The Sanctuary has attracted a lot of attention and incorporated a lot of international players such as Mirova-Althelia Sustainable Ocean Fund and Deloitte Social Finance; and
- In addition, "work has started on the design of the world class Visitor Centre. The search is on for an appropriate location and Blue Finance is partnering with Formula D Interactive of South Africa on this venture which will include both high tech and more natural elements to educate and entertain guests."[134]

Looking forward, "the livelihoods of 20,000 households in the Dominican Republic for example are expected to be improved as a result of the co-management there."[135] As for employment and investments, "more than 25 permanent staff will be recruited overall. Annual operational expenses are expected to run to about $1 million, with capital expenditure for vessels, scientific equipment and the construction of a visitor centre adding another $2 million to $3 million."[136]

Similarly, the Project team has:

{…} directly hired 10 fishermen so far out of 35 to 40 in our area to work as scuba divers, boat captains and in coral restoration. We have trained their wives and family members to sell souvenirs and other products. Collectively we need to get better at hiring fishermen and their families to reduce pressure on the reef.[137]

In addition:

'We have vast amounts of seaweed coming into shore, for example, that will choke the coral. So, we have built floating barriers to protect the area, and the local fishermen have been hired to clean the barriers and repair them where needed,' says Kheel {…}. Grupo Puntacana has started implementing measures already. 'We have installed channel markers and mooring buoys, we implemented a membership programme with rules, regulations and user fees,' says Kheel. 'We even have a patrol boat to respond to issues in our area, both rescue and illegal activity.

Another important result is the collaboration between TNC and the Carnegie Airborne Observatory which will:

> yield high-resolution maps of the new sanctuary using data collected from satellites, aircraft, drones, and by scuba divers. These detailed maps will enable managers to make important marine spatial planning decisions that appropriately allocate conservation efforts throughout the sanctuary.[138]

Challenges and How They Were Met

Some of the main challenges faced were internal people, financing, and stakeholder engagement.

With respect to internal management, it is often a challenge to find the right people with the right capacity. While there is a lot of goodwill in marine conservation, management skills are often lacking. Similarly, NGOs are often not designed for ongoing management and do not have good day-to-day management or day-to-day operations. To address this challenge, it was a good idea to bring in the private sector via the establishment of the PPP.

The challenge of financing has to do with the fact that grant funding is different than impact investments, which can take the form of debt financing or equity financing. In the MPA world, it is challenging to contract a loan. One needs to be very repetitive and convincing, and if successful, the loan— as a risk management approach—will be staggered based on outcomes. A performance-based approach is important to impact investors and for Blue Finance. To address this challenge, the financing for the Sanctuary is mixed with a main revenue stream from compulsory user fees.[139]

With respect to stakeholder engagement, "the notion that our oceans and coastal areas should be managed – particularly for the benefit of private investors – is an uncomfortable topic for some."[140] Furthermore, "the first hurdle to creating managed MPAs is convincing governments of their value. That process can take years of working with the various parties impacted by their creation, such as the fishing, tourism and energy industries. {...} A harder task, however, is convincing governments of the relationship between ocean health and climate change and the subsequent impact on coastal areas."[141]

Beyond Results

One important point is that Blue Finance did not need a proof of concept in the Dominican Republic, because they already had it.[142] In addition, the "whole project is designed to be replicable – and investable."[143]

Looking to the future, there are several items that need to occur. For instance, there is:

- The need to prove the PPP can deliver in the Dominican Republic (i.e., there are other co-management agreements signed and run by NGOs in other island nations such as the Bahamas, Belize, and Bonaire);
- The need to prove the PPP as an investment is a viable business model (i.e., and bring some financing to the PPP where private finance is novel); and
- Blue Finance is now looking to expand to Southeast Asia.[144]

It is important to note that Blue Finance's "PPP model typically operates in sites with viable tourism options and moderate to high visitation, generating the consistent financial revenues required to pay back investors."[145]

The Dominican Republic has also been identified as one of the top fifty sites to target long-term coral reef conservation investments.[146] As noted by Nicolas Pascal, inside the Sanctuary, there is a small reserve with very high fish biomass and high live coral cover which demonstrates the gains achieved through proper protection and management. While there will be some bleaching, it is comparatively less.[147]

Furthermore, as part of the Coalition for Private Investment in Conservation, Blue Finance has developed a publicly available blueprint of Public–Private Partnership for Marine Protected Areas.[148]

Lessons Learned

According to Mr. Nicolas Pascal, Executive Director at Blue Finance, there are several lessons learned throughout his years working on marine conservation and particularly from his work on the Sanctuary. Likewise, the two most important attributes are to have passion and to be a good communicator. In a PPP, you are dealing with both private sector actors and public sector actors. It is often very clear whether or not the private sector is interested. In contrast, you cannot rush this stage or this approach when working with the public sector. Similarly, the management plan, with all the relevant regulations and zoning, along with a discussion of how to incorporate all the stakeholders, is not something that can be rushed.[149]

Similarly:

{…} support of the former minister of environment was crucial to getting the initiative off the ground, even when the country already had regulations for co-management of protected areas in place. Blue Finance was approached by the government through its partnership with the UN and, supported by their funding, researched and identified the key partners in the country to work with.[150]

Other Resources on the Arrecifes Del Sureste Marine Sanctuary

Blue Finance

- http://blue-finance.org/

Coalition for Private Investment in Conservation's Blueprint of Public–Private Partnership for Marine Protected Areas by Blue Finance

- http://cpicfinance.com/cpic-blueprint-public-private-partnership-for-mar ine-protected-areas-by-blue-finance-3/; and
- http://cpicfinance.com/blueprints/sustainable-coastal-fisheries/

Dominican Republic's Ministry of Environment and Natural Resources

- https://ambiente.gob.do/

Dominican Republic's Ministry of Tourism

- https://mitur.gob.do/; and
- http://www.godominicanrepublic.com/about-dr/flora-fauna/

Fundación Grupo Puntacana (Punta Cana Foundation)

- https://www.puntacana.org/

Fundación Dominicana de Estudios Marinos (FUNDEMAR; Dominican Foundation for Marine Studies)

- https://www.fundemardr.org/

<u>Protected Planet</u>

- https://www.protectedplanet.net/arrecifes-del-sureste-marine-sanctuary

Financial Analysis

The following financial analysis will look at return and risk.

Return

According to the 266 respondents to the Global Impact Investing Network's (GIIN) 2019 Annual Impact Investor Survey:

> Impact investors target financial returns along a continuum ranging from capital preservation to competitive market rate. Most respondents principally target risk-adjusted, market-rate returns (66%). A further 19% primarily seek below-market returns that are closer to market rate, and the remaining 15% target returns closer to capital preservation. Over 70% of foundations and not-for-profit fund managers pursue below-market returns. Most Private Debt-focused Investors also target below-market returns (59%), whereas a majority of Private Equity-focused Investors target market-rate returns (79%).[151]

Thus, it is important to know your investor audience when it comes to proposing the impact investment and its expected returns—both financial returns and non-financial returns.

Risk

There are a wide variety of risks associated with impact investing depending on the exact nature of the investment, including the specific underlying assets (i.e., tourism and/or fisheries), the actual structure of the investment (e.g., debt financing or equity investment), and the overall time horizon of the investment.

As outlined by Pascal et al., the challenges (or risks) to impact investments in marine conservation include: investment sizes are too small; a shortage of high-quality investment opportunities with track records; the lack of common standards and metrics; the lack of risk mitigation (i.e., de-risking) strategies; and the fact that the market remains niche and boutique.[152]

Business Risk

Business risks are based around supply and demand for a given product. This said, depending on the underlying investment, does the demand exist—now and into the future—for example, for an ecotourism lodge, visiting the MPA, or seafood sourced from a Marine Stewardship Council certified fishery?

Supply chain costs may also present a business risk. Is the cost to produce the sustainably sourced seafood cost-competitive, relative to its premium price, to compete against more conventionally sourced seafood? Likewise, is there even a premium price that can be charged?

Furthermore, there could be material risk of companies having to switch suppliers due to consumer pressure focused on their supplier's fishing operations. Switching suppliers can involve costs (e.g., unwinding contracts, search costs for new suppliers, etc.), higher costs of goods, and competitive pressure as other firms secure exclusive rights with more sustainable suppliers.

Strategic Risk

Strategic risks for an impact investor in the marine space can be mitigated through a diversified portfolio consisting of a range of different projects. Likewise, the Sustainable Ocean Fund did not invest into just a single project, in one country, or one region. There are also strategic risks associated with impact investors changing their investment philosophy or investment direction. In addition, strategic risks may arise when choosing the host country to invest in, the actual project and its partners, or when lining up financing. Furthermore, there can be strategic risks associated with asking for too much, or too little, funding.

Reputation Risk

Reputation risks can arise from investing in fraudulent and/or inexperienced organizations that are undertaking the conservation activities. This includes organizations that are unable to deliver on the financial and environmental expectations, along with the social concerns of fair labor practices, gender equity, and fair wages.

Liquidity Risk

Investments in large, underlying assets—such as undertaking a large marine spatial planning exercise with a significant sustainable fishery operation—can pose more of a liquidity risk, than purchasing shares in a publicly traded company. In addition, impact investments into new industries (e.g., marine biodiversity offsets) can also pose a liquidity risk. However, as these projects and companies become more mainstream, their liquidity will likely increase.

Operational Risk

Operational risks for impact investors include inadequate technology (e.g., inadequate in-country computer systems, intermittent energy or internet access, insufficient monitoring of fishery activity, etc.), incompetent management, and possible fraud (e.g., illegal sourced fish or falsely claimed, sustainability certification).

Legal and Regulatory Risk

Legal and regulatory risk is potentially significant in host countries where the ease of doing business is often more challenging and corruption is more significant. This can make getting fair court hearings, recovered investments from bankrupt companies, or export licenses approved more difficult. In essence, there can be a lack of legal recourse.

Credit Risk

Credit risk includes default risk, bankruptcy risk, downgrade risk, and settlement risk.

From the impact investors' perspective, there are credit risks, and specifically default or bankruptcy risks, associated with the company or project receiving the impact investment.

Market Risk

Market risk includes interest rate risk, equity price risk, foreign exchange risk, and commodity price risk.

For impact investments in an export-oriented product (i.e., the fishing industry), foreign exchange risk can be significant depending on the host country's currency relative to the importing country's currency. For instance, a commodity and foreign exchange risk could occur if buyers are willing to switch to cheaper lobsters from Belize as a result of the change in foreign exchange, instead of lobsters from Australia or Mexico. In turn, the lobster fisheries in Australia or Mexico may face bankruptcy risk, while increased fishing pressure may arise in Belize specifically because of the foreign currency exchange.

Risk, Return, Time (Horizon), Taxes, Liquidity, Legal, and Unique (RRTTLLU)

Risk and Return

Please see above for the risk and return associated with impact investing. The type of return required by the impact investor may change over time as the composition of their portfolio changes. This said, it is very important to understand both the financial and non-financial returns expected by the impact investors.

Time Horizon

The time horizon for impact investing is likely to be a short-term to medium-term investment. Likewise, TNC's Conservation Notes had one-, three-, and five-year terms,[153] Credit Suisse and Althelia's Nature Conservation Notes were six-year notes, and the Althelia Sustainable Ocean is an eight-year fund.

Taxes

There are likely taxes, such as taxes on capital gains. Depending on the structure of the investment, there could be sales taxes or value-added taxes levied on the underlying products and employment taxes, along with export and/or import taxes and duties.

Liquidity

Depending on the nature of the investment, the investment might be more or less liquid. For instance, the Nature Conservation Notes were fairly liquid, while investments in sustainable fishing operations in an unstable region are likely to be less liquid.

Legal

Legal considerations associated with impact investing will vary from country to country. This said, it is important to consider the jurisdiction of the impact investment fund and the jurisdiction of the host countries where the investments are taking place.

Unique

One unique aspect is that the environment is only one of the ESG (Environmental, Social and Governance) components and there are lots of other competing investments. Furthermore, marine conservation and sustainable fisheries are just one sector of the environment which also includes freshwater, avoided deforestation, efficient transportation, and clean energy.

In addition, impact investors need to assess the risk-adjusted return after deducting management fees and taxes.

Policy Analysis

The following policy analysis will look at: defining the problem; establishing goals; selecting a policy; implementing a policy; and evaluating the policy. This said, impact investing depends on a wide variety of public policies to be in place.

Defining the Problem

Choosing which country to make an impact investment depends on factors such as the ease of doing business, the level of corruption, and the rule of law. While impact investors might be willing to make more risky investments, such investors are not willing to throw away their money. In addition, some investors—particularly institutional investors—have a legal and fiduciary obligation to only invest in countries and/or companies that have an investment grade rating.

Establishing Goals

The goals of impact investing are to get one's investment back, plus a financial return, while also making a positive environmental and social impact. Thus, impact investments should ideally also contribute to the conservation of coral reefs and promote economic development for local communities.

Selecting a Policy

When choosing which country to make an impact investment, one should assess the countries in their totality. If the investment grade rating is the primary concern, then as of July 2019, for instance, the countries of Indonesia, the Philippines and Malaysia were rated as investment grade. In contrast, Papua New Guinea, Egypt, and Fiji were rated as non-investment grade.

The following coral reef countries were ranked in order of the Trading Economics Credit Rating (TE Rating), which "scores the credit worthiness of a country between 100 (riskless) and 0 (likely to default) (Table 10.1)."[154]

Table 10.1 Credit ratings of countries with largest coral reefs

	Country	S&P		Moody's		Fitch		TE rating
1	Solomon Islands	N/A	N/A	B3	Stable	N/A	N/A	25
2	Papua New Guinea	B	Stable	B2	Stable	N/A	N/A	30
3	Egypt	B	Stable	B2	Stable	B+	Stable	30
4	Maldives	N/A	N/A	B2	Negative	B+	Stable	32
5	Fiji	B+	Stable	Ba3	Stable	N/A	N/A	35
6	Bahamas	BB+	Stable	Baa3	Stable	N/A	N/A	52
7	India	BBB-	Stable	Baa2	Stable	BBB-	Stable	56
8	Indonesia	BBB	Stable	Baa2	Stable	BBB	Stable	58
9	Philippines	BBB+	Stable	Baa2	Stable	BBB	Stable	60
10	Malaysia	A-	Stable	A3	Stable	A-	Stable	66
1	Saudi Arabia	A-	Stable	A1	Stable	A+	Stable	78
2	France	AA	Stable	Aa2	Positive	AA	Stable	92
3	United Kingdom	AA	Negative	Aa2	Stable	AA	Negative Watch	92
4	United States	AA+	Stable	Aaa	Stable	AAA	Stable	98
5	Australia	AAA	Stable	Aaa	Stable	AAA	Stable	100
6	Marshall Islands	N/A	N/A	N/A	N/A	N/A	N/A	N/A
7	Micronesia	N/A	N/A	N/A	N/A	N/A	N/A	N/A
8	Vanuatu	N/A	N/A	N/A	N/A	N/A	N/A	N/A
9	Tanzania	N/A	N/A	B1	Negative	N/A	N/A	N/A
10	Eritrea	N/A	N/A	N/A	N/A	N/A	N/A	N/A

Credit Trading Economics

If one's firm can make an impact investment in countries without an investment grade rating, then there are additional policies to consider such as policies related to the ease of doing business, corruption, the Human Development Index (HDI), and the rule of law.

For instance, the global rankings for ease of doing business, according to the World Bank,[155] for the top 20 countries with the largest coral reefs are as follows:

- #8: U.S.
  - #16: enforcing contracts;
  - #36: trading across borders;
  - #50: protecting minority investors; and
  - #53: starting a business.
- #9: U.K.
  - #15: protecting minority investors;
  - #19: starting a business;
  - #30: trading across borders; and
  - #32: enforcing contracts.
- #15: Malaysia
  - #2: protecting minority investors;
  - #33: enforcing contracts;
  - #48: trading across borders; and

  - #122: starting a business.
- #18: Australia
  - #5: enforcing contracts;
  - #7: starting a business;
  - #64: protecting minority investors; and
  - #103: trading across borders.
- #32: France
  - #1: trading across borders;
  - #12: enforcing contracts;
  - #30: starting a business; and
  - #38: protecting minority investors.
- #73: Indonesia
  - #51: protecting minority investors;
  - #116: trading across borders;
  - #134: starting a business; and

  - #146: enforcing contracts.
- #77: India
  - #7: protecting minority investors;
  - #80: trading across borders;
  - #137: starting a business; and

- #108: Papua New Guinea
  - #89: protecting minority investors;
  - #140: trading across borders;
  - #143: starting a business; and

  - #173: enforcing contracts.
- #115: Solomon Islands
  - #98: starting a business;
  - #110: protecting minority investors;
  - #156: enforcing contracts; and
  - #160: trading across borders.
- #118: Bahamas
  - #84: enforcing contracts;
  - #105: starting a business;
  - #132: protecting minority investors; and
  - #161: trading across borders.
- #120: Egypt
  - #72: protecting minority investors;
  - #109: starting a business;
  - #160: enforcing contracts; and

  - #171: trading across borders.
- #124: Philippines
  - #104: trading across borders;
  - #132: protecting minority investors;
  - #151: enforcing contracts; and
  - 166: starting a business.
- #139: Maldives
  - #71: starting a business;
  - #125: enforcing contracts;
  - #132: protecting minority investors; and

  - #155: trading across borders.
- #144: Tanzania
  - #64: enforcing contracts;
  - #131: protecting minority investors;
  - #163: starting a business; and

(continued)

(continued)

– #163: enforcing contracts.	– #183: trading across borders.
• #92: Saudi Arabia	• #150: Marshall Islands
– #7: protecting minority investors;	– #75: trading across borders;
– #59: enforcing contracts;	– #75: starting a business;
– #141: starting a business; and	– #103: enforcing contracts; and
– #158: trading across borders.	– #180: protecting minority investors.
• #94: Vanuatu	• #160: Micronesia
– #110: protecting minority investors;	– #61: trading across borders;
– #132: starting a business;	– #170: starting a business;
– #136: enforcing contracts; and	– #184: enforcing contracts; and
– #147: trading across borders.	– #185: protecting minority investors.
• #101: Fiji	• #189: Eritrea
– #79: trading across borders;	– #103: enforcing contracts;
– #97: enforcing contracts;	– #164: trading across borders;
– #99: protecting minority investors; and	– #174: protecting minority investors; and
– #161: starting a business.	– #187: starting a business.

The global rankings by Transparency International of the Corruption Perceptions Index 2018[156] for the top 20 countries with the largest coral reefs were as follows:

• #11 (Tied): U.K.;	• #89 (Tied): Indonesia;
• #13: Australia;	• #99 (Tied): Philippines;
• #21: France;	• #99 (Tied): Tanzania;
• #22: U.S.;	• #105 (Tied): Egypt;
• #29: Bahamas;	• #124 (Tied): Maldives;
• #58: Saudi Arabia;	• #138 (Tied): Papua New Guinea;
• #61 (Tied): Malaysia;	• #157: Eritrea;
• #64 (Tied): Vanuatu;	• Unranked: Fiji;
• #70 (Tied): Solomon Islands;	• Unranked: Marshall Islands; and
• #78 (Tied): India;	• Unranked: Micronesia.

The global rankings by UNDP the 2017 Human Development Index,[157] which combines life expectancy, education, and per capita income indicators, for the top 20 countries with the largest coral reefs were as follows:

• #3: Australia (HDI of 0.939);	• #113 (Tied): Philippines (0.699);
• #13: U.S. (0.924);	• #115: Egypt (0.696);
• #14: U.K. (0.922);	• #116 (Tied): Indonesia (0.694);
• #24: France (0.901);	• #130: India (0.640);
• #39 (Tied): Saudi Arabia (0.853);	• #131: Micronesia (0.627);
• #54: Bahamas (0.807);	• #138: Vanuatu (0.603);

(continued)

(continued)

• #57: Malaysia (0.802);	• #152: Solomon Islands (0.546);
• #92 (Tied): Fiji (0.741);	• #153: Papua New Guinea (0.544);
• #101 (Tied): Maldives (0.717);	• #154: Tanzania (0.538); and
• #106 (Tied): Marshall Islands (0.708);	• #179: Eritrea (0.440).

The World Justice Project has designed a rule of law ranking which primarily utilizes nine factors (e.g., Constraints on Government Powers, Absence of Corruption, Open Government, Fundamental Rights, Order and Security, Regulatory Enforcement, Civil Justice, and Criminal Justice, and Informal Justice).[158] Of the 126 countries assessed, the following global rankings are for the top 20 countries with the largest coral reefs:

• #11: Australia (Overall Score of 0.8);	• #121: Egypt (0.36);
• #12: U.K. (0.8);	• Unranked: Eritrea (N/A);
• #17: France (0.73);	• Unranked: Fiji (N/A);
• #20: U.S. (0.71);	• Unranked: Maldives (N/A);
• #39: Bahamas (0.61);	• Unranked: Marshall Islands (N/A);
• #51: Malaysia (0.55);	• Unranked: Micronesia (N/A);
• #62: Indonesia (0.52);	• Unranked: Papua New Guinea (N/A);
• #68: India (0.51);	• Unranked: Saudi Arabia (N/A);
• #90: Philippines (0.47);	• Unranked: Solomon Islands (N/A); and
• #91: Tanzania (0.47);	• Unranked: Vanuatu (N/A).

Other possible tools including looking at the "Fragile States Index" developed by the Fund for Peace and the Carnegie Endowment,[159] the "World Press Freedom Index" by Reporters Without Borders,[160] and/or the "Happy Planet Index" which calculates "sustainable well-being."[161]

Implementing a Policy

With these problems in mind, host countries need to implement a suite of public policies to better position themselves for impact investments. The reason impact investments are more prominent in Australia, the EU, and the United States is because of the strong rule of law, transparency, relative price discovery, the range of financial intermediaries, scalability of projects, and both managerial and technical experiences, along with other local capacities.

Evaluating the Policy

Despite mixed signals based off their rankings on the ease of doing business, corruption, the HDI, and the rule of law, the countries of Indonesia, the Philippines and Malaysia have investment grade ratings and seem to be better choices for impact investments. In addition, Australia, along with the overseas territories of France, the United States, and the United Kingdom could be good choices for impact investments.

Working with finance ministries and structuring underwriting policies can help mitigate risk and allow impact investment funds to attract investors. For instance, the Overseas Private Investment Corporation (OPIC), now known as the U.S. International Development Finance Corporation, provided political risk insurance for a forest carbon project in Cambodia,[162] and the Althelia Climate Fund received a risk-sharing loan guaranteed from USAID's Development Credit Authority for up to USD$133.8 million.[163]

Future Outlook for Instrument

It is quite possible that impact investments will increase in size in the future. For instance, it has been estimated that the "value of impact investments could swell to $30 trillion within a decade."[164]

With respect to insurance, Dr. Melanie McField, the Founder and Director of the Healthy Reefs for Healthy People program, notes that it is quite plausible for a scenario to arise where insurance companies offer their subscribers a discount in their insurance if they are protecting, for instance, their mangroves and offshore coral reefs. This said, Restoration Insurance Service Company (RISCO) for Coastal Risk Reduction, which was selected as a top idea for 2019 by the Global Innovation Lab for Climate Finance, is "an enterprise that restores and conserves mangrove habitats in vulnerable coastal areas. It contracts with coastal asset owners, who receive reduced insurance premium rates due to reduced property damage risk as a result of restoration efforts, and who, over time, use these cost savings to pay back the RISCO for the restoration and conservation work."[165]

As noted in the *Innovations for Coral Finance* report:

> However, such insurance does not seem replicable in all coastal contexts, in terms of both ecological and financial factors.
>
> First, the 'test' of this first example of parametric insurance is taking place in one of the world's most famous touristic places. It may be harder in other contexts to find hotel owners willing to get involved in such process. The

question of how to finance the process is of primary importance in ensuring the success of parametric insurance for coral ecosystem conservation. The first step is to assess the wide range of stakeholders who would be prepared to pay for such insurance: those who would take out the policy and pay the premium; and those who would pay the compensation alongside the insurance company if the compensation is not covered in full by the latter.

Second, some ecosystems seem to be more suited to restoration than others. Although there are a lot of academic papers and experimental projects on coral reef restoration, parametric insurance may not be efficient when it comes to protecting touristic assets through coral restoration, for two main reasons. On the one hand, there is no evidence of a successful coral restoration project on a broad scale so far. On the other hand, even though the literature about this is still in development, it seems that it is the geographical and physical shape of the barrier reefs that play a major role in mitigating the energy and height of waves reaching the shore, more than the health of coral reefs themselves. In contrast, mangrove reforestation projects could feasibly be financed through such a mechanism {…}.[166]

Further, "globally, the examples of conservation impact investing in developing economies are limited. However, there remains a significant interest in finding ways to channel private investment to conservation enterprises. Incubators provide a key way to develop these markets and businesses. Creating a regional or global incubator would be one way to diversify risk, achieve economies of scale, and leverage technical expertise."[167]

Yet, there have been many recent developments related to impact investments. For instance, BNP Paribas Asset Management "launched the first exchange-traded fund focused on the blue economy, with the €35 million fund investing in listed equities that contribute to the sustainable economic use of the ocean resources,"[168] Credit Suisse "announced its impact fund addressing ocean health and engaging to create a sustainable blue economy has raised 212 million,"[169] S2G Ventures "launched a fund with about 100 million in commitments to invest in alternative food sources from the world's oceans,"[170] and HSBC and Pollination have recently formed the "world's largest natural capital manager."[171]

Other Resources on Impact Investing

BlueSwell

- blueswell.sea-ahead.com

Case Foundation's Impact Investing Network Map

- https://casefoundation.org/networkmap/

Coalition for Private Investment in Conservation (CPIC)

- http://cpicfinance.com/

Conservation Enterprise Development Program by Wildlife Conservation Society (WCS)

- https://www.wcs.org/our-work/solutions/business; and
- https://www.youtube.com/watch?v=qIjtcVVYQ80&feature=youtu.be

Conservation Finance Alliance Incubator

- https://www.conservationfinancealliance.org/incubator

CI Ventures, LLC

- https://www.conservation.org/projects/conservation-international-ventur es-llc

EcoEnterprises Fund

- https://www.ecoenterprisesfund.com/

Eco-Business Fund

- https://www.ecobusiness.fund/en/

Five Ways to Advance Conservation Entrepreneurship

- https://ssir.org/articles/entry/five_ways_to_advance_conservation_entrep reneurship

Global Impact Investing Network

- https://thegiin.org/

ImpactAssets 50: An Annual Showcase of Impact Investment Fund Managers

- http://www.impactassets.org/ia50_new/?filters=.

Impact Finance Center

- http://www.impactfinancecenter.org/

Incubator for Nature Conservation by the IUCN

- https://www.iucn.org/theme/environmental-law/our-work/protected-areas-pas/incubator-nature-conservation-inc

iPAR

- https://iparimpact.com/

IRIS, an initiative of the Global Impact Investing Network

- https://iris.thegiin.org/

Mission Investors Exchange

- https://www.missioninvestors.org/

Motif Impact Portfolios

- https://www.motifinvesting.com/products/impact-portfolio

OpenInvest

- https://www.openinvest.co/how-it-works/

Planet Tracker

- https://planet-tracker.org/tracker-programmes/soft-commodities/seafood/

Private Capital for Nature Conservation: Could Impact Investments be a Solution?

- www.globalnature.org/bausteine.net/f/8525/Privatecapitalfornatureconservation_WEB.pdf?fd=3

Rockefeller Foundation's Bellagio Center

- https://www.rockefellerfoundation.org/our-work/bellagio-center/

Sustainable Blue Economy Finance Principles

- http://awsassets.panda.org/downloads/sustainable_blue_economy_finance_principles_2018_brochure_final.pdf

Toniic SDG Impact Theme Framework

- http://www.toniic.com/sdg-framework/

Notes

1. The Forum for Sustainable and Responsible Investment. "SRI Basics." Accessed February 28, 2020.http://www.ussif.org/sribasics.
2. Daily, Gretchen C. and Katherine Ellison. *The New Economy of Nature*. 116.
3. Pascal et al. "Impact Investing in Marine Conservation." 203.
4. Ibid. 203.
5. GIIN. "Who Leads the GIIN?" Accessed January 3, 2020. https://thegiin.org/giin/senior-management-team#amit-bouri.
6. WAVES. "About Us." Accessed December 13, 2016. https://www.wavespartnership.org/en/about-us.
7. NatureVest. "About Us." Accessed December 13, 2016. http://www.naturevesttnc.org/about-us/.
8. Ecosystem Marketplace. "Althelia Raises $80 Million For REDD and Ecosystem Services." Accessed December 13, 2016.http://www.ecosystemmarketplace.com/articles/althelia-raises-80-million-for-br-redd-and-ecosystem-services/.
9. Natural Capital Finance Alliance. "The NCD and NCFA Story." Accessed December 13, 2016.http://www.naturalcapitalfinancealliance.org/about-the-natural-capital-declaration/.
10. Natural Capital Coalition. "History, Vision & Mission." Accessed December 13, 2016.http://naturalcapitalcoalition.org/who/history-vision-mission/.
11. NatureVest. "Frequently Asked Questions." Accessed December 13, 2016.http://www.naturevesttnc.org/about-us/faq/.

12. Credit Suisse. "Sustainable Products & Services: Conservation Finance: Our Role in Conservation Finance." Accessed March 29, 2017.https://www.credit-suisse.com/us/en/about-us/responsibility/banking/sustainable-products-services.html.

13. Roumpis, Nick. "Sustainable Blue Economy Finance Principles Unveiled." *Environmental Finance.* March 9, 2018. Accessed December 31, 2019. https://www.environmental-finance.com/content/news/sustainable-blue-economy-finance-principles-unveiled.html.

14. WWF. "Introducing the Sustainable Blue Economy Finance Principles." Accessed December 31, 2019. http://awsassets.panda.org/downloads/sustainable_blue_economy_finance_principles_2018_brochure_final.pdf.

15. Hargreaves Lansdown. "The Global Sustainability Trust PLC: Launch postponed." Accessed December 31, 2019. https://www.hl.co.uk/shares/ipos-and-new-issues/the-global-sustainability-trust-plc#.

16. Cripps, Peter. "Impact Fund Eyes £200m London IPO." *Environmental Finance.* January 11, 2019. Accessed December 31, 2019. https://www.environmental-finance.com/content/news/impact-fund-eyes-200m-london-ipo.html.

17. IssueWire. "LIMEYARD, Equarius Risk Analytics and Thomas Schumann Capital Launch World's First Water Risk Index." December 11, 2018. Accessed April 13, 2020. https://www.issuewire.com/limeyard-equarius-risk-analytics-and-thomas-schumann-capital-launch-worlds-first-water-risk-index-1619586475837013.

18. Circulate Capital. "Press Release: Circulate Capital closes US $106M Fund to Protect Asia's Ocean from Plastic." December 4, 2019. Accessed January 15, 2020. https://1b495b75-5735-42b1-9df1-035d91de0b66.filesusr.com/ugd/77554d_149b9b5f9aae4a2ab409fdeb231eb3b3.pdf.

19. WWF. *Guide to Conservation Finance.* 36.

20. Pascal et al. "Impact Investing in Marine Conservation." 206.

21. Milder, Brian. "Conservation Finance in the Galápagos Islands: Capitalizing the 'Missing Middle'." In *Conservation Capital in the Americas,* edited by James N. Levitt. 120.

22. Gonzales, José and Hermilio Rosas. "The Caral Archeological Site and Supe Valley Restoration." In *Conservation Capital in the Americas,* edited by James N. Levitt. 92.

23. Meloy Fund. "About the Fund." Accessed April 13, 2020. https://www.meloyfund.com/about.

24. DFC. "Political Risk Insurance." Accessed April 13, 2020. https://www.dfc.gov/what-we-offer-our-products/political-risk-insurance.

25. CI. "CI Ventures, LLC." Accessed April 13, 2020. https://www.conservation.org/projects/conservation-international-ventures-llc.

26. Danigelis, Alyssa. "American ESG Investing Reaches $12 Trillion Assets, Says US SIF Foundation." *Environmental Leader.* November 2, 2018. Accessed

January 15, 2020. https://www.environmentalleader.com/2018/11/american-esg-investing/.

27. Global Impact Investing Network. "Annual Impact Investor Survey, 2019, Ninth Edition." June 2019. https://thegiin.org/assets/GIIN_2019%20Annual%20Impact%20Investor%20Survey_webfile.pdf. VIII.

28. United Nations' Principles for Responsible Investment. "About the PRI." Accessed March 18, 2020. https://www.unpri.org/about-the-pri.

29. Alliance to End Plastic Waste. "The Answers." Accessed January 15, 2020. https://endplasticwaste.org/answers/.

30. Circulate Capital. "Press Release: Circulate Capital closes US $106M Fund to Protect Asia's Ocean from Plastic." December 4, 2019. Accessed January 15, 2020. https://1b495b75-5735-42b1-9df1-035d91de0b66.filesusr.com/ugd/77554d_149b9b5f9aae4a2ab409fdeb231eb3b3.pdf.

31. Banca March. "News: March A.M. Launches the Mediterranean Fund." September 30, 2019. Accessed January 15, 2020. https://www.bancamarch.es/en/us/news/march-am-launches-the-mediterranean-fund.html.

32. Sky Ocean Rescue. "About Us." Accessed January 15, 2020. https://www.skyoceanrescue.com/about.

33. Althelia Ecosphere. "Our Team." Accessed February 20, 2017. https://althelia.com/about-us/our-team/#all.

34. Cripps, Peter. "First-of-Its Kind Sustainable Ocean Fund Hits the Market." *Environmental Finance*. February 6, 2017. Accessed December 31, 2019. https://www.environmental-finance.com/content/news/first-of-its-kind-sustainable-ocean-fund-hits-the-market.html.

35. Althelia Ecosphere. "Althelia Merges with Mirova to Invest €1 Billion for Nature." October 9, 2017. Accessed April 13, 2020. https://althelia.com/2017/10/09/mirova-completed-acquisition-althelia-ecosphere-first-step-invest-e1-billion-natural-capital-2020/.

36. Cripps, Peter. "First-of-Its Kind Sustainable Ocean Fund Hits the Market." *Environmental Finance*. February 6, 2017. Accessed December 31, 2019. https://www.environmental-finance.com/content/news/first-of-its-kind-sustainable-ocean-fund-hits-the-market.html.

37. Ibid.

38. Ibid.

39. Chen, James. "Société d'investissement à Capital Variable (SICAV)." *Investopedia*. June 25, 2019. Accessed December 31, 2019. https://www.investopedia.com/terms/s/sicav.asp.

40. Marchant, Christopher. "Althelia Sustainable Ocean Fund Raises Additional $50m." *Environmental Finance*. November 19, 2019. Accessed December 31, 2019. https://www.environmental-finance.com/content/news/althelia-sustainable-ocean-fund-raises-additional-$50m.html.

41. Kiss, Edit. Interviewed by Brian McFarland. March 2017.

42. Althelia Ecosphere. "White Paper: Investing for Impact and Value in the Marine Environment: The Sustainable Ocean Fund." 2.

43. Kiss, Edit. Interviewed by Brian McFarland. March 2017.
44. Althelia Funds. "ESG Approach." Accessed March 31, 2017. https://althelia.com/esg-approach/.
45. Althelia Funds. "Our Impacts Monitoring." Accessed March 31, 2017.https://althelia.com/our-approach/our-impacts-monitoring/.
46. Bank, David. "Beyond Carbon: Althelia Climate Fund Attracts Conservation Investors." *ImpactAlpha*. Last modified October 24, 2014.http://impactalpha.com/beyond-carbon-althelia-climate-fund-attracts-conservation-investors/.
47. Marchant, Christopher. "Althelia Sustainable Ocean Fund Raises Additional $50m." *Environmental Finance*. November 19, 2019. Accessed December 31, 2019. https://www.environmental-finance.com/content/news/althelia-sustainable-ocean-fund-raises-additional-$50m.html.
48. Mirova. "Mirova's Sustainable Ocean Fund reaches a final close at $132 m of commitments exceeding its target." August 5, 2020. Accessed October 20, 2020. Available: https://www.mirova.com/en/news/mirova-sustainable-ocean-fund-reaches-final-close-132m-commitments.
49. Ibid.
50. Ibid.
51. Ibid.
52. Althelia Ecosphere. "White Paper: Investing for Impact and Value in the Marine Environment: The Sustainable Ocean Fund." 7.
53. Ibid. 7.
54. Blue Finance. "Blue Finance Brings Together Public & Private Sectors in the Dominican Republic to Sustainably Manage 8000 km^2 of the Marine Sanctuary 'Arrecifes del Sureste'." Accessed February 7, 2019. http://blue-finance.org/?page_id=961.
55. Fitzgerald, Tim. Interviewed by Brian McFarland. April 2020.
56. Althelia Ecosphere. "White Paper: Investing for Impact and Value in the Marine Environment: The Sustainable Ocean Fund." 4.
57. Ibid. 5.
58. Ibid. 5–6.
59. Fitzgerald, Tim. Interviewed by Brian McFarland. April 2020.
60. Environmental Defense Fund, Rare/Meloy Fund and Encourage Capital. "Principles for Investment in Sustainable Wild-Caught Fisheries." 2018. www.fisheriesprinciples.org.
61. Fitzgerald, Tim. Interviewed by Brian McFarland. April 2020.
62. Ibid.
63. Marchant, Christopher. "Brazilian Biodiversity Fund Reaches First Close." *Environmental Finance*. October 29, 2019. Accessed December 31, 2019. https://www.environmental-finance.com/content/news/brazilian-biodiversity-fund-reaches-first-close.html.
64. Althelia Funds. "Althelia Climate Fund." Accessed December 31, 2019. https://althelia.com/althelia-climate-fund/.

65. Ibid. "Madagascar Fund." Accessed December 31, 2019. https://althelia.com/initiatives/madagascar-fund/.

66. Ibid. "Conservation Notes." Accessed December 31, 2019. https://althelia.com/initiatives/conservation-notes/.

67. Althelia Ecosphere. "White Paper: Investing for Impact and Value in the Marine Environment: The Sustainable Ocean Fund." 2.

68. Ibid. 3.

69. Althelia Funds. "Cordillera Azul National Park REDD+ Project." Accessed December 31, 2019.https://althelia.com/investment/cordillera-azul-national-park-redd-project/.

70. Ibid. "Guatemalan Conservation Coast." Accessed December 31, 2019. https://althelia.com/investment/guatemalan-caribbean-forest-corridor/.

71. Ibid. "Sumatra Merang Peatland." Accessed December 31, 2019. https://althelia.com/investment/sumatra-merang-peatland/.

72. Ibid. "Taita Hills Conservation and Sustainable Land Use Project." Accessed December 31, 2019. https://althelia.com/investment/taita-hills-conservation-and-sustainable-land-use-project/.

73. Ibid. "Tambopata-Bahuaja Biodiversity Reserve." Accessed December 31, 2019. https://althelia.com/investment/tambopata-bahuaja-redd-and-agroforestry-project/.

74. Fitzgerald, Tim. Interviewed by Brian McFarland. April 2020.

75. Ibid.

76. Ibid.

77. Rivera-Planter et al. "Economic Instruments for Sustainability in Mexico's Marine Protected Areas and the Perverse Subsidy Challenge." In *Economic Incentives for Marine and Coastal Conservation*, edited by Essam Yassin Mohammed. 230.

78. Vertigo Lab. "Innovations for Coral Finance, ICRI Publication." 51.

79. Ibid. 52.

80. Ibid. 52.

81. Ibid. 52.

82. Ibid. 52.

83. Ibid. 50.

84. Ibid. 50.

85. Secaira, Fernando. Interviewed by Brian McFarland. February 2019.

86. Iyer, Venkat et al. *Finance Tools for Coral Reef Conservation*. 39.

87. Secaira, Fernando. Interviewed by Brian McFarland. February 2019.

88. Secaira, Fernando. "Insuring Nature: An Insurance Policy for the Mesoamerican Reef." Online webinar, Reef Resilience Network and the EBM Tools Network, location unknown, February 28, 2018.

89. Ibid.

90. Ibid.

91. Ibid.

92. Secaira, Fernando. Interviewed by Brian McFarland. February 2019.

93. Ibid.

94. Ibid.

95. Secaira, Fernando. "Insuring Nature: An Insurance Policy for the Mesoamerican Reef." Online webinar, Reef Resilience Network and the EBM Tools Network, location unknown, February 28, 2018.

96. Ibid.

97. Ibid.

98. Secaira, Fernando. Interviewed by Brian McFarland. February 2019.

99. Secaira, Fernando. "Insuring Nature: An Insurance Policy for the Mesoamerican Reef." Online webinar, Reef Resilience Network and the EBM Tools Network, location unknown, February 28, 2018.

100. Secaira et al. "A Guide on How to Insure a Natural Asset." 2019. *The Nature Conservancy.* https://www.researchgate.net/publication/335703419_GUIDE_ON_HOW_TO_INSURE_A_NATURAL_ASSET.

101. Secaira, Fernando. Interviewed by Brian McFarland. February 2019.

102. Ibid.

103. Ibid.

104. Ibid.

105. Klein, Asher. "'Nature as an Asset': Coral Reef Insurance May Be a Sea Change for Conservation." *NBC Philadelphia.* April 18, 2018. Accessed February 22, 2019. https://www.nbcphiladelphia.com/news/national-international/Mexico-Coral-Reef-Parametric-Insurance-479834133.html.

106. Ibid.

107. Ibid.

108. Ibid.

109. Evans, Steve. "Nature Conservancy Targets Florida & Hawaii for Parametric Reef Insurance." *Artemis.* September 26, 2019. Accessed January 28, 2020. https://www.artemis.bm/news/nature-conservancy-targets-florida-hawaii-for-parametric-reef-insurance/.

110. Iyer, Venkat et al. *Finance Tools for Coral Reef Conservation.* 40.

111. Environmental Finance. "Mexican coral reef insurance attracts new underwriters." 2020. Available: https://www.environmental-finance.com/content/news/mexican-coral-reef-insurance-policy-renewed.html.

112. Secaira, Fernando. Interviewed by Brian McFarland. February 2019.

113. Blue Finance. "Blue Finance Brings Together Public & Private Sectors in the Dominican Republic to Sustainably Manage 8000 km^2 of the Marine Sanctuary 'Arrecifes del Sureste'." Accessed February 7, 2019. http://blue-finance.org/?page_id=961.

114. Protected Planet. "Arrecifes del Sureste in Dominican Republic." Accessed February 28, 2020. https://www.protectedplanet.net/arrecifes-del-sureste-marine-sanctuary.

115. Blue Finance. "Blue Finance Brings Together Public & Private Sectors in the Dominican Republic to Sustainably Manage 8000 km^2 of the Marine Sanctuary 'Arrecifes del Sureste'." Accessed February 7, 2019. http://blue-finance.org/?page_id=961.

116. Pascal et al. "Impact Investing in Marine Conservation." 217–218.

117. Avery, Helen. "Blue Finance: Why Marine PPPs Could Be a Win-Win-Win." *Euromoney*. June 5, 2018. Accessed December 30, 2019. https://www.euromoney.com/article/b18hg7vjwmy9hn/blue-finance-why-marine-ppps-could-be-a-win-win-win.

118. Ibid.

119. Pascal, Nicolas. Interviewed by Brian McFarland. January 2019.

120. Ibid.

121. Ibid.

122. Trading Economics. "Dominican Republic Tourist Arrivals." Accessed March 18, 2020. https://tradingeconomics.com/dominican-republic/tourist-arrivals.

123. Pascal, Nicolas. Interviewed by Brian McFarland. January 2019.

124. Blue Finance. "Blue Finance Brings Together Public & Private Sectors in the Dominican Republic to Sustainably Manage 8000 km^2 of the Marine Sanctuary 'Arrecifes del Sureste'." Accessed February 7, 2019. http://blue-finance.org/?page_id=961.

125. Pascal et al. "Impact Investing in Marine Conservation." 212.

126. Blue Finance. "Blue Finance Brings Together Public & Private Sectors in the Dominican Republic to Sustainably Manage 8000 km^2 of the Marine Sanctuary 'Arrecifes del Sureste'." Accessed February 7, 2019. http://blue-finance.org/?page_id=961.

127. Blue Finance. "The Right Person for the Job." Accessed February 7, 2019. http://blue-finance.org/?p=1261.

128. Asner, Greg and Clare LeDuff. "Bold Initiative Aims to Protect Coral Reefs in the Dominican Republic." *Mongabay*. July 16, 2018. Accessed April 13, 2020. https://news.mongabay.com/2018/07/bold-initiative-aims-to-protect-coral-reefs-in-the-dominican-republic/.

129. Pascal, Nicolas. Interviewed by Brian McFarland. January 2019.

130. Ibid.

131. Blue Finance. "Blue Finance Brings Together Public & Private Sectors in the Dominican Republic to Sustainably Manage 8000 km^2 of the Marine Sanctuary 'Arrecifes del Sureste'." Accessed February 7, 2019. http://blue-finance.org/?page_id=961.

132. Iyer, Venkat et al. *Finance Tools for Coral Reef Conservation*. 46.

133. Blue Finance. "Blue Finance Brings Together Public & Private Sectors in the Dominican Republic to Sustainably Manage 8000 km^2 of the Marine Sanctuary 'Arrecifes del Sureste'." Accessed February 7, 2019. http://blue-finance.org/?page_id=961.

134. Blue Finance. "The Right Person for the Job." Accessed February 7, 2019. http://blue-finance.org/?p=1261.

135. Avery, Helen. "Blue Finance: Why Marine PPPs Could Be a Win-Win-Win." *Euromoney*. June 5, 2018. Accessed December 30, 2019. https://www.euromo ney.com/article/b18hg7vjwmy9hn/blue-finance-why-marine-ppps-could-be-a-win-win-win.

136. Ibid.

137. Ibid.

138. Asner, Greg and Clare LeDuff. "Bold Initiative Aims to Protect Coral Reefs in the Dominican Republic." *Mongabay*. July 16, 2018. Accessed April 13, 2020. https://news.mongabay.com/2018/07/bold-initiative-aims-to-protect-coral-reefs-in-the-dominican-republic/.

139. Pascal, Nicolas. Interviewed by Brian McFarland. January 2019.

140. Avery, Helen. "Blue Finance: Why Marine PPPs Could Be a Win-Win-Win." *Euromoney*. June 5, 2018. Accessed December 30, 2019. https://www.euromo ney.com/article/b18hg7vjwmy9hn/blue-finance-why-marine-ppps-could-be-a-win-win-win.

141. Ibid.

142. Pascal, Nicolas. Interviewed by Brian McFarland. January 2019.

143. Avery, Helen. "Blue Finance: Why Marine PPPs Could Be a Win-Win-Win." *Euromoney*. June 5, 2018. Accessed December 30, 2019.

144. Pascal, Nicolas. Interviewed by Brian McFarland. January 2019.

145. Iyer, Venkat et al. *Finance Tools for Coral Reef Conservation*. 46.

146. Beyer et al. "Risk-Sensitive Planning for Conserving Coral Reefs Under Rapid Climate Change." 3 of 10.

147. Pascal, Nicolas. Interviewed by Brian McFarland. January 2019.

148. Coalition for Private Investment in Conservation. "CPIC Blueprint Public-Private Partnership for Marine Protected Areas by Blue Finance." Accessed February 7, 2019. http://cpicfinance.com/cpic-blueprint-public-private-partne rship-for-marine-protected-areas-by-blue-finance-3/.

149. Pascal, Nicolas. Interviewed by Brian McFarland. January 2019.

150. Avery, Helen. "Blue Finance: Why Marine PPPs Could Be a Win-Win-Win." *Euromoney*. June 5, 2018. Accessed December 30, 2019. https://www.euromo ney.com/article/b18hg7vjwmy9hn/blue-finance-why-marine-ppps-could-be-a-win-win-win.

151. Global Impact Investing Network. "Annual Impact Investor Survey, 2019, Ninth Edition." 2019. https://thegiin.org/assets/GIIN_2019%20Annual% 20Impact%20Investor%20Survey_webfile.pdf. 4.

152. Pascal et al. "Impact Investing in Marine Conservation." 209.

153. TNC. "The Conservation Note Prospectus." January 24, 2014. Accessed April 13, 2020. https://www.nature.org/media/aboutus/tnc-prospectus-jan uary-2014.pdf.

154. Trading Economics. "Credit Rating." Accessed July 23, 2019. http://www.tra dingeconomics.com/country-list/rating.

155. World Bank Group. "Doing Business: Economy Rankings." Accessed July 23, 2019. http://www.doingbusiness.org/rankings.

156. Transparency International. "Corruption Perceptions Index 2018." Accessed July 24, 2019. https://www.transparency.org/cpi2018.

157. UNDP. "Human Development Data (1990–2017)." Accessed July 23, 2019. http://hdr.undp.org/en/data#.

158. World Justice Project. "Rule of Law Index 2019." Accessed July 24, 2019. http://data.worldjusticeproject.org/#table and https://worldjusticeproject.org/our-work/research-and-data/wjp-rule-law-index-2019.

159. Fragile States Index. "Home." Accessed April 13, 2020. https://fragilestatesindex.org/.

160. Reporters without Borders. "The World Press Freedom Index." Accessed April 13, 2020. https://rsf.org/en/world-press-freedom-index.

161. Happy Planet Index. "Home." Accessed April 13, 2020. http://happyplanetindex.org/.

162. Environmental Finance. "Sustainable Forestry Deal of the Year." 2012. Accessed April 6, 2020. https://www.dfc.gov/sites/default/files/2019-08/EF0712_Awards_forest_web1.pdf.

163. USAID. "U.S. Government, Althelia Climate Fund Mobilize $133.8 Million for Forest Conservation and Alternative Livelihoods." Last modified May 28, 2014. Accessed April 6, 2020.https://2012-2017.usaid.gov/news-information/press-releases/may-28-2014-us-government-althelia-climate-fund-mobilize-1338-million-forest-conservation.

164. Cripps, Peter. "Impact Investments 'to Hit $30trn within Decade'." *Environmental Finance*. November 5, 2018. Accessed December 31, 2019. https://www.environmental-finance.com/content/news/impact-investments-to-hit-$30trn-within-decade.html.

165. The Global Innovation Lab for Climate Finance. "Investor Initiative Chooses Ideas That Will Mobilize Investment for Sustainable Cities, Energy Access, Blue Carbon, and Sustainable Agriculture." March 5, 2019. Accessed April 13, 2020. https://www.climatefinancelab.org/news/lab-selection-2019/.

166. Vertigo Lab. "Innovations for Coral Finance, ICRI Publication." 2017. http://vertigolab.eu/wp-content/uploads/2018/04/rapport-innovation-for-coral-finance.pdf. 53.

167. Iyer, Venkat et al. *Finance Tools for Coral Reef Conservation*. 64.

168. Environmental Finance. "BNP Paribas launches first blue economy ETF." 2020. Available: https://www.environmental-finance.com/content/news/bnp-paribas-launch-first-blue-economy-etf.html.

169. Environmental Finance. "Credit Suisse ocean health impact fund raises 212m." 2020. Available: https://www.environmental-finance.com/content/news/credit-suisse-ocean-health-impact-fund-raises-212m.html.

170. Environmental Finance. "100m raised for oceans and seafood impact fund." 2020. Available: https://www.environmental-finance.com/content/news/100m-raised-for-oceans-and-seafood-impact-fund.html.

171. Environmental Finance. "HSBC and Pollination form 'world's largest natural capital manager'." 2020. Available: https://www.environmental-finance.com/content/news/hsbc-and-pollination-form-worlds-largest-natural-capital-manager.html.

11

Payments for Ecosystem Services

Introduction

Payments for ecosystem services come in several different conservation financing instruments such as biodiversity offsets, species banking, watershed protection payments, mitigation banking, and forest carbon markets (i.e., particularly blue carbon offsets).

Historical Overview

In the late 1970s, the concept of payments for ecosystem services (PES, historically referred to as payments for environmental services) first appears.[1] For instance, "first used in the United States in the 1970s to mitigate damage to wetlands, biodiversity offset programmes have more recently been introduced in a number of countries."[2]

Other notable dates for PES include:

- "First used in the United States in the 1970s to mitigate damage to wetlands, biodiversity offset programmes have more recently been introduced in a number of countries. More than 100 countries {now} have laws or policies in place that require or enable the use of biodiversity offsets, or are currently considering their use."[3]
- 1994: "The market-based approach for irrigation was first set out in the communiqué of the 1994 Council of Australian Government for the Murray-Darling Basin, and was fully developed in the National Water Initiative in 2004 (ABS, 2008)."[4]

© The Author(s) 2021
B. J. McFarland, *Conservation of Tropical Coral Reefs*,
https://doi.org/10.1007/978-3-030-57012-5_11

- 1996: Costa Rica "earned an international reputation for reforestation when it introduced a unique piece of legislation, the 1996 Forestry Law, that promotes conservation through payments for environmental services, reforestation, and conservation programs."[5]
- 1998: The "New South Wales state legislature had taken an aggressive step to forward this vision in November 1998, adopting the world's first law establishing 'carbon rights'."[6]
- 2003: The Climate, Community & Biodiversity Alliance (CCBA) is founded.[7]
- 2004: First Clean Development Mechanism (CDM) Project (Brazil NovaGerar Landfill Gas to Energy Project) is registered.[8]
- 2004: The Business and Biodiversity Offsets Programme (BBOP) is established by Forest Trends.[9]
- 2005: Verified Carbon Standard Association (now known as Verra), which manages the Verified Carbon Standard (the world's largest voluntary carbon market certification standard) and the Climtate, Community & Biodiversity Standards (CCBS), is founded.[10]
- 2006: Boulder, Colorado passes the first carbon tax in the United States.[11]
- 2006: First CDM Forestry Project (Facilitating Reforestation for Guangxi Watershed Management in Pearl River Basin) is registered.[12]
- 2007: First tropical Climate, Community & Biodiversity Alliance (CCBA) Project (Small-scale Reforestation for Landscape Reforestation Project) is validated against the Climate, Community & Biodiversity Standards (CCBS).[13]
- 2007: In October 2007, "the first such measure – the use of the CCAR {California Climate Action Registry} Forest Project Protocols – was adopted. The protocols are the first rigorous governmental accounting standards in the United States for climate projects that embrace forest management and avoid deforestation."[14]
- 2007: In December 2007, "the Van Eck Forest became the first project from any emissions sector to register with the CCAR {California Climate Action Registry} as a carbon emissions reduction project."[15]
- 2007: The "concept of REDD was taken up by environmental groups and some tropical countries, the latter informally organized into a Coalition for Rainforest Nations, and in 2007 at the United Nations Climate Change Conference in Bali, Indonesia, the Bali Road Map was adopted as a two-year plan to finalize a binding international agreement in December 2009 in Copenhagen."[16]
- 2008: First tropical forest conservation CCBA Project (Reducing Carbon Emissions from Deforestation in the Ulu Masen Ecosystem) is validated against the CCBS.[17]

- 2011: Wildlife Works Carbon's Kasigau Corridor REDD Project—Phase I Rukinga Sanctuary becomes the first VCS REDD+ project to complete verification and issuance.[18]
- In 2012, "{...} the UNFCCC provided final approval of a mangrove conservation and rehabilitation project in Senegal. This was the first such project to receive accreditation under the UNFCCC's Clean Development Mechanism."[19]
- 2013: The Green Climate Fund "opened its doors" in South Korea.[20]
- 2015: The UNFCCC's Paris Agreement is signed by almost every country on Earth.[21]
- 2016 (September): The world's first global policy on biodiversity offsets is adopted.[22]

Mechanisms of Instrument

Whether speaking of biodiversity offsets or carbon offset credits, PES schemes often use a "polluter pays" approach and all PES schemes have a "buyer" or "beneficiary" and a "seller" or "provider:"

Providers of ES {ecosystem services} are normally the stewards of the resource base. Beneficiaries are agents whose economic performance or livelihoods are affected by those services (which often are appropriated 'free of charge'). Providers and beneficiaries normally belong to different social groups, but their welfares are interconnected through ES, the provision of which depends on decisions taken by multiple agents. The possibilities of coordination between these two groups (to align their interests) are very much conditioned by the allocation of rights and the types of ES (how they are produced, provided and consumed).[23]

Regarding Payments for Marine and Coastal Ecosystem Services (PMCES):

In many ways, PMCES {Payments for Marine and Coastal Ecosystem Services} can be seen as a response to the problem of undervaluation. They provide a means of recognizing, capturing and internalizing the ecosystem values that have traditionally been excluded from the prices, markets and policies that drive land and resource use decisions in coastal areas.[24]

{Furthermore} most PMCES schemes are built upon two premises: that ecosystem services have quantifiable economic value, and that this value can be used to entice investment in their restoration and maintenance. (Forest Trends and Katoomba Group, 2010)[25]

Although PES instruments work differently, essentially all PES instruments compensate the owner and/or steward of natural resources for its services. One could categorize PES in two categories: those ecological assets which already have a market value (e.g., water) and those ecological assets without a market value (e.g., the reduction of greenhouse gas emissions from mangrove forests). Water needs some regulatory support to establish a price that convinces a hydroelectric company, a large bottling company, or a local government to support the preservation of an upstream watershed. In contrast, GHGs—particularly carbon dioxide sequestered and stored in mangrove forests or seagrass—are often undervalued or considered to have no market value. This situation creates unclear and/or unstable pricing, and thus, regulation creates the potential for monetization.

Similarly, as explained by Marcello Hernandez-Blanco, there are two different types of payments for ecosystem services known as Coasean and Pigouvian. In a pure Coasean PES scheme, the government plays no role. For instance, if a water bottling company pays upstream farmers to protect the watershed. In contrast, a Pigouvian PES scheme is one that is characterized by government involvement, such as implementing a carbon price.[26]

Biodiversity Offsets

Biodiversity offsets, known also as species banking or conservation banking, "represents a free market mechanism for mitigating development impacts on endangered species' habitats."[27] The mitigation hierarchy starts with a firm avoiding impact to an endangered species' habitat, then minimizing any impacts if avoidance is impossible, followed by rehabilitating or restoring degraded habitats, and concluding with the use of biodiversity offsets as a last resort mechanism.[28]

Biodiversity offsets can be developed for a particular compliance market or as voluntary offsets. With respect to financing, there are several structures including upfront payments for the restoration work or staggered payments.

A comparable mechanism is known as mitigation banking, where:

Once all parties involved approve the plans and the bank meets its ecological success criteria (defined by the government, not the landowner), the mitigation banker earns credits that regulatory agencies will recognize. The landowner can then sell these mitigation credits to permittees and others who must compensate for adversely affected wetlands or other natural areas. The sale of wetland credits legally transfers the liability for wetland mitigation from the permitee to the wetland banker.[29]

With respect to biodiversity offsets:

They usually come in three different forms, depending on national legislation about whether they are set up on a voluntary or obligatory basis:

On-demand offset, in which the project owner takes entire financial and legal responsibility for compensation. Offsetting measures are usually restoration, rehabilitation, or habitat creation or preservation mechanisms.

Biodiversity banking, in which a public or private third party, specialized in compensation, sets up, without any obligations, the offsetting measures. Those actions generate offset 'credits' that the third party can sell to a project owner who has to compensate for the impacts of his project. Offsetting measures are realized before the implementation of the project.

Financial compensation, in which the project owner has to compensate for his damaging project by making a financial transfer to a third-party organization. This can be transferred to a compensation fund managed by the government or an environmental organization, to an environmental NGO or foundation, or to a territorial collectivity. Usually, financial transfers are used when on-demand offset is impossible.[30]

Another approach is to categorize biodiversity offsets as: one-off offsets; in-lieu fees; and biobanking:

One-off offsets: once (predicted) adverse impacts have been evaluated, the biodiversity offset is carried out by the developer or by a subcontractor (e.g. a conservation NGO). The developer assumes financial and legal liability. Verification is normally undertaken by a government agency or an accredited third party. One-off approaches are typically used in voluntary offsets and are common under regulatory programmes (e.g., vegetation management offsets in Queensland, Australia; Species Mitigation and Wetland Compensatory Mitigation in the United States).

In-lieu fees: a government agency stipulates a fee that a developer has to pay to a third party, to compensate for residual adverse biodiversity impacts. The third party (i.e. the offset provider) takes on the financial and legal responsibility for the offsets. In-lieu fee arrangements have been employed in the US Species Mitigation and Wetland Compensatory Mitigation, and in forest compensation schemes in India and Mexico.

Biobanking: once (predicted) adverse impacts are evaluated, the developer can purchase offsets directly from a public or private biobank. A biobank refers to a repository of existing offset credits, where each credit represents a quantified gain in biodiversity resulting from actions to restore, establish, enhance and/or preserve biodiversity. As under the in-lieu fee arrangement, financial and legal liability is transferred from the developer to the provider. Examples of biobanking include the US Wetland Compensatory Mitigation, the New

South Wales Biobanking scheme in Australia, and compensation pools under the German Impact mitigation Regulation.[31]

Regardless of the categorization:

Key design and implementation features that must be considered to ensure offset schemes are environmentally and cost effective, as well as distributional equitable include: thresholds and coverage; equivalence; additionality; permanence; monitoring, reporting and verification; transaction costs; and compliance and enforcement. {…} a distinct issue for offsets is how to ensure equivalence between the biodiversity loss at the development site, and the biodiversity gain at the offset site.

Yet, it is important to recognize that:

In some cases, adverse impacts to biodiversity may not be able to be fully compensated. This is because: the affected biodiversity is irreplaceable or extremely vulnerable; there are no available offset sites; or there are no known conservation approaches to achieve the offset outcomes required.[32]

Biodiversity offsets should result in no net loss and ideally, result in net gain.

Watershed Payments

According to the *State of Watershed Investment 2016* Report, a watershed investment is defined as "any transaction between a buyer and a seller where financial value is exchanged for activities or outcomes associated with the maintenance, restoration, or enhancement of watershed services or natural areas considered important for watershed services."[33]

Watershed payments are often structured between downstream users of water (e.g., hydroelectric facilities, downstream bottlers, and agricultural interests) and upstream watershed stewards (e.g., farmers and/or forest landowners). Similarly, "much of the funding for the 'green' solutions to flood control and stormwater quality comes from existing stormwater management monies, in addition to a stormwater utility fee within water bills that is based on the amount of impervious surface on a property."[34]

The Mechanisms of Developing Forest Carbon Projects

Voluntary forest carbon projects follow these general steps from the initial project idea through project implementation and ending with the issuance of verified emission reductions (also known as carbon offset credits). This process is similar whether the project is being designed and implemented against a compliance carbon standard or a voluntary carbon standard. Similarly, this process is relatively the same whether you are developing a more traditional forest carbon project (i.e., reforestation or forest conservation project) or a blue carbon project (i.e., mangrove conservation project). In addition to mangroves, blue carbon includes seagrass and salt marshes and could include kelp beds.

1. *Due Diligence and Assessment of Project Potential*

It is vitally important to undertake an initial assessment of the project potential and conduct sufficient due diligence. Some aspects include:

- Review of stakeholders and evaluation of potential partners;
- Hold initial stakeholder and partner meetings;
- Conduct financial calculations;
- Undertake risk assessment, legal analysis, and potentially draft contracts; and
- Analysis of landownership titles and carbon ownership (if applicable).

2. *Selection of Certification Standard and Methodology*

All compliance carbon projects and nearly all voluntary forest carbon projects are designed and audited against third-party certification standards.

The leading forest carbon certification standard in the voluntary carbon market is the Verified Carbon Standard (VCS) which is often coupled with the Climate, Community & Biodiversity Standards (CCBS) to demonstrate net positive community and biodiversity benefits. Additional voluntary carbon certification standards that accept forest carbon projects are the Gold Standard, the American Carbon Registry (ACR), and Plan Vivo.

Here are the most frequently used carbon certification standards:

- ACR: http://americancarbonregistry.org/;
- CAR: http://www.climateactionreserve.org/;
- CCBS: https://verra.org/project/ccb-program/;
- Gold Standard: http://www.goldstandard.org/;
- Plan Vivo: http://www.planvivo.org/; and
- VCS: https://verra.org/project/vcs-program/

A methodology, also known sometimes as a protocol, is essentially the blueprint for how various parameters of a project are calculated and/or determined. This includes guidance on how to design the forest carbon inventory, how to determine the project's start date and longevity, and how to develop a deforestation baseline. Each carbon certification standard will have a range of approved methodologies for use under the given standard. The approved methodologies and protocols under the CAR, the CDM, and the VCS are:

- CAR Protocols: http://www.climateactionreserve.org/how/protocols/;
- CDM Methodologies: http://cdm.unfccc.int/methodologies/index.html; and
- VCS Methodologies: https://verra.org/methodologies/

More specifically, there are a few VCS methodologies related to blue carbon:

- VCS' Methodology for Coastal Wetland Creation[35]; and
- VCS' Methodology for Tidal Wetland and Seagrass Restoration.[36]

Technical forestry firms capable of working through the complex methodologies include:

- Carbon Decisions International: http://carbondecisions.com/;
- Climate Focus: http://www.climatefocus.com/;
- ecoPartners: https://www.epcarbon.com/;
- Finite Carbon: http://www.finitecarbon.com/;
- Ostrya Conservation: https://www.ostryaconservation.com/;
- Silvestrum: http://www.silvestrum.com/;
- Restore America's Estuaries: https://estuaries.org/;
- TerraCarbon: http://terracarbon.com/;
- Wildlife Works Carbon: http://www.wildlifeworks.com/; and
- Winrock International: https://www.winrock.org/

3. *Selection of Independent Auditors*

The next step is to choose independent auditors, known as validation/verification bodies (VVBs). The auditors are subject matter experts. For example, forest carbon projects are often audited by professional foresters, agriculturalists, or community development experts.

Auditors are often accredited to the American National Standards Institute (ANSI), which is the U.S. representative of the International Organization for Standardization (ISO). This accreditation helps to ensure auditors have proper management systems in place including a conflict of interest policy, document retention policies, and proper document version control.

Auditors, such as Aster Global Environmental Solutions or SCS Global Services, then must be formally accepted by the certification standard. This acceptance is based off an application process and possibly includes a more rigorous training specific to the given certification standard. Many auditors are accepted across several carbon certification standards. Approved VVBs for the CAR and the VCS can be found here:

- CAR: http://www.climateactionreserve.org/how/verification/connect-with-a-verification-body/; and
- Gold Standard: https://www.goldstandard.org/resources/approved-aud itors

4. *Implementation of Project Activities*

After a forest carbon project has passed due diligence, chosen a certification standard and a methodology or protocol, and lined up an independent auditor, it is time to implement the on-the-ground activities (that is, if such activities are not already underway).

For an afforestation or reforestation blue carbon project, these activities include:

- Conducting a carbon sequestration/forest growth curve analysis;
- Choosing the composition of tree species or seagrasses to be planted;
- Identifying a source for seeds, saplings, or grasses (i.e., such as establishing a nursery); and
- Planting the seeds, saplings, or grasses at the project site.

For a blue carbon forest conservation project, the on-the-ground activities include:

- Designing and implementing a forest carbon inventory;
- Developing a deforestation and degradation baseline; and
- Starting to implement the specific activities which will help reduce deforestation and degradation at the project site. This might include the

landowner foregoing conversion of the forest to another land use such as a shrimp farm or developing alternative activities for local communities.

5. *Write Initial Project Documents*

While the project activities are being implemented, it is a wise to start writing the initial set of project documents. These project documents are comprehensive and require an elaboration on various aspects of the project including:

- Background information on the project and project proponents;
- Incorporation of the technical measurements derived from the chosen methodology;
- Maps of the project location and project activities;
- Demonstration of Free, Prior and Informed Consent and engagement with stakeholders;
- Discussion of all relevant laws and regulations; and
- Risk assessment of the project.

Many of the carbon certification standards offer templates for these project documents:

- VCS (see *Templates & Forms*): https://verra.org/project/vcs-program/rules-and-requirements/; and
- Gold Standard (see *Templates*): http://www.goldstandard.org/resources/aff orestation-reforestation-requirements

6. *Register/List Project; Undertake Validation Audit*

The project, upon completion of the initial project documents and after a sufficient amount of activities are implemented, will register or list with an approved registry. A registry for carbon projects is somewhat comparable to an online bank account, and upon logging in, you can view your current holdings and then either receive or initiate a transfer of VERs.

The validation audit will be conducted by the previously secured auditor. The validation audit involves the auditor reading through the initial project documents, checking all calculations, interviewing stakeholders, and visiting the project site.

Upon successful completion of the validation audit, the auditor will issue a validation report and validation statement. These documents essentially confirm that, in the auditor's professional opinion, the project has been designed and thus far implemented in accordance with all of the requirements under the chosen carbon certification standard and its chosen methodology or protocol.

7. *Conduct Monitoring, Reporting, and Verification with Issuance of Carbon Offset Credits*

The project, after achieving validation, will continue to implement the on-the-ground project activities and will periodically conduct monitoring, reporting, and verification (MRV).

Monitoring involves the monitoring or measurement of certain metrics associated with the project. For a blue carbon conservation project, this mostly involves monitoring deforestation and degradation relative to the established baseline, along with monitoring the project's impact on local communities and biodiversity.

Reporting involves the preparation of another set of project documents. This next set of project documents are prepared for a specific, historical time period—usually the previous calendar year.

Verification involves another audit, known as the verification audit. This audit involves a similar process of the auditor reading through the new project documents, checking all new calculations derived from the monitoring, (re)interviewing stakeholders, and (re)visiting the project site.

Upon successful completion of the verification audit, the auditor will issue a verification report and verification statement confirming the project has mitigated or sequestered a specific quantity of carbon dioxide equivalent emissions (CO_2e), in addition to community and biodiversity benefits. The project proponents would then deliver the project documents and auditor reports to the registry, and the registry will issue the specific quantity of offsets—each with a distinct serial number—into the project proponent's registry account.

8. *Sales*

Sales should not be left to the very end, but a common structure in the forest carbon market is for buyers to only purchase offsets after the project proponents have successfully completed the verification audit and are ready to transfer the issued VERs.

Size of Instrument

The following outlines the approximate sizes of the biodiversity offsets market, watershed payments, and the forest carbon sector.

As noted by Marcello Hernandez-Blanco, in reference to Salzman et al (2018) paper entitled, "The global status and trends of Payments for Ecosystem Services," there are 555 active PES programs around the world with combined annual payments over USD$36 billion. Of these 555 active PES programs, a total of 387 (i.e., 70%) are watershed PES schemes, whereas 120 (i.e., 21%) are biodiversity and habitat programs, and 48 (i.e., 9%) are forest and land-use carbon programs. Yet, there are no active PES programs for coastal marine ecosystems.[37]

Biodiversity Offsets

According to the OECD, "more than 100 countries {now} have laws or policies in place that require or enable the use of biodiversity offsets, or are currently considering their use."[38] In addition, "biodiversity offset programmes mobilised between USD 2.4 and 4 billion in 2011 and have substantial potential to be scaled-up."[39]

According to Holly J. Niner et al., "biodiversity offsetting policies applicable to marine environments were found to exist in six countries (US, Canada, Australia, France, Germany, Colombia) and have been actively considered in at least 27 others."[40]

With respect to the United States:

> The number of mitigation banks in the US has grown exponentially from an estimated 46 authorized mitigation banks in 1992 after the passage of the Clean Water Act to over 2,000 approved mitigation banks in 2015. Many more mitigation banks are undergoing review before final approval. Although precise figures are difficult to ascertain, private estimates indicate a total yearly credit sales volume in the US of over $2 billion.[41]

Although a bit outdated, the last *State of Biodiversity Markets* Report dated 2010, reported:

> {...} 39 existing compensatory mitigation programs around the world, ranging from programs with active mitigation banking of biodiversity credits to programs channeling development impact fees to policies that drive one-off offsets. There are another 25 programs in various stages of development or

investigation. Within each active offset program, there are numerous individual offset sites, including over 600 mitigation banks worldwide.

The global annual market size is $1.8-$2.9 billion at minimum, and likely much more, as 80% of existing programs are not transparent enough to estimate their market size. And the conservation impact of this market includes at least 86,000 hectares {212,000 acres} of land under some sort of conservation management or permanent legal protection per year.[42]

It is important to note that not all, in fact probably very few, of these commitments went to mitigation programs and biodiversity offsets in tropical coral reef ecosystems and some of this funding may also be included as overlapping figures reported in the section on domestic government expenditures.

Watershed Payments

According to the *State of Watershed Investment* 2016 Report, approximately USD$25 billion was spent on green infrastructure water projects across 62 countries and 419 distinct programs in 2015. Of this USD$25 billion, the vast majority (USD$23.7 billion) came in the form of direct subsidy payments from governments "to landowners to protect and restore water-critical landscapes and promote a green economy."[43] These landowners received an estimated USD$16 billion to reward their stewardship of watersheds. In addition, these projects "protected, rehabilitated, or created new habitat on more than 486 million hectares of land around the world, an area nearly 1.5 times the size of India."[44] Again, not all of these commitments went to watersheds adjacent to offshore coral reefs and some of this funding may also be included as overlapping figures reported in the section on domestic government expenditures. Yet, "programs in Queensland and the Australian Capital Territory stepped up payments to farmers to manage polluted runoff to the Great Barrier Reef. Municipal water service providers in Queensland and Melbourne also continued pilot offset projects for nutrients and stormwater."[45]

A recent development involved Ecosystem Investment Partners, which specializes in wetland and stream mitigation banks throughout the United States, which closed a USD$455 million wetlands fund in February 2020, bringing its total raised to nearly USD$1 billion.[46]

Forest Carbon

In 2018, the voluntary forest carbon market transacted 50.7 million tons of carbon dioxide equivalent emissions (mtCO2e) and was valued at approximately USD$171.9 million. The largest subsector of voluntary forest carbon in 2018 was Reducing Emissions from Deforestation and forest Degradation (REDD+) projects, which transacted 30.5 million mtCO2e and was valued at USD$46.6 million.[47] As of April 2020, the California compliance carbon market, which is one of the few compliance markets that accepts forest carbon offsets, saw approximately 80% of its offsets (137.9 million of 173.3 million total offsets) issued from U.S. forestry projects.[48]

While there are not as many blue carbon forest projects, there are still several mangrove restoration or conservation projects, such as:

- The Loru Forest Project, which is "managed by the Nakau Programme and protects 166 ha of tropical {coastal} rainforest on eastern Espiritu Santo, Vanuatu"[49];
- The Mangrove restoration and coastal greenbelt protection in the East coast of Aceh and North Sumatra Province, Indonesia Project, which aims to "increase the environmental carrying capacity of mangrove ecosystems in the east coast of Northern Sumatra for carbon sequestration, natural disaster risk reduction and local livelihood improvement"[50];
- The Mikoko Pamoja Project is "a community-led mangrove conservation and restoration project based in southern Kenya"[51]; and
- The Nakau Programme, which "is a landscape-level programme in the Pacific Islands which uses 'Payments for Ecosystem Services' (PES) for community-based rainforest & mangrove protection."[52]

Introduction to Case Studies

The following case study will examine the Reef Credit Initiative. The Initiative is setting up the infrastructure for local Queensland landowners to develop projects which issue credits representing a reduction in nutrients, pesticides, or sediments. This infrastructure includes, for example, a publicly available standard and methodologies. The two main problems facing the Great Barrier Reef, asides from global climate change, that the Initiative is attempting to address is the runoff from agricultural lands throughout Queensland and the clearing of coastal wetlands.

Case Study #1: Reef Credit Initiative

Introduction

Reef Credits is a quite simple concept—we have known the problems for a long time (climate change, pollution); the world needs to solve the climate change problem, but only Australia can solve its water quality problem. If you want the reef to be resilient or adapt to climate change, the reefs need clean water. While there are grants given by government to farmers, who are the main problem, to improve practices, buy machines, etc., the outcomes are not well known by Australia's National Audit Office. In contrast, the Reef Credit Initiative ("Initiative") will be a practice change that is verifiable. This work will not entirely displace NGOs, as you will still need research and additional help (Fig. 11.1).[53]

Essentially:

Fig. 11.1   Landscape Map of Australia (*Credit* Rainer Lesniewski)

A Reef Credit represents a quantifiable volume of nutrient, pesticide or sediment prevented from entering the Great Barrier Reef catchment. The relative value of pollutant reduction from nutrient, sediment or pesticide is set using the reef wide pollution reduction targets described in the Reef 2050 Water Quality Improvement Plan (2018). These values will be periodically amended by the Reef Credit Secretariat to reflect changes to the pollution reduction targets. A Reef Credit can then be sold to those seeking to invest in water quality improvements, such as government, private industry and philanthropists.[54]

More specifically:

The Reef Credit Scheme is designed to operate similarly to other successful voluntary and compliance ecosystem services markets around the world. This involves the development of a Reef Credit Standard, Reef Credit Methodologies and Reef Credit Projects. The administration of Reef Credits (Reef Credit Secretariat) will be overseen by an independent not-for-profit organisation {Eco-Markets Australia} with a skills-based board (Reef Credit Governance Body).[55]

The original, foundational methodologies were:

- Dissolved inorganic nitrogen (DIN) reduction, including "Reduction in nutrient run-off through managed fertiliser application" (*practice change)* and "Reduction in nutrient run-off through wetland systems" (*ecosystem repair)*; and
- Sediment reduction, including "Reduction in sediment run-off through gully remediation" (*ecosystem repair)* and "Reduction in sediment run-off through improved grazing practice" (*practice change*).[56]

As of March 2020, three foundational methodologies were "Nitrogen Use Efficiency, Wetlands and Gully Repair," and the first "Reef Credit methodology, Nitrogen Use Efficiency, has been approved after a rigorous peer review and consultation period."[57]

As of now, these are voluntary nutrient, pesticide, or sediment reduction offsets.

Natural resource management organisations, Terrain NRM and NQ Dry Tropics, have teamed up with environmental markets investor, GreenCollar, to develop the Reef Credit Scheme. The Scheme has been developed in collaboration with industry groups, research organisations and regional communities."

{…} With financial support from the Queensland Government, the partnership is focused on setting up a framework to support the Reef Credit Scheme.[58]

An independent not-for-profit company {Eco-Markets Australia} established in Queensland will administer the Reef Credit Scheme. The company will be governed by a skills based Board of Directors, with the day-to-day administration of the Reef Credit Standard delegated to the Reef Credit Secretariat (staff). The Members and Partners of the not-for-profit company will be organisations and agencies that endorse the principles and objectives of the Reef Credit Scheme, Guide and Standard. Partnership is offered to organisations that are unable to become members due to legal or other constraints. The Board and Reef Credit Secretariat will be supported on technical aspects of the Reef Credit Standard and Methodologies by a Technical Advisory Committee.[59]

Through this process, Reef Credits will be assured to be:

- Real;
- Measurable;
- Permanent;
- Additional;
- Independently audited;
- Unique;
- Transparent; and
- Conservative.[60]

The earliest possible Reef Credit Project start date is July 2017 and the project crediting period will be between five and twenty-five years.[61]

The National Australia Bank provided funding to Terrain NRM who, in partnership with GreenCollar, secured USD$325,000 funding over two years to support farmers' access to the Initiative.[62]

Identify the Problem

The two main problems, asides from climate change, that the Initiative is attempting to address is the runoff from agricultural lands throughout Queensland and the clearing of coastal wetlands. From 2000 to 2017, approximately 7.7 million hectares (19 million acres) of land were cleared in Australia.[63]

With respect to the runoff from agricultural lands, it is important to understand the scale of the problem. Likewise, "the agricultural land use in the catchments adjacent to the Great Barrier Reef lagoon, employs over 35,000 people and contributes approximately $3.7 billion annually in gross value of production."[64]

To address water quality, the Queensland and Federal governments' Reef 2050 Long Term Sustainability Plan and the Reef 2050 Water Quality Improvement Plan (2018) set water quality targets. The targets are to reduce dissolved inorganic nitrogen loads by 60%, fine sediment loads by 25%, particulate nutrients by 20% and reducing pesticide to protect at least 99% of aquatic species at the end of the catchment. The 2017 Reef Scientific Consensus statement recognised that current initiatives cannot meet the water quality targets and that exploration of management alternatives and increased support and resources will be required.[65]

Second, Queensland has lost approximately 90% of its coastal wetlands. These floodplains used to collect water and the various forms of nitrogen would collect there and then gas off into the atmosphere. Over the years, there has been heavy development along the coast where one of the goals is to drain the water out as quickly as possible. In contrast, floodplains naturally try to retain water as long as possible. Stakeholders have always known that if we are going to get the water clean again, we need to improve both agricultural practices and restore the floodplains.[66]

Why the Problem Is Important

The problems of agricultural runoff and the clearing of coastal wetlands are critically important. First, farmers need to adopt better practices. Second, "as well as capturing carbon and improving water quality, the project is restoring critical habitat for unique native wildlife including the endangered Southern Cassowary and more than 70 threatened flora and fauna species."[67]

How Problem Was Identified

The Initiative, along with the problems it was designed to help address, was collaboratively designed by GreenCollar, along with the organizations Terrain NRM and NQ Dry Topics.

Effectiveness of Process for Identifying Problem

The process of identifying the problems seems effective. The "whole thing – from governance oversight, auditing, verification – all needs to be done independently – otherwise, there could be a conflict."[68] Furthermore, the public display and the peer review process is all done by the Secretariat. GreenCollar and its other members make up the Secretariat. However, this is an interim model. The members will select new board members and then there is a 12-month phase in process until at which point the Secretariat goes independent.[69]

Steps Taken to Address the Problem

There were many steps taken to address the problems. These include:

- A "key conclusion in the 2016 Queensland Water Quality Science Task-force report was the need to consider incentives and market mechanisms 'to complement and integrate with regulation extension and education'."[70]
- In "early 2017, work began on designing the Reef Credit Scheme by foundation partners Terrain NRM, NQ Dry Tropics and GreenCollar as part of two reef water quality Major Integrated Projects (MIPs) funded by the Queensland Government. The initial concept and early consultations on mechanisms underlying the Reef Credit Scheme were first held in the in Wet Tropics and later introduced to the Burdekin Dry Tropics."[71]
- An "initial feasibility study was conducted as part of a consultation phase of the Major Integrated Projects. The study demonstrated the viability of the Reef Credit Scheme."[72]
- In mid-2017, "the Queensland Government committed support to the development of the Reef Credit Scheme, funding the establishment of a Reef Credit Secretariat for the start-up phase. Work then commenced on establishing the Secretariat and drafting of the Reef Credit Standard, Program Guide and foundation Reef Credit Methodologies."[73]
- In late 2017, "a Reef Credit Interim Steering Committee, comprising the CEOs of Terrain NRM, NQ Dry Tropics and GreenCollar, was established to guide the development of the Reef Credit Scheme."[74]
- In 2018, the Reef Credit Interim Steering Committee completed a final Options Paper for the design of the Reef Credit Scheme.[75]
- Throughout 2018, "a number of project sites have been identified across the catchments of the Reef to test application of the foundation Reef Credit Methodologies. At these sites, management activities will be undertaken to generate Reef Credits including agricultural and grazing practice change, wetland restoration and gully restoration."[76]
- "We are currently considering comments and expect to have a consultation summary and final Reef Credit Standard and Guide available on the website by March 2019."[77]
- "The role of the Interim Steering Committee will cease once the Reef Credit Scheme has transitioned from Beta phase to fully operational. Governance of the Reef Credit Scheme will then transfer to an independent not-for-profit company {Eco-Markets Australia} established in Queensland."[78]

As of March 2020, the Reef Credit Initiative was seeking to:

- Present the draft Reef Credit Standard and Guide to key stakeholders;
- Present the finalized Reef Credit Standard and Guide by March 2019;
- Undertake per review of first methodologies in mid-to-late 2019;
- Register the first Reef Credit Projects by early 2020; and
- Have the first issuance of Reef Credits by early 2020.[79]

Results

One of the main results, thus far, of the Initiative is setting up the program's infrastructure. This includes designing the standard, developing the methodologies, and engaging the variety of stakeholders (i.e., project partners, government officials, farmers, credit buyers, etc.). In addition, Mr. Mike Berwick of GreenCollar spoke to approximately sixty farmers (i.e., representing potentially sixty reef credit projects), with approximately ten of the farmers entering into signed contracts. Most of these potential projects had an interim assessment done, although some rework (i.e., redo the questions asked, inquire again about the farmers' practices, etc.) was needed due to the revised methodology.[80]

The first Reef Credit Initiative project, known as the Babinda Reef project near Cairns, Queensland, is a wetlands project. As of March 2020, "Green-Collar has nine fully contracted pilot projects and up to 40 other projects at various stages of development."[81] In July 2020, it was reported that the large private equity firm KKR had invested an undisclosed amount in Green-Collar; this will likely lead to GreenCollar scaling its environmental markets business.[82]

Challenges and How They Were Met

The challenges associated with the Initiative have been a little bit of everything.

First, the cost of setting up something like the Initiative should not be underestimated. The focal, subset area of Queensland where the Initiative is targeting is the size of Germany and this is where the Initiative is trying to change the behaviors of local farmers which is no small undertaking. In addition, the earlier methodology had serious issues and was too simplistic. For instance, the initial thought was that 17% of nitrogen goes out to the reef and for every kilo reduced, you would get 17% of it in a reef credit. However, that value varies dramatically due to lots of variables including soil type and climate. This said, the revised methodology has gone through the peer review process, a public comment period, and the standard has been approved.

We know what to do, but regulating farmers is not a viable, effective solution. For a farmer to adopt the necessary practices to improve water quality, it happens to be the very same practices that will improve their bottom lines (i.e., less inputs, improving soil health, etc.). However, the cost of getting there is a challenge. Imagine a farmer asking what do I do at my age, who is 65 years old and who does not believe climate change is an issue, but knows adopting is still good for their bottom line (i.e., such as retooling equipment)? In contrast, the "tin pot" farmers simply do not have the cash flow to adopt the changes. However, "we are just about there – regarding trust."[83]

Another challenge is the sheer complexity of the oversight, the auditors, government, and the methodology which is probably the hardest aspect. The reason the methodology, and getting it right, is such a challenge is because nitrogen loss is very hard to track. One's water table is often mixed with someone else's water table, some nitrogen runs across the surface while some nitrogen goes to the atmosphere, and the movement of nitrogen varies with the rains. All of these variables make measuring and monitoring near impossible, so you have to model all of this. Further, you need good science, government backing, and audits in order for people to trust it. To get this level of trust is no small task. To overcome this challenge, it has been important to have public comment periods and peer reviews for the methodologies.

There have also been challenges associated with program delays due to:

- responding to feedback from farmers and graziers, science, government, industry and community about the three foundation methodologies to ensure Reef Credit projects lead to verifiable water quality outcomes;
- building a team of people with skills and resources to work closely with {stakeholders} to create projects, collect and analyse data and trade Reef Credits; and
- creating and trialling efficient data collection systems and processing capacity.[84]

While there is a theoretically endless supply of reef credits, the next challenge will be buyers.[85]

Beyond Results

The Queensland State Government and science community tabulated farm practices into four risk categories. The best practices in agriculture, according to an agronomist, include, but are not limited to: no soil compaction; using GPS precision farming; practicing no-till; undertaking regular rotation and not always planting the same monoculture; and not overwatering. All of this will reduce the need for nitrogen and thus, reduce nitrogen loss.

The highest risk farmers drive all over the place, compact the soil, till it, grow sugarcane year in and year out, and there are more pathogens. The sugarcane farms' productivity was declining (i.e., despite better soil testing) and it was concluded that it was a soil health problem. This said, "we are just in the transition of the high-risk farmers to realize this is not the way to go – in my view, the secret to reduce loss is 'soil health' – to have good microbial activity and symbiotic relationship between soils and microorganisms."[86] This is a transformational change and it applies to any cultivated crop. Looking forward, there is a shifting attitude from the old way of thinking to the new way of thinking. The more progressive farmers get it, while the others are still suspicious. You need the progressive farmers to do it, to make more money, and to get public credibility, and then the others will follow.[87]

In addition to better agricultural practices, restoring the landscape is the other half of the problem. We can restore landscapes. In fact, wetlands can naturally abate a lot of nitrogen. There are quite a lot of farmers interested in this idea of restoration. For instance, as a farmer, I might make more money restoring wetlands than continuing to grow sugarcane on these marginal, sandy, old soils.[88]

Furthermore, there is the possibility of adding in other ecosystem services—such as fisheries and biodiversity values—to the carbon benefits.[89]

Another aspect of sustainability is that while the nitrogen methodology lacks permanence, the wetlands methodology has a permeance clause. However, the reef credits—assuming project success—would be issued every year for ten years based off the baseline and this should provide a relatively stable cash flow. Although it is true that after ten years, the farmers could go back to their old practices, it is important to note that the new practices will be making the farmers more money and thus, should be sustainable.[90]

Furthermore, "to ensure success, GreenCollar has firstly, assembled a dedicated team working on Reef Credits; is supporting the process to establish an independent Reef Credit Secretariat and authored the first formally approved methodology. {GreenCollar is} now focused on ensuring long term agreements to purchase Reef Credits."[91]

Lessons Learned

According to Mr. Mike Berwick, it is important to understand that you do not change these agricultural practices overnight and the practices will become long-term changes. The government's long-term sustainability plan is very good with science-based targets by catchment basins and by rivers. However, the government has not had a chance to meet those targets (i.e., goal of 80% nitrogen reduction by 2025, but only 3% achieved so far). For instance, to put in permanent beds, do proper drainage, and to go paddock by paddock because you have all different crops, will take time and expertise. Second, you cannot force someone to spend money to meet a standard if they do not have the money in the first place. Such regulations are often detested by farmers and in contrast, you need an incentive to more properly change practices.

Lastly, what GreenCollar has done is basically setup a bit of a blueprint system. While water quality markets are not new, what is new is turning a water quality credit into a locally specific and relevant reef credit. The practice of measuring, verifying, and monetizing water quality improvements is quite possibly transferable. Heavy metals, for example, could be the main pollutant from an industry intensive region and a comparable reef credit scheme could be developed for that particular catch basin.[92]

According to Mr. James Shultz, the Great Barrier Reef is a solvable problem. As opposed to a drop in the bucket with carbon offset credits, with a reduction in sediment and nutrient runoff, you can literally see the change. From a marketing standoff, it is important to make an immediate and significant impact that one can see.[93]

Other Resources on the Reef Credit Initiative

Australia Government, Department of the Environment and Energy's Reef 2050 Long-Term Sustainability Plan

- https://www.environment.gov.au/marine/gbr/publications/reef-2050-long-term-sustainability-plan-2018

Australia Government, Reef 2050: Water Quality Improvement Plan (2017–2022)

- https://www.reefplan.qld.gov.au/about/assets/reef-2050-water-quality-improvement-plan-2017-22.pdf

Great Barrier Reef Water Science Taskforce's Final Report (May 2016)

- https://environment.des.qld.gov.au/assets/documents/reef/gbrwst-finalr
  eport-2016.pdf

GreenCollar Group

- https://greencollar.com.au/reef-credits/

NQ Dry Tropics

- http://www.nqdrytropics.com.au/

Queensland Trust for Nature

- https://www.alca.org.au/across-australia/queensland/

Reef Credit

- https://www.reefcredit.org/

Reef Credit: Design Options Summary

- https://www.reefcredit.org/wp-content/uploads/2018/11/Reef-Credit-Sch
  eme-Design-Options-Summary.pdf

Reef Credit Briefing

- https://www.reefcredit.org/wp-content/uploads/2018/11/Presentation-on-
  the-Reef-Credit-Scheme.pdf

Reef Credit Guide

- Draft: https://www.reefcredit.org/wp-content/uploads/2018/11/Draft-
  Reef-Credit-Guide.pdf; and
- Adopted Version 1.0: https://www.reefcredit.org/wp-content/uploads/
  2019/03/Reef-Credit-Guide-Version-1.0.pdf

Reef Credit Scheme Overview

- https://www.reefcredit.org/wp-content/uploads/2018/11/Reef-Credit-Sch
  eme-Overview.pdf

<u>Reef Credit Standard</u>

- Draft: https://www.reefcredit.org/wp-content/uploads/2018/11/Draft-Reef-Credit-Standard.pdf; and
- Adopted   Version   1.1:   https://www.reefcredit.org/wp-content/uploads/2020/04/Reef-Credit-Standard-Version-1.1.pdf

<u>Terrain NRM</u>

- https://terrain.org.au/

Financial Analysis

The following financial analysis will look at return and risk.

Return

Compensation payments—whether it is supporting marine biodiversity offsets, blue carbon offsets, or watershed protection payments—may be voluntary or mandated by law. Depending on the market, there will be different buyers, different prices, and ultimately different returns.

Risk

There are a variety of risks facing PES, most notably policy risk and clarity over the ownership of the underlying asset (e.g., biodiversity offsets and/or blue carbon offsets). In addition, as noted in the *Innovations for Coral Finance* report:

> 'The scale and degree of connectivity between and within marine ecological units in three dimensions (Crowder et al., 2008), high biological and physical heterogeneity of both habitats and species on widely varying spatial and temporal scales (Crowder et al., 2008), poorly define property rights and the remote nature of governance relative to population centers (Vaissière et al., 2014)' are the many challenges marine biodiversity offsets are facing.[94]

As highlighted by Roldan Muradian, risks may include the fact that "{…} the effects of incentives on individual behaviour are far from universal. Even within the same cultural background, different social groups may react in a different way to incentives. {…} The effects of incentives are also mediated by group dynamics. This includes reputation, imitation and mutual reinforcement."[95] In addition, PES can also "undermine moral motives and social preferences for cooperation."[96]

Furthermore, theoretical challenges include:

- Currency (choosing metrics for measuring biodiversity);
- No net loss (defining requirements for demonstrating no net loss of biodiversity);
- Equivalence (demonstrating equivalence between biodiversity losses and gains);
- Longevity (defining how long offset schemes should endure);
- Time lag (deciding whether to allow a temporal gap between development and offset gains);
- Uncertainty (managing for uncertainties throughout the offset process);
- Reversibility (defining how reversible development impacts must be); and
- Thresholds (defining threshold biodiversity values beyond which offsets are not acceptable).[97]

Business Risk

The core business risk of forest carbon projects, including mangrove conservation projects, is whether "the price of carbon is lower than the opportunity costs, and {how to overcome the fact that} funding comes later than typical farming alternative uses."[98]

From the supplier and buyer's perspectives, there is business risk associated with how to price the asset. In addition to nature's intrinsic value, it is difficult to put an extrinsic value on nature such as coral reefs and iconic marine species. Likewise, what is the proper price of a marine biodiversity offset, or the fair price of a forest carbon offset bundled with a water quality credit?

Another risk is based around payment structure. For instance, "when companies decide to undertake offsets but prefer to pay over time – either through annual payments or through smaller lump-sum payments over a set period of years (e.g., five-year tranches) – some system should be considered to guarantee that those payments will be forthcoming over time."[99]

Strategic Risk

Strategic risks associated with PES projects include choosing the appropriate methodology and certification standard (assuming they exist); choosing the appropriate partners, especially those who are able to meet all the technical requirements and deliver verified offsets; and strategically choosing the right, successful project to support.

Strategic risks associated with marine biodiversity offsets include identifying a suitable area for the biodiversity offset project and negotiating with the sponsoring company and/or host government to commit to the terms of the project. It is also important to understand that there are a lot of reasons why a company may later back out of its biodiversity commitment (e.g., company bankruptcy, internal policy changes, or lack of profits) and create an orphan site, so project developers need to think through strategic options (i.e., diversifying revenue sources or engaging shareholders and host governments to pressure the company to maintain its commitment).

Strategic risks to blue carbon are that they are very dependent on location and accounting for belowground biomass is more challenging than traditional terrestrial carbon projects. Regarding location, being within the narrow range of the intertidal zone—such as for salt marshes—is necessary. If the project is too high out of the zone, then the roots will dry out and there will not be a salt marsh. In contrast, if the project is too low, then the area is going to be wet the whole time.[100] In addition, only certain species of seagrass grow in certain substrate.[101]

Belowground accounting is more complicated for blue carbon. With a mangrove forest, the root system is massive, it is constantly growing, and thus, is a significant carbon pool.[102] Further, take a mangrove project, where some amount of leaf litter stays in the mangrove forest and some is lost. Tidal fluctuations influence where this leaf litter material goes. However, does the material go to another mangrove forest? Is the material buried for a long time in the ocean? Or does the material go to shore and rot in the sun? Some material will also come into the project area from elsewhere. So, do you earn blue carbon offsets for what goes out and/or for what comes in?[103]

Reputation Risk

With respect to biodiversity offsets, there is also a potential reputation risk for project developers to partner with large-scale port construction, offshore gas and oil companies, or cruise line companies.

Liquidity Risk

There are relatively high upfront costs associated with developing PES projects (e.g., project documentation, technical studies, site visits, independent audits, etc.) and then ongoing costs (i.e., monitoring, reporting, and verification) need to be considered. This said, it is difficult to secure upfront financing, so there is a liquidity risk of not being able to convert the offsets into cash to recover sunk project costs.

Operational Risk

Operational risks associated with blue carbon projects include ensuring permanence and undertaking the actual restoration or conservation work. For instance, it is difficult to replant seagrass, as "you are essentially planting a prairie underwater with snorkelers or SCUBA divers."[104]

Furthermore, climate change and its impact on blue carbon projects is something that needs to be quantified for permanence. For instance, one must somehow ensure the salt marsh is not going to wash away due to sea-level rise or due to increasing hurricanes. The project developer needs to ensure the accretion level is greater than the lost and yet, "a hurricane is akin to a forest fire for blue carbon."[105]

One of the biggest hurdles to overcome is the technical challenges with designing and implementing a blue carbon project. This includes developing a new methodology or navigating the complexities of the chosen methodology or protocol.

The costs and time required to develop blue carbon projects pose another operational risk. However, if a firm is too academic-heavy and lacks the necessary marketing, or vice versa, this will present significant operational risks.

With respect to biodiversity offsets:

The marine environment, however, presents unique difficulties including the scale and degree of connectively between and within ecological units operating in three dimensions, high biological and physical heterogeneity of both habitats and species on widely varying spatial and temporal scales, poorly defined property rights and the remote nature of governance relative to population centres.[106]

Legal and Regulatory Risk

Legal and regulatory risks are significant for PES initiatives.

> Given the regulatory risk that environmental markets {such as a marine biodiversity offset program} can be altered or ended with the 'stroke of a pen' from policy makers, lenders are generally unwilling to value revenue from environmental markets when underwriting projects. With this heavy, or complete, discounting of environmental market revenues, these markets fail to entice the level of investment needed for projects to generate environmental credits and provide verifiable conservation benefits. How can we reduce the risks associated with environmental credits and keep these critically important projects moving forward?[107]

With respect to blue carbon, there is a legal and regulatory risk associated with ownership. There are a lot of potential blue carbon areas that are owned by a federal government. Thus, federal lands often do not allow private individuals to claim private ownership of a federal asset (i.e., imagining a carbon asset as being a federal asset). It should be noted that oil and gas companies get around this, in part by using a lease agreement. Nevertheless, with respect to blue carbon, there is also the issue of fragmented ownership across state and/or national borders.[108]

Establishing the rights and ownership of blue carbon is a concern in many developing countries, especially when it comes to community-based and Indigenous Peoples-based projects.

Around the world, there are few, if any, blue carbon projects that have undertaken work to nest into higher levels of carbon accounting at the national or subnational level. Yet, how exactly projects will fit into the jurisdiction oftentimes remains a challenge. Furthermore, how will blue carbon projects fit into the Paris Agreement?

Taxation on offsets, including blue carbon offsets, is often unclear and the specific regulations on the sale of such offsets can also present bureaucratic barriers to trade. This said, are blue carbon offsets to be taxed as a commodity, a service (i.e., ecosystem service), or as a derivative (i.e., its value is derived based off the underlying forest)?

Credit Risk

Credit risk includes default risk, bankruptcy risk, downgrade risk, and settlement risk.

Many project developers face default and bankruptcy risk due to the decline in PES prices. These risks may be mitigated, in part, by diversifying project revenues from sustainable fisheries, ecotourism user fees, or by bundling other ecosystem service offsets.

Market Risk

Market risk includes interest rate risk, equity price risk, foreign exchange risk, and commodity price risk.

If commodity prices increase, then the risk of conversion of tropical mangroves to an alternative land use, such as a shrimp farm, can also increase.

Risk, Return, Time (Horizon), Taxes, Liquidity, Legal, and Unique (RRTTLLU)

Risk and Return

Please see above for the risk and return associated with PES.

Time Horizon

The time horizon for PES is a long-term commitment. For instance, many forest carbon projects have a minimum Project Lifetime of 30 years and many projects have Project Lifetimes up to 100 years. The minimum Project Lifetime of a carbon offset project is determined by the certification standard. From a buyer's perspective, short time horizons often exist for the purchase of carbon offsets because they have no ongoing requirements (e.g., project management or monitoring, reporting, and verification of the project).

Taxes

There can be a variety of taxes associated with payments for ecosystem services. This includes a possible sales tax on the underlying PES asset, along with property taxes and employment taxes.

Liquidity

There appears to be a low level of liquidity for transferring the entire PES project—whether it be a biodiversity, a blue carbon, or watershed payment project. Furthermore, the liquidity of a given offset will depend on whether the offset is transacted in the voluntary market or the compliance market. For instance, there is a lower level of liquidity for voluntary forest carbon offsets when compared to the more liquid market for compliance forest carbon offsets in the California market.

Legal

Legal considerations associated with payments for ecosystem services will vary from country to country. In addition, these legal considerations may vary from state to state, and may even vary depending on whether the project proponent is a private citizen or, for example, an Indigenous Peoples' association.

Unique

One unique aspect of PES projects is the potential ability to stack or bundle several offset types. Bundling is when a single offset includes several ecosystem services (i.e., biodiversity and blue carbon offset), while stacking is when the ecosystem services are differentiated and sold separately.

Another unique aspect is that arguably there are compliance markets and there are voluntary mechanisms. Likewise, it has been argued that voluntary PES instruments are more of a voluntary transfer payment as opposed to an actual market. This said, markets require—among many things—price discovery and liquidity, which is often lacking in PES instruments such as voluntary blue carbon and biodiversity offsets.

Policy Analysis

The following policy analysis will look at: defining the problem; establishing goals; selecting a policy; implementing a policy; and evaluating the policy.

Defining the Problem

One problem is "owing to a lack of legislation in many countries (Niner et al., 2017b), most coral ecosystem biodiversity offsets have so far been based on voluntary schemes, such as in the Fiji case study {…}."[109] Other potential problems facing PES are most notably policy risk and clarity over the ownership of the underlying asset (e.g., forest carbon offsets). For instance:

- Is a private project developer entitled to ownership of PES assets, particularly if the PES assets are publicly owned? Does it matter whether this private project developer is a domestic or foreign entity?
- Are Indigenous Peoples and other local communities the owners of the PES assets within their territory? Under what conditions are such Indigenous Peoples and local communities able to sign contracts and sell, or transfer, their underlying PES assets?
- What, if any, government entity grants permission—such as a distinct license, permit, or certification—to undertake, develop, finance, and sell PES assets?
- What, if any, government charge (e.g., taxes, royalties, and/or registration costs) will be levied on the PES assets?
- Does it make a difference—and if so, how?—if the underlying PES asset is for biodiversity, water, and/or blue carbon?
- If a country has a water regulatory agency, a ministry of fisheries, and a ministry of environment, who then approves a bundled offset consisting of water, blue carbon, and biodiversity?

Policies which address these risks will help expand the market for PES projects, programs, and investments while simultaneously conserving tropical coral reefs and blue carbon, preserving biodiversity, and providing local sustainable economic opportunities.

Establishing Goals

The goals for establishing desirable policies for PES are to:

- Establish robust, long-term, transparent markets with price discovery and liquidity;
- Clarify how exactly landownership can be demonstrated and the process for effectively and efficiently resolving overlapping title claims; and

- Clearly state who can own the underlying PES assets and the additional, if any, costs to be levied by the government.

Selecting a Policy

There have been a wide range of policies that have had varying impacts on the development of PES. For instance, Ecuador has been a leader in watershed PES schemes, while several countries including Madagascar and Panama have enabled biodiversity offset markets, and several countries (e.g., India, Indonesia, and Kenya) host mangrove carbon offset projects. There is also the Philippine Blue Carbon Initiative.[110]

On the international level and with respect to policies designed to create a market for forest carbon projects, and potentially blue carbon, is the UNFCCC's Paris Agreement.

When selecting a marine biodiversity offset policy, it is important to note:

National (or supranational) policies {for marine biodiversity offsets} exist in the US, Canada, Australia, the EU, France, Germany and Colombia. These policies support the application of the five principles essential to biodiversity offsetting success with the exception of Colombia where detail relating to additionality was not found. Only one of these national policies, the Magnuson-Stevens Fishery Conservation and Management Act, has been developed specifically for marine application and with the exception of French, German and Colombian policy, all have application restricted to 'listed' marine habitats, species or protected areas.[111]

In addition to the U.S. Magnuson–Stevens Fishery Conservation and Management Act, there is also the allowance of mitigation banks in the United States under Section 404 of the Clean Water Act.[112] Australia's national policy is called the Environment Protection and Biodiversity Conservation Act 1999. Australia also has a Reef Trust and there is the Environmental Offsets Act 2014 for Queensland.[113]

On a subnational level:

In the US and Australia sub-national policy has been developed for specific marine application of biodiversity offsets in the instance of impacts to eelgrass in California, fish habitat in New South Wales and specifically for the Great Barrier Reef in Queensland. In Australia, sub-national policy supporting the application of biodiversity offsetting exists in five of its six states.[114]

In Australia's New South Wales, the Biodiversity Banking and Offsets Scheme (BioBanking) was a voluntary biodiversity offset scheme. This BioBanking Scheme was "replaced by the Biodiversity Offsets Scheme (BOS) under the Biodiversity Conservation Act 2016 (BC Act) which commenced on 25 August 2017. The BOS is also a market-based scheme that operates in a similar way to BioBanking."[115] Changes included:

- 1. Replacement of the BioBanking Assessment Methodology (BBAM) with the Biodiversity Assessment Methodology (BAM);
- 2. Replacement of BioBanking Agreements with Biodiversity Stewardship Agreements; and
- 3. Establishment of the Biodiversity Conservation Trust (BCT).[116]

Then again, it can also be possible to structure projects in the absence of specific policies:

> There is no biodiversity offset legislation in Fiji, nor any administrative recognition of biodiversity offsets in the Environmental Impact Assessment (EIA) process or alternative mitigation measures. However, an integrated resort development project (97 residential lots, 5 tourism lots and a marina) near Nadi international airport in the south of Viti Levu in 2006 led to the conservation of 110 ha of mangroves in the southwest of the main island to compensate the destruction of 8.8 ha due to the dredging of a canal.[117]

Implementing a Policy

Marine PES programs are often overseen at the national level. For instance, the Australian Government Department of the Environment is responsible for administering the Environment Protection and Biodiversity Conservation Act of 1999,[118] while the U.S. Magnuson–Stevens Fishery Conservation and Management Act is a Federal Congressional Act.

The UNFCCC Paris Agreement, in contrast to the Kyoto Protocol, has taken a bottom-up approach and the responsibility for implementing the Nationally Determined Contributions (NDCs) rests with each member country.

Evaluating the Policy

According to Holly J. Niner et al., the five key principles for biodiversity offsetting success are:

- 1. Use of mitigation hierarchy;
- 2. Demonstration of equivalence;
- 3. Ensure additionality;
- 4. Continuity; and
- 5. Compliance success.[119]

Thus, one way to evaluate a country's biodiversity offsetting policies is to see whether all five principles are addressed.

Almost 200 countries signed onto the UNFCCC Paris Agreement in December 2015, which now commits countries to carry out their NDCs. The Agreement has language on Internationally Transferred Mitigation Outcomes (ITMOS), which could be interpreted as a role for international blue carbon offsets:

> The use of internationally transferred mitigation outcomes to achieve nationally determined contributions under this Agreement shall be voluntary and authorized by participating Parties. {…}
>
> Recognizes the importance of adequate and predictable financial resources, including for results-based payments, as appropriate, for the implementation of policy approaches and positive incentives for reducing emissions from deforestation and forest degradation, and the role of conservation, sustainable management of forests and enhancement of forest carbon stocks; as well as alternative policy approaches, such as joint mitigation and adaptation approaches for the integral and sustainable management of forests; while reaffirming the importance of non-carbon benefits associated with such approaches; encouraging the coordination of support from, inter alia, public and private, bilateral and multilateral sources, such as the Green Climate Fund, and alternative sources in accordance with relevant decisions by the Conference of the Parties.[120]

Alongside COP21, 18 countries signed a declaration on international carbon markets. This group, which is being led by New Zealand, is also comprised of Australia, Canada, Chile, Colombia, Germany, Iceland, Indonesia, Italy, Japan, Mexico, Netherlands, Panama, Papua New Guinea, South Korea, Senegal, Ukraine, and the United States.[121]

Future Outlook for Instrument

There are a lot of potential, future developments with PES and it appears likely the use of PES projects and programs will increase:

In summary, PES's rapid rise has been the result of a combination of factors; (i) a window of opportunity opened up by the fall of integrated approaches; (ii) the appealing simple discourse and conceptualization that have dominated so far the analytical background of PES; (iii) the promises of win-win solutions, and in particular the possibility of reconciling rural development and environmental conservation through a single policy tool; and (iv) the strong support by influential organizations.[122]

There is increasing interest in blue carbon and it is likely more projects will be developed as methodologies get approved. For instance:

- "Swiss-based MSC Cruises aims to achieve carbon neutrality in its marine fleet operations by Jan. 2020 through offsetting, and plans to increasingly rely on blue carbon projects to maintain the goal {...}.[123]
- TNC and XL Catlin developed "Blue Carbon Resilience Credits," with potential pilot sites in Belize and Papua New Guinea.[124]
- There have been several studies conducted on the Turneffe Atoll of Belize to assess its PES values[125]; and
- Million Mangroves, a project of Natural Capital Partners, was launched.[126]

Costa Rica's Blue Fund could signal the future design of marine and coastal PES programs. The focus of Costa Rica's Blue Fund would be mangroves and coral reefs. The structure would involve the Costa Rican Government as the seller and the buyers would be many different sectors, such as tourists, private businesses, and fisheries.[127] In addition to Costa Rica, Australia may be a location for more blue carbon work:

Australia's mangroves, tidal marshes, and seagrass meadows are absorbing about 20 MtCO2 every year, according to a major new study that is the first to measure in detail the climate benefits of the coastal ecosystems. But the study, published in the journal Nature Communications, warns that degradation of these "vegetated coastal ecosystems" was already seeing 3 MtCO2 per year being released back into the atmosphere. The study reveals Australia's vast coastlines represent between 5% and 11% of all the so called "blue carbon" locked up in mangroves, seagrasses, and tidal marshes globally.[128]

One future location of marine biodiversity offsets is Belize:

Threats posed to the coastal marine environment have been directly addressed in Belize through the development of a marine biodiversity offset framework which is hoped to progress to a more formal state. This has been developed through a partnership with the Australia-Caribbean coral reef collaboration and the Belize Coastal Zone Management Authority and Institute.[129]

There are some other, possible ideal locations that may see more blue carbon projects and programs. Some locations, such as edges of certain wetlands in California are more peat-like than the salty sea marshes on the East Coast, and this is where one could get the biggest bank for your buck. Likewise, the USFWS's National Wetlands Inventory Program could be reviewed. Other ideal locations including restoring shrimp farms in Southeast Asia. Take, for instance, a place where mangroves were once deforested for a shrimp farm. Over time, for whatever reason, the shrimp farm failed and now the project activity would be to restore the area back to mangroves. Furthermore, some of the small island states that have the resources and could demonstrate co-benefits (i.e., like benefits to fisheries) may host future blue carbon projects and programs.[130]

However, one must recognize that blue carbon is often not going to fund all of the expensive restoration work, but rather, blue carbon could finance a part of it. Blue carbon may generate revenue to cover say 10–20% of the total costs and this revenue could, for example, go toward ongoing monitoring costs.

Then again, there are emerging carbon compliance mechanisms that may include international forestry offsets—possibly to include blue carbon—such as the California cap-and-trade program, the Carbon Offsetting and Reduction Scheme for International Aviation (CORSIA) of the International Civil Aviation Organization (ICAO), and the Paris Agreement.

There has also been a lot of work around storm protection and flood reduction values of coral reefs and mangroves. This includes modeling that has been developed based off engineering models. For instance, when considering the construction of a sea wall, there are models to examine different storms (i.e., wind speeds, wave heights, etc.) and estimate their impacts (e.g., based off different topographies, impacts with and without structures, where does water propagate, etc.). We can now do these model exercises with coral reefs and mangroves to better understand how they reduce wave height, wave energy, and inland flooding, along with how many people are impacted and how many dollars of damage result. This adaptation value of ecosystems is likely more valuable than the mitigation value. This might lead to combining the mitigation and adaptation values. Looking to the insurance industry, who knows how to price risk, the presence of offshore reefs or restored mangroves may factor into reduced insurance premiums and such values may be easier for policymakers to understand.[131]

Although not really a payment for ecosystem service, there is the emergence of plastic offsets. For instance, there is the Plastic Bank,[132] and Verra, the parent association of the Verified Carbon Standard, is assessing plastic offsets under the Plastic Recovery and Recycling Project Accounting Program.[133]

Ultimately:

{...} if PMCES {Payments for Marine and Coastal Ecosystem Services} are to reach their full potential, decision makers must be willing to rethink the way in which coastal developments are planned and implemented: to explicitly recognize the values associated with ecosystem services and to factor them into their calculations.[134]

Other Resources on Payments for Ecosystem Services

Artificial Intelligence for Ecosystem Services (ARIES)

- http://aries.integratedmodelling.org/

Blue Natural Capital Financing Facility

- https://bluenaturalcapital.org/

Bundling and Stacking for Maximizing Social, Ecological, and Economic Benefits: A Framing Paper for Discussion at the "Bundling and Stacking Workshop"

- https://rmportal.net/library/content/translinks/translinks-2012/Wildlife%20Conservation%20Society/2012%20PES%20Bundling%20and%20Stacking%20Workshop%20(Washington,%20DC,%20USA)/Paper_PES BundlingandStacking.pdf

California Air Resources Board's Cap-and-Trade Program

- https://www.arb.ca.gov/cc/capandtrade/capandtrade.htm

Carbon Market Institute

- http://carbonmarketinstitute.org/

Climate Finance Tracking at AfDB

- http://www.afdb.org/en/topics-and-sectors/sectors/climate-change/climate-finance-tracking-at-afdb/

Co$ting Nature

- http://www.policysupport.org/costingnature

ECOSTAR's Market Outlooks on European Voluntary Carbon, Biodiversity Offsets, and Watershed Investments

- http://www.ecostarhub.com/market-outlooks/

Forest Trends

- Business and Biodiversity Offsets Programme (BBOP): http://bbop.forest-trends.org/;
- BBOP Standard: http://bbop.forest-trends.org/pages/our_work_standard; and
- Coastal and Marine Initiative: https://www.forest-trends.org/who-we-are/initiatives/#ssection-2

Global Mangrove Alliance

- http://www.mangrovealliance.org/

IUCN

- https://www.iucn.org/theme/business-and-biodiversity/our-work/business-approaches-and-tools/biodiversity-offsets; and
- https://portals.iucn.org/offsetpolicy/

Million Mangroves, A Natural Capital Partners' Project

- https://millionmangroves.com/the-projects

Multi-scale integrated models of ecosystem services

- http://www.afordablefutures.com/; and

- https://ipbes.net/policy-support/tools-instruments/multi-scale-integrated-models-ecosystem-services-mimes

Natural Capital Project's InVEST (Integrated Valuation of Ecosystem Services and Tradeoffs) Software

- https://naturalcapitalproject.stanford.edu/software/invest

Regulatory In-Lieu Fee and Banking Information Tracking System

- https://www.usace.army.mil/Missions/Civil-Works/Regulatory-Program-and-Permits/techbio/

Stacking Ecosystem Services Payments Risks and Solutions

- https://nicholasinstitute.duke.edu/sites/default/files/publications/stacking-ecosystem-services-payments-paper.pdf

The Blue Carbon Initiative

- https://www.thebluecarboninitiative.org/

The International Partnership for Blue Carbon

- https://bluecarbonpartnership.org/

Toolkit for Ecosystem Service Site-Based Assessment

- http://tessa.tools/

USFWS's Guidance on Conservation Banking

- https://www.fws.gov/endangered/landowners/conservation-banking.html; and
- https://www.fws.gov/endangered/esa-library/pdf/Conservation_Banking_Guidance.pdf

USFWS's National Wetlands Inventory Program

- https://www.fws.gov/wetlands/

Notes

1. Gómez-Baggethun et al. "The History of Ecosystem Services in Economic Theory and Practice: From Early Notions to Markets and Payment Schemes." https://doi.org/10.1016/j.ecolecon.2009.11.007.
2. OECD. "Policy Highlights: Biodiversity Offsets: Effective Design and Implementation." Accessed October 29, 2018. http://www.oecd.org/environment/resources/Policy-Highlights-Biodiversity-Offsets-web.pdf. 2.
3. Ibid. 2.
4. Maraseni, Tek N. and Munir A. Hanjra. "Payments to Landholders for Managing Water, Land and Ecosystems (WLE) in Coastal Agricultural Catchments for Protecting the Great Barrier Reef." In *Economic Incentives for Marine and Coastal Conservation*, edited by Essam Yassin Mohammed. 194.
5. Martin. *On the Edge*. 127.
6. Daily, Gretchen C. and Katherine Ellison. *The New Economy of Nature*. 29.
7. Climate. Community & Biodiversity Alliance. "About the CCBA." Accessed April 14, 2020. http://www.climate-standards.org/about-ccba/.
8. UNFCCC. "CDM: Project Search." Accessed November 23, 2016. http://cdm.unfccc.int/Projects/projsearch.html.
9. Ten Kate, Kerry, Hase, Amrei and Maguire, Patrick. "Principles of the Business and Biodiversity Offsets Programme." March 2018. https://doi.org/10.1007/978-3-319-72581-9_3.
10. Verra. "About Verra." Accessed April 14, 2020. https://verra.org/about-verra/who-we-are/.
11. World Wildlife Fund. *Guide to Conservation Finance*. 9.
12. UNFCCC. "CDM: Project Search." Accessed November 23, 2016. http://cdm.unfccc.int/Projects/projsearch.html.
13. Climate. Community & Biodiversity Alliance. "Small-scale Reforestation for Landscape Reforestation." Accessed November 23, 2016. http://www.climate-standards.org/2007/01/01/small-scale-reforestation-for-landscape-reforestation/.
14. Wayburn, Laurie. "The Van Eck Forest: Carbon Markets and the New Economic Paradigm for Forest Sustainability." In *Conservation Capital in the Americas*, edited by James N. Levitt. 188.
15. Ibid. 189.
16. Elliott, Chris. "REDD+ Is a Little More Complex Than 'Not Cutting Down a Tree'." In *On the Edge*, by Claude Martin. Berkeley: Greystone Books, 2015. 235.
17. Climate. Community & Biodiversity Alliance. "Reducing Carbon Emissions from Deforestation in the Ulu Masen Ecosystem." Accessed November 23, 2016. http://www.climate-standards.org/2007/11/02/reducing-carbon-emissions-from-deforestation-in-the-ulu-masen-ecosystem/.

18. Warner, Maud and Molly Peters-Stanley. "Kenya Carbon Project Earns First-Ever VCS REDD Credits." *Ecosystem Marketplace*. Accessed November 29, 2016.http://www.ecosystemmarketplace.com/articles/kenyan-carbon-pro ject-earns-first-ever-vcs-redd-credits/.

19. Benkenstein, Alex and Romy Chevallier. "Africa's Mangrove Habitats: Prospects and Challenges of Payment for Coastal and Marine Ecosystem Services." In *Economic Incentives for Marine and Coastal Conservation*, edited by Essam Yassin Mohammed. 216.

20. Green Climate Fund. "Behind the Fund." Accessed November 23, 2016. http://www.greenclimate.fund/the-fund/behind-the-fund.

21. UNFCCC. "Paris Agreement." Accessed April 11, 2017. http://unfccc.int/ paris_agreement/items/9485.php.

22. IUCN. "Biodiversity Offsets." Accessed April 14, 2017. https://www.iucn.org/ theme/business-and-biodiversity/our-work/business-approaches-and-tools/bio diversity-offsets.

23. Muradian, Roldan. "Payments for Marine and Coastal Ecosystem Services and the Governance of Common Pool Natural Resources." In *Economic Incentives for Marine and Coastal Conservation*, edited by Essam Yassin Mohammed. 53.

24. Emerton, Lucy. "Using Valuation to Make the Case for Economic Incentives: Promoting Investments in Marine and Coastal Ecosystems as Development Infrastructure." In *Economic Incentives for Marine and Coastal Conservation*, edited by Essam Yassin Mohammed. 18.

25. Ibid. 20.

26. Hernandez-Blanco, Marcello. "Costa Rica's Blue Fund: A Research and Policy Agenda for Establishing a 'Blue Fund' for the Conservation and Restoration of Marine and Coastal Ecosystems in Costa Rica." *Conservation Finance Alliance*. January 15, 2020. https://www.conservationfinancealliance.org/ news/2019/12/20/upcoming-cfa-webinar-a-research-and-policy-agenda-for- establishing-a-blue-fund-for-the-conservation-and-restoration-of-marine-and- coastal-ecosystems-in-costa-rica.

27. Meyer, Shannon. "Markets for Ecosystem Services and Reclaiming the Great Dismal Swamp." In *Conservation Capital in the Americas*, edited by James N. Levitt. 198.

28. BBOP. "Mitigation Hierarchy." Accessed December 13, 2016.http://bbop.for est-trends.org/pages/mitigation_hierarchy.

29. Davis, Adam. "Mainstreaming Environmental Markets." In *From Walden to Wall Street*, edited by James Levitt. 158.

30. Vertigo Lab. "Innovations for Coral Finance, ICRI Publication." 23.

31. OECD. "Policy Highlights: Biodiversity Offsets: Effective Design and Imple- mentation." Accessed October 29, 2018. http://www.oecd.org/environment/ resources/Policy-Highlights-Biodiversity-Offsets-web.pdf. 5.

32. Ibid. 6.

33. Bennett, Genevieve and Franziska Ruef. "Alliances for Green Infrastructure: State of Watershed Investment 2016." 1.

34. McQueen, Mike and Ed McMahon. *Land Conservation Financing*. 152.
35. Verra. "VM0024 Methodology for Coastal Wetland Creation, v1.0." Accessed January 17, 2020. https://verra.org/methodology/vm0024-methodology-for-coastal-wetland-creation-v1-0/.
36. Verra. "VM0033 Methodology for Tidal Wetland and Seagrass Restoration, v1.0." Accessed January 17, 2020. https://verra.org/methodology/vm0033-methodology-for-tidal-wetland-and-seagrass-restoration-v1-0/.
37. Hernandez-Blanco, Marcello. "Costa Rica's Blue Fund: A Research and Policy Agenda for Establishing a 'Blue Fund' for the Conservation and Restoration of Marine and Coastal Ecosystems in Costa Rica." *Conservation Finance Alliance*. January 15, 2020. https://www.conservationfinancealliance.org/news/2019/12/20/upcoming-cfa-webinar-a-research-and-policy-agenda-for-establishing-a-blue-fund-for-the-conservation-and-restoration-of-marine-and-coastal-ecosystems-in-costa-rica.
38. OECD. "Policy Highlights: Biodiversity Offsets: Effective Design and Implementation." Accessed October 29, 2018. http://www.oecd.org/environment/resources/Policy-Highlights-Biodiversity-Offsets-web.pdf. 2.
39. Ibid. 6.
40. Niner, Holly J., et al. "A Global Snapshot of Marine Biodiversity Offsetting Policy." 1.
41. Iyer, Venkat, et al. *Finance Tools for Coral Reef Conservation*. 17.
42. Madsen, Becca, et al. "State of Biodiversity Markets 2009." IV.
43. Bennett, Genevieve and Franziska Ruef. "Alliances for Green Infrastructure: State of Watershed Investment 2016." 2.
44. Ibid. 2.
45. Ibid. 45.
46. Pothering, Jessica. "Ecosystem Investment Partners Closes Fourth Conservation Fund at $455 Million." *ImpactAlpha*. February 10, 2020. Accessed February 11, 2020. https://impactalpha.com/ecosystem-investment-partners-closes-fourth-conservation-fund-at-455-million/.
47. Donofrio, Stephen, et al. "Financing Emission Reductions for the Future: State of the Voluntary Carbon Markets 2019."
48. California Air Resources Board. "Compliance Offset Program." Accessed April 9, 2020.https://ww3.arb.ca.gov/cc/capandtrade/offsets/offsets.htm.
49. Plan Vivo. "Loru Forest Project, Vanuatu—The Nakau Programme." Accessed January 17, 2020. https://www.planvivo.org/project-network/loru-forest-project/.
50. VCS Project Database. "Mangrove Restoration and Coastal Greenbelt Protection in the East Coast of Aceh and North Sumatra Province, Indonesia." Accessed January 17, 2020. https://www.vcsprojectdatabase.org/#/project_details/1493.
51. Plan Vivo. "Mikoko Pamoja—Kenya." Accessed January 17, 2020. https://www.planvivo.org/project-network/mikoko-pamoja-kenya/.

52. Plan Vivo. "The Nakau Programme—Pacific Island Rainforest Carbon & Conservation." Accessed January 17, 2020. https://www.planvivo.org/project-network/the-nakau-programme/.
53. Berwick, Mike. Interviewed by Brian McFarland. September 2019.
54. GreenCollar. "Reef Credits." Accessed September 19, 2019. https://greenc ollar.com.au/reef-credits/.
55. Ibid.
56. Ibid.
57. GreenCollar. "The Reef Credit Update: Issue 1—March 2020." Accessed April 1, 2020.
58. Reef Credit. "Home." Accessed July 10, 2019. https://www.reefcredit.org/.
59. Terrain NRM, NQ Dry Tropics, and GreenCollar. "Reef Credit Guide: Beta Version 1.0." November 12, 2018. Accessed July 1, 2019. https://www.reefcr edit.org/wp-content/uploads/2018/11/Draft-Reef-Credit-Guide.pdf. Page 5.
60. Ibid. 8.
61. Ibid. 5.
62. Bentley, James. Interviewed by Brian McFarland. October 2019.
63. Steele, Jennie. "Australian authorities blind to Habitat Destruction." *Climate Action.* September 12, 2019. Accessed April 13. 2020. http://www.climateac tion.org/news/australian-authorities-blind-to-habitat-destruction.
64. Terrain NRM, NQ Dry Tropics, and GreenCollar. "Reef Credit Guide: Beta Version 1.0." November 12, 2018. Accessed July 1, 2019. https://www.reefcr edit.org/wp-content/uploads/2018/11/Draft-Reef-Credit-Guide.pdf. 3.
65. Ibid. 3.
66. Berwick, Mike. Interviewed by Brian McFarland. September 2019.
67. Carbon Pulse. "Qantas, GreenCollar Group Team Up on Carbon, Reef Credits." June 8, 2018. Accessed December 30, 2019. http://carbon-pulse. com/53654/.
68. Bentley, James. Interviewed by Brian McFarland. October 2019.
69. Ibid.
70. Terrain NRM, NQ Dry Tropics, and GreenCollar. "Reef Credit Guide: Beta Version 1.0." November 12, 2018. Accessed July 1, 2019. https://www.reefcr edit.org/wp-content/uploads/2018/11/Draft-Reef-Credit-Guide.pdf. 3.
71. Reef Credit. "Home." Accessed July 10, 2019. https://www.reefcredit.org/.
72. Ibid.
73. Ibid.
74. Ibid.
75. Ibid.
76. Ibid.
77. Ibid.
78. Ibid.
79. Reef Credit. "Home." Accessed April 1, 2020. https://www.reefcredit.org/.
80. Berwick, Mike. Interviewed by Brian McFarland. September 2019.

81. GreenCollar. "The Reef Credit Update: Issue 1—March 2020." Accessed April 1, 2020.

82. Environmental Finance. "KKR has global ambition for Australian environmental markets platform." 2020. Available: https://www.environmental-finance.com/content/news/kkr-has-global-ambition-for-australian-environmental-markets-platform.html.

83. Berwick, Mike. Interviewed by Brian McFarland. September 2019.

84. GreenCollar. "The Reef Credit Update: Issue 1—March 2020." Accessed April 1, 2020.

85. Berwick, Mike. Interviewed by Brian McFarland. September 2019.

86. Ibid.

87. Ibid.

88. Ibid.

89. Ibid.

90. Ibid.

91. GreenCollar. "The Reef Credit Update: Issue 1—March 2020." Accessed April 1, 2020.

92. Berwick, Mike. Interviewed by Brian McFarland. September 2019.

93. Shultz, James. Interviewed by Brian McFarland. March 2019.

94. Vertigo Lab. "Innovations for Coral Finance, ICRI Publication." 2017. http://vertigolab.eu/wp-content/uploads/2018/04/rapport-innovation-for-coral-finance.pdf. 24.

95. Muradian, Roldan. "Payments for Marine and Coastal Ecosystem Services and the Governance of Common Pool Natural Resources." In *Economic Incentives for Marine and Coastal Conservation*, edited by Essam Yassin Mohammed. 63.

96. Ibid. 64.

97. Bull, Joseph W., et al. "Biodiversity Offsets in Theory and Practice." 372.

98. Vitale et al. "Ecuador ChoCO$_2$ Conservation Carbon Project: Conservation International." In *Conservation Capital in the Americas*, edited by James N. Levitt. 171.

99. Iyer, Venkat, et al. *Finance Tools for Coral Reef Conservation.* 32.

100. Eaton, James. Interviewed by Brian McFarland. February 2020.

101. Settelmyer, Scott. Interviewed by Brian McFarland. March 2020.

102. Ibid.

103. Eaton, James. Interviewed by Brian McFarland. February 2020.

104. Ibid.

105. Ibid.

106. Niner, Holly J., et al. "A Global Snapshot of Marine Biodiversity Offsetting Policy." 4.

107. Carbon Pulse. "Ecosystem Marketplace: Can the World Bank Model Show Us How to De-Risk US Environmental Markets?" Last Updated September 28, 2018. Accessed July 31, 2019. www.carbon-pulse.com/48338/.

108. Eaton, James. Interviewed by Brian McFarland. February 2020.

109. Vertigo Lab. "Innovations for Coral Finance, ICRI Publication." 24.

110. Office of the President of the Philippines' Climate Change Commission. "The Philippine Blue Carbon Initiative." Accessed April 6, 2020. https://climate.gov.ph/our-programs/the-philippine-blue-carbon-initiative.

111. Niner, Holly J., et al. "A Global Snapshot of Marine Biodiversity Offsetting Policy." 6.

112. USEPA. "Section 404 of the Clean Water Act." April 16, 2019. https://www.epa.gov/cwa-404/mitigation-banks-under-cwa-section-404.

113. Niner, Holly J., et al. "A Global Snapshot of Marine Biodiversity Offsetting Policy." Appendix.

114. Ibid. 6.

115. New South Wales Government. "Biodiversity Banking and Offsets Scheme (BioBanking)." Accessed April 13, 2020. https://www.environment.nsw.gov.au/topics/animals-and-plants/biodiversity/biobanking.

116. Ibid.

117. Vertigo Lab. "Innovations for Coral Finance, ICRI Publication." 25.

118. Australian Government Department of the Environment. "About the EPBC Act." Accessed April 6, 2020. https://www.environment.gov.au/epbc/about.

119. Vertigo Lab. "Innovations for Coral Finance, ICRI Publication." 3.

120. UNFCCC. "Paris Agreement." Accessed April 6, 2020. https://unfccc.int/resource/docs/2015/cop21/eng/l09.pdf.

121. Ali, Hamza. "18 Countries Sign Declaration on International Carbon Markets." December 14, 2015. *Environmental Finance.* Accessed April 8, 2017. https://www.environmental-finance.com/content/news/18-countries-sign-declaration-on-international-carbon-markets.html.

122. Muradian, Roldan. "Payments for Marine and Coastal Ecosystem Services and the Governance of Common Pool Natural Resources." In *Economic Incentives for Marine and Coastal Conservation,* edited by Essam Yassin Mohammed. 59.

123. Carbon Pulse. "Cruise Line MSC Seeks CO2 Neutrality Through Blue Carbon Offsets." November 8, 2019. Accessed February 28, 2020. https://carbon-pulse.com/86121/.

124. The Global Innovation Lab for Climate Finance. "Blue Carbon Resilience Credit." Accessed April 6, 2020. https://www.climatefinancelab.org/project/blue-carbon-resilience-credit/.

125. Turneffe Atoll Trust. "Home: Related Reports." Accessed April 6, 2020. https://www.turneffeatoll.org/.

126. Natural Capital Partners. "The Projects." Accessed April 6, 2020. https://millionmangroves.com/the-projects.

127. Hernandez-Blanco, Marcello. "Costa Rica's Blue Fund: A Research and Policy Agenda for Establishing a 'Blue Fund' for the Conservation and Restoration of Marine and Coastal Ecosystems in Costa Rica." *Conservation Finance Alliance.* January 15, 2020. https://www.conservationfinancealliance.org/news/2019/12/20/upcoming-cfa-webinar-a-research-and-policy-agenda-for-establishing-a-blue-fund-for-the-conservation-and-restoration-of-marine-and-coastal-ecosystems-in-costa-rica.

128. Readfearn, Graham. "Australia's Vast Carbon Sink Releasing Millions of Tonnes of CO2 Back into Atmosphere." *The Guardian.* October 1, 2019. Accessed February 28, 2020. https://www.theguardian.com/environment/2019/oct/01/australias-vast-carbon-sink-releasing-millions-of-tonnes-of-co2-back-into-atmosphere.
129. Niner, Holly J., et al. "A Global Snapshot of Marine Biodiversity Offsetting Policy." 8.
130. Eaton, James. Interviewed by Brian McFarland. February 2020.
131. Settelmyer, Scott. Interviewed by Brian McFarland. March 2020.
132. Plastic Bank. "Home." Accessed April 6, 2020. https://plasticbank.com/.
133. Verra. "Plastic Recovery and Recycling Project Accounting Program." Accessed April 6, 2020. https://verra.org/project/plastic-accounting-program/.
134. Emerton, Lucy. "Using Valuation to Make the Case for Economic Incentives: Promoting Investments in Marine and Coastal Ecosystems as Development Infrastructure." In *Economic Incentives for Marine and Coastal Conservation,* edited by Essam Yassin Mohammed. 17.

12

Ecotourism

Introduction

Tropical beaches and their offshore coral reefs are some of the most popular, idyllic destinations in the world and provide a range of ecotourism activities from diving and snorkeling, to shell collecting, sunbathing, and recreational fishing.

For an excellent book, see *Coral Reefs: Tourism, Conservation, and Management*, edited by Bruce Prideaux and Anja Pabel.

Historical Overview

Low-impact tourism likely began earlier, but in 1983, the term "ecotourism" was first coined by Mexican architect Héctor Ceballos-Lascuráin.[1]

Other notable dates throughout the history of ecotourism include:

- 1923: In 1923, Barro Colorado Island (BCI), which is located within the Panama Canal's Gatun Lake, "became one of the first biological reserves in the New World, and is now the most intensively studied area in the tropics."[2]
- 1990: The International Ecotourism Society (TIES) was established and was "launched by a team organized by Megan Epler Wood at a conference in Florida in 1989 as the world's first international non-profit dedicated to ecotourism as a tool for conservation and sustainable development."[3]
- 1990: The Adventure Travel Trade Association (ATTA) is established.[4]

© The Author(s) 2021
B. J. McFarland, *Conservation of Tropical Coral Reefs*,
https://doi.org/10.1007/978-3-030-57012-5_12

- 1993: "That reputation has brought definite payoffs, particularly in the tourism industry, which in 1993 surpassed bananas as the country's {i.e., Costa Rica} leading revenue source."[5]
- 2002: The year is "declared International Year of Ecotourism. Quebec (Canada) holds the World Ecotourism Summit, which adopts the Quebec Declaration on Ecotourism."[6]
- 2003: Center for Responsible Travel (CREST) is established.[7]
- 2007: Partnership for Global Sustainable Tourism Criteria was formed. In 2009, the Sustainable Tourism Stewardship Council was created. Both entities merged in August 2010 to form the Global Sustainable Tourism Council (GSTC).[8]

Mechanisms of Instrument

There are a wide variety of revenue sources from ecotourism:

> Tourism has the potential to generate sustainable funding for conservation in MPAs through tourism-based user fees (e.g., protected area entry fees, diving fees, and yachting fees); revenues from commercial activities of protected area agencies; airport, cruise ship, or hotel taxes; and voluntary donations of tourism operators or tourists. For tourism-based revenue sources to be viable, tourism sites need to be both attractive and accessible to tourists. The most successful revenue generation strategies are built on strong market research and collaboration between government agencies, conservation organizations, and private operators. Revenue generation strategies also need to address the additional infrastructure costs and environmental impacts of increasing numbers of tourists. Environmental impacts can be mitigated through the imposition of fines and taxes, and voluntary tourism certification programs provide a mechanism for tourism operators to be recognized for their investment in sustainable operations.[9]

With this in mind:

> Entry fees are the most common kind of MPA user fees and have the advantage that only those who use the protected area need to pay the fee. In some cases, entry fees can generate enough revenue to pay for most of a protected area's operating costs, especially in cases where visitor numbers are high and entry fees are also relatively high. Many protected areas in developing countries charge entry fees that are far lower than what international visitors would be willing to pay. The introduction of two-tiered pricing, with substantially

higher rates for tourists than for local residents, can greatly increase the total amount of fees collected.[10]

Income might be generated from operating tourist facilities such as visitor lodges, campsites, stores and restaurants, activities (i.e., rides and tours), concessions, and through the hire and rental of equipment and facilities (i.e., diving gear).

Some countries require all foreign tourists (i.e., not just divers or not just people who visit the MPA) to pay a small conservation fee when they enter or leave the country. Such visitor fees might be collected directly by the MPA or via a third-party such as a tour operator.

The destination—whether it be an MPA or a UNESCO World Heritage Site—is often public. While there are private islands which may contribute to the conservation of nearby coral reefs, these are relatively small scale.

Oftentimes, ecotourism operators are licensed by the subnational or federal government to operate within a given MPA. For instance, in Mexico, ecotourism operators are allowed to bring tourists to snorkel with whale sharks (*Rhincodon typus*),[11] and in the Dominican Republic, there are three operators for swimming with humpback whales (*Megaptera novaeangliae*).[12]

It is important to note that the visitors to such places are not necessarily tourists but may also be students and/or researchers. For instance, this includes students participating in the School for Field Studies' Turks & Caicos program[13] and researchers visiting the Smithsonian Tropical Research Institute in Panama.[14]

Ideally, the destinations of ecotourism will have low impacts due to a restricted number of visitors that are both enforced and updated according to the sustainability of the location. Similarly, you often need a permit to visit the MPA and there are limits on the number of people who can visit.

Size of Instrument

The UN World Tourism Organization (UNWTO) estimates there were approximately 1.4 billion international tourist arrivals in 2018.[15] In addition, the World Travel & Tourism Council reported that the "global travel and tourism sector grew at 3.9% to contribute a record $8.8 trillion and 319 million jobs to the world economy in 2018."[16] More specific to coral reefs, Mark Spalding et al.'s research estimates "some 30% of the world's reefs are of value in the tourism sector, with a total value estimated at nearly US$36 billion, or over 9% of all coastal tourism value in the world's coral reef countries."[17]

Introduction to Case Studies

The following case studies will examine the Galápagos Islands, ecotourism in Belize, the Bonaire National Marine Park, the Palau National Marine Sanctuary, and Lady Elliot Island. While the Galápagos Islands do not have much coral reef cover, the Islands are one of the most iconic places on Earth, is often on the bucket list for international travelers, and is home to a tremendous amount of endemic species. Belize, although a relatively small country located in Central America, has an extraordinary collection of terrestrial and marine biodiversity. The Bonaire National Marine Park generates a relatively substantial sum of revenue from reef-based tourism and the reefs are an important part of Bonaire's culture. The Palau National Marine Sanctuary, which encompasses 475,077 km^2, helps to fund Palau's marine conservation. Palau has many leading accomplishments including the world's first shark sanctuary and the first nation-led, crowdsource funding campaign called "Stand with Palau." Lady Elliot Island, the "home of the manta ray," is a spectacular ecotourism lodge located on the Southernmost island of the Great Barrier Reef and within the well-preserved Green Zone.

Case Study #1: The Galápagos Islands

Introduction

While the Galápagos Islands do not have much coral reef cover, the Islands are one of the most iconic places on Earth and is often on the bucket list for international travelers. The Galápagos Islands, which are made up of approximately "127 islands, islets and rocks, of which 19 are large and 4 are inhabited,"[18] are owned by Ecuador and are located approximately 600 miles (966 kilometers) off the coast of Ecuador (Fig. 12.1).

Tourism is one of the major revenue sources for Ecuador, and particularly for the Galápagos National Park Service. Few countries have entry taxes, while in contrast, the Galápagos Islands have an entry tax of USD$100 for every foreign tourist who is over the age of 12 and who is not a resident of Ecuador. This entry tax is split between the various actors such as the navy and the municipality as follows:

- 50%: Galápagos National Park;
- 25%: Galápagos municipalities;
- 20%: Government Council of Galápagos; and
- 5%: Galápagos parishes.[19]

Fig. 12.1 Map of the Galápagos Islands (*Credit* Open Source; Courtesy of WildAid and the Galápagos National Park Service)

Thus, the Galápagos Islands have the financing mechanism in place and with a budget of approximately USD$15 million to USD$16 million a year, it is fairly well funded.[20] There are a few other funding mechanisms in place. For instance, in the past, there was the Lindblad Expeditions Galápagos Conservation Fund[21] and the Galápagos Emergency Response Fund.[22] However, it is safe to say that a vast majority of the funding comes from tourism.

Identify Problem

There are some coral reefs located in the North, such as around Darwin Island. However, by 1997–1998, there were very few corals left around the Galápagos Islands. The University of San Francisco in Quito did an old study on coral reefs around San Cristobal island where they started to measure and monitor for any diseases and bleaching. Starting in 2009, and then particularly in 2010–2011 as a result of El Niño, some coral bleaching took place. In the central archipelago, the coral reefs were completely wiped out as the coral reefs could not sustain the extreme temperature fluctuations.[23]

Prior to 1998, and before the "Special Law for Conservation and Sustainable Development of the Galápagos" was enacted, industrial tuna fishing fleets (i.e., long liners) were also a major problem facing the Galápagos Islands. Once the special law was created, commercial tuna fishing fleets were no longer a major threat. The artisanal fishing fleets, with an estimated 1000 fisherfolk and 400 vessels, became a focus. Fishing is still prized and some of the activity is illegal.[24]

Around 2001, the Reserve got an airplane, which until today, has been relatively useful in monitoring for encroachment and illegal fishing. In 2006, legislation was starting to be written and then in 2009, the program went online with transceivers that essentially enabled one to view where the fishing vessels are and such vessels started to respect the Reserve areas. However, the fishing tactics started to change and you began to see the hauling of 6–8 small boats that would do the fishing for the mother ship. The mother ship would stay outside of the 40 nautical mile buffer zone and the smaller fishing boats would then unload the fish onto the mother boat and the mother boat would either go into port or do an exchange at sea.[25]

Asides from illegal fisherfolk and poachers, the Galápagos Islands are facing problems associated with the impacts of tourism and invasive species.[26] From prior to 1998 to 2018, tourism has exploded. For instance, in 2017, a total of approximately 241,800 tourists arrived in the Galápagos Islands and in 2018, a further 275,817 tourists arrived (i.e., a 14% increase with approximately 34,000 additional tourists).[27] Tourism has a footprint because the Galápagos Islands are located 600 miles off the coast of mainland Ecuador and nearly everything is imported. Alongside these imports comes the introduction of invasive species such as feral cats and goats, which for instance, eat the local floral and the tortoises are left with next to nothing to eat. Unfortunately, there are a lot of marine invasive species around the Galápagos Islands such as mussels and African clams, but the most difficult invasives are the ones not easily seen (i.e., parasites).[28]

Also, with respect to tourism, there is still poor management in many areas of the Reserve and the number of tourists is far too high. Dr. Judith Denkinger and the Universidad San Francisco de Quito (University of San Francisco in Quito) did a study back in 2015 that showed the levels of tourism the islands could support—150,000 annual visitors—had already been reached and that it should be frozen at this level. However nowadays, the number of visitors is far higher and is well above 225,000 annual visitors. The problem is not just the number of tourists, but everything they need (e.g., gasoline, food, etc.) and the effect of island hopping. A somewhat related problem of tourism is that the liveaboard operations primarily support international companies. The rise of ground-based tourism, although contributing to the increased number of tourists, at least supports local communities.[29]

Why the Problem Is Important

From the giant Galápagos tortoise (*Chelonoidis nigra*), to the Galápagos marine iguana (*Amblyrhynchus cristatus*), to Darwin's finches, to being one of best places in the world to dive with scalloped hammerhead sharks (*Sphyrna lewini*), the Galápagos Islands are home to some of the most unique species and some of the highest levels of endemism on Earth. As explained by Mr. Marcel Bigue of WildAid, "I do not think there is one that possesses such high levels of endemism and that is as well preserved as the Galápagos."[30]

The Galápagos Islands provides opportunities for fishing, tourism, local incomes, and foreign revenue for the Ecuador Government. Furthermore, the Galápagos Islands is listed as a World Heritage Marine Programme.

How Problem Was Identified

The problems facing the Galápagos Islands have been identified by several actors, including the work of NGOs such as WildAid and the work of universities such as the University of San Francisco in Quito.

Effectiveness of Process for Identifying Problem

The process for identifying the problems appears effective as there is a lot of attention paid to the Galápagos Islands due to their historical importance and iconic stature.

Steps Taken to Address the Problem

There are numerous steps taken to establish the Galápagos Islands National Park and to address the problems associated with overfishing, tourism, and invasive species. For instance:

- A total of "97% of the total emerged surface (7,665,100 ha) was declared a National Park in 1959. Human settlements are restricted to the remaining 3% in specifically zoned rural and urban areas on four islands (a fifth island only has an airport, tourism dock, fuel containment, and military facilities). The islands are surrounded by the Galapagos Marine Reserve which was created in 1986 (70,000 km^2) and extended to its current area (133,000 km2) in 1998"[31]; and

- Also, in 1998, the Special Law for Conservation and Sustainable Development of the Galápagos was passed.[32]

To help address problems associated with tourism, there is relatively strict regulation of the Galápagos Islands, including cutting down the amount of time tourists can spend from three months to six weeks. In addition, tourists need to apply for, and be granted, a visa.[33]

WildAid also took steps to address the problems surrounding illegal fishing. WildAid first started by asking what were the biggest threats and what could WildAid do to help. Planning with the Marine Resources Department of the Park Service was undertaken. Ultimately, WildAid started to work on improving the management of the fisheries and to stop illegal fisherfolk. This was done through ranger training, along with an internal analysis and needs assessment of the human and financial gaps including looking at education levels and potential user errors. Then, WildAid provided training, helped get the right equipment, and implemented proper systems. Although its focus was on the fisheries, WildAid adapted to fill in gaps and this is how its biosecurity work came up. In addition, Sea Shepherd has helped finance patrols for shark poachers.[34]

To address the issue of invasive species, there was a huge eradication project undertaken and the project was fairly successful on one of the islands. The process for WildAid was largely through its work with the Galápagos National Park System and later, WildAid worked with the Biosecurity Agency which was created. More specifically, WildAid worked with both senior level officials and field staff. WildAid also helped the field staff to lobby their senior staff. WildAid saw a need for a quarantine system as this was a huge gap and this is when WildAid started working with the Biosecurity Agency. Likewise, invasive species became more of a problem as the controls that were around in 1988 were the same controls being used in 2008. WildAid's first step was to look at what the top three invasives were and also looked at how biosecurity was done in the United States, Chile, and other places that had the same level of resources as the Galápagos Islands. Unfortunately, there were more problems than the funds WildAid had to address the problems.[35] Likewise, according to Mr. Marcel Bigue with respect to invasive species, "ideally the entire project would be completed in an additional five years, however, it is difficult to say with certainty, as we are already four years into the process. The total project cost is estimated at $40 M and the Government of Ecuador will be picking up 95% of all costs."[36]

Results

WildAid, which aims is to reduce global consumption of wildlife products, in part through use of celebrities and which also works to secure marine protected areas such as the Galápagos Islands, accomplished quite a bit. According to WildAid's 2017 Annual Report:

- Using technology provided by WildAid, "park rangers apprehended 10 shark poaching vessels, including a Chinese cargo vessel with over 6,000 frozen sharks in its hold, and an Ecuadorian fishing vessel and its eight support boats with 300 juvenile sharks. The Chinese vessel's crew was sentenced to 1-3 years in prison and owner charged a $6 million fine. The Ecuadorian vessel's crewmembers were sentenced with 1.5-3 years in prison, and the owner charged a $300,000 fine. Galápagos rangers conducted 478 patrols and 1,001 inspections throughout the year {2017}, which resulted in 68 infractions."[37]
- WildAid "developed a 10-year fleet renewal plan that will increase surveillance in the new Darwin and Wolf shark sanctuary and update the Galápagos National Park fleet to better fit the park's needs. Ultimately, this will reduce annual operating expenses by over $2 million."[38]
- WildAid "helped Galápagos National Park launch a pilot plastic-reduction campaign at an elementary school to promote the use of reusable bottles, resulting in a 95% reduction in the use of disposable bottles. In 2018, we will expand to three more schools."[39]
- WildAid also "launched a canine unit with three canine detection teams to prevent invasive species from entering the Galápagos. The teams can detect Giant African Land Snails {*Achatina fulica*}, one of the most destructive snail species in the world, and up to 10 prohibited products. Their work led to 40 detections of prohibited items in passenger baggage."[40]
- Furthermore, WildAid "invested in equipment for biosecurity labs and specialized training for 170 staff, which improved invasive species detection and response capacity. The team inspected over one million pieces of luggage and 17 million kg of cargo at entry points, resulting in 5,956 confiscations of prohibited or contaminated products. The lab analyzed over 3,000 samples suspected to contain invasive species."[41]

Similarly, according to WildAid's 2018 Annual Report, WildAid also:

- Published "a fleet renewal plan for Galapagos National Park (GNP), a 10-year project that will decrease GNP operating costs by millions annually, while providing greater protection for the reserve."[42]
- WildAid, together with the GNP, "hosted the first regional enforcement workshop in Latin America, with attendees representing 30+ MPAs and seven countries. They shared lessons learned, successes and common challenges in MPA enforcement to strengthen protection in the region."[43]
- Galapagos rangers "conducted more than 8,800 patrol hours, inspecting 1,760 artisanal/tourism vessels and finding 70 infractions, including fishing without required permits and the use of illegal fishing gear."[44]
- Furthermore, "using electronic surveillance tools, rangers detected 31 industrial fishing vessels illegally entering the reserve. All vessels now face substantial fines."[45]

According to Dr. Judith Denkinger, results are still materializing. Dr. Denkinger works with sea lions and the sea lions had problems with pathogens to the degree where 60% of the pups were lost. Researchers found the sea lions had distemper/Leptospira, which is believed to have been from people, dogs, or cows and then possibly transmitted to sea lions.[46] Vaccines are not allowed in the Galápagos Islands and there is little control over bringing non-vaccinated animals to the Islands. Likewise, there is a concern these vaccines could be reactivated (i.e., if it's a live culture, it could be spread to wild populations). Despite the ongoing problems with rats and cats, Dr. Denkinger believes their invasive species program is successful. Yet on San Cristobal Island, wild cats are still a big problem due to the cats eating the eggs or eating young birds.[47]

Dr. Denkinger also thinks the Galápagos Islands is doing a very good job of addressing the illegal fishing problem as evidenced by the recently seized Chinese boat. Unfortunately, the pressure is very high as shark fins get turned over to the international boats that are setup over the border of the marine reserve. In addition, large drift nets with satellite transmitters are let loose and they sometimes drift through the actual reserve because there is no control where they go and all the bycatch (i.e., sharks, turtles, sea lions, etc.) can be devastating. This said, commercial fishing in the reserve is not allowed and nor is the use of drift nets in the reserve allowed—but it still happens.[48]

Challenges and How They Were Met

There are approximately 85 vessels that tour the Galápagos Islands and, in the past, you would go on a liveaboard. The archipelagos are very well regulated (i.e., habited versus uninhabited islands) and you need a guide to visit. Then around 2010, there was a shift. The Islands had maxed out vessel tours and a handful of operators were getting the lion share of revenue with very little of the revenue trickling down. There were populist mayors raising issues about people versus parks, and how we the people need to benefit. This is when land-based tourism started to explode in the Islands with lots of people starting to offer something akin to Airbnb. Now people are taking day trips from the Islands in fast boats and this is the big, new problem with tourism. There are now lots of people, making walking trails and doing locally based tourism.[49] A related challenge is that there are powerful families in Ecuador who are a powerful force in the country that both run the country and who make a lot of money from mass tourism. This mass tourism versus controlled, good tourism is "like squeezing a balloon."[50]

Another challenge is that at one point, park management was very decentralized and had lots of autonomy (i.e., and it still does to some degree today). However, as of recent, this could be changing and there is a greater likelihood the entry tax will go to the general fund of Ecuador as the price of oil declines (i.e., this is a similar situation as in Mexico).[51] Likewise, Ecuador's crude oil basket was above USD$80 per barrel from approximately 2011 to 2014, and was then under USD$50 per barrel for most of 2016 and 2017.[52]

Galápagos Islands is a challenging place. For instance, how do you prevent people from moving to the Islands if, for example, they have a friend living there, who is making $5\times$ more money than on the mainland. How do you control population? Not only are salaries higher on the Islands than on mainland Ecuador, but a lot of living expenses and other items (e.g., plane travel, gasoline, etc.) are subsidized. This system encourages more people to move to the Islands. In addition, the system encourages people to buy cars as training cars are shipped from the mainland to the Islands to teach people how to drive.[53]

Beyond Results

It is important to note that, looking back at 1998, WildAid started with nothing as the marine reserve was not in the wheelhouse of the National Park Service; rather the Park Service was just focused on the terrestrial environment and this tends to be similar for other park services. As described by Marcel Bigue, every site is different, but there is general stuff needed for all compliance such as capacity building, training systems, monitoring systems, outreach to stakeholders, and effective prosecution. Similarly, things are definitely scalable. There are templates and one does not need to replicate everything from scratch. For instance, boats may be different, but most marine parks are still using boats for monitoring and enforcement. These results are now being applied to other MPAs throughout Ecuador. WildAid, for example, helped setup a peer-to-peer network to show others in country how local staff at the Galápagos Islands are working. WildAid, and its partners, also took the results beyond the Galápagos Islands by, for instance, hosting a regional workshop for law enforcement who came to the Islands around July 2018 to learn how it was/is done there.[54]

In contrast, Dr. Judith Denkinger and colleagues fear the future will be even more dire. From 2002 to now, Dr. Denkinger has seen dramatic changes: more people, more cars, less fish, and less sea lions. While, the terrestrial side is kind of well organized—for example, with the control of invasive species—the marine reserve is very difficult. The marine reserve is a huge area, monitoring and enforcement are expensive, and only a few small areas are strictly conserved. A lot of the overall area is used for tourism or small-scale fisheries, while the prohibition is only on industrial fishers. Yet, tuna and grouper populations are still declining. Sea cucumbers and lobsters are kind of coming back, but they have not had the necessary time to recover.[55]

Lessons Learned

Marcel Bigue's extensive experience working with WildAid, and at the Galapagos Islands, has resulted in numerous lessons learned. For instance, you can have the best technical assistance and the best technology package, but if there is not the political will to enforce fisheries regulations, you are not going to be very successful. Likewise, since 2004, Mr. Bigue has been connected to the Galápagos Islands and has seen the figuratively "high tide and low tide"—some administrations are very passionate about protecting and enforcing the fisheries, while other administrations "could care less." Nowadays in Ecuador, along with Argentina, Brazil and the United States, the national governments are all tending to the right politically speaking. Rafael Vicente Correa Delgado, the former president of Ecuador, did an about face and the country has now entered a turbulent time. In conservation, you have your victories and you have backtracks, and right now Ecuador is in back-track. However, the Park and the government have built a good foundation with lots of dedicated civil servants. In addition, tourism will continue as the Galápagos Islands, mainly, still remains a high-end destination. Another important lesson learned is to recognize that no one technology does it all.[56]

Dr. Judith Denkinger thinks that while piracy and drift nets are hard to manage, one of the major problems is corruption in the institutions and the lack of political will. All the positions in the Park are political positions and all come from the same political background. However, the best should be chosen, not someone handpicked by the environmental minister. This said, there were recently three changes in the environmental minister and the current environmental minister is working with oil companies to explore how to drill in Yasuní National Park. Yet, the park director, who is handpicked by the environmental minister, will then assign everyone else (e.g., biologists, rangers; etc.). Instead, you need to hire someone from the Galápagos Islands, but if no such qualified person is available, then you can hire from the mainland afterward.[57]

Other Resources on the Galápagos Islands

Charles Darwin Foundation

- https://www.darwinfoundation.org/en/

Galápagos Conservation Trust

- https://galapagosconservation.org.uk/; and
- https://galapagosconservation.org.uk/what-we-do/projects/

Galápagos Conservancy

- https://www.galapagos.org/; and
- https://www.galapagos.org/wp-content/uploads/2019/05/Annual-Report-2018-web.pdf

Galápagos National Park Service

- https://www.galapagos.gob.ec/en/national-park/;
- 2017 Annual Report: http://www.galapagos.gob.ec/wp-content/uploads/downloads/2018/02/informe_visitantes_anual_2017.pdf; and
- 2018 Annual Report: http://www.galapagos.gob.ec/wp-content/uploads/downloads/2019/01/INFORME-ANUAL-DE-VISITANTES-A-LAS-ÁREAS-PROTEGIDAS-DE-GALÁPAGOS-2018.pdf

The Galapagos Science Center, A Joint Effort Between the University of North Carolina at Chapel Hill and The Universidad San Francisco de Quito in Ecuador

- https://Galápagos.unc.edu/gsc/;
- http://galapagosscience.org/;
- http://www.usfq.edu.ec/Galápagos/Paginas/about_Galápagos.aspx; and
- http://www.usfq.edu.ec/opi/stations/Paginas/Galápagos-science-center.aspx

WildAid

- https://wildaid.org/expeditions-galapagos/; and
- https://wildaid.org/programs/marine-protection/

Case Study #2: Ecotourism in Belize

Introduction

Belize, a relatively small country located in Central America, has an extraordinary collection of terrestrial and marine biodiversity. Belize is "home to almost 1,400 species, {and} the Belize Barrier Reef is one of the most diverse ecosystems in the world. It has been recognized as a UNESCO World Heritage site since 1996"[58] (Fig. 12.2).

Ecotourism is a major contributor to the Belize economy, in large part due to the fact that Belize has some incredible reefs and dive sites such as the Hol Chan Marine Reserve, Ambergris Caye, and the Great Blue Hole, which is rated as one of the top dives in the world.[59] Furthermore, Belize has some relatively large reserves including the Southwater Caye Marine Reserve and the Corozal Bay Wildlife Sanctuary.[60]

From a financing perspective, in the beginning stages, "funding support for the MPA system {was} being provided by donor agencies such as the GEF / UNDP Coastal Zone Management Project, the EU, the GEF / World Bank Mesoamerica Barrier Reef System Project, TNC, WWF, and the Summit Foundation."[61]

Today:

Belize's system involves collecting the equivalent of a $3.75 conservation tax from all foreign tourists at the same time that they pay the $15 airport departure tax, and earmarking all of this revenue for conservation projects

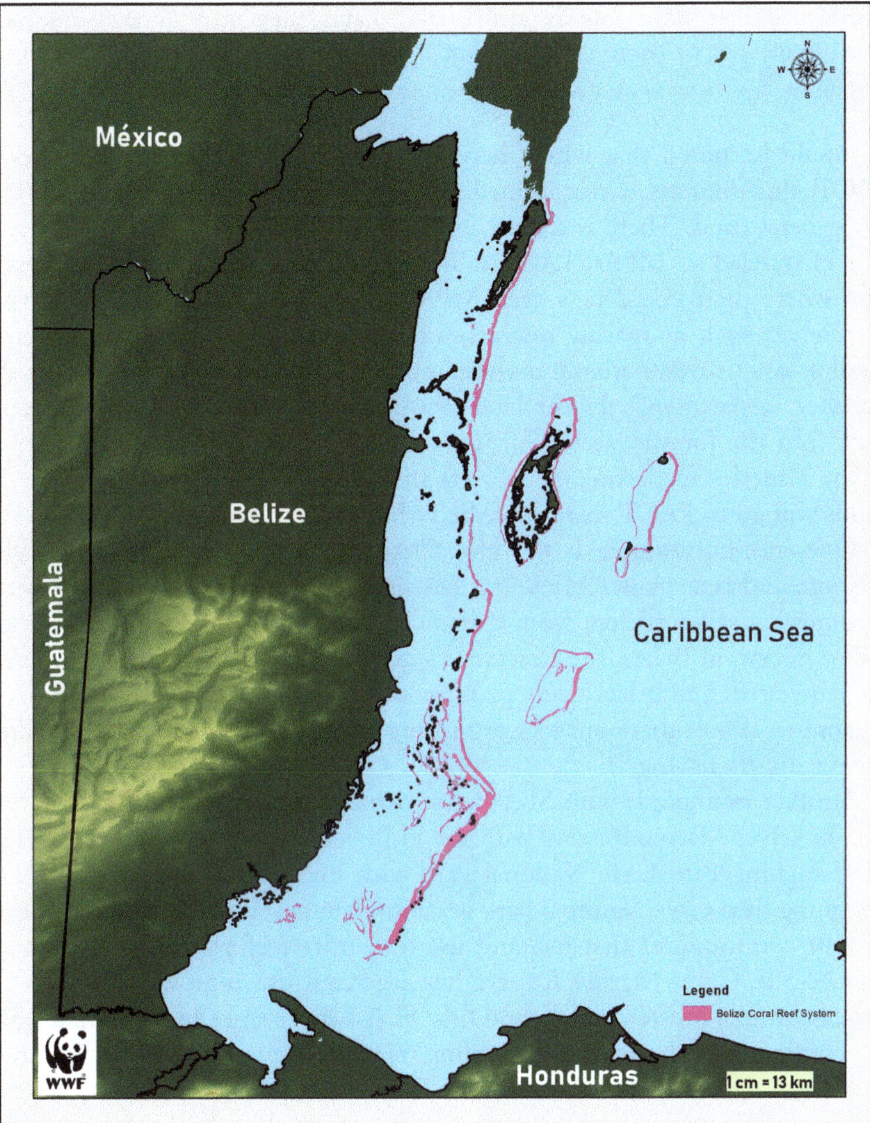

Fig. 12.2 The Coral Reef System of Belize (*Credit* WWF Mesoamerica)

administered by {Protected Areas Conservation Trust} PACT. Cruise ship passengers also pay the fee. Most foreign visitors to Belize are ecotourists who go there either to see the rainforests, or to swim, dive, snorkel, and fish in the world's second longest barrier reef. A survey done before the fee was imposed showed that most foreign visitors were even willing to pay $20 as a conservation fee. However, the tourism industry feared, without any corroborating evidence, that setting the fee at that level might cause many foreign tourists to decide not to come to Belize, but to visit cheaper neighboring countries instead.[62]

It should be noted that while there is a departure tax that is channeled to PACT, this amount is not immediately redirected to reserves as it goes to the general fund. There is a need for more work on oversight to make sure it is channeled to MPAs. PACT is doing landscape and seascape conservation where there could be several MPAs. In Belize, there are different types of reserves such as marine reserves (i.e., under the Fisheries Department), wildlife sanctuaries, national monuments, and national parks (i.e., under the Forestry Department). In the land sector, user fees are kept and reinvested back into the forestry sector. Visitor fees from marine areas are now going to the Fisheries Department. Belize is almost at the point in time where the funds will go to PACT instead of the Fisheries Department.[63]

One specific example is the Hol Chan Marine Reserve, which is right offshore and is a smaller MPA, but has significant biodiversity. This Reserve is somewhat self-sufficient from a financial perspective with their fees. As the oldest reserve in Belize, the Reserve is a good example as it has a trust, funds are reinvested, and it has a very good model of outreach. Because the Reserve is close to shore, there are lots of visitors and activities include snorkeling, diving and fly fishing.[64]

Another example is with SEA Belize, which comanages the Gladden Spit & Silk Cayes Marine Reserve with the Fisheries Department and comanages the Laughing Bird Caye National Park with the Forestry Department. The Laughing Bird Caye National Park is a complete no-take zone and is adjacent to four communities that fish and use the entire reef system. The Gladden Spit & Silk Cayes Marine Reserve has a general use zone which allows for the catching of lobster, conch, and fish. SEA Belize's work includes managing the commercial fisheries and upholding regulations for lobster, conch, finfish, and outer reef fish. In addition, SEA Belize uses protocols that are aligned with regional and national programs to monitor the health of corals and coral bleaching. Furthermore, SEA Belize makes sure that rangers are out at both marine protected areas and that there is monitoring taking place. Ultimately, there are lots of dynamics at play with an outreach program working directly with schools, fishing communities, private beach owners, and tour operators about marine conservation in general.[65]

SEA Belize's source of funding is primarily two sources. SEA Belize collects ticket fees from visitors, mainly foreigners, visiting both parks. The Fisheries and Forestry Departments are reworking the co-management agreements where the ticket fees will stay with SEA Belize. These funds then go directly to the day-to-day management, particularly payment of staff, of both MPAs. Asides from ticket fees, at Laughing Bird Caye National Park there is a gift shop that sells local arts and crafts, along with personalized t-shirts and mugs that say SEA Belize or the name of the MPA. There are opportunities to expand this gift shop and to increase revenues.[66] Second, traditional grant writing is another revenue stream. This involves finding thematic ways and different calls for proposals, then submitting a response to the funder, and going through the revision process. This grant funding is mainly project-based and for example, the Belize Marine Fund is funding the installation of new demarcation barriers, signage, and overall beautification of the areas.[67] Similarly, the Protected Areas Conservation Trust (PACT), a subsidiary of the Belize Government, was looking for application submissions for conservation investments strategies. SEA Belize sought to improve Little Water Caye with better monitoring and to build a dormitory (i.e., to provide housing, food, etc.). There are also a few other donors, which reached out to SEA Belize as they wanted to specifically fund conservation activities in Southern Belize.[68]

Furthermore, a "{…} loan from the Inter-American Development Bank to the Ministry of Tourism for implementing Belize's Sustainable Tourism Plan" was extended.[69]

Identify the Problem

The Belize Barrier Reef Reserve System "was added in 2009 to UNESCO's List of World Heritage in Danger and to date {2017} remains on this list based on threats related to the removal of mangrove cover, unsustainable coastal development, and offshore oil exploration."[70]

Historically, illegal fishing was a problem facing Belize in general and particularly within the Hol Chan Marine Reserve. Hol Chan was historically a fishing community and its designation was championed by fishers. Nowadays, there are different problems facing both Belize and the Hol Chan Marine Reserve. Illegal fishing is no longer an issue as many fisherfolk transitioned into tourism. For example, hotels and resorts hire people, who were once fisherfolk, for tourism-related jobs such as tour guides, boat drivers, and fishing guides.

Currently, the problems facing Belize and Hol Chan include overtourism, clearing of mangroves, disease, pollution, climate change and *Sargassum*. The impact of tourism, which is related to the clearing of mangroves for development, has had a detrimental impact on the local environment due to the boost in tourism. Disease, particularly white disease, is becoming a problem. White disease started off in Florida and in early 2019, the white disease was seen in Mexico. It was then positively identified in Belize in 2019. Climate change is a current problem due to increasing sea temperatures, and coupled with pollution runoff which brings nutrients to the sea, a perfect growth environment has been created for *Sargassum*.[71]

Problems inside Laughing Bird Caye National Park and Gladden Spit & Silk Cayes Marine Reserve include the changing climate, the changing seas, changing currents, the severity of storms, and the erosion. There is severe degradation, for instance, at South Silk Caye Island inside of the Gladden Spit & Silk Cayes Marine Reserve. Tourists are allowed to visit the island, but the island is suffering from severe erosion and "you have an island that is basically washing away."[72] Tour guides and tour operators are reaching out to SEA Belize and saying that we need to do something and we need do something fast. At Laughing Bird Caye National Park, there is also a lot of erosion. The ranger house is being exposed to the elements, such as wave action, and there is a need to maintain the integrity of the island for both visitors and staff. SEA Belize recently had a consultant do an analysis of the carrying capacity of the islands and there is now a need to ensure that access to the islands is controlled.[73]

Another problem is illegal fishing, which not only occurs in the MPAs, but also in adjacent areas. The Southern Barrier Reef of Belize is a complex ecosystem and is one of the most productive places in the country. While the coral reef has been categorized as doing very well and is a fairly healthy system, a lot of fisherfolk tend to congregate and there is a need to bump up enforcement (i.e., make sure rangers have enough fuel). There are three types of illegal fishing. First, there is small-scale fishing, which usually involves local, licensed fisherfolk in skiffs (i.e., 20-foot long boats with an outboard motor and typically with 3–6 people on board). While many respect the regulations, some will fish during seasonal closures and will not respect the size limitations. SEA Belize needs to make sure they have the manpower to regularly monitor. Second, there are sailboats, mainly licensed fisherfolk from Northern Belize communities, that will go out for a two-week trip (i.e., as opposed to the previously mentioned small scale fisherfolk who go out for a day trip) to range the reef from North to South. Third, there are foreigners who mainly fish at night and are coming from Honduras and Guatemala. These fisherfolk also use skiffs, but they will break into the no-take zones and take as much fish as possible.[74]

Another challenge is that the two MPAs are not contiguous; one is about 45 minutes from the SEA Belize office and the other is about one hour away.[75]

Why the Problem Is Important

These problems are important, in part, because Belize has some incredible reefs and dive sites such as the Hol Chan Reserve, Ambergris Caye, and the Great Blue Hole.

> Despite a land area of just less than 23,000 km^2 and a population of about a quarter of a million people, Belize possesses a spectacular coastline nearly 400 km long and is home to a trove of coastal and offshore resources. Three largely undeveloped and awe-inspiring "atolls" (Turneffe, Lighthouse, and Glover's Reefs) lie offshore the internationally acclaimed barrier reef. Belize continues to develop one of the world's most advanced and visionary systems of marine protected areas.[76]

If a significant die off of corals occur throughout Belize due to the white disease, or tourism visitation numbers are reduced due to the *Sargassum* problem, then Belize could lose a lot of tourism-related revenue. This said, Belize earned an estimated USD$427 million in tourism revenue in 2017,[77] and in 2019, there were 500,000+ overnight visitors and one million cruise line passengers. The latest statistics show that Belize surpassed the half a million mark, welcoming a total of 503,177 overnight visitors in 2019. This represents more than 100% growth in the sector in the past decade (up from 241,119 in 2010). Cruise arrivals also surpassed the 1 million mark again for the fourth consecutive year.[78]

In addition to reefs, Belize has several atolls (Glover's Reef), which are not very common in the Caribbean. These atolls are far enough away (~40 miles) and thus, not very accessible to tourists and are relatively, quite pristine.[79] Furthermore, the Belize Barrier Reef Reserve System is listed as a World Heritage Marine Programme.

How Problem Was Identified

The aforementioned problems were identified in different ways. For *Sargassum*, it is quite visible so everyone can see it and nearly everyone is aware (e.g., private sector hotels, government agencies, etc.) of the *Sargassum* problem.

With respect to illegal fishing, fisherfolk were complaining about catching less, while the stock assessment—particularly for finfish—by the Fisheries Department identified the catch per unit effort (CPUE) was decreasing.

White disease was originally identified by scientists in the Florida Keys and then it took approximately 1–2 years before it appeared in Mexico and Belize.

With respect to the Laughing Bird Caye National Park and Gladden Spit & Silk Cayes Marine Reserve, stakeholder engagement is very strong and there is an open-door policy. When, for instance, the tour operators association was facing problems and felt threatened (i.e., with the erosion at South Silk Caye), the association felt comfortable approaching SEA Belize and saying that intervention was needed. Likewise, communities notice the illegal fisherfolk, particularly the foreigners, and the communities will come to SEA Belize.[80]

As a ten-year old organization, SEA Belize has the data and can see changes in some of the species (i.e., different abundance and dynamics). There are also ongoing surveys such as the national conch survey which is done every two years. When generating these reports, whether for conch or lobsters, you can tell what are the problems. In addition, SEA Belize works closely with the Fisheries Department.[81]

Effectiveness of Process for Identifying Problem

Overall, while there is always room for improvement, the processes have been effective. There is always an attempt to get good consultation before the passage of a new regulation. For instance, MPA managers will consult with NGOs and local communities.[82]

The process for identifying the *Sargassum* problem has so far been effective. There has been effective government collaboration as the Belize Government has organized a committee, is now chairing the committee, and has involved the private sector (i.e., beachfront hotels and resorts), along with other representatives from the government, private sector, and NGO community.[83]

With respect to the effectiveness for identifying the problem of white disease, it has been challenging because practitioners are not really sure what is contributing to the disease. For instance, the disease could be related to sediment runoff and those have a land-based source.[84] Alternatively, the disease could be related to something in the water such as an invasive parasite.

The Coastal Zone Management Authority and Institute (CZMAI) "reached out directly to stakeholders through regional CACs {Coastal Advisory Committees}, television and radio ads, announcements in print news and media, convened public meetings in coastal regions, and shared educational materials related to the process."[85]

Steps Taken to Address the Problem

There are numerous steps taken to address the aforementioned problems and to position Belize as a world leader in ecotourism.

For instance, "the first marine habitat to be included in a protected area in Belize was at Half Moon Caye, on Lighthouse Reef atoll, which was designated as a natural monument in 1982 (a portion of the caye itself having been protected since 1928 due to its booby colony)."[86]

To address illegal fishing and overfishing, Belize announced in 2019 a drastic expansion of its MPAs and no-take zones from 4.5 to 11.6% of its waters.[87] Nowadays, Belize is seeking to put limitations on the catch and to establish shorter fishing seasons for shellfish and conch.[88]

The steps taken to address the *Sargassum* problem essentially are the Belize Government put together a *Sargassum* committee and has begun exploring how to collect *Sargassum* and use it in an economical manner.

Furthermore, "the Conservation Compliance Unit of the {Belize} Fisheries Department enforces fishery regulations. When first established, with funding from the USAID, it was well equipped and provided an enforcement capability considerably greater than that found in most other Caribbean countries."[89]

Results

In 1996, the Belize Barrier Reef Reserve System (BBRRS) was inscribed on the World Heritage List which "{…} is comprised of seven protected areas; Bacalar Chico National Park and Marine Reserve, Blue Hole Natural Monument, Half Moon Caye Natural Monument, South Water Caye Marine Reserve, Glover's Reef Marine Reserve, Laughing Bird Caye National Park and Sapodilla Cayes Marine Reserve. The largest reef complex in the Atlantic-Caribbean region, it represents the second largest reef system in the world. The seven protected areas that constitute the BBRRS comprise 12% of the entire Reef Complex."[90]

More specific to the Hol Chan Marine Reserve, which is one of the oldest MPAs in Belize, one of the most significant results is that you can actually see the results of protection as several large species of fish can been seen in Hol Chan which you would not be able to see elsewhere.[91]

The Laughing Bird Caye National Park was designated a national park in 1991 and became a World Heritage Site in 1996. There are a lot of additional results at the Laughing Bird Caye National Park, which is one of the biggest no-take zones in Belize. For instance, SEA Belize has ensured monitoring is enforced and ensured finances are safeguarded for the MPA. The communities can see that SEA Belize is continuing their mission to completely preserve the area and the local fisherfolk know the benefit of the no-take zone due to the spillover effect. SEA Belize also supports Fragments of Hope, which is doing coral restoration in Laughing Bird Caye National Park and one can see an exponential increase in coral cover over the last seven years. Furthermore, both the local economy and the national economy have benefitted from tourism revenue.

Challenges and How They Were Met

Current challenges facing ecotourism in Belize, and particularly the Hol Chan Marine Reserve, include: climate change; managing tourism demand; community expansion; deforestation of mangroves; contamination due to erosion; and working with stakeholders.

Global climate change is creating a higher chance of storms hitting Belize and harming local residents, infrastructure, and the ecosystems.[92] Starting in 2007, WWF Mesoamerica started to use the topic of climate change as an opportunity to formulate new partnerships with the local communities and private sector. WWF Mesoamerica started by increasing local communities and the private sector's awareness of climate change, mapped their vulnerability, looked at adaptation strategies, and then went in and offered to help them implement measures.[93]

In order to address the challenges of overtourism, community expansion, and contamination due to erosion and sediment runoff, Belize enacted its Coastal Zone Management Plan ("the Plan") in 2016. The Plan, which seeks a blend of conservation and development, helps define the scale of development and the type of development. However, it is important to note the Plan is not a legal instrument and an entity can still go against the Plan. The Coastal Zone Management Authority and Institute, which is the governmental entity responsible for overseeing the Plan, is now in the process of revising the Coastal Zone Management Act. The challenge is that ideally you should have the act first, and then have a plan right afterward. While the Plan is good (i.e., it involves ecosystem services, establishes regional committees, and is based off lots of consultations), the Act is outdated. The first act was done in 1998, but there was no plan until 2016 and there is a need to update the act. Belize is now seeking to get the Plan incorporated into the new act and make it official law, which should be done in 2020.[94]

Whale sharks are a popular attraction in Belize. In fact, Gladden Spit & Silk Cayes Marine Reserve "{…} attracts the most predictable and dense aggregation of whale sharks in the world."[95] The whale sharks pass through Gladden Spit & Silk Cayes Marine Reserve as there is a spawning aggregation zone and this is where the whale sharks go. However, Belize is at the point where the whale shark sightings are starting to decrease. There is an increasing number of visitors, but the spawning aggregation still allows for mutton snapper fishing. While the tour guides are trying to look at fish and see whale sharks, the fisherfolk are trying to catch snapper. The question is, how does one move forward? Do you close tourism and/or fishing? Both groups are thinking in opposition. Looking at visitation rates, in 2017, Ms. Denise Garcia recalls talking to tour operators and fisherfolk who were saying they see nothing. In 2018, the number of visitors fell. Then in 2019, the number of visitors basically doubled. The question was, why do you sell tours, if you complain about the poor number of sightings? The reason is that the operators are still breaking even and they did not want to close the zone, or have stricter measures implemented.[96]

Another challenge has been some unsound development on the island of Hol Chan, such as the clearing of mangroves and filling in of wetlands.[97] Belize revised its mangrove regulation in June 2018, that assesses higher fees for applications and for licensing to clear mangroves. The previous mangrove act had a relatively small penalty of USD$250, that did not dissuade bad actors. A related challenge is that the government did not have the capacity to monitor everywhere—especially areas out of sight—and the mangrove clearing could have major impacts before it is detected. The new mangrove regulation levies fines for mangrove restoration if you illegally cleared a mangrove forest.[98]

Working with stakeholders, particularly in the beginning, is challenging.[99] For instance, SEA Belize has not been able to place a number on Laughing Bird Caye National Park regarding quantifying the benefits to fisheries of the no-take zone (i.e., how much fish is being replenished). WWF, particularly Nadia Bood and her team, are exploring what tools to use to put a number of the tourism figures and fisheries benefits, but they remain a challenge.[100]

At Gladden Spit & Silk Cayes Marine Reserve, there is a different framework because there is a no-take zone and a general zone that allows some extraction. The question is, how do you maintain the preservation of the no-take zone and simultaneously regulate the extraction of commercial fish species in the general zone? There is also a delicate balance with allowing tourists to use South Silk Caye, but then to manage the other two cayes for protection. There are a lot of shorebirds using the islands and nesting sites for turtles. Yet, there is a constant challenge to balance the usage and to maintain the integrity.[101]

Beyond Results

The results for ecotourism in Belize, while impressive to date, are facing some serious threats to their sustainability. For instance, there has been pollution impacts from Belize and Southern Mexico, and impacts from bleaching and disease.

With respect to SEA Belize, there have recently been conversations with the board and staff when researching for the strategic plan with respect to long-term financing plans. In SEA Belize's perspective, there is a potential to maximize what SEA Belize already has out there. There is a goal to develop further financing mechanisms as the current status quo is not working as there is not as much money coming in. For instance, SEA Belize is rethinking how to use the Little Water Caye (i.e., SEA Belize owns a part of the Caye). The Caye has always been there for a ranger base; however, could SEA Belize use the island to divert some of the tourism from South Silk Caye? If so, how should SEA Belize put in place measures to make sure Little Water Caye is not impacted like South Silk Caye and to make sure no problems arise as a result of competition. Furthermore, SEA Belize, in collaboration with Fragments of Hope, are looking to expand the restoration efforts beyond Laughing Bird Caye National Park to other marine protected areas such as the Gladden Spit & Silk Cayes Marine Reserve.[102]

Further, regarding the coastal zone management plans:

> Belize's National Emergency Management Organization also used maps and graphs from the Plan to support their disaster risk reduction plans, specifically to illustrate where changes in coastal protection services would lead to more at-risk populations and infrastructure, including coastal access points, schools, and emergency services.[103]

Lessons Learned

According to Ms. Nadia Bood of WWF Mesoamerica, it has been a struggle and a battle—particularly for members of the conservation community, but "we are getting there." Likewise, Belize has a strong NGO and conservation community that is able to collaboratively work together. While it has not always gone smoothly, Belize is going in the right direction. Likewise, sometimes you need to call out the government, but you also need to work with the government. For instance, WWF Mesoamerica was a leading advocate for the Coastal Zone Management Plan and WWF Mesoamerica later became part of the task force to do the regulation.[104]

An additional lesson learned, according to Ms. Bood, is that if you really want to stem the problem of *Sargassum*, you need a multi-country, regional approach because you have a lot of nutrients coming off of South America and from the Southern United States which are also contributing to this problem.[105]

According to Ms. Denise Garcia of SEA Belize, there are a lot of lessons learned. The classic model in Belize is that you have an NGO who applies for the co-management of a protected area with the government. The NGO will then seek out the funding (i.e., apply for grants and work with donors to try to meet the gap in financing). After ten years of doing this traditional approach, it does not seem to be sustainable. NGOs have to be constantly evolving the manner in which they seek financing through creative fundraising and creative management of the protected areas. The tradition approach of grant writing and proposals does not take you far. Ticket fees do help with financing, but when there is a change in government, those fees could be removed. Instead, you need to have a very structured finance team to support your work. You need to develop a strategic plan to seek financing and to diversity revenues.[106]

Another lesson learned by Ms. Garcia is that one should not just focus only on their own protected areas. Rather, you are looking at a system. Ask, is what you are doing innovative? Is it replicable? Can other countries use it? NGOs need to consider all of this and oftentimes we are not collaborating enough and we are not sharing data enough. It is quite often the case that someone is working on a similar issue and if we openly share data on a systems level, then we can make a bigger impact.[107]

Lastly, access and visitor usage are ever evolving issues and we need to always rethink how we manage protected areas.[108]

According to Verutes et al.,

Based on findings from surveys and interviews conducted during the planning process, we learned that repeated engagement made stakeholders feel more committed to the process and optimistic about the potential for positive outcomes. We heard that the technical elements of our approach (e.g., use of maps and quantitative data, incorporation of stakeholder ideas and values into the assessment) were some of the main reasons stakeholders were continuing to participate in the process.[109]

Other Resources on Ecotourism in Belize

Association of Protected Areas Management Organizations

- https://apamobelize.org/

Belize Audubon Society

- http://www.belizeaudubon.org/

Belize Marine Fund

- https://marfund.org/en/belize-marine-fund/

Blue Solutions

- www.bluesolutions.info

Blue Ventures—Belize

- https://blueventures.org/volunteer/belize/

Coastal Zone Management Authority and Institute

- https://www.coastalzonebelize.org/

ECOMAR

- http://www.ecomarbelize.org/

Gladden Spit & Silk Cayes Marine Reserve

- http://www.fisheries.gov.bz/gladden-spit-silk-caye/

Green Reef

- https://ambergriscaye.com/greenreef/index.html

Laughing Bird Caye

- https://www.laughingbird.org/; and
- https://fragmentsofhope.org/laughingbirdcaye/

Protected Areas Conservation Trust

- https://www.pactbelize.org/

The Sarstoon Temash Institute for Indigenous Management

- https://www.satiim.org.bz/about-satiim/

Southern Environmental Association (SEA Belize)

- http://www.seabelize.org/

TIDE Tours (a subsidiary of TIDE, the Toledo Institute for Development and Environment)

- https://www.tidetours.org/

Toledo Ecotourism Association

- http://belizemayatourism.org/

Turneffe Atoll Sustainability Association

- http://www.turneffeatollmarinereserve.org/home

Ya'axché Conservation Trust

- https://ecotourismbelize.com/

Case Study #3: Bonaire National Marine Park

Introduction

Formerly part of the Netherlands Antilles and now an island municipality of the Netherlands, Bonaire is located in the southern Caribbean near Venezuela and is home to some of the best shore diving in the world (Fig. 12.3).

In addition to Bonaire, the Dutch Caribbean also consists of the islands of Aruba, Curaçao, Klein Bonaire, Sint Maarten, Saba, and Sint Eustatius. Bonaire was one of the first countries in the Caribbean to pass legislation to establish an MPA and one of the first to charge divers. Initially:

> The driving forces behind the creation of a Marine Park on Bonaire included Bonaire devotees Captain Don Stewart and Carel Steensma and The Netherlands Antilles National Parks Foundation. Carel Steensma was a personal friend of HRH Prince Bernhard of the Netherlands, Chairman of the WWF at the time. The project of the BNMP {Bonaire National Marine Park} started in 1979, with funding from WWF Netherlands, the Dutch Government, the Government of the Netherlands Antilles, and the Island Government of Bonaire.[110]

The Bonaire National Marine Park {"Marine Park"} was officially established in 1992. Shortly afterward, in December 1999, Klein Bonaire was added:

> Klein Bonaire had been in private hands for almost 130 years when a group of divers began working on assuring the islet's continued undeveloped condition. Their goal was realized in December of 1999, when the government purchased Klein Bonaire for US$ 5 million. The management of Klein Bonaire was assigned to the Bonaire National Marine Park; thus, Klein Bonaire may be the only terrestrial area in the world which is managed by a Marine Park.[111]

In the past:

> In the Caribbean islands of Bonaire and Saba, diving fees that were introduced more than 10 years ago now finance a large share of the costs of managing the MPAs. Divers in Bonaire are required to pay a flat annual fee of ten dollars, while divers in Saba pay a fee of three dollars per dive. On both islands, all of the revenue generated by these fees goes into a nonprofit conservation foundation that manages the protected areas, based on a long-term contract with the government. The admission fees have enjoyed widespread support from visiting divers, and the existence of well-managed and maintained parks has become a strong positive marketing tool for the islands themselves. The system

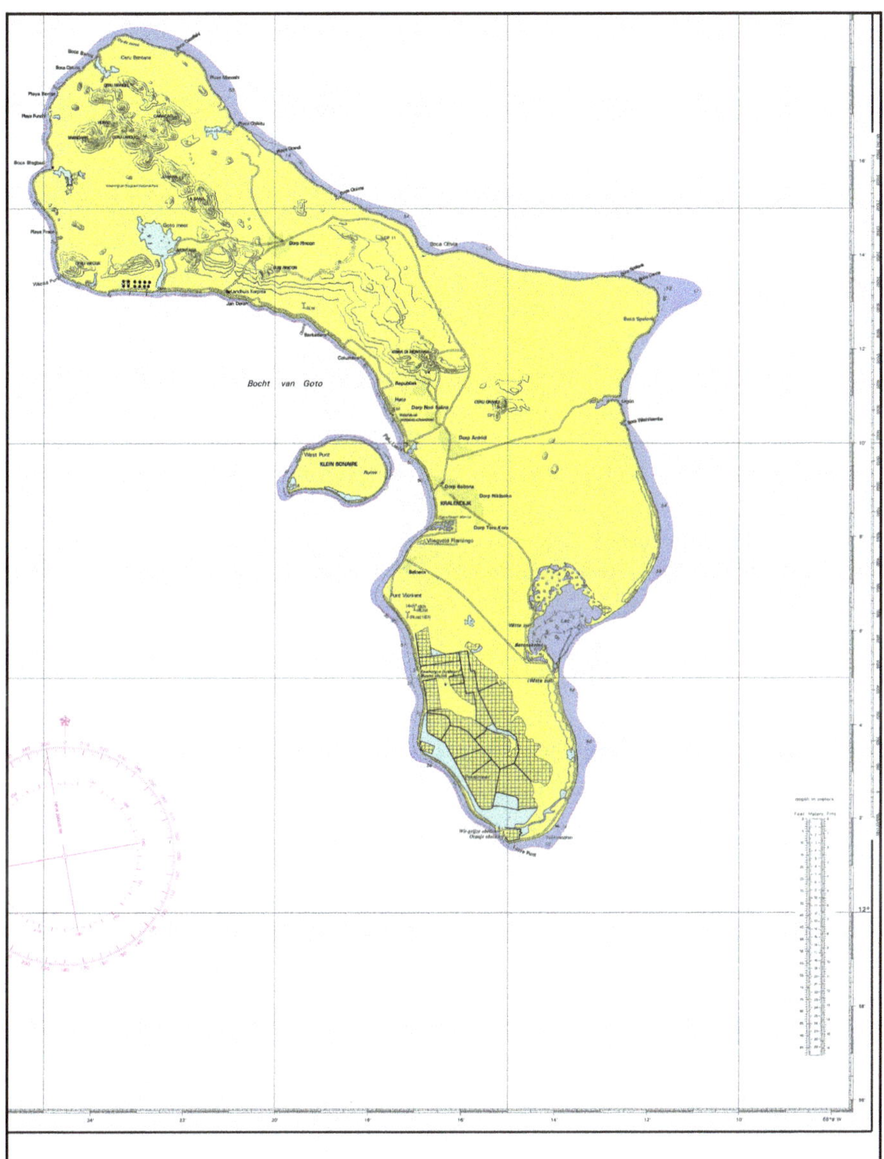

Fig. 12.3 Map of Bonaire National Marine Park (*Credit* Public Entity Bonaire, Department of Spatial Planning & Development)

is self-policing since divers are required to display a plastic tag, which has since become a collectors' item.[112]

Today:

In Bonaire, in the Netherlands Antilles, well known for its scuba diving, Bonaire National Marine Park (BNMP) has, since 1992, covered the cost of basic park operations through a fee charged to divers and other users of the park. BNMP receives no government funding. After park managers judged funding to be insufficient for sound management of the MPA, because of several issues (inflation, expansion of managing team and materials needed, reorientation of park policies, etc.), several ideas were tested to keep the BNMP self-sufficient financially. Among them, STINAPA (the BNMP's multistake-holder management body) started to charge mooring fees for boats, increasing self-generated incomes. However, BNMP remains at an 80-90% level of self-financing. BNMP has teamed up with the Coral Reef Alliance (CORAL), a US-based NGO, to accept donations through CORAL's tax-exempt status, to supplement its budget.[113]

In 2019, the fee was USD$45 per year for divers and USD$25 per year for other users.[114] Approximately 80–90% of the fees goes to management salaries, boats and vehicles, mooring maintenance, outreach materials, and law enforcement. The remaining 10–20% raised from grants goes to boat purchases, research, and monitoring.[115]

With respect to cruise ships, if you stay on the cruise ship, you do not pay the fee. Similarly, if you get off the cruise ship and, for instance, go to lunch onshore, then you also do not pay the fee. However, if you get off the cruise ship and go to a dive shop (i.e., to snorkel or dive), then you need to pay the fee and get the tag.[116]

Although a bit outdated, in 2001, the annual total economic value of Bonaire's natural environment was estimated at USD$110 million.[117]

Identify the Problem

Historically, one of the problems facing the terrestrial ecosystem on Klein Bonaire was the overgrazing by goats:

The flora of Klein Bonaire has undergone severe degradation in the past due to intensive grazing by goats. Historical photos from the 1930s show vegetation consisting of large, full-grown trees and the absence of a shrub layer. At the time the island can be presumed to have been heavily populated with goats, which while largely not affecting adult trees, affected regeneration of new plants. The absence of goats for over forty years has allowed the natural flora of Klein Bonaire to make a comeback, so that it has become home to many varieties of plants and animals, some not even present on Bonaire itself.[118]

There are perceived concerns about too many people using the reefs of the Marine Park and about sewage runoff.[119]

Another problem is that historically, agriculture was a bigger part of peoples' livelihoods and a lot of people live from fishing. Bonaire's ancestors already had a system in place to balance nature, where for instance, fisherfolk would voluntarily leave a place alone for fish to rebound or otherwise, there would be no food. Today, there are policies regulating these matters. For instance, lobsters need to be a certain size and only caught during the open season, and you are not allowed to catch queen conchs. Yet, queen conchs are still being caught.[120] In addition, some people get insulted a bit by the policies, but it is necessary because new migrants to Bonaire (i.e., like new investors or entrepreneurs) do not understand this ancestral system.[121]

A further problem is *Sargassum*.[122]

Why the Problem Is Important

Bonaire has some of the most beautiful coral reefs in the Caribbean. This said, "Bonaire's fringing coral reefs are home to virtually every species of hard and soft coral found in the Caribbean. More than 340 fish species live here, making it one of the healthiest and most bio-diverse reefs in the region."[123] Similarly, the Marine Park "is world famous for its easy access and has ranked in the top 5 shore diving destinations for many years."[124]

With respect to Klein Bonaire, it is "{...} an important stopover for migratory birds and includes the most important nesting grounds for endangered Hawksbill (*Eretmochelys imbricata*) and Loggerhead (*Caretta caretta*) sea turtles on Bonaire."[125] Furthermore, "Lac and Klein Bonaire are both a RAMSAR site and therefore internationally recognized as important wetlands areas."[126]

It should also be noted that the problems are important because Bonaire generates a relatively substantial sum of revenue from reef-based tourism and the reefs are an important part of Bonaire's culture.

How Problem Was Identified

Essentially, the Bonaire Government makes the policies and STINAPA does the management and supervision of the Marine Park. STINAPA collects the fees that are paid to use the park and if people are violating regulations, STINAPA is in charge.

Effectiveness of Process for Identifying Problem

The process for identifying the problems appears to be effective.

Steps Taken to Address the Problem

Bonaire has a long history of steps taken to conserve its coral reefs and to address the problems:

> Beginning with turtle protection in 1961, the prohibition of spear fishing in 1971, and protection for coral, dead or alive, in 1975. The driving forces behind the creation of a Marine Park on Bonaire included Bonaire devotees Captain Don Stewart and late Carel Steensma and The Netherlands Antilles National Parks Foundation. The Bonaire National Marine Park was established in 1979.[127]

Another important step took place in 1978 and 1980, when "{...} the park instituted a permanent mooring system, first in the world, for boats to moor instead of dropping anchors."[128]

With respect to addressing the problem of sewage runoff, new legislation was enacted around 2017 which requires septic tanks to be installed a minimum of 500 meters from the ocean, which should help.[129] In 1994, Bonaire adopted a planning and zoning ordinance,[130] and in 2008, a Nature Policy Plan was implemented by the Bonaire Government. In 2015, the Public Housing, Spatial Planning, and Environmental Protection Act BES (Wet vrom BES) of the Ministry of Infrastructure and Environment came into force.[131] There are also a Bonaire Nuisance Ordinance and a Bonaire Waste Ordinance.[132] Collectively, these are important steps as there are now different zones in Bonaire and likewise, different activities are allowed in different zones.[133]

Results

The Marine Park, which covers 2700 hectares (6672 acres), was established in 1979 and is currently managed by the nonprofit organization Stichting Nationale Parken Bonaire (STINAPA Bonaire). In 1992, Bonaire became "one of the first dive destinations to introduce admission fees for scuba divers."[134]

Challenges and How They Were Met

As described by STINAPA, the main challenge is balancing the interests of the multiple stakeholder groups:

> The primary challenge of managing the Bonaire National Marine Park is dealing with the varied groups and individuals who use the waters around Bonaire, and encouraging the sustainable use of natural resources. BNMP employees work with government departments, divers, boaters, fishermen, businesspeople, homeowners, contractors and other diverse groups who have, in some cases, nothing in common but their relationship to the Marine Park. The challenges to managing the ecosystems that form the Bonaire National Marine Park are dramatically increasing as the economy of Bonaire has grown in the past years.[135]

Another challenge, although not as pronounced as on the island of Sint Maarten, is the presence of cruise ships. With one cruise ship, let alone with two cruise ships, the beaches and reefs of Bonaire can quickly become full. This creates friction between the cruise tourists and the stayover tourists, as the stayover tourists (i.e., who are mainly from the Netherlands and the United States) cannot even book tours. In addition, the cruise ship tourists do not spend a lot of money on the islands as compared to the stayover tourists.[136]

When you have a proposed project that wants to develop a certain area on Bonaire, there needs to be a Management Effectiveness Report completed. This report is quite detailed (i.e., the way you are going to build, how you are going to put in infrastructure, etc.) and its quite costly, especially if looking at areas of mangroves. After spending a lot of money, the result might be that you cannot do the project and this presents a challenge.[137]

Managing tourism is another challenge. For instance, users of the Marine Park are not allowed to touch certain corals. However, humans do not always listen and tourists who visit the Marine Park on their own are not being supervised by a dive shop instructor. In addition, when it comes to cruise tourism, one must know: who are the actors, what is happening when the ship arrives, what do people do, how do they move, who is selling them services, and what services are being provided. The main challenge is how to do all of this in a coordinated manner and correctly organize it.[138] A related challenge is that Bonaire does not have a lot of statistics on tourists, except for the number of visitors and how much was collectively spent. However, it is unknown how many people visited the Marine Park in a given year, what those people did (i.e., diving and/or snorkeling), and there is no entry card (i.e., indicating how long people will stay, where they will stay, etc.). This said, an exit survey has been proposed to shed light on these questions.[139]

Beyond Results

Bonaire has developed its economy around the tourism sector as the island "does not have miles of shopping malls or miles of white beaches, and does not have 10,000 casinos. Instead, Bonaire has the sun, the sea, and nature."[140] Similarly, Bonaire does a lot of conservation because that is their source of income and there are a lot of regulations and policies concerning conservation so that visitors do not destroy nature.[141]

Lessons Learned

According to Ms. Helvig D. W. Cecilia-Thode, an important lesson learned is that you need to create international awareness. There is a lot of awareness in Bonaire, but it is mostly geared toward the locals. However, when tourists visit, they are the ones who are not really taking care of the reefs and of nature. The awareness needs to be broader and more international, so that before tourists arrive, they are aware (i.e., awareness provided on board the cruise ships—before the tourists arrive).[142] Another important lesson learned, shared by Ms. Paulina E. Martis-van Arneman, is that you can have development and you can maintain nature, but you need guidelines and regulations.

Other Resources on Bonaire

Dutch Caribbean Nature Alliance

- https://www.dcnanature.org/bonaire-national-marine-park/

<u>Stichting Nationale Parken Bonaire (STINAPA Bonaire)</u>

- https://stinapabonaire.org/stinapa/; and
- https://stinapabonaire.org/resources/

<u>Tourism Corporate Bonaire 2017–2027 Strategic Tourism Plan</u>

- https://www.tourismbonaire.com/strategictourismplan

<u>What's Bonaire's Nature Worth? Analysis by the Economics of Ecosystems and Biodiversity on Bonaire</u>

- https://www.ivm.vu.nl/en/Images/2001_TEEB_Bonaire_total_tcm234-310328.pdf

Case Study #4: Palau National Marine Sanctuary

Introduction

Palau is an archipelago nation in the Micronesia region of the South Pacific with some of the best corals in the world (Fig. 12.4).

The Palau National Marine Sanctuary ("Sanctuary") encompasses 475,077 km^2 (183,000 square miles) and to help fund Palau's marine conservation, there have been several initiatives:

Fig. 12.4 Map of Palau (*Credit* Peter Hermes Furian)

In 2015, Palau became the first country to close 80% of its EEZ to extractive activities. In 2009, it had already implemented a green fee of US$15 per tourist, used for financing local communities' conservation efforts under the Protected Areas Network in Palau. This initiative raised approximately US$2.6 million in 3 years.

In 2012, the green fee was increased to US$30 to improve the entire public water and sewerage system of Palau, and contribute to the endowment fund which will help Palau achieve its promise to effectively conserve at least 30% of the nearshore marine resources and 20% of the terrestrial resources by 2020 under the Micronesia Challenge. The US$30 green fee will be merged with the US$20 departure tax and an additional US$50 to form the US$100 Palau visitor's fee, according to a decision taken by the senate in March 2017 {...}.

The Palau visitor's fee will go into the Palau Security Fund, to be used first to finance projects related to aviation, transportation, immigration, and border security. The remainder of the fund will be disbursed at the discretion of the Palau Congress for 'projects and undertakings that will ensure the security and well-being of the national and state governments.' This announcement in 2017 created some doubts, as the Palau visitor's fee avoids specific earmarking. However, US$10 of the visitor's fee will be restricted and earmarked for the Fisheries Protection Trust Fund used to finance the Palau National Marine Sanctuary.[143]

Palau has a USD$100 visitor's fee, similar to Galápagos Islands' entry tax of USD$100 for every tourist. In addition, Palau set up the first nation-led, crowdsource funding campaign called "Stand with Palau."[144]

Palau is not a particularly representative as its case is a bit easier than other profiled countries. Likewise, the President of Palau was supportive of conservation, there has been a long history of conservation, there were only about 21,500 people living in Palau as of July 2018,[145] and there is high-end tourism with an estimated USD$4 million to $5 million a year from foreign fishing as compared to USD$90 million a year from tourism.[146]

Identify the Problem

Palau is facing several problems, "from climate change and ocean acidification to increasing pollution and illegal fishing."[147]

Many generations ago, there may have been some overfishing. However, Palauans as a people and through their traditional practices, have a deep conservation ethic. The chief would declare "blu," in order to help manage the local fisheries by closing off a species to fishing or to establish fishing seasons.

Palau gained independence in 1994 and it was during this same year that the Palau Conservation Society (PCS) was founded. Back in the early-to-mid 1990s, there was little tourism. However, as Palau opened up as a new nation, tourism significantly increased, primarily due to Japanese tourism. The overfishing, particularly of reef fish, that may be happening by local people is often in response to demand from outsiders (i.e., such as due to the increase in tourism over the last dozen years).

As part of the negotiation to become independent from the United States, there was a proposal to build a road on Babeldaob (i.e., the largest island of Palau). However, the topography of Babeldaob includes a lot of hills and there is very little topsoil. The soil that does exist is often clay and mud soils. Thus, the impact of building the road, in conjunction with erosion, has led to sedimentation being both a historical and current issue facing the nearby reefs of Palau.

These problems are being exacerbated by climate change and elevated temperatures, with some reported bleaching events.[148]

Why the Problem Is Important

The aforementioned problems are important because Palau:

> {…} is world renowned for its healthy and incredibly diverse marine ecosystem. Home to more than 1,300 species of fish and 700 species of coral, the Micronesian island nation has been called one of the seven underwater wonders of the world. The nutrient-rich waters are teeming with sharks, turtles, manta rays, dugongs and tropical fish.[149]

Such iconic locations include the Sanctuary and more specifically the Blue Corner Palau and Eil Malk Island. Furthermore, the problems are important because the reefs are important to Palau's culture, their generation of tourism revenue, and it is home to many rare and endangered species such as Napolean wrasse (*Chelinus undulates*) along with green (*Chelonia mydas*) and hawksbill turtles (*Eretmochelys imbricate*).[150]

Furthermore, Palau's Rock Islands Southern Lagoon is listed as a World Heritage Marine Programme.

How Problem Was Identified

With a conservation ethic and a close relationship to nature existing for hundreds of years, it is Ms. Abolade (Bola) Majekobaje's deduction that the problems were probably identified by local people (i.e., probably local fisher-folk) who were noticing that the fish populations were declining. However, it is hard to discern what came first: were fisherfolk telling NGOs that fish are declining, or did NGOs first conduct monitoring and then inform the communities?[151]

It was reported that fisherfolk from some of the islands were coming to the PCS and saying that the fish were smaller and that it took longer to catch what they needed. The science then came behind those stories, where scientists would catch, measure, and analyze the results. The local fisherfolk were originally blaming people from the other states, but it ended up being the local fisherfolk themselves.[152]

The role of nonprofit organizations, both local and international, have been important. For instance, the Governor of Ngermeskang, which now has the Ngermeskang Bird Sanctuary, did not know how unique the area was for birds until the PCS and BirdLife started doing the monitoring. Furthermore, there are a lot of scientists doing data collection, who are affiliated with NGOs such as TNC or with the U.S. NOAA through PCS.[153]

Effectiveness of Process for Identifying Problem

It appears as though the process for identifying problems has been effective.

Steps Taken to Address the Problem

There have been many steps taken to address the various problems facing Palau. This includes:

- In 1994, "the same year Palau became independent, it passed the Marine Protection Act, which included a moratorium on fishing for bumphead parrotfish {*Bolbometopon muricatum*}"[154];
- Also, in 1994, the PCS was established;
- The world's first shark sanctuary is established in Palau in 2009[155];
- Palau declares a marine mammal sanctuary in 2010[156];
- Rock Islands Southern Lagoon becomes a UNESCO World Heritage site in 2012[157];
- In 2014, Palau's Dugong Protection Act is signed into law[158]; and
- "On October 28, 2015, President Tommy E. Remengesau, Jr. signed into law the Palau National Marine Sanctuary Act, one of the world's most ambitious ocean conservation initiatives to date aimed at not only

protecting Palau's marine resources, but also at protecting the world's tuna stocks."[159]

Results

With a rich history of conservation, Palau and particularly the Sanctuary, has accomplished a lot. This includes:

- Establishing "{…} landmark legislation {which} creates a no-take Marine Sanctuary (approximately 500,000 km^2) covering 80% of Palau's Exclusive Economic Zone (EEZ), in which no-fishing will occur, and creates a Domestic Fishing Zone covering approximately 20% of Palau's EEZ in which traditional and domestic fishing activities will be allowed to provide fish solely for the domestic market."[160]
- Creating the world's first shark sanctuary.[161]
- Palau setup the first nation-led, crowdsource funding campaign called "Stand with Palau."[162]
- In 2019, the PSC celebrated its 25th anniversary and "as a Birdlife International Partner, this program also stewards Important Bird Areas (IBAs)."[163]

The PCS has helped to modernize conservation in Palau. Palau is still a culturally traditional society and having PCS as a moderator and trusted local group has been very helpful. PCS goes out, learns, and gets exposed to the big ideas, and then makes it have a local Palau twist. In this sense, PCS acts as a conservation catalyst, yet sometimes is acting behind the scene. In addition, another important result for PCS has been to move from a species conservation approach to more of a landscape conservation approach. PCS has established a research institute and individuals who were at PCS are able to access financing from GEF.[164]

Challenges and How They Were Met

Some of the challenges facing Palau have been cultural. For instance, there was no idea of private property until the Japanese came in and said we need to map the whole country. Another cultural challenge is that Micronesian society is maternal and women have a lot of power. Similarly, the resources are owned by the state and the clans. The Federal Government basically does not own anything, and nor do NGOs.[165]

Another challenge raised by Dr. Rob Dunbar is when you compare the Galápagos Islands with Palau. Most people immediately get on boats in the Galápagos Islands; in contrast, in Palau, with approximately 80,000–150,000 annual visitors, most people get on boats during the day and then everybody returns to the same small town in the evening (i.e., which puts pressure on the local ecosystem due to sewage, supplying food to restaurants, etc.). Likewise, the continued growth of reef fishing to supply high-end tourism is probably not sustainable.[166]

There have also been some political challenges. There was a somewhat controversial, recent legislation that banned foreign fishing exports. However, the only fish processing plant in Palau was a Taiwan-owned company which got its fish from Taiwan boats. Thus, Taiwan did not like this legislation and nor did Japan which imports a lot of tuna. In addition, Delta was earning revenue from this cargo (i.e., making money bringing tuna to Japan). Further, Palau recognized Taiwan as an independent country, which angered China, and China banned its citizens from going to Palau. This is a challenge because Chinese citizens were the largest group of tourists visiting Palau. While Palau wanted its citizens to get into domestic fishing, there were very few who wanted to buy boats, to learn how to fish, and to crew boats.[167]

For the PCS, competition for funding has been a main challenge as there are conservation organizations in each state in Palau. Human capacity is another broad issue. Palau is a small island country with approximately 20,000 people and 1000 foreigners in the tourism sector. Likewise, it is hard to find locals with the capacity and yet, there is a lot of demand for employees, so Palau is seeing people move to Palau. Yet in contrast, as a young nation, Palau's outward migration is also a challenge for good governance. It is difficult to keep human resources in place (i.e., keep the people after they have been trained), as many people are leaving to the United States where Palauans can live and work in the United States with just a passport. While Palauans live in paradise, things—such as housing and food—are getting very expensive.[168]

Tourism and development do not need to be in competition to one another; however, development is happening a lot faster than policy and regulations. This is particularly true with land use, and to a degree ocean use, and presents a huge threat. Palau wants its economy to grow and to be less dependent on foreign aid, but at the same time cannot undertake unsustainable coastal development. Likewise, the government needs to map out areas that are okay for development (i.e., maintain mangroves) and planners need to lead these efforts.[169]

Beyond Results

The Palau visitor's fee is divided up with a part of it going to the PCS trust fund, a part of it goes to the states, and a part of it goes to the Sanctuary. As long as tourism stays a core part of the Palau economy, it looks like there will be a sustainable, long-term funding mechanism for the Sanctuary.[170]

Even if the PCS were to disappear, the conservation sector had modernized and there are more NGOs now working in Palau such as TNC, BirdLife International (i.e., PCS is their local partner), Island Conservation, and CI.[171]

Furthermore, as Dr. Rob Dunbar notes, "the history of protection in Palau is long enough, that you can see the improvements. Take Blue Corner Palau, for example, where you are 'guaranteed' to see fifty sharks."[172]

Lessons Learned

As shared by Dr. Dunbar, tackling tough topics like, how to create a marine sanctuary, what use is allowed, and where is this use allowed, is not easy and it will take a lot of work to find out the answer because you just do not simply know the answer. Likewise, it is important to start early with smart, engaged people to start figuring out the answers. There is a group, including Dr. Dunbar, of about 15 people with permission from the Palau Government, tasked with looking at how do we think through this for the benefit of the Palauan people.[173]

An important lesson learned for Ms. Abolade (Bola) Majekobaje and the PCS is that having communities at the center of what you are doing and community buy-in is very important. The PCS really tries to live and breathe by this belief. PCS will not go into a community unless there is a community buy-in. Similarly, it is important to know that money is not everything.

Another lesson learned and what may not be the first thing you think of when you think of conservation, is that governance is very important. One can have lots of money for a project, but if you want it to be sustained beyond you, there needs to be enabling conditions for proper ongoing governance (e.g., community buy-in, community likes the project, resources, and ability to sustain the idea and project, etc.).[174]

Other Resources on Palau National Marine Sanctuary

Big Ocean Managers

- https://bigoceanmanagers.org/

Coral Reef Research Foundation

- https://coralreefpalau.org/; and
- http://www.glispa.org/glispa-bright-spots/163-palau-s-protected-areas-network-act

Global Island Partnership

- http://www.glispa.org/glispa-bright-spots/163-palau-s-protected-areas-network-act

Island Conservation

- https://www.islandconservation.org/palau/

PSC (Annual Reports, State of Birds, etc.)

- https://www.palauconservation.org/programs/resources/

Palau International Coral Reef Center (PICRC)

- http://picrc.org/

Palau Visitors Authority

- https://www.pristineparadisepalau.com/national-marine-sanctuary/

The Pew Charitable Trusts' Global Ocean Legacy

- https://www.pewtrusts.org/en/projects/archived-projects/global-ocean-legacy-palau

Case Study #5: Lady Elliot Island

Introduction

In 1816, "{…} a lesser-known sea captain, Thomas Stewart discovered and named the most southern coral isle of the Great Barrier Reef,"[175] Lady Elliot Island. Lady Elliot Island is the southernmost coral cay of the Great Barrier Reef and the only coral cay throughout the Great Barrier Reef where one can land a fixed-wing aircraft[176] (Figs. 12.5 and 12.6).

Located approximately 90 kilometers (56 miles) off the coast of Bundaberg, Lady Elliot Island is located within the Green Zone of the Great Barrier Reef (GBR) and is one of the most spectacular locations throughout the GBR. Further, Lady Elliot Island—which is managed in part with a lease agreement with the Lady Elliot Island Eco Resort—appears to be one of the operations that is most committed to ecotourism. For instance, upon arrival, a detailed orientation to the ecotourism resort includes discussions around:

- Do not waste food and all food waste is composted;
- Climate change and the island's carbon offset project with Greenfleet;
- History of the island's restoration activities and location of onsite tree nursery; and
- Solar-powered island, which is on track of 100% renewable energy generation by 2020.

In addition, Lady Elliot Island's dedication to ecotourism includes humpback whale fact sheets in guest rooms, recycling bins, a Reef Education Center, and Peter Gash and his team have developed an Environmental Management Plan. Overall, Lady Elliot Island is a smaller island with fewer people and "just" a fringing reef around the Island.

Identify the Problem

Historical problems associated with Lady Elliot Island include the impact of guano miners and "the destructive foraging of feral goats that had been brought to the island in the 1880s as a source of food {…}."[177]

From 1863 to 1873, "Mr. J. Askunas was principally responsible for the large-scale destruction of the natural environment of Lady Elliot Island. He was a guano miner."[178] Guano, which is essentially "{…} the cemented deposits formed by accumulations of bird droppings – and rock phosphate represent natural resources that have been extracted from some islands of the

Fig. 12.5 Map of Lady Elliot Island (*Credit* Lady Elliot Island Eco Resort)

Fig. 12.6 Close Up Map of Lady Elliot Island (*Credit* Lady Elliot Island Eco Resort)

Great Barrier Reef in order to supply phosphatic fertiliser for agriculture."[179] Other than guano mining, which was also good for gunpowder, there were really no other activities taking place at Lady Elliot Island. There is a bit of conflict as to what happened to the guano from Lady Elliot Island and there is some talk that some of the guano went to gunpowder.[180] This was a significant problem because "in addition to the disruption caused to seabirds, the removal of material threatened the stability of the cay and increased its susceptibility to erosion during storms."[181]

In addition, there were numerous shipwrecks including the Port St John in May 1939, the Vansittart in August 1975, the Tahuna in November 1975, the Huzure and Thisby in early 1980, and the Apollo I in May 1980.[182]

Why the Problem Is Important

The problem, particularly of guano mining, is important because it disrupts both the island's terrestrial ecosystem, along with the surrounding marine ecosystem. Take for instance:

The cycle starts with the nutrients in bird poo finding their way into the aquifer beneath the island, and the island's steady revegetation over the past 50 years has attracted thousands of noodies and other birds, re-starting and intensifying a process that had almost switched off during the century after the guano mining when the island was a bare rock; kept that way by a herd of goats. Water samples from boreholes show particularly high concentrations of nutrients in the aquifer near the bird rookeries in the Pisonia forest and the data show how very good the aquifer is at providing a constant and substantial drip-feed of nutrients to the surrounding waters, supporting the island's marine biodiversity, which in turn provides food for the birds, and so the cycle begins again.[183]

This said, Lady Elliot Island has some of the best visibility of the entire GBR with world-renowned dive sites and snorkeling sites offshore including the Lighthouse Bommie. Lady Elliot Island is known as the "home of the manta ray," and likewise, Lighthouse Bommie is a known cleaning station for manta rays (Fig. 12.7).

Furthermore, Lady Elliot Island is an important nesting site for green and loggerhead turtles and Lady Elliot Island is one of the most important seabird rookeries throughout the GBR. Case-in-point:

In 1973, a group from the Bundaberg branch of the Wildlife Preservation Society of Queensland visited Lady Elliot Island and during an 18-hour

Fig. 12.7 Picture of Manta Ray at Lady Elliot Island (*Credit* Brigitta Jozan)

observation period recorded 25 species of birds. These included wedgetailed shearwater (mutton bird), brown booby, greater frigate bird, reef heron, pied oyster catcher, sooty oyster catcher, pacific golden plover, red capped dotterel, Mongolian dotterel, whimbrel, bar tailed godwit, greenshank, grey tailed tattler, turnstone, sharp tailed sandpiper, silver gull, black naped tern, little tern, crested tern, roseate tern, bridled tern, common noddy, common tern, house sparrow and sacred kingfisher.[184]

In addition, migrating humpback whales often pass by the island.

How Problem Was Identified

Don Adams, "the man who gave Lady Elliot Island back its vitality," recognized the problem associated with a lack of transportation to and from the Queensland islands, including Lady Elliot Island.[185] Such problems were identified via flyovers. In addition, one could observe the former problems of guano mining as the island's vegetation was severely degraded.

Effectiveness of Process for Identifying Problem

The process for identifying the historical problem of Lady Elliot Island's guano mining and subsequent degradation of the island's vegetation was effective.

Steps Taken to Address the Problem

There were many historical steps taken at Lady Elliot Island, including:

- Messrs J & J Rooney completed the construction of the lighthouse and the keeper's cottage in 1873.[186]
- In the 1960s, Don Adams originally paved the way for a tourism lease on Lady Elliot Island when he received a one-year, renewable lease.[187]
- In 1969, the airstrip was constructed.[188]
- The "last of the goats was shot in 1969 by Don {Adams} and the lighthouse keepers."[189]
- A "functional, glass-bottom boat was brought to the island and on 28 September 1969, the airstrip and terminal building were officially opened by the Premier of Queensland the Hon. Joh. Bjelke-Petersen."[190]

- It "was during this time {1969} that Don commenced a programmed restoration of the island's flora. He collected coconuts from Dunk Island, pandanus, casuarinas and shrubs from the mainland. Gradually as the trees took hold and multiplied, the island once again became a haven for sea birds. In recognition of his wonderful conservation work on Lady Elliot Island, the Wildlife Preservation Society of Queensland presented Don Adams with a conservation award in February 1974."[191]
- "Around the year 2000, the Queensland National Parks and Wildlife Service and the island resort manager, Steve Heath and his staff planted about 50 pisonia trees on the western side of the runway as part of the continuing re-growth strategy."[192]
- The "Bundaberg News Mail reported on 16 July 2004 that Federal authorities had recognised the importance of the historic Lady Elliot Island lighthouse by having the structure placed on the Commonwealth Heritage List."[193]
- Also. in 2004, when Australia's zoning changed, Ms. Virginia Chadwick was very instrumental as she had the responsibility for putting in Green Zones and she initiated in 2004 for Lady Elliot Island to be a Green Zone.[194]

More recently, one of the major, initial steps taken to address the aforementioned problems was that Mr. Peter Gash, and his associates, received a lease in 2005 for Lady Elliot Island from the Commonwealth Government (i.e., federal government) of Australia. Likewise, "when the lease on the island came up for renewal in 2005, Peter invited business associates Gold Coast lawyer Michael Kyle and former Australian Surf Life Saving Champion Grant Kenny to be his partners in the LEI venture. This group became the successful leaseholder."[195]

While Lady Elliot Island, and a few other islands of the GBR such as Dent Island and Low Isles are owned by the Commonwealth Government, the islands of the GBR are predominantly managed and controlled by the Queensland Government. The old system was a 99-year lease and in the past, it was often known as a peppercorn lease (i.e., a very low-cost lease) and people would use the leased island for sheep, goats, and/or cows. Mr. Gash's lease is a ten-year block for three terms (i.e., thirty years), which is much more expensive than the old state island leases, and Mr. Gash, who is currently about halfway through the thirty-year lease, is seeking to request another fifteen-year extension. The lease for the island is to the low water mark and in addition, Mr. Gash and his associates have a permit to essentially do no harm and to look after the place.[196]

Additional steps taken with the new lessees included the removal of invasive species and weeds such as lantana, along with restoration including "the pisonia tree, pandanus palm, octopus bush, natural ground cover and the tournefortia tree."[197] In 2013, a larger reforestation project was launched called "The Re-Greening, Revegetation/Weed Eradication Programme." Furthermore, solar photovoltaic panels have been installed throughout the island and "all vessels on Lady Elliot Island have been fitted with jet drives, replacing the old style and un-environmentally friendly propellers to ensure the reduction of environmental impact on the reef and its inhabitants."[198]

Results

There have been numerous results achieved at Lady Elliot Island. It is important to note that when the trees were cut down, the birds disappeared, and the forest-reef-forest nutrient cycle was disrupted. Since then, trees have been planted, the birds have returned, and the cycle is reestablished. In addition, these restorations efforts have raised awareness and the coral reefs are recovering.[199]

In the past, you could fish or spear, but now with the Green Zone established, the wildlife is getting bigger around Lady Elliot Island.[200]

Another result has been the installation of solar power, which has helped reduce energy costs, reduced greenhouse gas emissions, and further helped to raise awareness.[201]

Yet, the single greatest result to date, according to Mr. Peter Gash, the Managing Director and Custodian of Lady Elliot Island, is that as he gets older, Lady Elliot Island has loaned itself and made itself a platform for people to experience nature, learn about nature, and highlight our ability to live in harmony with nature with a limited human footprint. Likewise, visitors have a great holiday at Lady Elliot Island and then take that passion back home and spread the awareness.[202]

Challenges and How They Were Met

As explained by Mr. Gash, we are in a challenging position because millions of tiny actions got us to where we are—not just one thing—by millions of people and the same thing is going to turn us around. Mr. Gash plays a large role in educating people and particularly younger people. Similarly, the biggest asset of Lady Elliot Island is the people and consequently, the biggest challenges are also the people. You need to get the right people, get them motivated, and keep them motivated—which can all be a challenge. For example, and keeping in mind that Lady Elliot Island is a small island, when one person gets sick, then everyone gets sick.

Other challenges included the fact that the old buildings were in terribly bad shape and getting a sustainable cash flow (i.e., Mr. Gash started with a small pot). For instance, Mr. Gash reinvested a significant amount of money back into the island, such as tree plantings. Big government and big business are driven by finances and the government asked, how much money is Lady Elliot Island making us? Yet, it was not even Mr. Gash's responsibility to look after the forest and the government was originally like, what are you doing? However, Mr. Gash did the plantings at his own expense and the government helped draw up a revegetation plan. The plantings started small, with a few trees initially planted and then got up to 1000 trees. After four years, the government came to Peter and said they still did not have money. However, the Government talked to the Great Barrier Reef Foundation and the Government indirectly helped by getting the Foundation to go see Lady Elliot Island. A total of five sites were targeted and approximately USD$15 million was raised around the middle of 2018. Lady Elliot Island started to get some real funding from the Foundation, in conjunction with approximately USD$1 million of Lady Elliot Island's own money, and this in part, helped to scale up the restoration activities to about 9000 trees with another 9000 trees in the nursery.

Whereas other resorts could borrow USD$50—USD$100 million and make the resort very nice, this comes with more tourists, more waste, more food, and more sunscreen—all of which, can impact the health of the coral reefs. At Lady Elliot Island, the resort is allowed to have 150 guests a night and 250 visitors during the day. However, the resort does not go anywhere close to this limit, as they usually have 80–90 guests at night and bring in a total of approximately 20–30 guests per day. This is less than half the overall permitted people and is needed for conservation.[203]

Don Adams "{…} first major problem was a logistical one. There was no jetty on which to land the heavy earthmoving equipment necessary to build the airstrip and, more importantly, no aircraft available at the time in the Maryborough district capable of carrying the machinery across the open stretch of Pacific Ocean to the island."[204]

Running an island resort 80 km out to sea presents a multitude of challenges on a daily basis. 'There is a 150-bed resort with around 30 staff on duty at any one time and the logistics are all our responsibility compared with a similar sized complex on the mainland. We generate our own power, desalinate the seawater for drinking purposes and ablutions, maintain a sewerage treatment plant and recycle most of our rubbish'.[205]

'The issues associated with transporting diesel to the island, handling it here, the on-going maintenance of the generators every two hundred hours and of course the quest to respond to climate change were the drivers towards increasing solar power,' Peter {Gash} recalled. {…} This on-going installation reached a significant milestone in March 2018 with 100 kilowatts of available power and the counter achievement of the need for the diesel falling from nearly 600 litres a day, down to around 70 litres a day, getting us closer to our goal of 100% renewable by 2020.[206]

The buildings are a very expensive component of the overall cost of operating this island because it is such a hostile and salty environment and added to that is a good dose of bird droppings on the roofs during the birdlife breeding season.[207]

Running a dive shop is also challenging. For instance, there are multiple types of insurance that must be acquired and annual maintenance costs, for a smaller outfitter, can cost upward of USD$50,000. In addition, it is hard to find help and even harder to keep good help. This is, in part, because there is not a long training process for dive masters and dive instructors as opposed to training required to become a marine biologist. This shorter training process and transferrable skills results in dive masters and dive instructors leaving for a competitor or traveling aboard to Mexico, the Maldives, Indonesia, etc. It is also important to note that what keeps dive shops operating is the snorkelers. The approximate ratio is 60–70% snorkelers to 30–40% divers. Likewise, dive shops can earn more money from snorkelers due to less insurance and less equipment needs.[208]

Beyond Results

The successes of Lady Elliot Island are reaching audiences beyond those who have visited the Island. For instance:

Produced for the BBC in 2015, the iconic three-part series, *The Great Barrier Reef with Sir David Attenborough*, featured Sir David in the second episode referring to the wonder of the Manta ray cleaning stations off Lady Elliot Island. {…} The series went on to win a British Academy of Film and Television Arts (BAFTA) Award in 2017.[209]

Lessons Learned

There are many lessons learned by Mr. Peter Gash, the Managing Director and Custodian of Lady Elliot Island. The big one, as explained by Mr. Gash, is that you need to believe in yourself, be who you are, and if you have a goal—for example—to convert a degraded island into an ecotourism destination, you need to trust yourself, trust your values, and trust your judgment. Likewise, if is not helpful to blame others, such as blaming the government for the current predicament. A related lesson learned that was shared by Mr. Gash is that when we are shown the right way, we often go the right way. This said, one should always be positive. Although the negativity flows so strongly—particularly from the media—and we can tend to gravitate to the negative, it helps to keep positive because people like positivity.[210]

Another lesson learned is that you must listen to others—whether they are the smallest, the youngest, and/or the perceived least educated. Everyone, such as guests and friends, can have contributions by relaying things they see or by offering up suggestions. Similarly, you should look for support in amazing places by doing amazing work. By doing amazing work and by acknowledging people for the effort they put in, you can receive support from local people, local councils, foundations and nongovernmental organizations, and governments.[211]

When making decisions, one should weigh how the decision is going to affect the environment (i.e., conservation and sustainability impacts) and how the decision is going to impact the finances. While Mr. Gash will take the environment over finances at any time, it is important to note that financial sustainability is critical. Ultimately, it is a balance and nature will pay you back.[212]

Lastly, "{…} it is worth remembering that no matter how many laws are made, how many research papers are produced, how many programmes planned or how many restrictions are enforced, the Great Barrier Reef heritage will only be passed on intact to the next generation if each individual who visits this marine wonderland takes nothing home but photographs and happy memories."[213]

Other Resources on Lady Elliot Island

Great Barrier Reef Marine Park Authority's History of Commonwealth Islands

- http://www.gbrmpa.gov.au/our-work/Managing-multiple-uses/managing-commonwealth-islands/history-of-commonwealth-islands

Lady Elliot Island

- http://www.ladyelliot.com.au/

Project Manta

- https://www.sites.google.com/site/projectmantasite/

The Great Barrier Reef with Sir David Attenborough

- https://www.bbc.co.uk/programmes/b06vbz1l; and
- https://www.smithsonianchannel.com/shows/david-attenboroughs-great-barrier-reef/1005189

Tourism and Events Queensland

- https://teq.queensland.com/

Walsh, Anthony. Lady Elliot Island: Eco-haven of the world's Great Barrier Reef. Australia: IPG Marketing Solutions Brisbane, 2018.

Financial Analysis

The following financial analysis will look at return and risk.

Return

The revenue stream is often visitor fees and licensing fees for operators which, depending on the site and the host country, will be returned to the relevant ministry, the general fund, or back to the particular MPA. There is also a huge job market associated with reefs from dive shops to boat operators.

Risk

Ecotourism, and especially conventional forms of tourism, can have both positive and negative impacts on tropical coral reefs. One major risk is having too many tourists, but this can be managed via a restriction on the number of tourists allowed to visit, say a particular MPA. Tourists can negatively impact reefs via their use of sunscreen, disturbing reefs with their snorkel and scuba fins, and wastewater discharge from their lodges, while park fees can go toward conservation of the same reefs.

For instance, the Seychelles was absolutely untouched and there was "a feeling that no one ever snorkeled there."[214] However, over the course of a decade, tourism started to fill up the empty places. Tourists were seeking sun and fun and many tourists would go snorkeling for ten minutes and then go to the beach. Yet, just dropping anchor started to change the environment. A "50-pound steel chunk (i.e., anchor) should not be dropped onto the reef every time you pull up to a site."[215] A mooring initiative is obvious and only takes a little organization. This said, the reefs that have the most impacts are those that have the most potential due to their proximity and accessibility.[216]

Business Risk

Many business risks associated with ecotourism are similar to other industries. Such business risks include:

- Theft of property or money from visitors or employees; and
- Staff morale, retention, recruitment, and training.

Furthermore, a significant external event—such as a global recession, terrorist attack, global pandemic such as the severe acute respiratory syndrome coronavirus 2 (SARS-CoV-2) and the coronavirus's COVID-19 disease, or a severe storm—can result in a business risk due to a drop in tourism and travel.

Strategic Risk

Strategic risks for ecotourism companies include:

- Ferocious competition from other ecotourism companies;
- Inability to pay higher commissions and lost revenue share; and
- How to structure tours which require upward of two years of advance planning, such as trips to remote atolls.

Another strategic risk is for tour operators to choose the right type of tourism offering and be able to respond to market demands; otherwise, a tourist staying at an ecotourism lodge may book offsite activities with another business.

Reputation Risk

The level of corruption and the ease of doing business in many countries can be very challenging. For instance, there could be a tendency to pay bribes to the general manager of a hotel or to the park rangers in order to receive preferential treatment.

Another reputation risk—although more a risk to the MPA—is the fact that:

In some countries, the fees collected do not always benefit conservation, as many park systems lack incentives for their staff to rigorously collect and account for entry fees. Additionally, in many countries entry fees are deposited into the general government treasury rather than allocated back to the park system. To ensure an effective financial stream, revenue from protected area entry fees should be channeled directly back into the protected area system to cover operational needs such as staff salaries or investment needs such as infrastructure.[217]

This lost revenue can result in delayed maintenance and upkeep at the MPA, which in turn can lessen the visual appeal to tourists. If visitors are harmed or lose their lives during the trip, which is possible in part due to a location's remoteness and its wildlife, this can present a reputational risk for both the tour operator and the MPA.

Greenwashing, in the sense of falsely stating or overstating one's sustainability practices, can also generate reputational risk. For instance, the cruise line industry has generated negative press due to dumping garbage at sea including a USD$40 million fine to Princess Cruise Lines Ltd.,[218] having numerous outbreaks of the illness norovirus,[219] and their generation of air pollution.[220] This said, "passengers often believe that the ship's waste is treated or stored for land removal, but in fact most of it is dumped at sea, sometimes illegally."[221]

Furthermore, overtourism poses a threat to local communities and biodiversity, along with presenting a reputation risk. For instance, the island of Boracay in the Philippines was temporarily closed to tourism due to excessive waste.[222] Similarly, bleached corals or having charismatic species go locally extinct can pose additional risks.

Liquidity Risk

There is a liquidity risk because ecotourism companies have few hard assets to sell. While lodges, dive boats, or websites may be transferred, it is unlikely that an ecotourism company could transfer their ecotourism operating license. This could be the case because the government may not allow the license, whether or not it is exclusive, to be transferred.

Operational Risk

In Mexico, a substantial amount of money is raised from visitors to protected areas and this money is collected by the federal government. The collected fees are then distributed to approximately 170 protected areas throughout Mexico. Thus, there is a disincentive for a particular protected area to increase visitation, when the funds are distributed throughout the country.

As explained by Mr. Kenneth Johnson, founder of Mexico-based ecotourism company EcoColors, it is necessary to learn lessons, you need passion to protect the reef, you need to have someone who understands business, and you need to have quality standards in place to protect the reef. Mr. Johnson has done some great outreach to SCUBA diver groups where he spoke to the economics (i.e., you make money by protecting the reef). In addition, Mr. Johnson has participated in "train the trainers" events where he briefed 20 SCUBA instructors who then taught 1000 diver guides. As explained by Mr. Johnson, there are operational risks as people are "too busy, have things to do, and people don't want to change the way they do things."[223]

Furthermore, wholesalers are pushing for too high of commissions and this is creating a significant operational risk. In fact, wholesalers are asking for 40–55% commission, if not 60% commission, and Mr. Johnson heard of commissions as high as 72% paid to wholesalers. Refusing to pay this high of commission can lead to lost business. However, paying the commission is tough and although it forces operators to be creative (which can be a good thing), it can also lead to less quality equipment, fewer staff benefits, less maintenance of boats or vehicles, and a higher ratio of customers to guides.[224]

Legal and Regulatory Risk

There are numerous legal and regulatory risks associated with ecotourism. This includes obtaining an operating license, obtaining the required permits to visit some marine protected areas, and following all the various rules, regulations, and policies.

For instance, offering differentiated fees depending on whether tourists are from the country or from abroad may be considered illegal discrimination.

Another risk is if a government decides to open up a previously designated MPA for recreational and/or commercial fishing, which then impacts the quality of the ecotourism experience.

Political risk, including social unrest and terrorism, can also be significant and have a detrimental impact on ecotourism. For example, it was reported in January 2020 by Stratfor, "the world's leading geopolitical intelligence platform," that the Al-Qaeda affiliate in East Africa Al Shabaab "is poised to surge in East Africa."[225] This could directly impact tourists' willingness to visit marine parks in Kenya and possibly in Tanzania.

Legislation or certification standards can assist to minimize risks associated with the physical impact of overfishing and overtourism where MPAs are poorly managed and where people are not educated. Likewise, one should not kick or stand on reefs. Thus, tour operators must uphold code of conducts, provide a certain level of engagement/education, and have a tour guide present as opposed to just "dumping people in the water."

Credit Risk

Credit risk includes default risk, bankruptcy risk, downgrade risk, and settlement risk.

Market Risk

Market risk includes interest rate risk, equity price risk, foreign exchange risk, and commodity price risk.

Market risks are particularly concerning for ecotourism, since travel—especially travel to distant tropical places like the Maldives or Palau—is oftentimes a luxury item.

Risk, Return, Time (Horizon), Taxes, Liquidity, Legal, and Unique (RRTTLLU)

Risk and Return

Please see above for the risk and return associated with ecotourism.

Time Horizon

The time horizon for ecotourism is a medium-term to long-term commitment. Reserves can involve capital intensive investments such as hotels, restaurants, and vehicles. Shorter time horizons may exist for smaller investments such as investments into the creation of a concession or for establishing ecotourism companies that facilitate tours.

Taxes

There are likely a variety of taxes such as property, business, sales, employment, and/or some form of tourism tax for visitors (e.g., if visitors stay at onsite facilities).

Liquidity

Liquidity might exist with concessions or other onsite facilities (e.g., tourist attraction or lodging) that could be transferred to another owner.

Legal

Legal considerations associated with ecotourism will vary from country to country. Similarly, different permits may be required to visit an MPA versus a marine sanctuary, and many countries require visitors to obtain tourist visas.

Unique

One unique aspect is that the assets of ecotourism can be hard to sell, especially as opposed to sustainable, certified commodities. In addition, factors outside of the control of the ecotourism operators (such as changing consumer behaviors or a global recession) can present a unique risk.

Policy Analysis

The following policy analysis will look at: defining the problem; establishing goals; selecting a policy; implementing a policy; and evaluating the policy. This said, ecotourism depends on a wide variety of policies to be in place.

Defining the Problem

Ecotourists have numerous destinations to choose from, are likely traveling with friends and/or family, and will be spending a relatively large sum of disposable income. With these factors in mind, most tourists want to travel to a place that is safe, easy to get to, and that is relatively affordable. Unfortunately, many countries that host tropical coral reefs, such as the Maldives, Palau, and the Seychelles, are far from the EU and the United States, are relatively difficult to travel to (e.g., due to distance, poor local infrastructure, and/or language barriers), and/or are relatively unsafe.

Establishing Goals

Some of the policy goals for ecotourism are to encourage travelers to visit your country, while simultaneously preserving the natural landscape and providing local economic opportunities.

Selecting a Policy

One of the world's leading ecotourism destinations is Costa Rica, which is due to the country's rich biodiversity, safe environment (e.g., low crime rates, less violence, and less poverty), high literacy rates, and relatively modern facilities including an international airport, good highway systems, and quality healthcare. Some of the leading ecotourism countries, in terms of supportive policies, are Australia, Belize, Bonaire, and Palau.

Furthermore, the global rankings of the top 20 countries with the largest coral reefs as of July 2019 for ease of doing business, according to the World Bank, were as follows[226]:

- #8: U.S.
  - #16: enforcing contracts
  - #36: trading across borders
  - #50: protecting minority investors
  - #53: starting a business
- #118: Bahamas
  - #84: enforcing contracts
  - #105: starting a business
  - #132: protecting minority investors
  - #161: trading across borders

- #9: U.K.
  - #15: protecting minority investors
  - #19: starting a business
  - #30: trading across borders
  - #32: enforcing contracts
- #15: Malaysia
  - #2: protecting minority investors
  - #33: enforcing contracts
  - #48: trading across borders
  - #122: starting a business
- #73: Indonesia
  - #51: protecting minority investors
  - #116: trading across borders
  - #134: starting a business
  - #146: enforcing contracts
- #77: India
  - #7: protecting minority investors
  - #80: trading across borders
  - #137: starting a business
  - #163: enforcing contracts
- #92: Saudi Arabia
  - #7: protecting minority investors
  - #59: enforcing contracts
  - #141: starting a business
  - #158: trading across borders
- #94: Vanuatu
  - #110: protecting minority investors
  - #132: starting a business
  - #136: enforcing contracts
  - #147: trading across borders
- #101: Fiji
  - #79: trading across borders
  - #97: enforcing contracts
  - #99: protecting minority investors
  - #161: starting a business
- #108: Papua New Guinea
  - #89: protecting minority investors
  - #140: trading across borders
  - #143: starting a business
  - #173: enforcing contracts
- #115: Solomon Islands
  - #98: starting a business
  - #110: protecting minority investors
  - #156: enforcing contracts
  - #160: trading across borders
- #18: Australia
  - #5: enforcing contracts
  - #7: starting a business
  - #64: protecting minority investors
  - #103: trading across borders
- #32: France
  - #1: trading across borders
  - #12: enforcing contracts
  - #30: starting a business
  - #38: protecting minority investors
- #120: Egypt
  - #72: protecting minority investors
  - #109: starting a business
  - #160: enforcing contracts
  - #171: trading across borders
- #124: Philippines
  - #104: trading across borders
  - #132: protecting minority investors
  - #151: enforcing contracts
  - #166: starting a business
- #139: Maldives
  - #71: starting a business
  - #125: enforcing contracts
  - #132: protecting minority investors
  - #155: trading across borders
- #144: Tanzania
  - #64: enforcing contracts
  - #131: protecting minority investors
  - #163: starting a business
  - #183: trading across borders
- #150: Marshall Islands
  - #75: trading across borders
  - #75: starting a business
  - #103: enforcing contracts
  - #180: protecting minority investors
- #160: Micronesia
  - #61: trading across borders
  - #170: starting a business
  - #184: enforcing contracts
  - #185: protecting minority investors
- #189: Eritrea
  - #103: enforcing contracts
  - #164: trading across borders
  - #174: protecting minority investors
  - #187: starting a business

The global rankings by Transparency International of the Corruption Perceptions Index 2018 for the top 20 countries with the largest coral reefs were as follows[227]:

- #11 (Tied): U.K.
- #13: Australia
- #21: France
- #22: United States
- #29: Bahamas
- #58: Saudi Arabia
- #61 (Tied): Malaysia
- #64 (Tied): Vanuatu
- #70 (Tied): Solomon Islands
- #78 (Tied): India
- #89 (Tied): Indonesia
- #99 (Tied): Philippines
- #99 (Tied): Tanzania
- #105 (Tied): Egypt
- #124 (Tied): Maldives
- #138 (Tied): Papua New Guinea
- #157: Eritrea
- Unranked: Fiji
- Unranked: Marshall Islands
- Unranked: Micronesia

The global rankings by the UNDP of the 2017 Human Development Index (HDI),[228] which combines life expectancy, education, and per capita income indicators, for the top 20 countries with the largest coral reefs were as follows:

- #3: Australia (HDI of 0.939)
- #13: United States (0.924)
- #14: United Kingdom (0.922)
- #24: France (0.901)
- #39 (Tied): Saudi Arabia (0.853)
- #54: Bahamas (0.807)
- #57: Malaysia (0.802)
- #92 (Tied): Fiji (0.741)
- #101 (Tied): Maldives (0.717)
- #106 (Tied): Marshall Islands (0.708)
- #113 (Tied): Philippines (0.699)
- #115: Egypt (0.696)
- #116 (Tied): Indonesia (0.694)
- #130: India (0.640)
- #131: Micronesia (0.627)
- #138: Vanuatu (0.603)
- #152: Solomon Islands (0.546)
- #153: Papua New Guinea (0.544)
- #154: Tanzania (0.538)
- #179: Eritrea (0.440)

Implementing a Policy

Palau has implemented several policies including a ban that goes into effect in 2020 on non-reef friendly sunblock,[229] and Palau has banned certain types of fishing (i.e., such as shark fishing). All of this would not be possibly without tourism. Whereas small island nations without a lot of options could choose to undertake resource exploitation, Palau has focused on resource conservation. The entity primarily responsible for tourism in Palau is the Palau Visitors Authority.[230]

Evaluating the Policy

Thus far, Palau appears to be quite effective in its policies. Non-reef friendly sunblock is really a bad thing and it a significant step for a nation to go to that level, where there is no such sunblock sold in Palau. Likewise, "the Government has signed a law that restricts the sale and use of sunscreen and skincare products that contain a list of ten different chemicals. The ban will be put into place in 2020."[231]

Furthermore, Palau has introduced the Palau Pledge, whereas "upon entry, visitors need to sign a passport pledge to act in an ecologically responsible way on the island, for the sake of Palau's children and future generations of Palauans."[232]

Future Outlook for Instrument

As explained by Ted Cheeseman, anything that creates stability—be it political, social, or economic stability—helps tourism. A key takeaway is that associations, when effective, on one hand can create much-needed clarity for operations and, on the other hand, can also be a major resource through which conservation issues can be addressed.[233]

One future direction is to utilize tourism revenue to promote greater sustainability including the deployment of renewable energy systems:

> 'Take the Caribbean,' he says. 'Half the GDP is in tourism, so investment in preserving beaches and marine life is essential. Yet here are economies that rely on diesel for electricity, which is both dirty and expensive, and at the same time have lots of wind and sun. They need investments in renewable energy projects, and the reduction in cost of electricity could provide the returns.'[234]

There is also the emerging concept of ridge-to-reef, which could link tourism activities from reefs to onshore forests and hillsides.

As predicted by Mr. Kenneth Johnson, the future of Mexican tourism, and which seems relevant to a host of other areas, is that there will likely be more construction of hotels and more tourism, particularly more massive tourism but also more specialized tourism. This increase in tourism will likely include more snorkelers and divers. With this in mind, if we cannot develop the awareness, people will just go with the cheapest option and the cheapest option is not always the best—as it could entail more pollution or more environmental impact. It is also important to note that what is cheap at the moment, is not always the cheapest in the long term.[235]

Another challenge, going forward, is that once something is nice and people like it, it gets publicized and massive amounts of people want to do it. For instance, Mr. Johnson met the director of a gorilla reserve in Rwanda. The director explained that the gorilla activity had too many people, so they set a capacity limit, put the price at USD$200 per person for the activity, and all the spots were full. They decided to further limit the amount of people, put the price up to USD$500 per person, and again all the spots were full. They upped the price to USD$1000 and still all the spots were full, so they finally decided to leave the price at USD$1500 per person and they had approximately thirty people sign up. The reserve now has a reduced environmental impact and there is a lot of money for local schools and hospitals, but the activity is limited to only rich people.[236]

Other Resources on Ecotourism

Center for Responsible Travel

- http://www.responsibletravel.org/

Global Sustainable Tourism Council

- https://www.gstcouncil.org/en/

International Institute for Peace through Tourism

- http://www.iipt.org/AboutUs.html

Professional Association of Diving Instructors (PADI) and PADI's ScubaEarth

- https://www.padi.com/; and
- https://www.scubaearth.com/

Scuba Schools International

- https://www.divessi.com/

The Adventure Travel Trade Association

- http://www.adventuretravel.biz/

The International Ecotourism Society

- http://www.ecotourism.org/

Travel and Tourism Satellite Account Program

- http://travel.trade.gov/research/programs/satellite/

UNWTO

- http://www2.unwto.org/en;
- http://media.unwto.org/content/infographics; and
- Word Tourism Barometer: http://www.unwto.org/facts/eng/barometer. htm

World Travel & Tourism Council

- http://www.wttc.org/research/

WWF's Pay for Nature View

- http://assets.panda.org/downloads/paypernatureviewphotos.pdf

Notes

1. EcoClub. "International Ecotourism Monthly." Year 7. Issue 85. Accessed December 14, 2016. http://ecoclub.com/news/085.pdf. 2.
2. Smithsonian Tropical Research Institute. "History." Accessed November 23, 2016. https://www.stri.si.edu/english/about_stri/history.php.
3. International Ecotourism Society. "Our Story." Accessed December 13, 2016. https://www.ecotourism.org/our-story.
4. Adventure Travel Trade Association. "Overview." Accessed December 14, 2016. http://www.adventuretravel.biz/about/.
5. Daily, Gretchen C. and Katherine Ellison. *The New Economy of Nature.* 178.
6. World Tourism Organization. "History." Accessed December 14, 2016. http://www.unwto.org/content/history-0.
7. Center for Responsible Travel. "History." Accessed December 14, 2016. http://www.responsibletravel.org/whoWeAre/aboutUs.php.

8. Global Sustainable Tourism Council. "Our History." Accessed December 14, 2016. https://www.gstcouncil.org/en/about/gstc-overview/our-history.html.

9. Spergel, Barry and Melissa Moye. "Financing Marine Conservation: A Menu of Options." 31.

10. Ibid. 31.

11. Johnson, Kenneth. Interviewed by Brian McFarland. October 2019.

12. Pascal, Nicolas. Interviewed by Brian McFarland. January 2019.

13. School for Field Studies. "Turks & Caicos Islands." Accessed February 28, 2020. http://fieldstudies.org/centers/tci/.

14. Smithsonian Tropical Research Institute. "Fees." Accessed February 28, 2020. https://stri.si.edu/fees.

15. UNWTO. "International Tourist Arrivals Reach 1.4 Billion Two Years Ahead of Forecasts." January 21, 2019. Accessed March 26, 2020. https://www.unwto.org/global/press-release/2019-01-21/international-tourist-arrivals-reach-14-billion-two-years-ahead-forecasts.

16. World Travel & Tourism Council. "Travel & Tourism Continues Strong Growth Above Global GDP." February 27, 2019. Accessed March 26, 2020. https://www.wttc.org/about/media-centre/press-releases/press-releases/2019/travel-tourism-continues-strong-growth-above-global-gdp/.

17. Spalding, Mark et al. "Mapping the Global Value and Distribution of Coral Reef Tourism." Marine Policy. Volume 82, August 2017, Pages 104–113. https://doi.org/10.1016/j.marpol.2017.05.014.

18. World Heritage Centre. "Galápagos Islands." Accessed September 16, 2019. https://whc.unesco.org/en/list/1/.

19. Galapagos National Park Directorate. "Tribute of Entry." Accessed September 16, 2019. http://www.galapagos.gob.ec/en/tribute-of-entry/.

20. Bigue, Marcel. Interviewed by Brian McFarland. October 2018.

21. Spergel, Barry and Melissa Moye. "Financing Marine Conservation: A Menu of Options." 37.

22. Ibid. 59.

23. Denkinger, Judith. Interviewed by Brian McFarland. December 2018.

24. Bigue, Marcel. Interviewed by Brian McFarland. October 2018.

25. Ibid.

26. Natural Habitat Adventures. "About Galapagos: Threats to the Marine Reserve." Accessed January 28, 2020. http://aboutgalapagos.nathab.com/conservation/marine-reserve/.

27. Ecuador Ministry of Environment and Galápagos National Park Service. "2018 Informe Anual: Visitantes a las áreas protegidas de Galápagos." 2019. Accessed January 28, 2020. http://www.galapagos.gob.ec/wp-content/uploads/downloads/2019/01/INFORME-ANUAL-DE-VISITANTES-A-LAS-ÁREAS-PROTEGIDAS-DE-GALÁPAGOS-2018.pdf. 4.

28. Bigue, Marcel. Interviewed by Brian McFarland. October 2018.

29. Denkinger, Judith. Interviewed by Brian McFarland. December 2018.

30. Kareus, Matt. "Interview: Marcel Bigue on WildAid's Ambitious Plans to Combat Invasive Species in the Galapagos." *About Galapagos.* October 10, 2014. Accessed December 30, 2019. http://aboutgalapagos.nathab.com/blog/interview-marcel-bigue-on-wildaids-ambitious-plans-to-combat-invasive-species/.

31. World Heritage Centre. "Galápagos Islands." Accessed September 16, 2019. https://whc.unesco.org/en/list/1/.

32. WWF. "The Galápagos." Accessed September 16, 2019. https://www.worldwildlife.org/places/the-galapagos.

33. Denkinger, Judith. Interviewed by Brian McFarland. December 2018.

34. Sea Shepherd. "Tiburon Martillo." Accessed January 28, 2020. https://seashepherd.org/galapagos/tiburon-martillo/.

35. Bigue, Marcel. Interviewed by Brian McFarland. October 2018.

36. Natural Habitat Adventures. "Interview: Marcel Bigue on WildAid's Ambitious Plans to Combat Invasive Species in the Galapagos." October 10, 2014. Accessed January 28, 2020. https://aboutgalapagos.nathab.com/blog/interview-marcel-bigue-on-wildaids-ambitious-plans-to-combat-invasive-species/.

37. WildAid. "2017 Annual Report." 2017. Accessed September 16, 2019. https://wildaid.org/wp-content/uploads/2018/04/Annual-Report-2017.pdf. 20.

38. Ibid.

39. Ibid.

40. Ibid.

41. Ibid.

42. WildAid. "2018 Annual Report." 2018. Accessed September 16, 2019. https://wildaid.org/wp-content/uploads/2019/06/WildAid_Annual-Report-2018_v13.pdf. 22.

43. Ibid.

44. Ibid.

45. Ibid.

46. Denkinger et al. "Pup Mortality and Evidence for Pathogen Exposure in Galapagos Sea Lions (*Zalophus Wollebaeki*) on San Cristobal Island, Galapagos, Ecuador." *J Wildl Dis.* 2017 July. Vol 53. Issue 3. 491–498. https://doi.org/10.7589/2016-05-092. Epub March 20, 2017; Also see: Morris Animal Foundation. "Distemper an Increasing Threat to Marine Mammals: Research Raises Concerns About Virus Impacting Seals, Sea Lions and Sea Otters." December 28, 2015. Accessed January 28, 2020. https://www.globenewswire.com/news-release/2015/12/28/798318/10158783/en/Distemper-an-increasing-threat-to-marine-mammals.html.

47. Denkinger, Judith. Interviewed by Brian McFarland. December 2018.

48. Ibid.

49. Bigue, Marcel. Interviewed by Brian McFarland. October 2018.

50. Denkinger, Judith. Interviewed by Brian McFarland. December 2018.

51. Bigue, Marcel. Interviewed by Brian McFarland. October 2018.

52. CEIC. "Ecuador Crude Oil Prices: 1984–2018." Accessed January 28, 2020. https://www.ceicdata.com/en/ecuador/crude-oil-prices.
53. Denkinger, Judith. Interviewed by Brian McFarland. December 2018.
54. Bigue, Marcel. Interviewed by Brian McFarland. October 2018.
55. Denkinger, Judith. Interviewed by Brian McFarland. December 2018.
56. Bigue, Marcel. Interviewed by Brian McFarland. October 2018.
57. Denkinger, Judith. Interviewed by Brian McFarland. December 2018.
58. WWF. "Safeguarding Belize's Reef for Future Generations." October 18, 2019. Accessed July 24, 2019. http://wwf.panda.org/?314230.
59. Scuba Diving. "The Best Dives in the World." November 9, 2018. Accessed July 24, 2019. https://www.scubadiving.com/best-dive-sites-in-world.
60. Atlas of Marine Protection. "Belize (BLZ)." 2020. Accessed February 28, 2020. http://www.mpatlas.org/region/country/BLZ/.
61. Gibson et al. "Belize's Evolving System of Marine Reserves." In: Sobel and Dahlgren. *Marine Reserves*. 293.
62. Spergel, Barry and Melissa Moye. "Financing Marine Conservation: A Menu of Options." 35.
63. Bood, Nadia. Interviewed by Brian McFarland. July 2019.
64. Ibid.
65. Garcia, Denise. Interviewed by Brian McFarland. August 2019.
66. Ibid.
67. MAR Fund. "Belize Marine Fund." Accessed April 13, 2020. https://marfund.org/en/belize-marine-fund/#resources_bmf.
68. Garcia, Denise. Interviewed by Brian McFarland. August 2019.
69. Verutes, Gregory M. et al. "Integrated Planning That Safeguards Ecosystems and Balances Multiple Objectives in Coastal Belize." 11.
70. Ibid. 2.
71. Bood, Nadia. Interviewed by Brian McFarland. July 2019.
72. Garcia, Denise. Interviewed by Brian McFarland. August 2019.
73. Ibid.
74. Ibid.
75. Ibid.
76. Gibson et al. "Belize's Evolving System of Marine Reserves." In: Sobel and Dahlgren. *Marine Reserves*. 287.
77. Statista. "International Tourism Revenue in Belize from 2005 to 2017." Accessed March 18, 2020. https://www.statista.com/statistics/814865/belize-tourism-revenue/.
78. Belize Tourism Board. "Statistics." Accessed March 18, 2020. https://belizetourismboard.org/belize-tourism/statistics/.
79. Wright, James. Interviewed by Brian McFarland. January 2018.
80. Garcia, Denise. Interviewed by Brian McFarland. August 2019.
81. Ibid.
82. Bood, Nadia. Interviewed by Brian McFarland. July 2019.
83. Ibid.

84. Ibid.
85. Verutes, Gregory M. et al. "Integrated Planning That Safeguards Ecosystems and Balances Multiple Objectives in Coastal Belize." 8.
86. Gibson et al. "Belize's Evolving System of Marine Reserves." In: Sobel and Dahlgren. *Marine Reserves.* 298.
87. WCS. "Government of Belize Expands Marine Protected Areas in Biodiverse Offshore Waters." April 3, 2019. Accessed April 13, 2020. https://newsroom.wcs.org/News-Releases/articleType/ArticleView/articleId/12150/Government-of-Belize-Expands-Marine-Protected-Areas-in-Biodiverse-Offshore-Waters.aspx.
88. Bood, Nadia. Interviewed by Brian McFarland. July 2019.
89. Gibson et al. "Belize's Evolving System of Marine Reserves." In: Sobel and Dahlgren. *Marine Reserves.* 290.
90. World Heritage Centre. "Belize Barrier Reef Reserve System." Accessed February 28, 2020. https://whc.unesco.org/en/list/764/.
91. Bood, Nadia. Interviewed by Brian McFarland. July 2019.
92. Reynolds, Aline. "Scientists Work to Build Climate Change Resilience in Caribbean Coral Reef." *Earth Institute.* January 8, 2019. Accessed February 28, 2020. https://blogs.ei.columbia.edu/2019/01/08/climate-resilience-mesoamerican-reef/.
93. GRID Arendal. "Nadia Bood from the World Wildlife Fund (WWF) Belize on Her Blue Solution." 2016. Accessed July 1, 2019. https://www.grida.no/resources/4901.
94. Bood, Nadia. Interviewed by Brian McFarland. July 2019.
95. Gibson et al. "Belize's Evolving System of Marine Reserves." In: Sobel and Dahlgren. *Marine Reserves.* 304.
96. Garcia, Denise. Interviewed by Brian McFarland. August 2019.
97. WWF. "Belize Gives Hope That Global Trend of Mangroves Loss Can Be Reversed." June 3, 2018. Accessed February 28, 2020. https://medium.com/@WWF/belize-gives-hope-that-global-trend-of-mangroves-loss-can-be-reversed-c79f384f9c0.
98. Bood, Nadia. Interviewed by Brian McFarland. July 2019.
99. GRID Arendal. "Nadia Bood from the World Wildlife Fund (WWF) Belize on Her Blue Solution." 2016. Accessed July 1, 2019. https://www.grida.no/resources/4901.
100. Garcia, Denise. Interviewed by Brian McFarland. August 2019.
101. Ibid.
102. Ibid.
103. Verutes, Gregory M. et al. "Integrated Planning That Safeguards Ecosystems and Balances Multiple Objectives in Coastal Belize." 10.
104. Bood, Nadia. Interviewed by Brian McFarland. July 2019.
105. Ibid.
106. Garcia, Denise. Interviewed by Brian McFarland. August 2019.
107. Ibid.

108. Ibid.
109. Verutes, Gregory M. et al. "Integrated Planning That Safeguards Ecosystems and Balances Multiple Objectives in Coastal Belize." 9.
110. Tourism Corporation Bonaire. "Marine Park." Accessed September 18, 2019. https://www.tourismbonaire.com/sightseeing/marine-park.
111. STINAPA. "Bonaire National Marine Park: Klein Bonaire." Accessed September 18, 2019. https://stinapabonaire.org/bonaire-national/.
112. Spergel, Barry and Melissa Moye. "Financing Marine Conservation: A Menu of Options." 33.
113. Vertigo Lab. "Innovations for Coral Finance, ICRI Publication." 36.
114. STINAPA. "Nature Tag." Accessed April 3, 2020. https://stinapabonaire.org/stinapa/nature-tags/; Also see: STINAPA. "Nature Fee 2019." Accessed April 3, 2020. https://stinapabonaire.org/nature-fee-2019/.
115. Vertigo Lab. "Innovations for Coral Finance, ICRI Publication." 36.
116. Martis-van Arneman, Paulina E. Interviewed by Brian McFarland. September 2019.
117. TEEB. "What's Bonarie's Nature Worth?" 2001. Accessed April 13, 2020. https://www.ivm.vu.nl/en/Images/2001_TEEB_Bonaire_total_tcm234-310328.pdf.
118. STINAPA. "Bonaire National Marine Park: Klein Bonaire." Accessed September 18, 2019. https://stinapabonaire.org/bonaire-national/.
119. van de Velden, Henk. Interviewed by Brian McFarland. September 2019.
120. STINAPA. "Marine Park News: Queen Conch is still being caught." July 5, 2018. Accessed September 18, 2019. https://stinapabonaire.org/marine-park-news/queen-conch-is-still-being-caught/.
121. van de Velden, Henk. Interviewed by Brian McFarland. September 2019.
122. STINAPA. "Marine Park News: Sargassum Clean Up Lagun." July 27, 2018. Accessed September 18, 2019. https://stinapabonaire.org/marine-park-news/sargassum-clean-up-lagun/.
123. Dutch Caribbean Nature Alliance. "Bonaire National Marine Park." Accessed September 18, 2019. https://www.dcnanature.org/bonaire-national-marine-park/.
124. STINAPA. "Bonaire National Marine Park: Klein Bonaire." Accessed September 18, 2019. https://stinapabonaire.org/bonaire-national/.
125. Dutch Caribbean Nature Alliance. "Bonaire National Marine Park." Accessed September 18, 2019. https://www.dcnanature.org/bonaire-national-marine-park/.
126. STINAPA. "Bonaire National Marine Park: Klein Bonaire." Accessed September 18, 2019. https://stinapabonaire.org/bonaire-national/.
127. Ibid.
128. Tourism Corporation Bonaire. "Marine Park." Accessed September 18, 2019. https://www.tourismbonaire.com/sightseeing/marine-park.
129. van de Velden, Henk. Interviewed by Brian McFarland. September 2019.

130. VanEps Kunneman VanDoorne. "Guide to Doing Business in Saba: Planning & Zoning Regulations." Accessed April 13, 2020. https://www.doingbusinessdutchcaribbean.com/saba/real-estate-construction-law-sab/planning-zoning-regulations-sab/.

131. Rijksdienst Caribisch Nederland. "Environmental Regulation." Accessed April 13, 2020. https://english.rijksdienstcn.com/infrastructure-and-water-management/environment-regulation.

132. VanEps Kunneman VanDoorne. "Guide to Doing Business in Bonaire: Nuisance, Nature & Environmental Regulations." Accessed April 13, 2020. https://www.doingbusinessdutchcaribbean.com/bonaire/real-estate-construction-law-bon/nuisance-nature-environmental-regulations-bon/.

133. Martis-van Arneman, Paulina E. Interviewed by Brian McFarland. September 2019.

134. Dutch Caribbean Nature Alliance. "Bonaire National Marine Park." Accessed September 18, 2019. https://www.dcnanature.org/bonaire-national-marine-park/.

135. STINAPA. "Bonaire National Marine Park: Klein Bonaire." Accessed September 18, 2019. https://stinapabonaire.org/bonaire-national/.

136. van de Velden, Henk. Interviewed by Brian McFarland. September 2019.

137. Martis-van Arneman, Paulina E. Interviewed by Brian McFarland. September 2019.

138. Ibid.

139. van de Velden, Henk. Interviewed by Brian McFarland. September 2019.

140. Martis-van Arneman, Paulina E. Interviewed by Brian McFarland. September 2019.

141. Ibid.

142. Cecilia-Thode, Helvig D.W. Interviewed by Brian McFarland. September 2019.

143. Vertigo Lab. "Innovations for Coral Finance, ICRI Publication." 20.

144. Shiffman, David. "Island Nation Sets Up World's First Crowdfunded Marine Protected Area." *Scientific American.* October 21, 2014. Accessed September 11, 2019. https://www.scientificamerican.com/article/island-nation-sets-up-worlds-first-crowdfunded-marine-protected-area/.

145. U.S. Central Intelligence Agency. "World Factbook: Australia—Oceania: Palau." Accessed September 11, 2019. https://www.cia.gov/library/publications/the-world-factbook/geos/print_ps.html.

146. Dunbar, Rob. Interviewed by Brian McFarland. April 2019.

147. Palau Visitors Authority. "National Marine Sanctuary." Accessed September 11, 2019. https://www.pristineparadisepalau.com/national-marine-sanctuary/.

148. Majekobaje, Abolade (Bola). Interviewed by Brian McFarland. September 2019.

149. Pew Charitable Trusts. "Archived Project: Global Ocean Legacy: Palau." Accessed September 11, 2019. https://www.pewtrusts.org/en/projects/archived-projects/global-ocean-legacy-palau.

150. PCS. "Priority Species." Accessed September 11, 2019. https://www.palaucons ervation.org/priority-species/.

151. Majekobaje, Abolade (Bola). Interviewed by Brian McFarland. September 2019.

152. Ibid.

153. Ibid.

154. Pew Charitable Trusts. "Palau National Marine Sanctuary: Building Palau's Future and Honoring Its Past." Accessed September 11, 2019. https:// www.pewtrusts.org/-/media/assets/2017/07/palau_update2017_v6.pdf?la= en&hash=4743F4F5B5593533FA12DB9E24FAFAF1F598802F.

155. Ibid. 6.

156. Ibid.

157. Ibid.

158. Ibid.

159. Palau Visitors Authority. "National Marine Sanctuary." Accessed September 11, 2019. https://www.pristineparadisepalau.com/national-marine-sanctuary/.

160. Ibid.

161. MarineBio Conservation Society. "World's First "Shark Sanctuary."" September 25, 2009. Accessed December 30, 2019. https://marinebio.org/wor lds-first-shark-sanctuary/.

162. Shiffman, David. "Island Nation Sets Up World's First Crowdfunded Marine Protected Area." *Scientific American.* October 21, 2014. Accessed September 11, 2019. https://www.scientificamerican.com/article/island-nat ion-sets-up-worlds-first-crowdfunded-marine-protected-area/.

163. PCS. "Conservation & Protected Areas Program." Accessed February 28, 2020. https://www.palauconservation.org/programs/conservation-and-pro tected-areas/.

164. Majekobaje, Abolade (Bola). Interviewed by Brian McFarland. September 2019.

165. Ibid.

166. Dunbar, Rob. Interviewed by Brian McFarland. April 2019.

167. Ibid.

168. Majekobaje, Abolade (Bola). Interviewed by Brian McFarland. September 2019.

169. Ibid.

170. Ibid.

171. Ibid.

172. Dunbar, Rob. Interviewed by Brian McFarland. April 2019.

173. Ibid.

174. Majekobaje, Abolade (Bola). Interviewed by Brian McFarland. September 2019.

175. Walsh. *Lady Elliot Island*. 19.

176. Ibid. 4.

177. Ibid. 57.

178. Ibid. 27.
179. Daley. *The Great Barrier Reef*. 156.
180. Gash, Peter. Interviewed by Brian McFarland. November 2019.
181. Daley. *The Great Barrier Reef*. 178.
182. Walsh. *Lady Elliot Island*. 72–75.
183. Ibid. 83.
184. Ibid. 106.
185. Ibid. 55.
186. Ibid. 39.
187. Lady Elliot Island. "Custodians." Accessed December 30, 2019. https://ladyel
liot.com.au/custodians/.
188. Walsh. *Lady Elliot Island*. 41.
189. Ibid. 57.
190. Ibid. 58.
191. Ibid.
192. Ibid. 79.
193. Ibid. 45.
194. Gash, Peter. Interviewed by Brian McFarland. November 2019.
195. Walsh. *Lady Elliot Island*. 78.
196. Gash, Peter. Interviewed by Brian McFarland. November 2019.
197. Walsh. *Lady Elliot Island*. 79.
198. Ibid. 85.
199. Gash, Peter. Interviewed by Brian McFarland. November 2019.
200. Ibid.
201. Ibid.
202. Ibid.
203. Ibid.
204. Walsh. *Lady Elliot Island*. 56.
205. Ibid. 78–79.
206. Ibid. 84.
207. Ibid. 86.
208. Wink, Jay. Interviewed by Brian McFarland. September 2019.
209. Walsh. *Lady Elliot Island*. 112.
210. Gash, Peter. Interviewed by Brian McFarland. November 2019.
211. Ibid.
212. Ibid.
213. Walsh. *Lady Elliot Island*. 116.
214. Cheeseman, Ted. Interviewed by Brian McFarland. January 2019.
215. Ibid.
216. Ibid.
217. WWF. *Guide to Conservation Finance*. 13.
218. U.S. Attorney General Office, Southern District of Florida. "Princess Cruise
Lines To Pay Largest-Ever Criminal Penalty For Deliberate Vessel Pollution."

Last modified December 1, 2016. https://www.justice.gov/usao-sdfl/pr/pri ncess-cruise-lines-pay-largest-ever-criminal-penalty-deliberate-vessel-pollution.

219. Centers for Disease Control and Prevention. "Outbreak Updates for International Cruise Ships." Accessed February 20, 2017. https://www.cdc.gov/nceh/vsp/surv/gilist.htm.

220. Friends of the Earth. "2016 Cruise Ship Report Card." Accessed February 20, 2017. http://www.foe.org/cruise-report-card.

221. Helvarg. *50 Ways to Save the Ocean.* 82.

222. McKirdy, Euan. "Philippines Closes 'Cesspool' Tourist Island of Boracay." *CNN*. April 5, 2018. Accessed February 28, 2020. https://edition.cnn.com/2018/04/04/asia/philippines-duterte-boracay-shutdown-intl/index.html.

223. Johnson, Kenneth. Interviewed by Brian McFarland. October 2019.

224. Ibid.

225. Abi-Hanna, Thomas. "Assessments: Al Shabaab Is Poised to Surge in East Africa." *Statfor*. January 15, 2020. Accessed January 24, 2020. https://worldv iew.stratfor.com/article/al-shabaab-militant-somalia-kenya-attack-us-troops.

226. World Bank Group. "Doing Business: Economy Rankings." Accessed July 23, 2019. http://www.doingbusiness.org/rankings.

227. Transparency International. "Corruption Perceptions Index 2018." Accessed July 24, 2019. https://www.transparency.org/cpi2018.

228. UNDP. "Human Development Data (1990–2017)." Accessed July 23, 2019. http://hdr.undp.org/en/data#.

229. Cooper, Rachel. "Pacific Island to Ban Sunscreen to Protect Coral Reefs." *Climate Action*. November 2, 2018. Accessed February 28, 2020. http://www.climateaction.org/news/pacific-island-to-ban-sunscreen-to-protect-coral-reefs.

230. Pristine Paradise Palau. "Palau Visitors Authority." Accessed April 6, 2020. https://www.pristineparadisepalau.com/palau-visitors-authority/.

231. Cooper, Rachel. "Pacific Island to Ban Sunscreen to Protect Coral Reefs." *Climate Action*. November 2, 2018. Accessed February 28, 2020. http://www.climateaction.org/news/pacific-island-to-ban-sunscreen-to-protect-coral-reefs.

232. Palau Pledge. "Home." Accessed April 6, 2020. https://www.palaupledge.com/.

233. Cheeseman, Ted. Interviewed by Brian McFarland. September 2016.

234. Avery, Helen. "Blue Finance: Why Marine PPPs Could Be a Win-Win-Win." *Euromoney*. June 5, 2018. Accessed December 30, 2019. https://www.euromo ney.com/article/b18hg7vjwmy9hn/blue-finance-why-marine-ppps-could-be-a-win-win-win.

235. Johnson, Kenneth. Interviewed by Brian McFarland. October 2019.

236. Ibid.

13

Debt Conversions

Introduction

Debt conversions, which were more commonly known as debt-for-nature swaps, have been used a handful of times for the conservation of tropical coral reefs. In the past, debt-for-nature conversions were implemented more for the conservation of tropical forests. This said, over the course of thirty years and in at least 25 countries, the U.S. Tropical Forest Conservation Act alone generated USD$218.4 million in local conservation funds.[1]

Countries that take on foreign debt are frequently required to earn export earnings to pay back foreign currency-denominated debt obligations. Such export earnings are often generated from promoting the development of commercial agriculture and livestock (i.e., which can lead to significant defor- estation, degradation, and erosion) or promoting large-scale tourism (i.e., such as building ports to host cruise liners and building hotel infrastructure). Some of this debt might be eligible for debt conversions.

Historical Overview

In 1987, the first debt-for-nature swaps were conducted in Bolivia and Ecuador.[2] It is important to note that part of the 1987 debt-for-nature swap in Ecuador was for the Galapagos National Park.[3]

A few, of the many, dates related to debt conversions include:

© The Author(s) 2021
B. J. McFarland, *Conservation of Tropical Coral Reefs*,
https://doi.org/10.1007/978-3-030-57012-5_13

- 1984: Thomas Lovejoy, then Vice President of WWF, applies the earlier concept of debt-for-equity swaps to conservation and is largely attributed with coming up with the concept of debt-for-nature swaps.[4]
- 1985: Chile becomes the first country "to undertake an institutionalised debt-to-equity swap programme {…}."[5]
- 1987: WWF and the Frank Weeden Foundation purchase debt with "assistance provided by Citicorp Investment Bank and Shearman and Sterling. {The swap was} structured as participation where Citicorp purchased Ecuador debt, on behalf of WWF, from Bankers Trust Co. The debt was then assigned to Fundación Natura (subject to WWF's participation) and exchanged for local currency bonds."[6] An additional beneficiary was the Charles Darwin Foundation.
- 1988: WWF purchases debt for the benefit of the Philippines including for the El Nido National Marine Park with Haribon Foundation and the Philippine Department of Environment and Natural Resources (DENR) as the beneficiaries.[7]
- 1989: WWF purchases debt from Bankers Trust for the benefit Ecuador including the Galapagos National Park, with Fundación Natura as the beneficiary.[8]
- 1990: WWF and the USAID purchase debt for the benefit of the Philippines, including for the El Nido Marine Sanctuary and the Tubbataha Reef National Park with Haribon Foundation and the Philippine Department of Environment and Natural Resources (DENR) as the beneficiaries.[9]
- 1991: TNC, USAID, and the Puerto Rican Conservation Trust purchase a debt-for-nature swap for an "endowment fund for JNPT {Jamaica Nationals Parks Trust}, a conservation trust fund dedicated to the conservation of biological diversity through support of Jamaica's national parks system: Montego Bay Marine Park and Blue and John Crow Mountains National Park."[10]
- 1991–1998: A series of debt-for-nature swaps are purchased by CI, Sequoia Foundation, MacArthur Foundation and/or USAID for the benefit of Mexico including for "fisheries (industrial and artisanal, wetlands) and protected areas (creating new ones and strengthening existing ones)" and with CIMEX as the beneficiary.[11]
- 1992: WWF and the USAID purchase debt for the benefit of the Philippines, including for the El Nido Marine Sanctuary with Haribon Foundation, the Philippine Department of Environment and Natural Resources, and the Foundation for the Philippine Environment as the beneficiaries.[12]
- 2006 (May): Mexico becomes "among the first sovereigns to successfully issue a catastrophe bond."[13]

- 2009: A "major initiative in the catastrophic bond market occurred in late 2009 when the World Bank announced the launch of its catastrophe bond issuance 'MultiCat' programme."[14]
- 2015: The Seychelles' Climate Adaptation and Impact Investment Debt Swap was officially completed.[15]

Mechanisms of Instrument

According to WWF:

Two primary debt relief instruments have provided funding for the environment: commercial debt-for-nature swaps and bilateral debt-for-nature swaps. In commercial swaps, a commercial creditor sells debt owned by a foreign government at a discount on the secondary market (usually to an NGO). Historically, commercial swaps generated $117 million for conservation around the world, but they are no longer common. Bilateral debt-for-nature swaps are similar to commercial swaps, but involve 'sovereign' debt owed by one government to another rather than commercial debt owed to a bank or commercial creditor.[16]

More specifically:

One option is debt reduction or redirection. Through this option, a country can choose to make interest and/or principal payments on treated debt, in local currency, to a forest fund {or a coral reef conservation fund}. The payments remain in the country rather than being paid to {for instance} the United States. A second option is the subsidized debt-for-nature swap, through which {for instance} the U.S. government and NGOs contribute funding to reduce or cancel a portion of the eligible host country's debt. Payments on treated debt are made in local currency for conservation activities as agreed by the U.S. government, the host country government, and the NGOs. Under the third option, debt buy-back, a country may purchase one or more eligible loans in U.S. dollars at a discount in exchange for a commitment to a tropical forest {or coral reef} fund.[17]

In addition to official external debt and commercial, private external debt, one looking to structure a debt conversion could also look into a host country or company's domestic (i.e., internal) debt. Interestingly, while debt conversions are no longer commonly structured for the conservation of tropical forests, these conservation financing instruments have recently been structured for marine conservation initiatives, including in the Seychelles.[18]

Advantages of this mechanism include: "mobilization of new resources for climate change mitigation or adaptation, financial benefits for the debtor country, poverty reduction and co-benefits, and predictability of funding and potential to attract further funding."[19] Such financial benefits include a better interest rate, debt relief, and the possibility to repay the debt in local currency. It should also be noted that debt-for-nature conversions should be in line with the host countries' national development plans.

Size of Instrument

Under the Enterprise for the Americas Initiative (EAI), the United States "restructured, and in one case sold, debt equivalent to a face value of over $1 billion owed by Latin American countries; {...} Nearly $177 million in local currency for environmental, natural resource, health protection, and child development projects within debtor countries were generated from these transactions."[20]

Subsequent to the EAI was the U.S. Tropical Forest Conservation Act (TFCA). Since 1998, "$233.4 million has been used under TFCA to restructure loan agreements in 14 countries (20 transactions), and over $339.4 million will be generated for tropical forest conservation at the conclusion of these agreements."[21]

According to the UNDP:

> The value of the debt under DNS {debt-for-nature swaps} agreements surpassed US$2.6 billion from 1985-2015 and resulted in transfers of circa US$1.2 billion to conservation projects worldwide. Bilateral DNS dominate, accounting for over 93 per cent of the total. Most transactions, about US$2 billion in value, were completed in the 1990s. From 2010-2015, transactions amounted to about US$150 million. Commercial DNS accounted for approximately US$200 million in restructured debt and US$123 million in allocations to conservation. These transactions came to an end after 2000 with the number of commercial DNS declining to just two.[22]

Yet:

> TFCA was authorized to receive appropriations through FY2007, but no funds have been appropriated for the program since FY2014. TFCA is being considered for reauthorization in the 115th Congress in S. 1023. This bill would expand the purpose of TFCA to include coral reefs and authorize $20 million in appropriations annually from FY2018 to FY2021, among other things.[23]

This said, the TFCA was reauthorized and signed into law in 2019 and in the authorization, the scope was expanded to coral reef ecosystems.[24]

While at least 25 host countries have participated in forestry-related debt-for-nature conservations,[25] several countries have participated in marine-related debt-for-nature conversions including Jamaica, Madagascar, the Philippines, and the Seychelles.

Introduction to Case Studies

The following case studies will examine the Seychelles' Climate Adaptation and Impact Investment Debt Swap, Debt-for-Nature Swaps in the Philippines, and the Jamaica Debt-for-Nature Swap and the Environmental Foundation of Jamaica. The Seychelles' Climate Adaptation and Impact Investment Debt Swap had numerous first of its kind, including the first debt-for-nature swap that generated cash flow to the marine environment, the first that used impact capital, the first that was negotiated in the Paris Club, and the first to incentivize policy commitments that were tied into the legal agreement. The Philippines, which is part of the Coral Triangle, in home to one of the largest and most diverse areas of tropical coral reefs. In the late 1980s and early 1990s, WWF helped orchestrate several debt-for-nature swaps in the Philippines. The swaps agreement called for the Philippines to invest into local conservation activities, including the direct implementation and management of the El Nido Marine Sanctuary and the Tubbataha Reef National Park. The Environmental Foundation of Jamaica (EFJ) was involved with one of the first debt swaps done between the U.S. Government and a territory in the Caribbean and the very first one in the English-speaking Caribbean. Since 1993, EFJ has received ~3700 proposals, funded 1300+ projects, and these projects have received a total of USD$43.93 million.

Case Study #1: Seychelles' Climate Adaptation and Impact Investment Debt Swap

The Seychelles, located off the coast of East Africa and relatively close to Madagascar, is comprised of 115 islands and manages an EEZ of approximately 1.4 million square kilometers[26] (Fig. 13.1).

As an island nation, the Seychelles is largely dependent on fishing (i.e., approximately 30% of the nation's GDP) and on tourism (i.e., approximately 60% of GDP).[27]

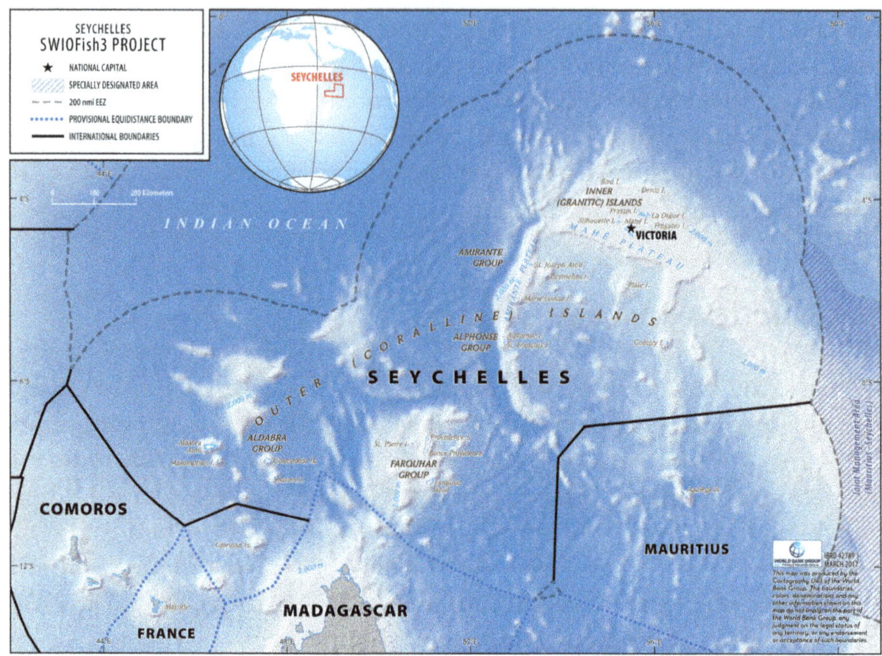

Fig. 13.1   Map of the Seychelles (*Credit* Cartography Unit, The World Bank)

In 2015, the Seychelles' Climate Adaptation and Impact Investment Debt Swap was officially completed.[28] In the arrangement, the debt swap is the financing option and the public–private trust fund known as the Seychelles' Conservation and Climate Adaptation Trust (SeyCCAT) is the transparent mechanism which gives confidence to both private investors and public officials.

With this in mind, the funders were the China Global Conservation Fund, the Jeremy and Hannelore Grantham Environmental Trust, the Lyda Hill and the Lyda Hill Foundation, the Oak Foundation, Oceans 5, the Turnbull Burnstein Family Charitable Fund, and the Waitt Foundation.[29] In addition, the main collaborators were the Governments of Belgium, France, Italy, South Africa, and the United Kingdom, along with UNDP, GEF, and the Global Island Partnership.[30]

The Seychelles debt swap had numerous first of its kind, including the first debt-for-nature swap that generated cash flow to the marine environment, the first that used impact capital, the first that was negotiated in the Paris Club, and the first to incentivize policy commitments that were tied into the legal agreement. The swap was almost the first South–South debt-for-nature swap; however, South Africa ended up pulling out in the end.[31]

This debt-for-climate-adaptation swap converts a portion of the Seychelles' debt to other countries into more manageable debt held by a local entity. To accomplish this refinancing, TNC provided $23 million in an impact capital loan and $5 million in grants to buy-back $29.6 million of Seychelles debt at a 5.4 percent discount.

The Government of Seychelles can now redirect this portion of their debt service to an independent, nationally based, public-private trust fund called the Seychelles Conservation and Climate Adaptation Trust (SeyCCAT).

Debt service payments will fund three distinct streams: one to repay impact investors, one to capitalize SeyCCAT's endowment, and one to fund work on the ground that advances marine and coastal conservation, including strategies for ecosystem-based climate adaptation and disaster risk reduction.

The estimated amount to be invested in these conservation activities and the endowment through the debt swap is $13 million, with nearly 70 percent of this payable in local currency rather than hard currency, averting the extra cost of conversion.

In addition, the period of debt payment will be extended from eight years to 20 years, reducing the government's annual debt service by over $2 million annually, freeing up funds for other needs of the citizens of Seychelles.

TNC will seek to transfer the impact investment loan into the holding of another investor and put recouped dollars into new projects, a proven approach that enables TNC to mobilize large sums quickly and extend conservation benefit globally.[32]

Identify the Problem

One significant problem, which led to the exploration of structuring a debt conversion, was that initially Seychelles had too much debt. The 2008 global financial crisis hit Seychelles particularly hard and in 2008, Seychelles was one of the most indebted countries in the world with a debt-to-GDP ratio of approximately 150%. Furthermore, Seychelles had minimal growth rates and was unable to invest in better management of their marine resources, which are extensive. Although Seychelles restructured their debt with creditors and received approximately a 40% haircut on this debt, it still left Seychelles with less then what they wanted to invest in marine resources.[33] For instance:

The Seychelles debt exchange came against the background of a severe balance of payments and public debt crisis which erupted in 2008. The authorities had pursued a decade of expansionary monetary and fiscal policies as part of their overarching development objective to double per capita income by 2017. However, persistent macroeconomic imbalances coupled with an inflexible exchange rate regime and complex exchange restrictions lead to a steady deterioration in Seychelles export competitiveness. Economic growth declined. From an average real growth rate of 7.5% in 2005-07, Seychelles growth rate fell by more than half to 3.1% in 2008. The economy shrank by 9.5% in 2009.

The global economic downturn exacerbated Seychelles weak economic performance as did the fuel and food price shocks that hit the economy in 2007-08. A growing shortage of foreign exchange prompted the authorities to liberalise the exchange regime and float the domestic currency, the rupee. The currency fell by almost 50% after the float though it regained some of its value in the following year. To sustain budgetary spending, the government borrowed in the international capital markets. Between 2006 and 2007, the authorities also issued three bonds: a US$200 million Eurobond maturing in 2011; a privately placed amortising note for €54.7 million; and a US$30 million supplement to the Eurobond.[34]

Yet,

> {…} these efforts were insufficient to stem the economy's downward spiral. During 2007-08, inflation surged, accelerating to 32% from 5.2% a year earlier, the current account widened and a substantial build-up of external payment arrears occurred. By 2008, the country's reserves were virtually exhausted and Seychelles defaulted on its bond payments due in July and October of that year.
>
> Seychelles default followed a rapid increase in the country's public debt and significant arrears accumulation. At the end of 2008, the country's public debt amounted to 151% of GDP and was estimated to have risen further to 175% at the end of 2009. Seychelles was ranked among the most highly indebted countries in the world.[35]

In addition to its debt profile, the Seychelles is also starting to witness the impacts of global climate change. Coral bleaching, as a result of warming sea surface temperatures, has impacted the Seychelles since 1998. Further, there is sea-level rise, along with increasing storm surges and saltwater intrusion.[36] Climate change is also increasing the cost of management as a result of bleaching, cyclones, and coastal erosion.

Why the Problem Is Important

The Seychelles' problem of not having sufficient financing for marine conservation, in part due to its heavy foreign debt load, is coupled with the impacts it is facing from overtourism, port construction, sea-level rise, and climate change.

These problems are important because the Seychelles is host to an extraordinary marine landscape including the Aldabra Atoll, a World Heritage Site. One unique aspect of the Aldabra Atoll is that it "retains some 152,000 giant tortoises, the world's largest population of this reptile."[37]

Other famous sites in the Seychelles include Coco Island and Anse Soleil Beach.

How Problem Was Identified

Collaboration and coordination are critical, in part, because as of 2019, the Seychelles only had approximately 100,000 inhabitants.[38]

Effectiveness of Process for Identifying Problem

It appears the process for identifying the problems facing Seychelles, along with the challenges of the debt swap, were effectively identified.

Steps Taken to Address the Problem

A series of major steps taken to address the problems facing the Seychelles and to initiate the debt swap was to undertake financial analyses. According to Dr. Andrew Rylance, Technical Advisor to the Government of Seychelles-UNDP-GEF project on Protected Area Finance,[39] one must demonstrate the business case, by proving its ability to:

- 1. Generate more revenue to finance conservation;
- 2. Improve cost-efficiency;
- 3. Strengthen ability to plan and utilize revenues effectively; and
- 4. Reinvest in infrastructure to improve attractions for tourism businesses to utilize.[40]

With respect to generating revenue, there were the following three proposed changes:

- 1. Adjust entrance fees according to evidence from willingness-to-pay surveys of tourists;
- 2. Guests staying at hotels within or bordering SNPAs MPAs {Seychelles National Parks Authority's Marine Protected Areas} were to contribute

entrance fees. The proposal was for hotel guests to pay one entrance ticket, irrespective of length of stay; and

- 3. Introduce multiple use tickets to encourage more visits per tourist.[41]

There was also management plan and business plan training, which involved developing a database of 70 business plans that were shared. This led to improved revenue, increased efficiency, and a professionalization of the business.[42]

With respect to a debt conversion, the following general steps are taken to finalize negotiations on a debt conversion:

- 1. Find a willing buyer (i.e., debtor country). This includes finalizing the conservation and policy commitments via a cabinet-level endorsement;
- 2. Find a willing seller (i.e., creditor country or commercial creditor); and
- 3. Get the financing in place (i.e., combination of grant, impact investments, and/or loan capital).[43]

Then, in order to close the transaction, the next steps include:

- 1. Create a trust fund and get the trust fund operational; and
- 2. Finalize all legal agreements.[44]

Results

There have been numerous results of the Seychelles' debt swap. This includes:

- Representatives met with Paris Club creditors, particularly France, the United Kingdom, Belgium, and Italy. The Paris Club is hosted in Paris by the French Treasury, which meets approximately ten times a year for creditor countries to meet with debtor countries.
- In 2015, the SeyCCAT was established. The Trust Fund is structured as a grant-making entity (i.e., for climate change adaptation work and for the marine spatial planning exercise), and not an implementing entity.
- Approximately USD\$9.4 million for marine and climate adaptation work (over 20 years) will be generated;
- It will also capitalize a USD\$6.6 million endowment; and
- 400,000 square kilometers (i.e., four million hectares) of new MPAs were created.[45] In fact, these 400,000 square kilometers of new MPAs is the second largest such network in the Indian Ocean as of 2019.[46]

Another major result of the Seychelles' debt swap was the marine spatial planning exercise for the Seychelles' EEZ. The marine spatial planning exercise can be envisioned as a cost–benefit tool where you lay hexagons and determine what is happening there and what are the conflicts. You then run, say 200 iterations, and those activities that are picked up 50% of the times are those that people can work together to decide what is done.

Initially, the plan was to identify where 30% of the most critical ecosystems were located within the Seychelles' EEZ and to get these specific areas into management. However, the Seychelles Government approached TNC and stated they wanted help with zoning/marine spatial planning for the entire EEZ. Since 2013, TNC has assisted with this exercise, along with the GEF, to help with:

- Mapping historical dive sites;
- Marking out maritime shipping zones and lanes;
- Zoning protected areas;
- Identifying potential areas for the deployment of renewable energy and areas for fossil fuel exploration; and
- Separating out where commercial fishing takes place (i.e., where are the pelagic tuna), where semi-industrial fishing takes place (i.e., mainly fishing for grouper and snapper), and where artisanal fishing takes place.

As of February 2019, and after five years of work, approximately 26% of the protected areas have been zoned and the partners have identified the remaining 4% of the initial 30% to be protected, which is located in nearshore water and is likely to be the most difficult. Most of the remaining 70% of the EEZ has also been zoned, including tourism areas, fishing zones, and shipping lanes.

Based off this marine spatial planning, there will be new and revised policies relating to coastal zone management and fisheries, along with other relevant marine polices. Likewise, the marine spatial planning exercise helps identify where all of the policy holes exist.

For instance, all boats are required to have electronic monitoring installed and turned on. However, if the coast guard sees a boat that does not have their monitoring turned on and is moving slowly (i.e., say 2 mph and is probably fishing), the coast guard can approach but cannot enter the boat. In this case, the boat may just say they were not fishing and claim their electronic monitoring device does not work.

Furthermore, the next step will be to cost out the implementation of monitoring all of these different zones and to put in place the appropriate regulations.[47]

It is also important to note that "this is the first comprehensive marine plan in the Western Indian Ocean. It is significant for improving ocean management for this global biodiversity hotspot, and for Seychelles' commitment to best practices in marine spatial planning. Once approved, the Seychelles marine spatial plan will be the second largest in the world."[48]

In addition to marine spatial planning, some additional activities that have been funded by the debt swap, include:

- 1. Expand and secure marine protected areas;
- 2. Coral and mangrove restoration projects;
- 3. Develop and/or reform fisheries, coastal zone management, and marine policies;
- 4. Economic diversification for affected users (sustainable fisheries, sustainable tourism); and
- 5. Social resiliency to climate change in coastal communities.[49]

A further result that arose from the Seychelles debt swap, is that the Seychelles was then able to move forward and structure the first ever blue bond, valued at USD$15 million, at the end of 2018 (see case study in Chapter 15).[50]

Challenges and How They Were Met

Some of the overall challenges for the Seychelles National Park Authority, included:

- Being under resourced;
- Being budget dependent;
- Having a lack of investment in infrastructure; and
- Entrance fees are not directed to conservation.

Entrance fees in 2015 generated approximately USD$3.9 million and extra budgetary funding was provided via GEF grants, endowment trusts, donations, and through corporate social responsibility.[51] To overcome this challenge of being underfunded, one huge result was that the Seychelles National Parks Authority was granted financial autonomy. The Authority was able to launch a new website and a new online ticket portal.[52] This online payment system, coupled with point of sale kiosks, contributed to a

"30% increase in revenue captured; increased ranger time doing conservation; {along with an} operator management system; and a customer database."[53] The Seychelles' customer database creates an option if people want to take a survey or to receive updates in the future about the park. In the future, a donation option (i.e., to become friends of the Seychelles Park or to adopt a tortoise or to adopt a reef) will be created. In addition, the customer database will allow more targeted marketing based off people's country of origin (i.e., German, French, and Italians are most common tourists), where they visited, when they visited, etc.[54]

Another challenge is that a lot of people visit the Seychelles for the beach and only a fraction visits the MPAs (i.e., which if visited, would increase the revenue for these areas). Similarly, there are not a huge number of cruise liners that visit the Seychelles, but there are a large number of people on each cruise liner. Such vessels often come from Kenya, visit the Seychelles, and then head onto Madagascar. To increase revenue from such sources, there are discussions about a cruise ship tax of USD$10 for each tourist onboard, which could be significant if you imagine 50,000 passengers. Such cruise ships do visit some of the Aldabra Atolls, along with some of the inner islands and some specific protected areas.[55]

There is also the potential challenge of overtourism, which is a conversation that is now taking place in the Seychelles. According to Dr. Rylance, with respect to overtourism, it is a matter of how visitors enter and what is their impact. For instance, if you have a good mooring system in a marine park, you have less impact with more boats than you do with less boats and no mooring system.[56] In the marine spatial planning, there will be sustainable use zones and aquaculture could be a possibility.[57]

Related to tourism, in the past there were piracy concerns, but no piracy events have recently taken place near the Seychelles.[58]

Another challenge highlighted by Dr. Rylance, was the need for "consistent, clear communication."[59] For instance, sign layouts were different in different places which were sometimes confusing and sometimes contradictory. Now, there is clear and consistent messaging on the trails and signs are made from recycled materials. In addition, the team is now working on a marketing and communication plan. This involves an increase in social media presence (e.g., YouTube, Instagram, etc.) with analytics and a targeted message, along with a diversification of media channels.[60]

Specifically, regarding the Seychelles' debt swap, some challenges included:

- The traditional first mover challenges, as this was the first debt-for-nature swap that generated cash flow to the marine environment, first that used impact capital, first that was negotiated in the Paris Club, and the first to incentivize policy commitments that were tied into the legal agreement;

- It can take a while to complete the transaction, especially if there is not a continuation of people you are working with on the transaction. For instance, you simply cannot do the deals without government buy-in. However, even with governments that do not change, people get swapped around among the different ministries or departments and oftentimes, you need to restart some of the process. Yet, at the end of the day, the government needs to make the decisions;
- Working with multiple stakeholders, as some do not have direct experience and some do not typically talk to each other. Likewise, the Initiative received input from all major sectors including commercial fishing, tourism and marine charters, biodiversity conservation, renewable energy, the port authority, maritime safety, and non-renewable resources.
- There was also a bit of a challenge of a foreign nonprofit (i.e., TNC) being involved. Likewise, in the past, debt-for-nature swaps, particularly when large-scale financing and local communities are involved, have been criticized for benefitting INGOs and not so much the local communities and debtor countries. This was mitigated via the establishment of SeyCCAT and the involvement of local NGOs. Outside of the Seychelles, the only group that has a board representative is TNC. Further, local communities were key stakeholders and were involved in the marine spatial planning;
- When creating the no-take zones, the proponents needed to identify where people can fish. There was a need to build the understanding, the trust, the relationships, and the methodology for how the stakeholders were to arrive at these decisions;
- An attempt to move away from using the phrase "debt-for-nature swaps," and instead use the phrase "debt conversions." This was because swaps— as in credit default swaps, which contributed to the 2007–2008 global financial crisis—have a bad connotation;
- Mapping out where fossil fuel exploration is to take place, which can be seen as a bit hypocritical. While TNC was not specifically advocating for fossil fuel extraction, you have to realize that Petro Seychelles, particularly as a state-owned company, needs to be at the table. Likewise, they are a stakeholder, they are part of the government, and fossil fuel extraction is a part of the Seychelles' development strategy; and
- Raising all the capital can be a challenge, and grant money is often the most difficult.[61]

There was a potential challenge associated with whether the Seychelles Government would actually use the proceeds for their originally intended use and whether the TNC faced a payback risk. To counter these potential challenges, the Seychelles signed a legal agreement with TNC essentially stating that the Seychelles Government would pay back TNC. In addition, the SeyCCAT was made under law and another governmental administration would have to change the law (i.e., would have to go to parliament) to abolish SeyCCAT.[62] Yet of the estimated USD$6 billion of debt swaps completed, there are no examples of a country defaulting on a debt-for-nature swap default (i.e., they would be defaulting on themselves). Such a default situation could possibly arise from an external shock, such as a hurricane.[63] As an additional component, the Trust will use parametric insurance to insure one year's worth of payments from the government. Thus, if a hurricane hits, the government's payments to the Trust will be relaxed and the Trust will receive the insurance payout. This will allow the government to pay for the cost of recovery, instead of making payments to the Trust.[64]

Furthermore, the TNC was expecting a 20% discount; but the Seychelles was doing very well with its macroeconomics and the discount on its debt ended up being less.[65]

Beyond Results

The Seychelles debt swap will likely be replicated and further scaled up. For instance, as of approximately April 2016, possible debt swap pipeline included:

- Grenada (USD$19.4 million in bilateral debt and USD$115 million in commercial debt);
- St. Kitts and Nevis (USD$29 million in commercial debt);
- Jamaica (USD$250 million in bilateral debt);
- Palau (USD$41 million in bilateral debt);
- The Marshall Islands (USD$27 million in bilateral debt); and
- The Federated States of Micronesia (USD$25 million in bilateral debt).[66]

There are many ongoing benefits to the Seychelles Government. For instance:

- Redirection of external debt services to investments in country: upward of USD$17 million (over 20 years), with two-thirds of this payable in local currency;

- Improved fiscal space. Maturities were extended on USD$27.3 million of debt from 8 years to 20 years on average; and
- Government entities are eligible to apply for funding from SeyCCAT.[67]

Looking beyond the results for the Seychelles and to other possible countries, similar benefits may be earned through a debt-for-nature swap. Likewise, Barbados has expressed interest in a "maritime debt-for-nature swap."[68] Furthermore, the benefits of the MPAs and the importance of investing in these MPAs are being communicated via nationally distributed posters and newspapers.[69]

Lessons Learned

There are many lessons learned from the Seychelles' debt swap and this includes:

- One must have patience. It took approximately four years to complete the Seychelles deal and hopefully future deals will go quicker;
- Scale matters. For all the time and effort needed to complete a debt swap, the deal should have a big impact and be worthwhile for everyone involved;
- The ministry of finance, especially the minister, is a key stakeholder. Historically and around the world, TNC has good working relationships with ministries of environment, particularly the Seychelles' Ministry of Environment. However, having both the Seychelles' Ministry of Finance and the Ministry of Environment working together was great. With respect to building a relationship with the Minister of Finance, they were interested in how the proceeds were to be used, but were most interested in their debt profile and their balance of payments. Likewise, the Ministry of Finance was interested in the USD$17 million of debt being redirected/reinvested into Seychelles (of which 2/3rds of the payment were to be made in local currency) and interested in the maturities extended from 8 to 20 years (i.e., which frees up USD$2 million per year)[70];
- Having broad stakeholder consensus at multiple levels is necessary and having high-level commitments is very helpful. For instance, the Seychelles Vice President was at Rio+20; and
- Overall, the concept of a marine debt-for-nature swap was very well received by SIDS, public and private donors, and impact investors.[71]

Additional lessons learned include:

- Such deals need to have an internal, government champion; and
- You need to understand how to raise significant loan capital and impact capital from commercial entities/banks, and how to de-risk (e.g., via a World Bank loan guarantee or via OPIC political risk insurance; OPIC is now the U.S. International Development Finance Corporation).[72]

Lastly, it is important to note that the President of France and the President of Seychelles were both involved (i.e., working at the highest level). Timing is also very important. The Seychelles was able to pull off the transaction right before the UNFCCC COP21 in Paris, as people seem to be more motivated if it is before a big event.[73]

According to Dr. Andrew Rylance, lessons learned include:

- There was pushback from the government for the park authority to have autonomy. There was also some pushback both inside and outside the organization about making a budget dependent entity (i.e., the Seychelles National Park Authority) into a revenue generator. Some were opposed to this concept, while others in the private sector thought the Authority should be revenue generators. In the end, the Ministry of Finance was surprisingly very supportive and helped answer their questions very well.
- Do not give an immense amount of information and lots of technical information; it is best to keep it simple.
- There are other conservation financing options such as increasing blue bonds and impact investments (i.e., which is probably one of the fastest growing areas with private capital going into sustainable business models within and/or near protected areas). Yet, there is always a concern about a pipeline of projects.
- A marketing and communication plan is very important and all the protected areas in the Seychelles received marketing training. For instance, Facebook was very important to reach residents of Seychelles (e.g., highlight problems with poaching and have teachers use the updates to share with students). Similarly, short videos are posted on YouTube that informs people about what is going on in the parks and these videos can then be shared on YouTube or via Facebook (i.e., with hotels sharing with visitors).
- Another lesson learned identified by Dr. Rylance is that no matter how much money is going into the park, if it is not being communicated well, it is not going to make a difference.[74]

Additional lessons learned that were highlighted by Dr. Rylance include:

- "Collect the evidence, communicate them clearly – to inform policy decisions, demonstrate the business case;
- Generating additional finance is just the start – how to effectively use the financing is just as important;
- Take a long-term perspective—changing the way financing conservation is viewed by all stakeholders—and not everyone will agree at the start;
- Long-term projects with external finance and technical support can help de-risk innovation by Protected Area Authorities and private businesses."[75]

Other Resources for Seychelles' Climate Adaptation and Impact Investment Debt Swap

Government of Seychelles' Conservation and Climate Adaptation Trust of Seychelles Act (2015)

- https://www.seylii.org/sc/sc/legislation/Act%2018%20of%202015%20Conservation%20and%20Climate%20Adaptation%20Trust%20of%20Seychelles%20Act%2C%202015.pdf

Marine Conservation Society Seychelles

- http://www.mcss.sc/

Programme Coordinating Unit's Protected Area Finance Project in Seychelles

- http://www.pcusey.sc/index.php/pcu-projects/ongoing/202-paf

Seychelles Conservation and Climate Adaptation Trust

- www.seyccat.org

Seychelles Island Foundation

- https://www.sif.sc/

Seychelles' Marine Spatial Planning Initiative

- http://seymsp.com/; and

- https://www.cbd.int/doc/meetings/mar/soiom-2016-01/other/soiom-2016-01-seychelles-02-en.pdf

Seychelles National Park Authority's 2017–2021 Strategic Plan and its 2019–2024 Marketing & Communications Plan

- https://www.snpa.gov.sc/index.php/downloads/documents

Seychelles' YouTube Video

- http://www.seychelles.cc/

The GEF

- https://www.thegef.org/project/seychelles-protected-areas-finance-project

TNC and NatureVest

- https://www.nature.org/en-us/about-us/who-we-are/how-we-work/finance-investing/naturevest/ocean-protection/; and
- https://www.nature.org/en-us/about-us/where-we-work/africa/stories-in-africa/seychelles-conservation-commitment-comes-to-life/

Case Study #2: Debt-for-Nature Swaps in the Philippines

Introduction

The Philippines, which is part of the Coral Triangle, is home to one of the largest and most diverse areas of tropical coral reefs (Fig. 13.2).

In the late 1980s and early 1990s, WWF helped orchestrate several debt-for-nature swaps:

From 1988-1993, WWF negotiated four commercial debt-for-nature swaps in the Philippines, which generated a total of $27.3 million in conservation funds. A large number of projects funded through debt-for-nature swaps were aimed at the conservation of marine biology. {…} In 1993, WWF (with funding provided by the U.S. Agency for International Development-USAID) was able to purchased debt owed by the Philippine government to international commercial banks that had a face value of $19 million for a price of

Fig. 13.2 Map of the Philippines (*Credit* Peter Hermes Furian)

only $13 million. In exchange for WWF's cancellation of the debt, the Philippine government allocated the equivalent of $17 million in Philippine pesos to establish a permanent endowment for the newly created Foundation for the Philippine Environment (FPE). The income earned by investing FPE's endowment has been used to make hundreds of grants to NGOs and local community groups. The Bank of Tokyo also donated debt, which was subsequently sold on the secondary debt market to generate additional funding for FPE.[76]

In addition, there were debt-for-nature swaps conducted in 1988, 1990, and 1992:

- 1988: WWF purchases debt for the benefit of the Philippines, including for the El Nido National Marine Park with Haribon Foundation and the Philippine Department of Environment and Natural Resources (DENR) as the beneficiaries.[77]
- 1990: WWF and USAID purchase debt for the benefit of the Philippines, including for the El Nido Marine Sanctuary and the Tubbataha Reef National Park with Haribon Foundation and the Philippine DENR as the beneficiaries.[78]
- 1992: WWF and the USAID purchase debt for the benefit of the Philippines, including for the El Nido Marine Sanctuary with Haribon Foundation, DENR, and the Foundation for the Philippine Environment as the beneficiaries.[79]

WWF-US, before WWF had a Philippines' office, was working with the Haribon Foundation. WWF-US and the Haribon Foundation then met with the DENR, particularly with the former president of the Haribon Foundation (who was then the undersecretary of DENR), about a proposed debt-for-nature swap. This relationship helped a lot and DENR positively responded to the idea of a debt-for-nature swap. Ultimately, WWF-US bought the Philippines debt at a 50% discount in the secondary market. There was no actual money exchange; rather, the debt-for-nature swap agreement called for the Philippines to invest the Central Bank Special Series Bills into local conservation activities, including the direct implementation and management by the Philippines Government of the El Nido Marine Sanctuary ("Sanctuary") and the Tubbataha Reef National Park ("Tubbataha"). As part of the debt-for-nature swap, another eight projects (approximately) were managed by NGOs. The subsequent debt-for-nature swaps in the Philippines built on the success of the original swap.[80]

Identify the Problem

Because of poverty, dynamite fishing, cyanide fishing, and poaching throughout the Philippines, there is a pressing problem to provide livelihoods for local communities.[81] In the Philippines, there are a lot of islands and tourism is a big focus; however, pollution and acidification are non-discriminate. While some of the coral reefs in the Philippines are quite pristine, others are very degraded and cyanide has become relatively common use.[82]

Going back to the late 1980s, there was a lack of financing as MPAs were far down the list of governmental priorities.[83]

At Tubbataha, which is "way in the middle of nowhere," the threats were less intense. This said, they are small islands and no facilities except for guard houses for the marine rangers. There are some incidences of poaching, but not as much as other places.[84] One major problem at the Tubbataha was the grounding of the USS Guardian on January 17, 2013.[85]

In the 1980s, the Sanctuary was pristine as tourism was not booming. However, since then, a lot of people discovered the place, which is a relatively short plane ride from Manilla, and because you have a choice of the beach and/or diving, there are lots of tourists. This said, there are more anthropogenic problems of tourism, waste management, and poaching at the Sanctuary. The El Nido local government is enforcing the law of waste management and encroachment of beach easements from tourism facilities such as hotels and restaurants. The problem of too much tourism in the Philippines can be perhaps best evidenced by the closing of the Philippines' Boracay Island, due to untreated waste and pollution, in April 2018.[86]

In addition to illegal fishing, illegal logging and the lack of financing were additional problems for the Sanctuary. Once the former manager of the Sanctuary, Nilda Baling, the "dragon lady," was no longer there, illegal fishing came back. The Sanctuary is one of the largest MPAs in the Philippines and the diversity of corals is still good, but some areas experienced bleaching and are starting to recover. For a while, illegal logging was controlled because there was a logging mortarium implemented about one year after Ms. Baling started. Ms. Baling lobbied for the ban due to siltation and its impact on the reefs. Now, it is disheartening, as the provincial government is undertaking a massive road construction project and there is likely to be increased siltation which will impact the reefs, mangroves, and the seagrass beds. Although the local communities understand these problems, the local provincial government does not seem to fully appreciate the impacts.[87]

Furthermore, there is an important population of dugongs (*Dugong dugon*) that feed on the sea grasses at the Sanctuary, but the sea grasses are being impacted (i.e., habitat destruction needs to be stopped) and there is poaching of dugongs. In addition, there is the tourist activity of swimming with dugongs, but the activity needs to be regulated and enforcement is always needed.[88]

Why the Problem Is Important

The Philippines is an important part of the Coral Triangle. In the words of Ms. Baling, the El Nido Marine Sanctuary is "a paradise."[89] There is an important population of dugongs at the Sanctuary, which are listed as a species vulnerable to extinction according to the IUCN Red List.[90] Regarding El Nido:

> In 1991, the Government of the Philippines proclaimed Bacuit Bay as a marine reserve. In 1998, the protected area was expanded to include terrestrial ecosystems and portions of the municipality of Taytay. It is now known as El Nido-Taytay Managed Resource Protected Area, which covers over 36,000 hectares of land and 54,000 hectares of marine waters.[91]

Tubbataha, which has relatively high SCUBA fees, is a World Heritage Marine Programme. Furthermore, numerous sites throughout the Philippines are listed as among the top fifty bioclimatic units (i.e., Palawan, Mindanao, and Negros) and the Philippines is one of seven key countries for the Coral Reef Rescue Initiative.

How Problem Was Identified

The Philippines is a mega diverse country and thus, there is a wide variety of ways to identify the problems affecting coral reefs. Most of the times, stakeholders are those who identify the problems. For instance, there are a lot of marine scientists and some of whom study dugongs. The studies are occasionally popularized to the point where the local and national news report on the studies and attention is given to the problems facing the dugongs. Similarly, the academic community—particularly those that study biodiversity—will occasionally give papers to the government. With respect to government, there are local government reports on local violations. The Philippines Government is involved with the national parks and their management plans to identify threats and incidences. Furthermore, there are a lot of local and international NGOs operating in the Philippines and these NGOs are another way of identifying the problems.[92]

Effectiveness of Process for Identifying Problem

In general, the process of identifying the problems has been effective. This said, the problem identification was quite accurate as the problems twenty years ago are the same problems today. The Philippines has a lot of diverse environmental laws, but the problem is enforcement.

With the population increasing, there is added pressure on the natural environment because people depend on these natural resources. There has to be other government programs that can assist the local communities to diversify their livelihoods. Tourism has been an answer for local fisherfolk, youth, and women as eco-guides, and because tourism has been heavily promoted by the government.[93]

With respect to the Sanctuary, there is always resistance, but you need to prove things. For instance, when you are able to catch a commercial fishing boat, which has done illegal fishing, all the fish are confiscated and given to the local communities and the community has a feast. The event is major news in the communities and has helped to get both community and government support.[94]

As explained by Ms. Nilda Baling, we bring the illegal fisherfolk (i.e., who are mainly coming from Cebu, other provinces in the Philippines, and occasionally from China) to a regional trial court. Unfortunately, there were still pending cases when Ms. Baling left, but there were no funds to bring in witnesses, policemen, etc.—so the cases might be dismissed.[95]

Steps Taken to Address the Problem

In the past, there were thirteen distinct steps to create a protected area in the Philippines. This was a process that could take up to ten years. This process generally included:

- An initial assessment of the area to identify biodiversity and biological significance (i.e., unique or endemic species are usually the basis);
- Next, go through a series of consultations with lots of back and forth with the legislative body and the President; and
- Finally, the area is then signed into law.

There is new legislation called "the Expanded National Integrated Protected Areas System (NIPAS) Act of 2018," which legislated 94 new areas be enacted into law (i.e., fast tracked). This new law is very good and now provides legal protection, because protected areas are not really protected unless enacted into law.[96]

Results

There have been a lot of results due to the debt-for-nature swaps and its impact on Tubbataha and the Sanctuary. First and foremost, the debt-for-nature swaps were successfully completed and funding was received for both MPAs. Tubbataha was declared a national marine park in 1988, the first of its kind in the Philippines,[97] and was then designated a World Heritage Site in 1993.[98] As a result of its distinction as a World Heritage Site, the government is forced to protect it due to its international recognition.[99]

With respect to the Sanctuary, environmental awareness has been heightened (i.e., it is not perfect, but a lot of people are now aware).[100] Similarly for Ms. Baling, the biggest result was getting the Sanctuary known—not just in the Philippines, but worldwide. The Sanctuary used to be such a remote, isolated area; however, now all kinds of tourists come to visit. Even outside of the country, Ms. Baling will hear others talking about the Sanctuary. Basically, Ms. Baling and her team helped "put that small place on the map."[101]

Another major result was building the capacity of the local community and getting them to know the benefits of conservation. This capacity building also includes the local government and getting other local governments to have their areas designated as MPAs. Likewise, the governments now recognize the benefits.[102]

Furthermore, developing the experience of the staff was a major milestone. The staff of the Sanctuary get invited to share their lessons on things such as how to do an apprehension. These invitations are often to other MPAs to share the challenges and issues faced regarding the management of their MPAs. The staff have established networks and even with tourists, who go back home, relate their experiences with others, and keep in contact with their guides. Similarly, the communities were trained and after the project, they were recruited into other jobs because of their specific skills. This includes guides who were trained on what to do and what not to do when diving and snorkeling, and these divers were recruited by resorts to cater to Taiwanese and Japanese tourists who need guides.[103]

Challenges and How They Were Met

Two major challenges faced at the Sanctuary were the sustainability of gains, which is the most challenging, followed by the challenge of working hard with the local communities to gain their trust.

With respect to the sustainability of gains, WWF was to pay off the debt and the Philippines Government would then come up with the equivalent in local conservation funds. The program started with ₱900,000 (Philippine Pesos) and with nine staff members. The initial focus was on a marine turtle sanctuary because the beaches were a nice nesting site for the turtles. The group then started doing education about conservation and trying to fit in with the communities. Whether the debt-for-nature swap ultimately worked good or bad, Ms. Nilda Baling is not sure, but Ms. Baling knows that they made good use of the funds, that there was good community support, and there was an expansion of the serious work of conservation. They were strong in enforcement and were attracting tourists because the corals are beautiful. Furthermore, the NIPAS Act of 2018 used El Nido's work to help inform legislation (i.e., such as the establishment of management boards) because when implementing the debt-for-nature swap, El Nido effectively started an advisor group made up of local leaders. However, there was not a sustainable plan (i.e., like long-term financing). The Sanctuary did not have a system of collecting user fees and there was no discussion of payments for ecosystem services. Implementation, in terms of conservation, was good, but after three years there was no plan for what to do next.[104]

With respect to working with the communities, it was originally hard for Ms. Baling to find her place in the community as the community was conservative and Ms. Baling was not from the communities. In fact, within three months, Ms. Baling was almost evicted. Ms. Baling earned the communities' lack of trust because Ms. Baling thought enforcement was already taking place as there was an office, but enforcement was not taking place. Instead, Ms. Baling had to initiate enforcement against illegal logging and illegal fishing. At the time, there were two big resorts in the area and Ms. Baling was questioned whether she was working on behalf of the resorts and not the communities. To help address this challenge, Ms. Baling's first assignments were to setup an office, see what was possible to accomplish, and to start recruiting staff from the local community. It was Ms. Baling's aspiration to help build the capacity of the local communities and Ms. Baling chose to not hire anyone from Manila. This said, it was very difficult to tap from the community as many people were into fishing and this was the first time they were employed in such a project.[105]

Another challenge is that it is difficult to win funding that goes into the local government. The money collected by the Sanctuary goes to the Integrated Protected Area Fund where 75% of the funds are left for the protected area and 25% of the funds go back to the National Government's Treasury so that other protected areas, which do not have as much funding, can access a part of the 25%. However, the 75% of the funds is retained by the local government and not specifically by the Sanctuary. There is now also a dispute between the national and local governments on how to divide this funding.[106]

Beyond Results

With the first debt-for-nature swap, it was the first time a significant amount of money was invested into park management in the Philippines, particularly for the Sanctuary, and this helped to improve governance and to professionalize park management. For instance, more rangers were hired, the park put in more infrastructure (i.e., staff and guard housing that provided a decent enough place to stay) and a small office.[107]

Similarly, the debt-for-nature swap was an innovative financing mechanism at the time. It was the first one in the country and it was followed by others, including one recently completed by the Philippines and Italy. The Forest Foundation Philippines was created due to US Tropical Forest Conservation Act and the Foundation for the Philippine Environment was also created. Both of these are grant funding organizations that are investing in conservation projects on a continuous basis from the interest off their endowment.[108]

Lessons Learned

According to Anabelle Plantilla, leadership (i.e., governance) is still very important. If you have poor governance, then poaching will happen left and right. Similarly, if there is a corrupt leader, he will serve his own interests. While there should be good governance in place (i.e., including stakeholder engagement, stakeholder participation, and accountability), it is difficult to achieve. However, there are pockets of success in the Philippines and these leaders need to be incentivized to replicate their types of governance.[109]

Ms. Nilda Baling has several lessons learned to share. First, from the manager of an MPA's perspective, it is not about what specific debts are being paid off with the debt-for-nature swap, but how the money is being used. Park managers do not necessarily understand the bank operations and how the exchanges of money take place; rather, what is most important to park managers on the ground is making good use of the money.

Second, if there are debt-for-nature swap programs again, the programs should look at the lessons learned. It should be very clear where the money went and how it was spent. This said, there were successes from the first debt-for-nature swaps, but it was unclear what happened to the other tranches of debt-for-nature swaps. The Philippines Environment Foundation was created and several small projects were funded from the second debt swap, but it was not clear what those projects achieved and where most of the money went as the Sanctuary did not get any of this funding. There must be trust between governments. If there will be another debt-for-nature swap program, it should also look at the Philippines' national program to see where we can fit in—as opposed to starting a new program that is different than the thrust of the Philippines Government.[110]

A third and related lesson learnt concerns the conduit of funds. Often times, the funds go first to an NGO and then the NGO disburses the funds to projects. It might be wise to look at the credibility of the NGOs when they are being used as a conduit. Some financial institutions give directly to the NGOs and the NGOs develop their own system of management. Instead, there should be close coordination between the national government and the NGO so there is not a scenario where the NGO just does it their own way (i.e., as opposed to looking at how they can help the national government pursue projects they cannot due to a lack of funding). Likewise, a responsible NGO should be chosen that works together with the national government to be more aligned, coordinate together, and specify specific roles and responsibilities with sustained financial plans in place.[111]

One more lesson learned shared by Ms. Baling is that we need good managers in protected areas. Women can be better, or good, managers of protected areas as they are often more into details and have a mother's instinct.[112]

Other Resources for Debt-for-Nature Swaps in the Philippines

Atlas of Biodiversity Conservation in the Coral Triangle

- http://uoa.maps.arcgis.com/apps/MinimalGallery/index.html?appid=e0ccfa15b54d49de9ea6a8db2079d8ac

El Nido Resorts

- https://www.elnidoresorts.com/

Foundation for the Philippine Environment

- https://fpe.ph/

<u>Philippines' Department of Environment and Natural Resources</u>

- http://mimaropa.denr.gov.ph/index.php/el-nido-taytay-managed-res
  ource-protected-area

<u>The Forest Foundation Philippines</u>

- http://new.forestfoundation.ph/; and
- http://www.forestfoundation.ph/our-story/

<u>Tubbataha Management Office</u>

- http://www.tubbatahareef.org/home

<u>World Heritage Site</u>

- El Nido-Taytay Managed Resource Protected Area: https://whc.unesco.
  org/en/tentativelists/5034/; and
- Tubbataha Reefs Natural Park: https://whc.unesco.org/en/list/653/

<u>WWF</u>

- http://wwf.panda.org/knowledge_hub/where_we_work/coraltriangle/cor
  altrianglefacts/places/tubbatahareefphilippines/

Case Study #3: Jamaica Debt-for-Nature Swap and the Environmental Foundation of Jamaica

Introduction

The Environmental Foundation of Jamaica (EFJ) was involved with one of the first debt swaps done between the U.S. Government and a territory in the Caribbean and the very first one in the English-speaking Caribbean (Fig. 13.3).

There were initial discussions in the 1990s, followed by a 1992 indicative agreement, and then EFJ was started in 1993. The first thing was that it was an exchange for PL480 funds, primarily dealing with agriculture. The first tranche of debt identified was just over USD$9 million. However, before all of this could be passed in the U.S. Congress, something needed to be

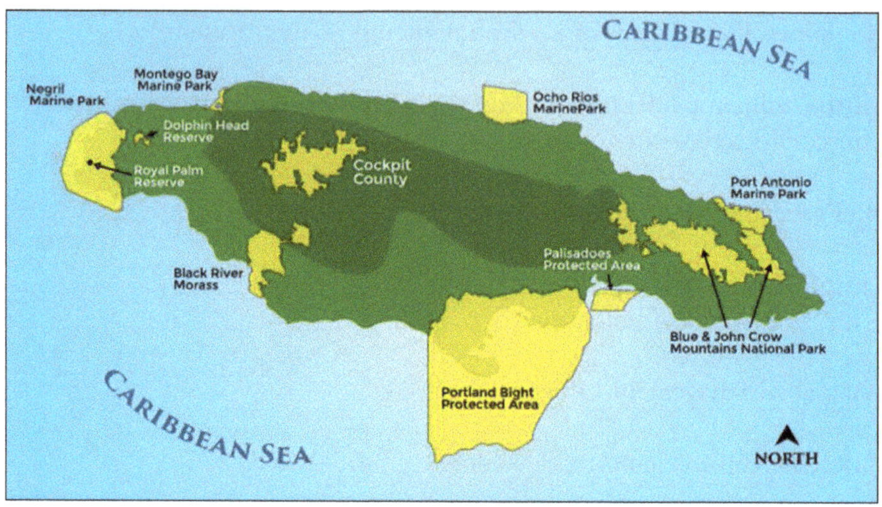

Fig. 13.3 Map of Jamaica (*Credit* Environmental Foundation of Jamaica)

specifically done for child survival and child development. Another tranche, totaling just over USD$12 million of USAID debt was identified, and this brought the total to just over USD$21 million that flowed into EFJ. The original financing was structured from 1993 to 2012, making the nature of the agreement approximately twenty years. Initially, the structure was a sinking fund that involved two separate income streams twice a year, with large sums deposited into the fund in the early years and gradually the amounts tapered off over time. EFJ was given the mandate to do environmental and social issues (i.e., child survival and child development) and broadly looked at conservation and sustainability.[113] One of many initiatives of the EFJ was to work with the Negril Marine Park, along with coastal zone and marine planning, and a queen conch (*Strombus gigas*) survey.

Technically, due to the nature of the sinking fund, EFJ should have closed its doors in 2012. However, EFJ merged with the Jamaica Protected Areas Trust (JPAT) which was established as a result of TNC and USAID under the Tropical Forest Conservation Act. The mandate required the JPAT had to look at forestry issues and the funds established JPAT for the years 2004–2024 (i.e., twenty years).[114]

EFJ is the largest grant-maker in Jamaica and works to catalyze on-the-ground change in environmental conservation and child survival and development. Born out of a creative "debt-swap" between the Governments of the United States and Jamaica, the EFJ received approximately US$21.39 million between 1993 and 2012 under this arrangement. In twenty four years (1993-2017), through a strong track record of financial and grant accountability systems, and impressive returns on investments, the EFJ has taken a corpus of US$21.39 million {…}. This has resulted in the EFJ having a significant impact on lives in Jamaica and the region.[115]

About 60% of the grants went to the environmental category. The maximum grant cap was USD$100,000 and the grants were normally given for twelve months.[116]

On a related note, in 1991, TNC, USAID, and the Puerto Rican Conservation Trust did a debt-for-nature swap for an "endowment fund for JNPT {Jamaica Nationals Parks Trust}, a conservation trust fund dedicated to the conservation of biological diversity through support of Jamaica's national parks system: Montego Bay Marine Park and Blue and John Crow Mountains National Park."[117] However, the JNPT was on a very small scale and is not the same as the JPAT or the EFJ.

Identify the Problem

Larger problems that faced Jamaica were overfishing, impacts from tourism, and natural disasters. The problem was so bad that:

After a series of natural and man-made disasters in the 1980s and 1990s, Jamaica lost 85% of its once-bountiful coral reefs. Meanwhile, fish catches declined to a sixth of what they had been in the 1950s, pushing families that depend on seafood closer to poverty. Many scientists thought that most of Jamaica's coral reef had been permanently replaced by seaweed, like a jungle overtaking a ruined cathedral.[118]

With respect to overfishing:

In the mid-20th century, fishing was one of Jamaica's key economic drivers. In a bid to expand the market, during the 1950s and 1960s, the government sponsored a boat mechanisation scheme to subsidise outboard engines and fuel, while also developing training programmes for fishermen in the use of improved equipment. These initiatives increased landings by some 50 percent. By the 1960s, Jamaica's fisheries sector was thriving, but with this came the inevitable adverse effects. According to the report Forgotten Fish: Jamaica's reef history, by 1969 grunts and large snappers were rare around shallow reefs, meaning that fishermen had to travel numerous miles to secure a profitable catch. After years of overfishing – together with other unsustainable practices, such as dynamite and longline fishing – Jamaica's fish stocks had declined massively. Additional government subsidies in the 1970s enabled fishermen to continue their activities; however, in spite of this, their catches continued to become smaller with each passing year.[119]

Similarly:

Population pressure in the 1960s led to an increased demand for food, which was in large part satisfied from the abundant life found on fringing reefs. The Jamaican fisheries began to overexploit the resource so that by the beginning of the 1970s the biomass had been reduced by 80 per cent. Composition of reef species changed (an indicator of overfishing) so that sea urchins became the dominant grazers on algae rather than herbivorous fish. This at least allowed the coral to remain dominant. Nevertheless, the change in composition of reef species, caused by the removal of competing herbivorous fish and urchin predators (such as trigger fish) created an over-reliance upon sea urchins to maintain that coral dominance.

Corals continued to survive whilst the sea urchins kept the competing algae in check. However, a water-born disease that appeared in 1982 spread through the sea urchin population, reducing numbers by 99 per cent. Consequently, algae came to dominate the shallow waters. Coral cover along the Jamaican coastline fell from 52 per cent in 1977, to 3 per cent by 1993, with algal cover increasing by up to 92 per cent at the end of the same period. Overfishing had ultimately caused the loss of the majority of Jamaica's coral reef ecosystems.[120]

Another major problem facing Jamaica is the impact from tourism:

The story of Jamaica's world-renowned tourism industry starts with one individual: Abraham 'Abe' Elias Issa. Considered the 'father of Jamaican tourism', in 1949, Issa opened the Tower Isle Hotel – the country's first all-season resort. Located just outside Ocho Rio – back then, a small fishing town – the hotel swiftly rose to fame, luring Hollywood stars, sporting legends and even British royalty.[121]

Since then, Jamaica has developed a large tourism footprint and hosts one of the largest cruise liner ports in the world.[122] This, in part, has contributed to the degradation of Jamaica's coral reefs:

The construction and subsequent presence of large-scale hotel resorts has further exacerbated the problem. Coastal developments have contributed to the country's rapid destruction of coral reefs and mangrove forests, while also causing other negative effects, such as increased sedimentation. Poor water quality, thanks to solid waste, chemical spills, nitrification and thermal discharges, has also disturbed marine life. Add in other factors, including public disregard and physical damage from marine users, as cited in a 2017 World Bank report, and the situation has reached dire straits.[123]

Natural disasters, particularly hurricanes, have also been a problem for the people and reefs of Jamaica:

The first calamity was 1980's Hurricane Allen, one of the most powerful cyclones in recorded history. 'Its 40-foot waves crashed against the shore and basically chewed up the reef,' Gayle says. Coral can grow back after natural disasters, but only when given a chance to recover – which it never got. That same decade, a mysterious epidemic killed more than 95% of the black sea urchins in the Caribbean, while overfishing ravaged fish populations. And surging waste from the island's growing human population, which nearly doubled between 1960 and 2010, released chemicals and nutrients into the water that spur faster algae growth. The result: Seaweed and algae took over.[124]

Why the Problem Is Important

The aforementioned problems are important because local communities in Jamaica are dependent on the sea, particularly for fishing. In addition, Jamaica generates significant revenue from tourism, in part due to tourists' desire to see coral reefs.

How Problem Was Identified

The aforementioned problems have been identified by researchers since the 1950s. Likewise, "studied continuously since the 1950s, Jamaican reefs are probably some of the best recorded in the world and local human impacts some of the best studied over a long period of time."[125]

Effectiveness of Process for Identifying Problem

The process appears to be quite effective, especially due to the longitudinal nature of the studies.

Steps Taken to Address the Problem

There were numerous steps taken to address the problems of overfishing, impacts from tourism, and natural disasters.

In 1965, Jamaica "became the site of the first global research hub for coral reefs, the Discovery Bay Marine Lab, now associated with the University of the West Indies."[126] In 1993, Jamaica and the United States entered into the debt-for-nature swap that led to the creation of the EFJ:

> In 1993, Jamaica became the first country in the Caribbean and Latin America to sign the EAI {Enterprise for the Americas Initiative} agreement. An essential component of the EAI agreement is that programmes be undertaken by non-governmental and community based organisations, and not by the state, with the intention of building a stronger civil society capacity for sustainable development.[127]

Decades later, "in 2012, the governments of Jamaica and the U.S. gave consent for the consolidation of the operations of the EFJ and the Forest Conservation Fund (FCF). The Forest Conservation Fund was established pursuant to the Tropical Forest Conservation Act (TFCA) of 1998."[128]

Another series of steps taken by Jamaica involved coral restoration.

> In Jamaica, more than a dozen grassroots-run coral nurseries and fish sanctuaries have sprung up in the past decade, supported by small grants from foundations, local businesses such as hotels and scuba clinics, and the Jamaican government. At White River Fish Sanctuary {...}, the clearest proof of early success is the return of tropical fish that inhabit the reefs – as well as hungry pelicans, skimming the surface of the water to feed on them.[129]

However, it is important to understand that:

> The delicate labor of the coral gardener is only one part of restoring a reef – and for all its intricacy, it's actually the most straightforward part. Convincing lifelong fishermen to curtail when and where they fish and controlling the surging waste dumped into the ocean are trickier endeavors.[130]

Results

In addition to structuring the debt-for-nature swap and managing funds, a major result of the EFJ has been grant giving. In fact, since 1993 (as of April 2020), EFJ received 3699 proposals, funded 1321 projects, and these "1,321 projects have received a total of US$43.93 million or J$2.52 billion."[131] For instance, this includes:

- In 2016, UWI and Discovery Bay Marine Laboratory were awarded USD$5 million for "establishing one raceway system for growing fragmented coral for ecosystem restoration with flowing water maintained by renewable energy"[132]; and
- Also, in 2016, White River Marine Association Limited was awarded USD$4,084,700 for "protecting and rebuilding reefs in a community-led project to counter effects of climate change."[133]

Additional recent interventions by EFJ have been around climate change adaptation (i.e., EFJ does not do climate change mitigation funding). More specifically, about USD$4.3 million has been granted to 135 climate change adaptation projects over three years. For instance, EFJ has funded about 40–50 greenhouses which are each producing 3–4 climate resilient crops (i.e., such as cassava) and farmers must have a sustainability plan.[134]

In addition, EFJ has funded approximately 30–40 projects around beekeeping over the last few years. Such alternative livelihoods can help local communities reduce their relience on fishing, to not cut down trees, and to reduce the practice of slash-and-burn agriculture. Rather, local communities plant flowering trees that are endemic to Jamaica and can manage bee boxes.[135]

Water harvesting is another series of projects funded by EFJ. This includes small water system projects, such as helping schools impacted by the drought in Jamaica with harvesting rainwater at the schools. EFJ has also provided funding to farmers to create small water catchments (e.g., small ponds and small dams) and have provided funding for solar pumps to pump out the collected water, along with funding for liners for the ponds and for hoses to take the water from ponds to distribute to adjacent farms.[136]

Furthermore, a few years ago and from which people are still reaping benefits, EFJ provided funding to fish sanctuaries. This includes funding to the Oracabessa Bay Fish Sanctuary in St. Mary, the White River Fish Sanctuary in St. Ann, and to the Caribbean Coastal Area Management Foundation (C-CAM) for the Portland Bight Protected Area. The model is for local fisherfolk to police the areas themselves.[137] Likewise, the "Oracabessa Bay Fish Sanctuary was the first of the grassroots-led efforts to revive Jamaica's coral reefs. Its sanctuary was legally incorporated in 2010, and its approach of enlisting local fisherfolk as patrols became a model for other regions."[138] The Oracabessa Bay Fish Sanctuary has had some notable successes:

In 2008, the Oracabessa Foundation teamed up with the St Mary Fishermen's Cooperative to create the Oracabessa Bay Fish Sanctuary, a no-fishing zone that stretches along the length of the town that was legally recognised in 2010.

Being a prominent hotelier in the area, and thanks also to his celebrity links, Blackwell has provided considerable assistance to Oracabessa, raising funds and awareness while also revitalising the marine area in the protected zone. But thanks to the involvement of local fishermen in particular, the sanctuary has shown impressive results: a 1,800 percent increase in fish biomass in the past seven years, along with a 150 percent increase in coral coverage. Oracabessa has inspired the fishermen of other areas to partner up with local hotels and conservationists to follow suit. Today, Jamaica has 17 Special Fishery Conservation Areas in total, and while their management and effectiveness varies, they are all showing promising results.[139]

Overall, the financing has been from the combination of both EFJ's principle and interest. There have been additional conversations, collaborations, and funding from partners such as the Jamaica Government (i.e., for EFJ's climate change adaptation work) and the Inter-American Development Bank.[140]

Challenges and How They Were Met

One challenge for EFJ, as an organization, is that there was not a lot of data collection done in the early years. Nowadays, there is more data, but it lacks the complete picture of the impact because not all of the data was captured. Similarly, figures since 2010 are not representative of all the work done as most of the funding, and the bulk of the money spent, was from 1999 to 2009.[141]

Another challenge has been the monitoring and enforcement of no-fishing zones:

Two years ago, the fishermen joined with local businesses, including hotel owners, to form a marine association and negotiate the boundaries for a no-fishing zone stretching two miles along the coast. A simple line in the water is hardly a deterrent, however – to make the boundary meaningful, it must be enforced. Today, the local fishermen take turns patrolling the boundary in the Interceptor. {...} Most of the older and more established fishermen, who own boats and set out lines and wire cages, have come to accept the no-fishing zone. Besides, the risk of having their equipment confiscated is too great. But not everyone is on board. Some younger men hunt with lightweight spearguns, swimming out to sea and firing at close-range. These men – some of them poor and with few options – are the most likely trespassers.[142]

Yet another challenge is with tourism, which can impact the local reefs and have limited economic benefits to the wider, local communities:

The all-inclusive-resort model, while extremely successful in attracting visitors to Jamaica, has an unfortunately limited impact on the local economy. With all food and drink included in a holiday package, an array of restaurants to pick from and a variety of in-house activities to sample, most tourists simply don't venture outside of their resort throughout their entire stay.[143]

According to Mr. Barrington Lewis, CEO of the EFJ, it took some time (i.e., approximately 2–3 years) to arrive at the final agreement. Because Jamaica had so much debt, it was not a big challenge to identify the debt. Also, there were no exchange rate issues as the exchanges were did in US dollars and the exchanges were conducted at a floating exchange rate. However, one major shortcoming of EFJ was how it was originally set up as a sinking fund. This said, there were many foundations that were started after the EFJ and because of the good work that EFJ did, but these foundations would likely suffer if EFJ was no longer in existence. While EFJ does not do the implementation of the work, but rather manage the funds, EFJ is still the largest fund maker in Jamaica. As explained by Mr. Lewis, when EFJ had very good years, they should have set up an endowment fund in order to grow the endowment to a sizable amount which would have enabled EFJ to make grants in perpetuity. In 2004, the EFJ board made an endowment fund on paper. However, it was not until 2011, when the endowment was first seeded. Being two years out from the last scheduled payment of the debt-for-nature swap, the endowment funding started too late. In contrast, the Jamaica Protected Areas Trust (JPAT), the organization EFJ merged with, set up an endowment fund right in the beginning with 50% put toward grants and 50% put toward administrative costs and funding the endowment. This model worked. Over time, JPAT was able to continue doing its work and now with the merger, EFJ will not be closing any time soon as there is a USD$6 million endowment in place and the partners are looking to raise additional funds.[144]

Beyond Results

Some of the work appears to be sustainable. For instance, EFJ is working closely with the White River Marine Park where partners are doing coral restoration near Ocho Rios. EFJ is actively, currently working with local partners and has provided approximately 50–60% of the funding. Local partners are working with USAID to take the initiative a step further. This has been successful as the area is now protected, overfishing reduced, and local partners are able to plant corals.[145]

Yet, "the impending expansion of the tourism industry is a fine example of this predicament. The government's 5x5x5 agenda involves a target of five million visitors generating \$4bn by 2021, which is tied with the goal of creating 15,000 new hotel rooms and 125,000 direct jobs. But with greater volumes come greater risks to the environment."[146]

Lessons Learned

Mr. Barrington Lewis, the former accountant for EFJ who was later promoted to CEO, has a lot of lessons learned to share. This includes:

- Establishing an endowment fund is an excellent idea;
- Capturing data is absolutely key, so the impact of the work can be showcased and particularly to the general public. This was previously a tremendous shortcoming;
- One recurring decimal is how to help develop the capacity of the organization with which you work, especially the smaller organizations. People may receive training, but then these people move on. Instead, you need to find a mechanism to help the organization to establish the project and to keep the project going. The original negotiation was that EFJ would only provide a specific amount to cover the organization's operating expenses. However, EFJ needs to look at how to do this;
- For the community and the local area where you are working, there absolutely needs to be buy-in. The community needs to own the project;
- Sustainability needs to be looked at seriously and it should be ensured that there is some financial sustainability plan in place. Likewise, these organizations need to have a combination of strategic and/or marketing plans in place;
- Governance structure is absolutely key, even if you are getting funding from multiple sources. Mr. Lewis' belief is that your board should always be skewed toward the NGO sector so that the NGO sector is heard. For instance, if your board is comprised of nine members, then for example, five of the nine seats should be from the NGO sector and four seats should be from other sectors (i.e., government, private sector, etc.). This helps to ensure the NGOs' interests are better served and it also gives credence to the donors that the cause they are funding is being served; and
- The final thing to add is that they need to have ownership of the project and something as simple has having sweat equity—I will put this funding in, but it is important for them to have co-funding and sweat equity.[147]

Other Resources for Jamaica Debt-for-Nature Swap and the Environmental Foundation of Jamaica

Enterprise for the Americas Initiative's Agreement

- http://www.efj.org.jm/wp-content/uploads/2018/01/EAIAgreement1.pdf

EFJ

- http://www.efj.org.jm/

Caribbean Coastal Area Management Foundation

- https://ccam.org.jm/research-conservation/projects/

International Coral Reef Initiative

- https://www.icriforum.org/about-icri/members-networks/jamaica

Jamaica Conservation & Development Trust

- https://www.jcdt.org.jm/

Negril area Environmental Protection Trust

- https://nept.wordpress.com/about/

Oracabessa Bay Fish Sanctuary

- http://www.oracabessafishsanctuary.org/

White River Fish Sanctuary

- https://whiteriverfishsanctuary.reefsupport.org/

Financial Analysis

The following financial analysis will look at return and risk.

Return

While the forgiver of debt may receive only a portion of the debt's face value, the forgiver may still receive some payment. This payment is better than if the borrower defaulted on the entire debt. In addition, the forgiver of debt may reduce its costs (e.g., legal and managerial costs) going forward that are associated with servicing the debt. The entity that has its debt relieved will generate a return based off the canceled debt's face value, including future interest payments that would have been due. Often, this forgiven debt is denominated in US dollars or Euros, and the entity that has its debt forgiven will have funds denominated in a local currency available for conservation activities. For NGOs and on-the-ground practitioners, this source of funds might be in addition to their other traditional revenue streams.

Risk

Disadvantages of this mechanism, particularly as a debt-for-climate conversion, include: "inverse relation between debt for climate swap potential and governance quality, transaction costs, budget squeeze, risk of inflation, and perceived sovereignty issues."[148]

There are numerous risks associated with such conversions with respect to the entity forgiving the debt and with respect to the entity whose debt is being forgiven.

Business Risk

Depending on the size of the debt conversion, there could be significant impacts to the income statement and balance sheet for the entity forgiving the debt. In addition, choosing a trusted, in-country facilitator of the debt conversion is a critical business risk.

Strategic Risk

From a debtor country's perspective, one strategic risk is:

> Large shares of foreign currency debt leave governments particularly vulnerable to depreciation of the local currency as well as cross-currency volatility. {...}
> What is desirable, therefore, in terms of a safe debt structure is for governments

to hold a significant share of long-term domestic currency debt in their portfolio. {…} Sharp rises in debt levels are frequently associated with shocks arising from adverse terms of trade, declines in imports by trading partners, adverse developments in commodity prices, weak growth and natural disasters.[149]

Furthermore, how the debt conversion is structured involves strategic risk. This includes structuring the amount to be forgiven, choosing the strategic partners (e.g., local or international NGOs, host country's finance and environment ministries, etc.) and the timing of the deal (e.g., with respect to currency exchanges and interest rates). For instance, "'many of these countries' debt is in the junk arena, so trying to raise money at lower interest rates with a discount is a challenge," says Weary. 'We need to have credit enhancements – perhaps by using the World Bank to guarantee the loans.'"[150]

From the sum of money raised by the debt conversion, there is a strategic risk associated with identifying the best projects within the most threatened biomes in the host country.

Reputation Risk

To most significant reputation risk could occur if the:

> {…} revenue from debt relief arrangements may not actually be put toward its originally intended use. To mitigate these risks, debt relief agreements should be designed to minimize market vulnerability and provide for contingency action in the event such a situation arises. Agreements can be struck with the government in advance to use a trust fund to manage the proceeds from the swap. Also, while debt relief for the environment usually supports a range of conservation priorities, agreements can be structured to promote the conservation of specific species, habitats or protected areas.[151]

Such a situation presents a reputation risk to both the entity forgiving the debt and the entity whose debt is forgiven. Further, this situation would likely result in fewer, if any, future debt conversions between the parties.

There is also a reputational risk for the debtor country if a "moral hazard" situation arises, where the debtor country completes a debt conversion, but shortly thereafter experiences an overall increase in its debt.[152]

There is a potential reputational risk if the debt being forgiven was originally taken out by corrupt officials.

With respect to the TFCA, there was "false media reports that warned that the U.S. Government would retain control over forests involved in the agreement."[153] Furthermore,

> Those who oppose debt-for-nature transactions often argue that they are not adequately enforced by debtor countries, generate insufficient funds to improve environmental problems, and may infringe on national sovereignty. Three-party debt transactions have historically had weak enforcement mechanisms; however, bilateral debt transactions such as those conducted under the EAI generally include safeguards and default provisions to protect the U.S. government from losing funds. National sovereignty became an issue with the first debt-for-nature swap in Bolivia when a conservation organization was reported to have obtained title to forested lands. There was a public outcry and ensuing political crisis when the Bolivian people thought a large part of their country had been given to a foreign organization. Consequently, conservation organizations involved in recent three-party transactions have generally refrained from directly buying land in debtor countries with conservation funds earned from debt-for-nature transactions.[154]

Liquidity Risk

It is difficult to issue debt conversions because relatively few marine debt conversions have occurred. This creates a lack of institutional knowledge with structuring the conversion and this poses a liquidity risk because few people know how to execute the deals. Furthermore, structuring debt relief for a distressed entity—whether a government or company—can also pose a liquidity risk. Liquidity issues could arise if there is an external shock, such as a global recession or a significant natural disaster.

Operational Risk

Operational risks related to debt conversions include having the technical and managerial experience to execute a debt conversion. Another operational risk is that for large companies and finance ministries, it may take too much time and money to develop a relatively small debt conversion. Similarly, there is the need to manage multiple stakeholders, possibly including multiple bond-holders, and there could be some bondholders who hold out. In contrast, the debt conversion may be too large for the in-country facilitator, especially if it is a local NGO, which again poses operational risk.

There also needs to be a monitoring component to ensure the proper use of funds, which could include a third-party trust fund whose financial statements are annually audited. Similarly, one must ask whether the debtor country has good financial record keeping.

Legal and Regulatory Risk

There are several legal and regulatory risks that could arise when structuring debt conversions. First, debt conversions can involve complicated negotiations between governments and across multiple departments or ministries. In addition, there is the chance that some government bureaucrats in the host country will not want to have NGOs telling them what to do and how to spend the money on conservation matters.

Depending on the jurisdiction of the debt and the parties involved, there are possible legal and regulatory risk. Government-to-government debt conversions will need to clear regulatory hurdles, which can involve bureaucracy and time delays. This is especially true if political parties change during the transaction and there are new appointees to the responsible departments or ministries. Likewise, another risk could occur if the debt conversion gets suspended by the debtor country.

Legal risks include legally closing out the debt, potentially writing a new contract or addendum related to debt forgiveness, and any legal ramifications associated with how the forgiven debt is to be used for conservation.

Credit Risk

Credit risk includes default risk, bankruptcy risk, downgrade risk, and settlement risk.

Debt conversions are likely structured for debt that faces a potential default, bankruptcy, or downgrade. This can jeopardize the closing of a transaction, especially if not completed in time.

Market Risk

Market risk includes interest rate risk, equity price risk, foreign exchange risk, and commodity price risk.

Debt conversions can generate "millions of local dollars for conservation but contains some risk. Local currency devaluation and inflation can reduce or even eliminate the cash value of the conservation commitment."[155]

Risk, Return, Time (Horizon), Taxes, Liquidity, Legal, and Unique (RRTTLLU)

Risk and Return

Please see above for the risk and return associated with debt conversions.

Time Horizon

The time horizon for debt conversions depend on the entities involved. The entity forgiving the debt—whether a commercial entity or a foreign government—is likely looking at a shorter time frame. In contrast, the government receiving the debt forgiveness and especially the on-the-ground entities responsible for implementing the terms of the debt conversion, are likely looking at longer time frames.

Taxes

There are likely limited, if any, taxes due to the government that is forgiving the debt or to the government that is receiving debt forgiveness. Commercial entities may be required to pay taxes, such as taxes on capital gains.

Liquidity

Distressed debt, including debt that is structured for a debt conversion, is likely to be less liquid than other traditional forms of debt.

Legal

Legal considerations associated with debt conversions will vary from country to country. It is important to consider both the legal framework of the host country receiving the debt forgiveness and the legal framework for the host country (i.e., whether a government involved in the swap, or where the company is headquartered) forgiving the debt, along with where the trust fund (if applicable) is registered.

Unique

Relatively few marine debt conversions have been executed. The conservation component of the debt conversions is unique with each transaction. Thus, it is important to identify the unique aspects (e.g., the specific in-country facilitator) of the various projects, which could include a combination of marine and/or forest conservation projects.

Policy Analysis

The following policy analysis will look at: defining the problem; establishing goals; selecting a policy; implementing a policy; and evaluating the policy.

Defining the Problem

As noted in the *Innovations for Coral Finance* report, according to the CBD:

> 13 creditor countries and 31 debtor countries have been involved in DFNS, while only one case of marine DFNS has been implemented so far, in the Seychelles. {...} Involvement of key actors and the financial and environmental guarantees they bring to the negotiations, as well as the evolution of both the international financial context and the national debt structure, are among the criteria to be considered when it comes to DFNS implementation. {...} In 2015, UNDP found that 15 of the 39 SIDS that are UN members had public debt-to-GDP ratios higher than 60%. Four SIDS – most of them in the Caribbean – reported debt-to-GDP ratios above 100%. Identifying those countries that have pre-existing commitments in terms of nature conservation, as well as international NGOs willing to get involved as a third party, and foundations ready to commit to granting a DFNS, could be the first steps towards generalizing such process. However, the global financial context, as well as the evolution of the national public budget in the short term, are to be closely watched, as they may impact both the process launch and the conduct and outcome of negotiations.[156]

Thus, one of the essential problems with debt-for-nature conversions is that you often need to find a host government with distressed debt.

Establishing Goals

A few of the public policy goals are:

- Enabling legislation for both the buyer and "seller" of debt; and
- Clear policies in place that the forgiven debt will be used specifically for the intended purposes (e.g., establishing or expanding an MPA, establishment of dedicated trust fund, etc.).

In addition, for a debt-for-nature conversion to be more easily structured, it is important to have a strong rule of law, strong in-country institutions (including the financial infrastructure), a robust civil society, and transparency over where the funds will be used.

High external debt might reveal opportunities for debt-for-nature conversions. However, some smaller countries with lower external debt levels, such as Belize (USD$1.33 billion), may present better opportunities for debt-for-nature conversions.[157]

Regarding the following classification, a country's income level was determined by the World Bank's 2019 fiscal year,[158] the list of heavily indebted poor countries (HIPC) designation is as of December 2019,[159] and the list of least developed countries (LDC) is as of December 2018.[160] The external debt stocks (total in 2018, DOD, current US$),[161] and the total debt service (% of exports of goods, services and primary income) were derived from the World Bank Group.[162] The change in the Corruption Perceptions Index from 2015 versus 2018 was derived from Transparency International.[163] The 2017 GAIN Country Index is from the Notre Dame Global Adaptation Initiative and is defined as "the ND-GAIN Country Index summarizes a country's vulnerability to climate change and other global challenges in combination with its readiness to improve resilience. It aims to help governments, businesses and communities better prioritize investments for a more efficient response to the immediate global challenges ahead"[164] (Table 13.1).

Further, programs under the U.S. Enterprise for the Americas Initiative (EAI) and Tropical Forest Conservation Act (TFCA) required eligible countries to:

- 1. Have a democratically elected government;
- 2. Not support terrorism;
- 3. Not fail to cooperate with the United States on drug control; and
- 4. Not engage in gross violations of human rights.[165,166]

Table 13.1 Country ranking on debt, Corruption and Adaptation (*Credit* Brian McFarland, Adopted in part from Warland and Michaelowa)

	Country	Classification	External debt stocks (total in 2018, DOD, current US$)	Debt/exports (%)	Change in Corruption Perceptions Index (2015 vs. 2018)	2017 GAIN Country Index from Notre Dame Global Adaptation Initiative
1	Australia	High-income	N/A	N/A	−2 (79 to 77)	71.0 (Ranked #5)
2	Bahamas	High-income; SIDS	N/A	N/A	N/A (65 in 2018)	52.7 (Ranked #66)
3	Egypt	Lower-middle income	98,704,878,246.4	15.0 (2018)	−1 (36 to 35)	46.1 (Ranked #101)
4	Eritrea	Low-income; HIPC; LDC	791,223,131.9	4.3 (2000)	+6 (18 to 24)	26.3 (Ranked #179)
5	Fiji	Upper-middle income; SIDS	851,797,658.2	2.0 (2018)	N/A	48.8 (Ranked #87)
6	France	High-income	N/A	N/A	+2 (70 to 72)	66.6 (Ranked #17)
7	India	Lower-middle income	521,390,564,017.4	11.4 (2018)	+3 (38 to 41)	42.2 (Ranked #120)
8	Indonesia	Lower-middle income	369,840,320,430.1	10.9 (2018)	+2 (36 to 38)	45.9 (Ranked #102)
9	Malaysia	Upper-middle income	N/A	N/A	−3 (50 to 47)	57.3 (Ranked #42)
10	Maldives	Upper-middle income; SIDS	2,332,419,859.0	9.2 (2018)	N/A (31 in 2018)	40.5 (Ranked #126)
11	Marshall Islands	Upper-middle income; SIDS	N/A	N/A	N/A	N/A

(continued)

Table 13.1   (continued)

Country	Classification	External debt stocks (total in 2018, DOD, current US$)	Debt/exports (%)	Change in Corruption Perceptions Index (2015 vs. 2018)	2017 GAIN Country Index from Notre Dame Global Adaptation Initiative
12 Micronesia	Lower-middle income; SIDS	N/A	N/A	N/A	36.5 (Ranked #152)
13 Papua New Guinea	Lower-middle income; SIDS	17,718,401,681.9	26.1 (2018)	+3 (25 to 28)	35.2 (Ranked #161)
14 Philippines	Lower-middle income	78,824,362,849.6	8.7 (2018)	+1 (35 to 36)	43.1 (Ranked #112)
15 Saudi Arabia	High-income	N/A	N/A	−3 (52 to 49)	54.3 (Ranked #56)
16 Solomon Islands	Lower-middle income; LDC; SIDS	388,511,174.2	5.6 (2018)	N/A (44 in 2018)	38.1 (Ranked #144)
17 Tanzania	Low-income; HIPC; LDC	18,584,988,400.3	8.4 (2017)	+6 (30 to 36)	37.0 (Ranked #149)
18 United Kingdom	High-income	N/A	N/A	−1 (81 to 80)	69.1 (Ranked #12)
19 United States	High-income	N/A	N/A	−5 (76 to 71)	67.9 (Ranked #15)
20 Vanuatu	Lower-middle income; LDC, SIDS	402,236,822.4	2.1 (2016)	N/A (46 in 2018)	39.3 (Ranked #135)

Selecting a Policy

The U.S. TFCA, which was passed in 1998 and was expanded from the previous 1991 EAI, has facilitated a relatively significant number of debt-for-nature conversions.[167] This said, there have been numerous multiparty debt-for-nature conversions, and several countries other than the United

States (e.g., Canada, Finland, Italy, Norway, and Sweden) have structured bilateral debt-for-nature conversions.[168]

When reviewing a country and selecting a policy, countries that previously participated in a debt-for-nature conversion (even if for a terrestrial project) could make for a good candidate. One could also look to see if the debtor country participates in debt exchanges. Further, one could look at a country's trading partners to see if the creditor country has a preferred region, shared language, etc., with potential candidates.

As noted above and as reasoned by Linde Warland and Axel Michaelowa,

> {…} the following four general criteria can be derived in order to identify countries for which debt for climate swaps would be appropriate:
>
> - Country is either an LDC {least developed country}, HIPC {highly indebted poor country}, or SIDS {small island developing states};
> - High level of indebtedness (external debt $\geq$ 150% of exports);
> - Governance improvement (improvement of at least 5 percentage points in the Corruption Perception Index by Transparency International between 2005 or a later year and 2014 {or more recently available year}; and
> - Highly vulnerable (above average vulnerability rating in the Notre Dame Global Adaptation Index Rating).[169]

According to this analysis, the following countries meet three or four of the criteria and host coral reefs: Bangladesh; Barbados; Dominica; Mauritius; Seychelles; and St. Vincent and the Grenadines. It should also be noted that several countries hosting coral reefs, such as Belize, Cook Islands, Fiji, Maldives, and Palau, were missing data and were thus excluded from the list of countries meeting three or four of the criteria.[170]

Implementing a Policy

The U.S. Department of Treasury and Department of State (including USAID) all have responsibilities related to the implementation of the TFCA.[171]

Evaluating the Policy

The EAI, which preceded the TFCA, facilitated debt-for-nature conversions with at least eight countries including Argentina, El Salvador, Jamaica, and Uruguay. These debt conversions generated USD$177.8 million in conservation funds and helped lay the groundwork for the larger TFCA.[172] The

TFCA was reauthorized and signed into law in 2019 and in the authorization, the scope was expanded to coral reef ecosystems. This expands the number of eligible countries and USD$15 million was appropriated for TFCA in FY2020.[173] For a more complete, internal evaluation of the TFCA, see the U.S. Government's assessment.[174]

Future Outlook for Instrument

There has been a decrease in terrestrial debt-for-nature conversions. As noted by the Congressional Research Service, this could be due to:

- 1. Eligible debt to conduct these transactions has decreased, making these transactions either insignificant or not allowed for some debtor countries;
- 2. The amount involved in the transactions is too small for some eligible countries to show interest in participating;
- 3. There is a lack of appropriations to support debt-for-nature transactions; and/or
- 4. The focus on tropical forests (i.e., through TFCA) might be too narrow for many eligible countries.[175]

It is possible that commercial debt-for-nature conversions may become more plausible in the future than sovereign debt-for-nature conversions. Furthermore, there could be an evolution where South–South debt-for-nature conversions are structured between large developing countries such as China, Brazil, and India and their regional counterparts. Regarding other future sites:

Grenada and Jamaica have expressed interest in similar deals. In the South Pacific, Palau and the Marshall Islands also wish to follow suit. The World Bank is helping to bring the debt swap product to Jamaica. The country carries almost US$ 700 million in structural bilateral debt, but at this stage there is no available information on what proportion of that will be subject to a conversion, or whether it will concern the face value or only the interest rate.[176]

With respect to the future of marine debt-for-nature conversions, there could be up to USD$500 million worth of deals.[177] According to a report by Linde Warland and Axel Michaelowa on the potential for climate debt swaps (although not necessarily for a marine investment):

Developing an assessment approach applying indicators for these four criteria {highly indebted countries, least developed countries, and small island states with recently improved governance and a high level of vulnerability}, we find that nine countries with a total debt stock of 22.3 billion USD would be priority candidates for debt for climate swaps. Relaxing the criteria in a way that either high indebtedness or high vulnerability is required beyond good governance, another 19 countries with a debt stock of 97.3 billion USD could be targeted.[178]

Further, could a debt-for-equity be structured for an impact investor looking to invest in a local marine investment such as improving fisheries or ecotourism?

Other Resources on Debt-for-Nature Conversions

Buckley, Ross P. "Debt-for-Development Exchanges: An Innovative Response to the Global Financial Crisis."

- https://papers.ssrn.com/sol3/papers.cfm?abstract_id=1583197

Caribbean Climate-Smart Accelerator

- https://www.caribbeanaccelerator.org/our-work

Conservation Finance Guide's Understanding Debt-for-Nature Swaps

- http://www.conservationfinance.org/guide/guide/indexba4.htm

Moye, M. (2011). Overview of Debt Conversion. Debt Relief International Ltd.

Reilly, W. K. (2006). "Using International Finance to Further Conservation: The First 15 Years of Debt-for-Nature Swaps." In Jochnick, C. and Preston, F. A. (Eds.), *Sovereign Debt at the Crossroads: Challenges and Proposals for Resolving the Third World Debt Crisis* (197–214). New York, NY: Oxford University Press.

Robinson, Michele. "Debt Restructuring Initiatives Paper for the Commonwealth Secretariat."

- https://michelerobinson.net/yahoo_site_admin/assets/docs/Debt_Restructuring_Initiatives_paper_-_Final_Version_June_2010.43100215.pdf.

Sheikh, Pervaze A. "Debt-for-Nature Initiatives and the Tropical Forest Conservation Act: Status and Implementation." Updated July 24, 2018. *Congressional Research Service.*

- https://crsreports.congress.gov

Warland, Linde and Axel Michaelowa. "Can Debt for Climate Swaps Be a Promising Climate Finance Instrument? Lessons from the Past and Recommendations for the Future."

- https://www.perspectives.cc/fileadmin/Publications/Can_debt_for_cli mate_swaps_be_a_promising_climate_finance_instrument_Warland_ Linde__Michaelowa_Axel_2015.pdf

WWF Center for Conservation Finance's Commercial Debt-for-Nature Swap

- https://www.cbd.int/doc/external/wwf/wwf-commercial-swaps-en.pdf

Notes

1. Sheikh, Pervaze A. "Debt-for-Nature Initiatives and the Tropical Forest Conservation Act: Status and Implementation." March 30, 2010. 5–7.
2. Ibid. 3–7.
3. WWF Center for Conservation Finance. "Commercial Debt-for-Nature Swaps." 1.
4. Lovejoy, Thomas E. "Aid Debtor Nations' Ecology." *New York Times.* October, 4, 1984. A31.
5. Robinson, Michele. "Debt Restructuring Initiatives Paper for the Commonwealth Secretariat." 39.
6. WWF Center for Conservation Finance. "Commercial Debt-for-Nature Swaps." 3.
7. Ibid. 6.
8. Ibid. 3.
9. Ibid. 7.
10. Ibid. 4.
11. Ibid. 5.
12. Ibid. 7.
13. Robinson, Michele. "Debt Restructuring Initiatives Paper for the Commonwealth Secretariat." 65.
14. Ibid. 66.

15. SeyCCAT. "Seychelles' Conservation and Climate Adaptation Trust (SeyCCAT): Achieving Conservation Through Innovative Finance and Creative Collaborations." Accessed August 7, 2019.

16. WWF. *Guide to Conservation Finance.* 41.

17. Ibid. 42.

18. TNC. "Seychelles." Accessed April 15, 2017. https://www.nature.org/ourinitiatives/regions/africa/wherewework/seychelles.xml.

19. Warland, Linde and Axel Michaelowa. "Can Debt for Climate Swaps Be a Promising Climate Finance Instrument?" 6–7.

20. Sheikh, Pervaze A. "Debt-for-Nature Initiatives and the Tropical Forest Conservation Act (TFCA): Status and Implementation." Updated July 24, 2018. Summary Page.

21. Ibid. Summary Page.

22. UNDP. "Debt for Nature Swaps." January 17, 2017. Accessed April 2, 2020. http://www.undp.org/content/dam/sdfinance/doc/Debt%20for%20Nature%20Swaps%20_%20UNDP.pdf. 2.

23. Sheikh, Pervaze A. "Debt-for-Nature Initiatives and the Tropical Forest Conservation Act (TFCA): Status and Implementation." Updated July 24, 2018. Summary Page.

24. Sheikh, Pervaze A. Interviewed by Brian McFarland. January 2020.

25. Sheikh, Pervaze A. "Debt-for-Nature Initiatives and the Tropical Forest Conservation Act: Status and Implementation." March 30, 2010. 5–7.

26. So Seychelles. "The Geography of the Seychelles." Accessed April 9, 2020. http://www.seychelles.org/seychelles-info/geography-seychelles.

27. Weary, Rob. Interviewed by Brian McFarland. February 2019.

28. SeyCCAT. "Seychelles' Conservation and Climate Adaptation Trust (SeyCCAT): Achieving Conservation Through Innovative Finance and Creative Collaborations." Accessed August 7, 2019. https://seyccat.org/wp-content/uploads/2017/09/infographic_talking_points_bat_web-1.pdf.

29. TNC. "Seychelles Fact Sheet: Investing for Resiliency." 2016. Accessed August 7, 2019. https://www.cbd.int/doc/meetings/mar/soiom-2016-01/other/soiom-2016-01-seychelles-01-en.pdf.

30. Ibid.

31. Weary, Rob. Interviewed by Brian McFarland. February 2019.

32. TNC. "Seychelles Fact Sheet: Investing for Resiliency." 2016. Accessed August 7, 2019. https://www.cbd.int/doc/meetings/mar/soiom-2016-01/other/soiom-2016-01-seychelles-01-en.pdf.

33. Weary, Rob. Interviewed by Brian McFarland. February 2019.

34. Robinson, Michele. "Debt Restructuring Initiatives Paper for the Commonwealth Secretariat." 31.

35. Ibid., 31.

36. Dogley, Didier. "Financing Marine Conservation and Adaptation to Climate Change in the Seychelles via a Debt Swap." Online Presentation, Hosted by

Yale Center for Business and the Environment, New Haven, CT, April 21, 2016. https://www.youtube.com/watch?v=Pn8yc_2TZJY.

37. World Heritage Centre. "Aldabra Atoll." Accessed November 23, 2019. http://whc.unesco.org/en/list/185.

38. Worldometers. "Seychelles Population." Accessed November 23, 2019. https://www.worldometers.info/world-population/seychelles-population/.

39. Fujairah International Conference for Coastal Areas and Marine Environment. "Speakers—Dr. Andrew Rylance." Accessed November 23, 2019. http://fimec.ae/en/?page_id=115.

40. Rylance, Andrew. "BIOFIN Webinar: Marine Protected Areas Finance—The Seychelles Case." October 31, 2018. Available: https://www.youtube.com/watch?v=CsVqoUvSXsY.

41. Ibid.

42. Ibid.

43. Weary, Rob. "Financing Marine Conservation and Adaptation to Climate Change in the Seychelles via a Debt Swap." Online Presentation, Hosted by Yale Center for Business and the Environment, New Haven, CT, 21 April 2016. https://www.youtube.com/watch?v=Pn8yc_2TZJY.

44. Ibid.

45. Ibid.

46. TNC. "Seychelles Fact Sheet: Investing for Resiliency." 2016. Accessed August 7, 2019. https://www.cbd.int/doc/meetings/mar/soiom-2016-01/other/soiom-2016-01-seychelles-01-en.pdf.

47. Weary, Rob. Interviewed by Brian McFarland. February 2019.

48. TNC. "Seychelles Marine Spatial Planning Fact Sheet." 2016. Accessed August 7, 2019. https://www.cbd.int/doc/meetings/mar/soiom-2016-01/other/soiom-2016-01-seychelles-02-en.pdf.

49. Dogley, Didier. "Financing Marine Conservation and Adaptation to Climate Change in the Seychelles via a Debt Swap." Online Presentation, Hosted by Yale Center for Business and the Environment, New Haven, CT, April 21, 2016. https://www.youtube.com/watch?v=Pn8yc_2TZJY.

50. Weary, Rob. Interviewed by Brian McFarland. February 2019.

51. Rylance, Andrew. "BIOFIN Webinar: Marine Protected Areas Finance—The Seychelles Case." October 31, 2018. Available: https://www.youtube.com/watch?v=CsVqoUvSXsY.

52. Ibid.

53. Ibid.

54. Ibid.

55. Ibid.

56. Ibid.

57. Ibid.

58. Bahadur. *The Pirates of Somalia.* 141, 271–275.

59. Rylance, Dr. Andrew. "BIOFIN Webinar: Marine Protected Areas Finance—The Seychelles Case." October 31, 2018. Available: https://www.youtube.com/watch?v=CsVqoUvSXsY.

60. Ibid.

61. Weary, Rob. Interviewed by Brian McFarland. February 2019.

62. Dogley, Didier. "Financing Marine Conservation and Adaptation to Climate Change in the Seychelles via a Debt Swap." Online Presentation, Hosted by Yale Center for Business and the Environment, New Haven, CT, 21 April 2016. https://www.youtube.com/watch?v=Pn8yc_2TZJY.

63. Weary, Rob. "Financing Marine Conservation and Adaptation to Climate Change in the Seychelles via a Debt Swap." Online Presentation, Hosted by Yale Center for Business and the Environment, New Haven, CT, April 21, 2016. https://www.youtube.com/watch?v=Pn8yc_2TZJY.

64. Weary, Rob. Interviewed by Brian McFarland. February 2019.

65. Weary, Rob. "Financing Marine Conservation and Adaptation to Climate Change in the Seychelles via a Debt Swap." Online Presentation, Hosted by Yale Center for Business and the Environment, New Haven, CT, April 21, 2016. https://www.youtube.com/watch?v=Pn8yc_2TZJY.

66. Ibid.

67. Ibid.

68. Bowen, Pearson. "Barbados Interested in Debt Swap with US-Based Group." *Caribbean Broadcasting Corporation*. April 23, 2019. Accessed April 13, 2020. https://www.cbc.bb/2019/04/23/barbados-interested-in-debt-swap-with-us-based-group/.

69. Rylance, Dr. Andrew. "BIOFIN Webinar: Marine Protected Areas Finance—The Seychelles Case." October 31, 2018. Available: https://www.youtube.com/watch?v=CsVqoUvSXsY.

70. Weary, Rob. "Financing Marine Conservation and Adaptation to Climate Change in the Seychelles via a Debt Swap." Online Presentation, Hosted by Yale Center for Business and the Environment, New Haven, CT, April 21, 2016. https://www.youtube.com/watch?v=Pn8yc_2TZJY.

71. Ibid.

72. Weary, Rob. Interviewed by Brian McFarland. February 2019.

73. Dogley, Didier. "Financing Marine Conservation and Adaptation to Climate Change in the Seychelles via a Debt Swap." Online Presentation, Hosted by Yale Center for Business and the Environment, New Haven, CT, April 21, 2016. https://www.youtube.com/watch?v=Pn8yc_2TZJY.

74. Rylance, Dr. Andrew. "BIOFIN Webinar: Marine Protected Areas Finance—The Seychelles Case." October 31, 2018. Available: https://www.youtube.com/watch?v=CsVqoUvSXsY.

75. Ibid.

76. Spergel, Barry and Melissa Moye. "Financing Marine Conservation: A Menu of Options. 16–17.

77. WWF Center for Conservation Finance. "Commercial Debt-for-Nature Swaps." 6.

78. Ibid. 7.

79. Ibid.

80. Plantilla, Anabelle. Interviewed by Brian McFarland. August 2019.

81. Ibid.

82. Wright, James. Interviewed by Brian McFarland. January 2018.

83. Plantilla, Anabelle. Interviewed by Brian McFarland. August 2019.

84. Ibid.

85. LaGrone, Sam. "U.S. Pays Philippines $1.97 Million for Reef Damage from Guardian Grounding." *U.S. Naval Institute*. February 18, 2015. Accessed August 27, 2019. https://news.usni.org/2015/02/18/u-s-pays-philippines-1-97-million-reef-damage-guardian-grounding.

86. Plantilla, Anabelle. Interviewed by Brian McFarland. August 2019.

87. Baling, Nilda. Interviewed by Brian McFarland. August 2019.

88. Plantilla, Anabelle. Interviewed by Brian McFarland. August 2019.

89. Baling, Nilda. Interviewed by Brian McFarland. August 2019.

90. Marsh, H. and S. Sobtzick. 2015. *Dugong Dugon. The IUCN Red List of Threatened Species* 2015: e.T6909A43792211. http://dx.doi.org/10.2305/IUCN.UK.2015-4.RLTS.T6909A43792211.en. Downloaded on September, 16 2019.

91. World Heritage Centre. "El Nido-Taytay Managed Resource Protected Area." Accessed April 13, 2020. https://whc.unesco.org/en/tentativelists/5034/.

92. Plantilla, Anabelle. Interviewed by Brian McFarland. August 2019.

93. Ibid.

94. Baling, Nilda. Interviewed by Brian McFarland. August 2019.

95. Ibid.

96. Plantilla, Anabelle. Interviewed by Brian McFarland. August 2019.

97. Tubbataha Reefs Natural Park. "History." Accessed December 26, 2019. http://www.tubbatahareef.org/wp/history.

98. World Heritage Centre. "Tubbataha Reefs Natural Park." Accessed December 26, 2019. https://whc.unesco.org/en/list/653/.

99. Plantilla, Anabelle. Interviewed by Brian McFarland. August 2019.

100. Ibid.

101. Baling, Nilda. Interviewed by Brian McFarland. August 2019.

102. Ibid.

103. Ibid.

104. Ibid.

105. Ibid.

106. Ibid.

107. Plantilla, Anabelle. Interviewed by Brian McFarland. August 2019.

108. Ibid.

109. Ibid.

110. Baling, Nilda. Interviewed by Brian McFarland. August 2019.

111. Ibid.
112. Ibid.
113. Barrington, Lewis. Interviewed by Brian McFarland. October 2019.
114. Ibid.
115. EFJ. "Who We Are." Accessed October 25, 2019. http://www.efj.org.jm/?q= node%2F332.
116. Barrington, Lewis. Interviewed by Brian McFarland. October 2019.
117. WWF Center for Conservation Finance. "Commercial Debt-for-Nature Swaps." 4.
118. Larson, Christina. "'The Coral Are Coming Back': Reviving Jamaica's Elegant and Essential Coral Reefs." *Associated Press*. September 20, 2019. Accessed October 25, 2019. https://amp.usatoday.com/amp/2372475001.
119. Matsangou, Elizabeth. "The Precarious Balancing Act Facing Jamaica." *World Finance*. October 22, 2018. Accessed October 25, 2019. https://www.worldf inance.com/strategy/the-precarious-balancing-act-facing-jamaica.
120. Nybakken et al. in: Goodwin. *International Environmental Law and the Conservation of Coral Reefs*. 18–19.
121. Matsangou, Elizabeth. "The Precarious Balancing Act Facing Jamaica." *World Finance*. October 22, 2018. Accessed October 25, 2019. https://www.worldf inance.com/strategy/the-precarious-balancing-act-facing-jamaica.
122. WorldAtlas. "The 10 Busiest Cruise Ports in The World." 2019. Accessed December 26, 2019. https://www.worldatlas.com/articles/the-10-busiest-cru ise-ports-in-the-world.html.
123. Matsangou, Elizabeth. "The Precarious Balancing Act Facing Jamaica." *World Finance*. October 22, 2018. Accessed October 25, 2019. https://www.worldf inance.com/strategy/the-precarious-balancing-act-facing-jamaica.
124. Larson, Christina. "'The Corals Are Coming Back': Reviving Jamaica's Elegant and Essential Coral Reefs." *Associated Press*. September 20, 2019. Accessed October 25, 2019. https://amp.usatoday.com/amp/2372475001.
125. Nybakken et al. in: Goodwin. *International Environmental Law and the Conservation of Coral Reefs*. 18–19.
126. Larson, Christina. "'The corals are coming back': Reviving Jamaica's elegant and essential coral reefs." *Associated Press*. September 20, 2019. Accessed October 25, 2019. https://amp.usatoday.com/amp/2372475001.
127. EFJ. "History." Accessed October 25, 2019. http://www.efj.org.jm/history/.
128. Ibid.
129. Larson, Christina. "'The Coral Are Coming Back': Reviving Jamaica's Elegant and Essential Coral Reefs." *Associated Press*. September 20, 2019. Accessed October 25, 2019. https://amp.usatoday.com/amp/2372475001.
130. Ibid.
131. EFJ. "Impact of Our Work." Accessed April 13, 2020. http://www.efj.org.jm/ impact-of-our-work/.
132. EFJ. "Grant Awards 2016." Accessed December 26, 2019. http://www.efj.org. jm/wp-content/uploads/2018/01/Grant-Awards-Call-2016.pdf.

133. Ibid.
134. Barrington, Lewis. Interviewed by Brian McFarland. October 2019.
135. Ibid.
136. Ibid.
137. Ibid.
138. Larson, Christina. "'The Coral Are Coming Back': Reviving Jamaica's Elegant and Essential Coral Reefs." *Associated Press.* September 20, 2019. Accessed October 25, 2019. https://amp.usatoday.com/amp/2372475001.
139. Matsangou, Elizabeth. "The Precarious Balancing Act Facing Jamaica." *World Finance.* October 22, 2018. Accessed October 25, 2019. https://www.worldf inance.com/strategy/the-precarious-balancing-act-facing-jamaica.
140. Barrington, Lewis. Interviewed by Brian McFarland. October 2019.
141. Ibid.
142. Larson, Christina. "'The Coral Are Coming Back': Reviving Jamaica's Elegant and Essential Coral Reefs." *Associated Press.* September 20, 2019. Accessed October 25, 2019. https://amp.usatoday.com/amp/2372475001.
143. Matsangou, Elizabeth. "The Precarious Balancing Act Facing Jamaica." *World Finance.* October 22, 2018. Accessed October 25, 2019. https://www.worldf inance.com/strategy/the-precarious-balancing-act-facing-jamaica.
144. Barrington, Lewis. Interviewed by Brian McFarland. October 2019.
145. Ibid.
146. Matsangou, Elizabeth. "The Precarious Balancing Act Facing Jamaica." *World Finance.* October 22, 2018. Accessed October 25, 2019. https://www.worldf inance.com/strategy/the-precarious-balancing-act-facing-jamaica.
147. Barrington, Lewis. Interviewed by Brian McFarland. October 2019.
148. Warland, Linde and Axel Michaelowa. "Can Debt for Climate Swaps Be a Promising Climate Finance Instrument?" 6–7.
149. Robinson, Michele. "Debt Restructuring Initiatives Paper for the Commonwealth Secretariat." 54.
150. Avery, Helen. "Blue Finance: Why Marine PPPs Could Be a Win-Win-Win." *Euromoney.* June 5, 2018. Accessed August 12, 2019. https://www.euromo ney.com/article/b18hg7vjwmy9hn/blue-finance-why-marine-ppps-could-be-a-win-win-win.
151. WWF. *Guide to Conservation Finance.* 41.
152. Warland, Linde and Axel Michaelowa. "Can Debt for Climate Swaps Be a Promising Climate Finance Instrument?" 2.
153. Sheikh, Pervaze A. "Debt-for-Nature Initiatives and the Tropical Forest Conservation Act (TFCA): Status and Implementation." Updated July 24, 2018. 13.
154. Ibid., 15.
155. WWF. *Guide to Conservation Finance.* 41.
156. Vertigo Lab. "Innovations for Coral Finance, ICRI Publication." 28.

157. World Bank Group. "World Development Indicators: External Debt Stocks, Total (DOD, Current US$)." Accessed July 31, 2019. http://databank.worldbank.org/data/reports.aspx?source=2&Topic=20.

158. World Bank Group. "World Bank Country and Lending Groups." Accessed December 27, 2019.https://datahelpdesk.worldbank.org/knowledgebase/articles/906519-world-bank-country-and-lending-groups.

159. World Bank Group. "Heavily Indebted Poor Countries (HIPC)." Accessed December 27, 2019. https://data.worldbank.org/region/heavily-indebted-poor-countries-hipc.

160. UN Committee for Development Policy. "List of Least Developed Countries (as of December 2018)." Accessed December 27, 2019. https://www.un.org/development/desa/dpad/wp-content/uploads/sites/45/publication/ldc_list.pdf.

161. World Bank Group. "World Development Indicators: External Debt Stocks, Total (DOD, Current US$)." Accessed December 26, 2019. http://databank.worldbank.org/data/reports.aspx?source=2&Topic=20.

162. World Bank Group. "Total Debt Service (% of Exports of Goods, Services and Primary Income)." Accessed December 27, 2019. https://data.worldbank.org/indicator/DT.TDS.DECT.EX.ZS.

163. Transparency International. "Corruption Perceptions Index 2018." Accessed December 26, 2019. https://www.transparency.org/cpi2018.

164. Notre Dame Global Adaptation Initiative. "ND-GAIN Country Index." Accessed December 27, 2019. https://gain.nd.edu/our-work/country-index/rankings/.

165. Sheikh, Pervaze A. "Debt-for-Nature Initiatives and the Tropical Forest Conservation Act (TFCA): Status and Implementation." Updated July 24, 2018. 9.

166. Warland, Linde and Axel Michaelowa. "Can Debt for Climate Swaps Be a Promising Climate Finance Instrument?" 13.

167. USAID. "Financing Forest Conservation: An Overview of the Tropical Forest Conservation Act." Accessed April 9, 2017. https://www.usaid.gov/biodiversity/TFCA.

168. Sheikh, Pervaze A. "Debt-for-Nature Initiatives and the Tropical Forest Conservation Act: Status and Implementation." March 30, 2010.

169. Warland, Linde and Axel Michaelowa. "Can Debt for Climate Swaps Be a Promising Climate Finance Instrument?" 14.

170. Ibid. 14.

171. USAID. "Financing Forest Conservation: An Overview of the Tropical Forest Conservation Act." Accessed April 9, 2017. https://www.usaid.gov/biodiversity/TFCA.

172. Sheikh, Pervaze A. "Debt-for-Nature Initiatives and the Tropical Forest Conservation Act: Status and Implementation." March 30, 2010.

173. Sheikh, Pervaze A. Interviewed by Brian McFarland. January 2020.

174. White House. "Detailed Information on the Tropical Forest Conservation Act Assessment." Accessed April 3, 2020. https://www.whitehouse.gov/sites/whitehouse.gov/files/omb/assets/OMB/expectmore//detail/10002240.2007.html.

175. Sheikh, Pervaze A. "Debt-for-Nature Initiatives and the Tropical Forest Conservation Act (TFCA): Status and Implementation." Updated July 24, 2018. 18.

176. Vertigo Lab. "Innovations for Coral Finance, ICRI Publication." 29.

177. Weary, Rob. Interviewed by Brian McFarland. February 2019.

178. Warland, Linde and Axel Michaelowa. "Can Debt for Climate Swaps Be a Promising Climate Finance Instrument?" 2.

14

Blue Procurement Models

Introduction

Blue procurement models come in several different structures. First, is the structure of obtaining marine aquarium products which are sustainable and beneficial to local communities. Second, is the structure of obtaining reef-dependent seafood that has been independently verified against a sustainability certification such as the Marine Stewardship Council (MSC). Third, is working with small-scale fisheries which may be subsistence-based and/or have a commercial aspect. Relatedly, a fourth potential form of blue procurement is the establishment of concessions, such as conservation or community-managed concessions, which produce sustainably sourced products.

Historical Overview

The sustainable procurement of locally produced commodities—particularly by Indigenous Peoples—has existed for thousands of years. However, the origin of commercial green procurement likely started with the creation of the Forest Stewardship Council (FSC) in 1994,[1] and the launch of the Sustainable Forestry Initiative (SFI) in 1995.[2] The Marine Stewardship Council (MSC) was launched shortly thereafter in 1997 after a statement of intent was signed between WWF and Unilever in 1996. The MSC Fisheries Standard was then "born" in 1998.[3]

Other important historical dates include:

© The Author(s) 2021
B. J. McFarland, *Conservation of Tropical Coral Reefs*,
https://doi.org/10.1007/978-3-030-57012-5_14

- 1999: "By 1999, with its standard in place and with growing interest from fisheries, fish buyers, and retailers, the MSC became independent of WWF and Unilever, and a year later {2000} certified its first fishery – for rock lobsters off the coast of Western Australia."[4]
- 2000: Carbon Disclosure Project (now known as CDP) is founded.[5]
- 2004: WWF initiated the Aquaculture Dialogues which "were a multi-stakeholder roundtable {that} formally started in 2004," and later "the Netherlands based Sustainable Trade Initiative IDH joined forces with WWF Netherlands to create the Aquaculture Stewardship Council {ASC} in 2010."[6]
- 2006: "The watershed came in February 2006, when Walmart, the world's largest retailer, declared that it was aiming to sell only MSC-certified seafood within five years. {…} Retail, however, is only half the story: in the United States, over 50 percent of all seafood is eaten in restaurants, so the catering sector is vital too. But then, just weeks after the Walmart breakthrough, the Compass Group – the largest catering company in the Americas, which supplies thousands of restaurants, schools, and hospitals – made a similar announcement. Sustainable seafood had entered the mainstream."[7]
- 2008: In 2008 "US Congress followed suit and passed an amendment to the Lacey Act of 1900, which prohibits trafficking in illegally harvested plants and wildlife."[8]
- 2009: CDP launched the global Water Disclosure Project.[9]
- 2015: Forest Trends' new project and platform Supply-Change.org is launched.[10]
- 2017: Indonesia "became the first nation to make its fishing fleet visible to the world via GFW {Global Fishing Watch}. Since then, Peru, Chile, Panama, Mexico, Namibia, and Costa Rica have either published their vessel data to the GFW map or have publicly committed to do so."[11]
- 2019 (October): Seafood Stewardship Index is launched by the World Benchmarking Alliance (WBA), "ranking 30 of the most influential seafood companies in the world on their commitments, transparency and performance to meet the UN SDGs."[12]
- 2019 (October): Global fishing data from the Global Fishing Watch becomes available on the Bloomberg Terminal.[13]

Mechanisms of Instrument

Commodity supply chains—whether for pelagic tuna, farm-raised oysters, or MSC-certified rock lobsters—often involve local producers, aggregators, processors, traders, manufacturers, and buyers, along with extensive transportation, marketing and financing networks, and auditors. Consumer pressure (e.g., product boycotts) and investor pressure (e.g., filing proxy statements) are helping to influence these supply chains to adopt more sustainable procurement practices and to address climate change. For instance:

> BlackRock, the world's largest asset manager, put corporate boards on notice that it expects them "to have demonstrable fluency in how climate risk affects the business" and how management is managing that risk. Specifically, the world's largest asset manager is encouraging companies, "in due course," to follow the recommendations of the Financial Stability Board's Task Force on Climate-Related Financial Disclosures, launched in December by Michael Bloomberg and Mark Carney, governor of the Bank of England.[14]

As stated by WWF,

> By implementing mechanisms such as tradable fishing quotas, levies, and product certification, fisheries managers can provide economic incentives for sustainable fishing practices, thereby improving both the environmental and economic conditions of the industry. Governments can raise revenues to manage fisheries by charging fishing access payments, license fees, excise taxes and fines.[15]

In addition to an independent certification such as MSC, there is also the use of individual fishing quotas (IFQs) and individual transferable quotas (ITQs):

> Under a system of IFQs, a government fisheries agency or an industry-wide association of fishers allocates specific shares of the total allowable catch of a given fish species within a defined area to specific individuals, groups, or companies. This is often done on the basis of their current or historical shares of a particular fishery, although lotteries and auctions have also been used as the basis for allocating quotas. Because fishers have the property right to secure future benefit from the resource, they are prepared to wait, and practice conservation, in order to optimize their long-term return.
>
> Under a system of ITQs, these shares or quotas are made freely transferable and divisible, and can be bought and sold to other fishers or fishing companies. ITQs can also be leased or mortgaged, like other types of property rights. Without any expenditure of public funds (other than for administrative

costs), the private sector thereby ends up achieving the same result as expensive government programs designed to buy out excess fishing capacity, either by scrapping excess fishing vessels or by paying individual fishers not to fish.[16]

An important mechanism of sustainable commodity supply chains is a certification standard, and numerous procurement models involve independent verification against a third-party certification. There are also internal certifications, which are viewed as potentially less transparent and less stringent. Some of the leading independent certification systems include Aquaculture Stewardship Council (ASC)-certified and MSC-certified seafood.

Obtaining certification may yield a premium price on the commodity when compared to its traditional counterpart, or certification may open up markets for certified products which would not otherwise exist. In addition to certification standards, there are reporting mechanisms such as reporting to the CDP, particularly its Water Program.[17] Companies reporting to the CDP Water Program include both publicly traded companies and privately held companies.

Several other innovative, technological advancements have enabled better supply chain management and traceability including the Global Fishing Watch and the Sea Around Us. The Global Fishing Watch is "promoting ocean sustainability through greater transparency. We use cutting-edge technology to visualise, track and share data about global fishing activity in near real-time and for free."[18] Similarly, the Sea Around Us "makes a wide range of data analyses and indicators available."[19]

Size of Instrument

There are several ways to assess the size of the blue procurement sector and its potential.

Aquaculture Stewardship Council (ASC)

The ASC is an independent certification standard used to assess social and environmental practices of farm-raised (i.e., aquaculture) seafood. The ASC, as of January 2020, certified a total of 1134 farms and another 273 farms were in assessment. Collectively, the certified production volume was approximately 1.9 million tons of farm-raised seafood.[20]

As noted by TNC, "the growth of aquaculture presents an opportunity that is both environmental and financial in nature. Current forecasts suggest producers will need $150 to $300 billion in capital expenditures in the next ten years to build out the infrastructure required to accommodate consumer demand."[21]

Marine Aquarium Trade

There is a global market for the marine aquarium trade:

> In terms of target species, Elizabeth Wood notes that for fish, 1,000 species are caught and an estimated 25-35 million specimens are traded annually. The most commonly traded are low individual value species such as butterflyfish, anemonefish, damselfish, gobies, angelfish and wrasse, with this being supplemented by low-volume trade in high-value fish such as clown triggerfish. In terms of identifying the major exporters, Colette Wabnitz and others indicated that between 1997-2002 there were the Philippines (43 per cent), Indonesia (26 per cent), and the Solomon Islands (12 per cent), the remaining trade being sourced from Australia, the Maldives, Fiji, Palau and Sri Lanka. Over the same time period, specimens from these states were mainly destined for the USA, the UK, France, Germany and the Netherlands.[22]

According to the FAO, "between 2000 and 2011, global exports of ornamental fish increased from US$181 million to US$372 million. Total trade in live marine ornamentals is estimated at around US$44 million annually."[23]

Marine Stewardship Council (MSC)

According to the MSC's 2018–2019 Annual Report, there were in 2019:

- 361 certified fisheries (including 13 fisheries currently suspended) and another 109 fisheries in assessment;
- 11.8 million tonnes of MSC certified catch, representing 15% of the overall global catch;
- The 109 fisheries in assessment represent another 1% of the global catch; and
- A total of 41 countries had MSC certified fisheries.[24]

In addition, there were approximately 37,000 products sold with the blue MSC label, which included one million tons of seafood sold to consumers with the MSC label, and an estimated USD$9.1 billion was spent by consumers on seafood with the MSC label.[25]

Introduction to Case Studies

The following case studies will examine LINI's Sustainable Ornamental Fish Aquaculture, Western Australia's MSC-Certified Rock Lobster Fishery, Rare's Fish Forever Work in the Philippines and Indonesia, and Blue Ventures' work in Madagascar. LINI is based in Bali, Indonesia and uses the sustainable capture of reef fishes as a driver of stewardship, including the Banggai cardinalfish. The Western Australia Rock Lobster Fishery was the first MSC-certified fishery in the world. The Western Rock Lobster is Western Australia's most valuable fishery and historically was Australia's most valuable single species wild capture fishery. Rare, a global nonprofit organization, has developed the Fish Forever Program, which is being used throughout the Philippines and Indonesia. Fish Forever is under the umbrella of coastal fisheries reform using fully protected areas to protect essential fish habitat and to both replenish and sustain fish production. OurFish is an application applied to the program and a signature program of Rare, which is also applied to the Fish Forever Program, as is the use of Pride Campaigns to shift key behaviors. Blue Ventures, in collaboration with local communities, has helped develop the concept or approach known as locally managed marine areas (LMMAs). Blue Ventures' LMMAs program focuses on three zones along Madagascar's west coast, which together include over 75 communities, a combined coastal population of more than 40,000 people, and a total marine area of almost 6000 km^2.

Case Study #1: LINI's Sustainable Ornamental Fish Aquaculture

Introduction

LINI is based in Bali, Indonesia and uses the sustainable capture of reef fishes as a driver of stewardship. LINI, which means "to line up," is an early initiative, small project but one that is a complex undertaking with many components. For instance, LINI is working on sustainable supply chains with a part of it including endemic species, there are financial aspects related to working with local communities whose livelihoods are dependent on the reefs, LINI undertakes coral reef restoration, and LINI has been working over the years to influence policy and to educate the government on how to see the coral reefs differently (i.e., the government sees the reefs primarily for tourism, but reefs also provide for sustainable fisheries)[26] (Fig. 14.1).

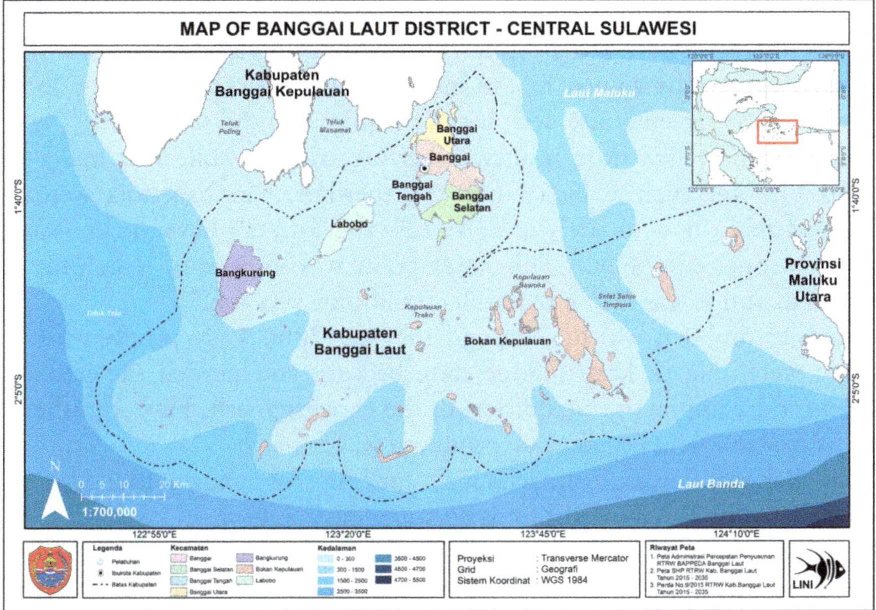

Fig. 14.1 Map of Banggai Laut District, Central Sulawesi, Indonesia (*Credit* LINI)

More specifically, the waters off Banggai Island are home to the Banggai cardinalfish (*Pterapogon kauderni*), which is an endemic species only found in this tiny area in Central Sulawesi, Indonesia. In addition, LINI also works with the octopus trade on Banggai Island and LINI works in North Bali and Banda Island, Indonesia with mainly the artisanal, handline tuna fisheries.[27]

Identify the Problem

One of the main problems is that while there are many organizations working within MPAs, there are fewer organizations—such as LINI—that are working outside of the MPAs. Likewise, there are very few organizations looking at the issues outside of the MPAs, in the areas where there are little-to-no regulations or enforcement.[28]

With respect to the supply chain for sustainable ornamental marine fish, there were a host of problems. Essentially, actors throughout the supply chain did not know how to do the business and there was overexploitation of the natural resource and fair trade was not happening. For instance, there were destructive fishing practices, practitioners did not know how to export fish nor how to hold the fish, and the exporters wanted to pay the fisherfolk as little as possible.

Regarding destructive fishing practices, cyanide has been shown to be harmful to reefs, fish, and people. Net-catching is more sustainable when fishers are properly trained in the method; however, where fishers do not know how to efficiently use nets, it can be quite harmful to the reef.[29]

Because there was little money, fisherfolk tried to collect as many fish as possible (i.e., such as blue tangs). Furthermore, the fisherfolk were not informed about how to safely dive, because they did not have the knowledge nor the proper equipment. This latter point was so serious, that fisherfolk even died, and it was like "a blood diamond but with fish."[30]

Another problem was that around the year 2000, only people who had money were able to keep marine tang because you needed the necessary gadgets and equipment, in addition to the expensive fish. However, the less expensive nano tang began to be produced in China.[31]

The Banggai cardinalfish, which is endemic to Banggai Island, is also classified on the IUCN Red List as endangered.[32] Likewise:

Since 1995 it {the Banggai cardinalfish} has increasingly been exploited for the international aquarium trade. In its proposal to have the species inscribed on Appendix II in 2007, the USA quoted figures of 600,000–700,000 specimens traded in 2001. The FAO assessment relating to this proposal was unsympathetic as it was felt the population levels had not reduced enough to merit inscription. Given the species' high productivity, the FAO interpreted the marine species annotation to the Fort Lauderdale Criteria as requiring an 80-90 per cent reduction in population size to justify Appendix II listing. At COP14, the proposal was withdrawn after a number of contracting parties concurred with the FAO opinion, whilst Indonesia reassured the USA that it had worked with the EU to implement sustainable management plans for the species."[33]

Why the Problem Is Important

Indonesia is an archipelago nation with a large number of people living by the coast and approximately 80% of the reefs are being exploited by small-scale fisheries. Yet, from the coast to the borders of the EEZs, the Indonesian Government does not see the activities outside of the MPAs as much of a concern and does not provide enough attention. Yet, these areas provide food security for Indonesians and provide income for the country through exports. In addition, it is often marginalized communities, that not enough organizations work with, who are dependent on these areas. Yet through sustainable fisheries, these local communities can be guardians of the reefs and you can link the fisherfolk with the hobbyists.[34]

Another reason why the problem is important and why the Marine Aquarium Council (MAC) wanted to start with Indonesia is because after the Philippines, Indonesia is one of the main exporters of aquarium fish.[35]

Furthermore, Central Sulawesi is amongst the fifty most important bioclimatic units.

How Problem Was Identified

The problems facing areas outside of the MPAs were identified mainly by NGOs. For instance, the founder of LINI, Ms. Gayatri Reksodihardjo-Lilley, previously worked with WWF Indonesia for nine years before LINI. It was during her time with WWF Indonesia, that Ms. Reksodihardjo-Lilley first worked on the issues which included identifying the problem of corals being mined for construction purposes and that corals were being unsustainably collected for ornaments and for the export trade. Ms. Reksodihardjo-Lilley then started talking to local fisherfolk and the coral collectors and advised them to collect dead corals rather than taking a big chunk of live corals. Ms. Reksodihardjo-Lilley and her team, back in 1998, then started lobbying the government to stop these practices. However, Ms. Reksodihardjo-Lilley and her team saw the problem, but did not see the problem from the business side, and just wanted it to all stop and "there were a lot of nos and a lot of do not do this or do that."[36]

Ms. Reksodihardjo-Lilley then joined the Marine Aquarium Council (MAC) and learned a lot about the problems facing the supply chain and realized that you can really make the aquarium trade sustainable and there are businesses who want to do the right thing.[37]

Ms. Reksodihardjo-Lilley then founded LINI to put it all together—from the local fisherfolk to the exporter, importer, and wholesaler—to achieve a situation where the private sector is benefitting conservation.[38]

Effectiveness of Process for Identifying Problem

Ms. Reksodihardjo-Lilley started seeing through the supply chain that the buyers from the retail shops wanted to help, but they did not know how to help. The MAC, as a standard-setting organization, started working through the issues with the belief that an independent certification could help demonstrate that fish coming from the producing country are able to maintain optimal health until the retailers. During this time, the MAC spread throughout Indonesia to provide training on better fish collection methods.

This included helping fisherfolk understand there were other options to catch fish and to do so without harming the environment. The thinking was that if the buyer could appreciate the changing ways, then they could reward the fisherfolk. Similarly, the fisherfolk could then see there was an incentive for them to change.[39]

Steps Taken to Address the Problem

There were several steps taken, as previously mentioned:

- Ms. Reksodihardjo-Lilley joined the MAC and learned about the problems facing the supply chain and realized that you can really make the aquarium trade sustainable and there are businesses who want to do the right thing.[40]
- Ms. Reksodihardjo-Lilley then founded LINI to put it all together—from the local fisherfolk to the exporter, importer, and wholesaler—to achieve a situation where the private sector is benefitting conservation.[41]

As explained by PhD candidate Ms. Shannon Switzer Swanson, what is great about LINI is that the main part of their work, that is productive, is that they are so attuned to the local context. LINI is very open to adapting their process to the communities' needs, the interest there, and their local resources. In Banggai, for example, there were a handful of people who are buying and selling blue tangs and Banggai cardinalfish. LINI worked with these people and used their skills to give them opportunities to raise and sell fish, and not just catch them.[42]

Results

In terms of policy, one of the most important results is that Ms. Gayatri Reksodihardjo-Lilley and her team were able to get the government to start collecting data on reef fish for the aquarium trade. It took ten years to achieve this, but it is now part of the agenda.[43]

In terms of the traditional fisherfolk and the local communities, Ms. Reksodihardjo-Lilley does not think they have achieved what they wanted to achieve. For instance, in the Banggai Islands, there are still fisherfolk collecting fish (i.e., particularly blue tangs) with cyanide. However, LINI just does not have enough people, time, and funding.[44]

Another achievement is that there are industries, particularly the retail shops in the buying countries, who are starting to recognize and support

LINI. For example, LINI has started to collaborate with Maidenhead Aquatics, which has over 160 stores across the United Kingdom.[45] Representatives from Maidenhead Aquatics come each year to meet with LINI to learn about the industry, to meet the fishers, to appreciate what the fishers do, and to help LINI do reef restoration in North Bali and in Banggai Island.[46]

Furthermore, Fair Trade certification just started in Indonesia for tuna and Safeway has started to offer this Fair Trade tuna from Indonesia.[47]

Challenges and How They Were Met

One challenge is that you cannot do premium pricing with sustainable, wild-caught marine aquarium fish. The premium pricing and payments are not very difficult to work on throughout the supply chain within the country. However, when you start talking with the exporters and especially with the importers, particularly in the United States, they do not want to increase their prices. It is like "greenwashing whereas people in the developed countries say yeah it has this certification, but as long as it is cheap." A related problem is that many fisherfolk have stopped collecting Banggai cardinalfish because the value is too small and the fisherfolk have other occupations such as tending to clove gardens or collecting octopus. This said, Indonesia is one of the top ten countries to export reef octopus for the food trade.[48]

Another challenge is that the MAC did not work well and in fact, it no longer exists. While people in Indonesia were eager to use the MAC, from the business and financial aspect, the exporters and importers did not want the cost to be added. Similarly, it is challenging to implement a labelling system—whether Fair Trade or MAC—with aquarium fish because you are dealing with a live specimen. Likewise, you can label coffee or chocolate, but you are dealing with a live animal and how do you put a label on a fish? It is also hard to trace and you need to put a big investment into the infrastructure.

A further challenge was that the United States, under CITES, sought to ban the trade in Banggai cardinalfish around 2008. The U.S. withdrew its proposal, but then approximately ten years later, the same scientists proposed a similar ban through the EU. However, there needs to be local stakeholders who properly manage the fishery, as opposed to a complete ban or moratorium. This is because the Banggai cardinalfish are easy to breed and easy to identify in the wild, so you just need to do enforcement.[49]

While 99% of marine aquarium fish cannot spawn in captivity, the reverse is true with freshwater fish. This said, if you can buy captive bred, buy captive bred, but a lot of marine fish are wild caught.[50] Thus, another challenge is that some areas, such as Hawaii and Fiji, are closing the marine aquarium trade. In Hawaii, there was a suspension of collecting marine aquarium fish. Green chromis, yellow tangs, and clownfish are popular tropical fish, with yellow tangs, in particular, coming from Hawaii. However, organizations, including the SCUBA industry, saw that tangs were disappearing and believed it was a result of collection. In Fiji, "everything being traded is stopped," including live rock (i.e., reef rock with life in it, which is not coral but helps in filtration) which was a big trade in Fiji. Unfortunately, some would mine live rock for roads and buildings. Around May 2018, the Indonesian Ministry of Marine Affairs and Fisheries stopped issuing the health certification for corals for the aquarium trade as they did not believe the aquarium industry was complying with the regulations. There are no corals being exported (i.e., both wild and cultured corals), but more and more species are coming from marine cultures. There is a lot of potential for marine cultures for coral farmers and there is a conservation aspect as farmers are basically creating coral habitat. However, instead of stopping just the bad actors, the Government stopped the entire trade, including those who were doing good for the environment. While the trade of wild-caught corals should be stopped, you need to do marine culture on a case-by-case basis as you are adding more corals when you do marine culture.[51]

Beyond Results

Researchers with Roger Williams University, including Dr. Andrew Rhyne, are developing a gadget to test the cyanide levels in fish and LINI is going to collaborate with the researchers.[52] This should help with the sustainability of the trade.

Lessons Learned

Ms. Gayatri Reksodihardjo-Lilley, with 20+ years of experience, has several lessons learned to share. First, there is a need for more organizations that are interested in this work and who are working on marine aquarium industry issues to get involved. Likewise, Ms. Reksodihardjo-Lilley has worked for the Big International NGOs (BINGOs) and they work a lot with dive tourism and food fish. However, the BINGOs—whether it is not cool or whether there is a mindset that fish should be used for food or swimming in the ocean and not in aquariums—are not working on sustainable aquarium fish.[53]

There is still a lot of work that needs to be done, but the marine aquarium trade can be sustainable for many of the species and the communities can be the guardian of the reefs. The fish being collected need healthy reefs and those communities that understand this need will self-protect the reef because they are dependent on the reefs for fish.[54]

It is not easy to get funding. Funding is needed so local communities can be trained about managing their resources. Likewise, the communities need help from other organizations on how to manage the resources, such as doing a resource assessment and monitoring of the resources by a government entity and/or an NGO. In addition, the industry can help itself throughout the supply chain such as the way that Maidenhead Aquatics makes a regular contribution or donation of £1 to LINI for every box of marine aquarium fish that is purchased through its supply chain.[55]

Other Resources for LINI's Sustainable Ornamental Fish Aquaculture

"Expanding our understanding of the trade in marine aquarium animals" paper by Andrew L. Rhyne et al.

- https://www.ncbi.nlm.nih.gov/pmc/articles/PMC5274522/

LINI

- http://lini.or.id/; and
- https://www.facebook.com/LINIfoundation

Maidenhead Aquatics

- https://www.fishkeeper.co.uk/

Marine Aquarium Biodiversity and Trade Flow

- https://www.aquariumtradedata.org

Ornamental Fish International

- https://www.ofish.org/

Reef to Aquarium

- https://reeftoaquarium.com/

"Wild caught ornamental fish: a perspective from the UK ornamental aquatic industry on the sustainability of aquatic organisms and livelihoods" paper by Tracey A. King

- https://onlinelibrary.wiley.com/doi/full/10.1111/jfb.13900

Case Study #2: Western Australia's MSC-Certified Rock Lobster Fishery

Introduction

The Western Australia Rock Lobster Fishery was the first Marine Stewardship Council (MSC) certified fishery in the world. The Fishery, which consists of zones A, B, and C, is located off the coast of Western Australia from the city of Augusta on the Southern boundary to the city of Onslow on the Northern boundary[56] (Fig. 14.2).

The Fishery had an "initial limited-entry system {which} evolved into a sophisticated TAE {Total Allowable Effort} control system with ITE {Individual Transferable Effort} units issued to each vessel, which allowed the fishery to operate close to maximum sustainable yield (MSY) levels and to generate significant licence values for fishers. More recently (2010), this fishery was converted to TAC {Total Allowable Catch} management with a strategy of targeting maximum economic yield.[57] In other words, the Fishery swapped from an input control regime to an output control regime.[58]

Another interesting feature of the Fishery is that "in a more general context, these prices paid for pot units traded through the licence register set up under the Fisheries Act are an equivalent to 'stock market' prices where the fishery as a whole is equivalent to a listed 'company' and the ITE or ITQ units are effectively 'shares' in that company."[59]

Identify the Problem

Historically, there were problems associated with limiting the number of fishing vessels, particularly vessels from out of state, and problems associated with the potential drastic increase in the number of lobster traps

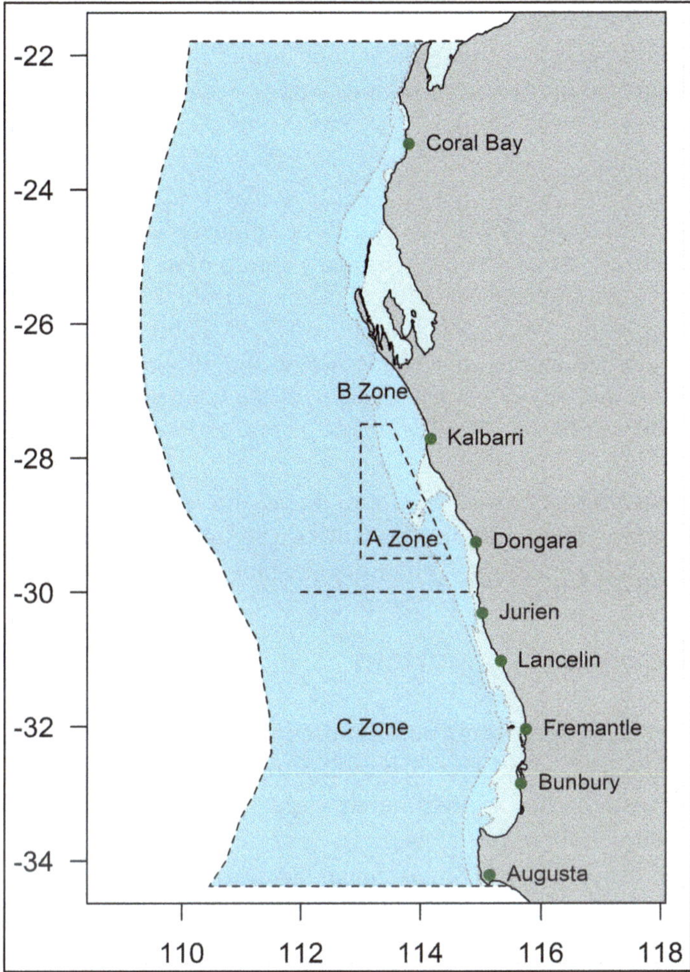

Fig. 14.2 Map of Western Australia Rock Lobster Fishery (*Credit* Simon de Lestang, DPIRD)

(known as pots). This said, "although the vessel and pot numbers were controlled, nominal fishing effort measured as 'pot lifts' continued to increase through the 1970s and 1980s and the effectiveness of fishing gear was also increasing."[60] Thus, there were concerns about "recruitment overfishing" and "effort creep."

Climate change is another problem facing the Fishery. For instance:

The south west of Western Australia is predicted to be heavily influenced by the impacts of climate change (e.g. increasing sea temperatures, declines in rainfall). Some climate change information has been taken into account in the rock lobster stock assessment process and the effect of the marine heat wave in 2010/11 on fisheries has been assessed but further information is required to examine potential impacts on this bioregion.[61]

This said:

> At a longer time scale, WRL {Western Rock Lobsters} have been rated a high risk to the effects of climate change as many aspects of its life history are highly sensitive to environmental conditions (Caputi et al., 2010). The economic performance of the WCRLMF {West Coast Rock Lobster Managed Fishery} is strongly affected by the value of the Australian dollar (affecting the price of lobsters), fuel and labour costs and status of the Chinese economy as China imports nearly all of the WRL.[62]

Bycatch is another potential problem. In addition to past issues with the bycatch of marine mammals and sea turtles, the "main bycatch species landed in the WCRLMF are octopus, champagne crabs, and baldchin grouper."[63]

Why the Problem Is Important

These problems are important because the Fishery provides jobs and revenue to the commercial fishery. The Fishery is also a large recreational fishery.

Western Australia has approximately eight species of rock lobster, but the "most abundant by far is the western rock lobster {*Panulirus cygnus*.}."[64] The Western Rock Lobster "{…} forms an important sector of the State's economy. It is Western Australia's most valuable fishery, with an estimated value of \$359 million in 2014, and has historically been Australia's most valuable single species wild capture fishery."[65]

Similarly, the Fishery "{…} is important for regional employment with 234 commercial vessels operating in 2017 with most of the catch handled by four main processing establishments. The rock lobster fishery is also a major recreational activity and provides a significant social benefit to the Western Australian community with over 60,000 recreational fishers holding rock lobster licences in 2017."[66]

It is also important to conserve seagrasses and shallow reefs, where approximately 95% of recruits live before the lobsters, upon nearing puberty, migrate West.

Further, Western Australia's Shark Bay and Ningaloo Coast are both listed as World Heritage Marine Programmes.

How Problem Was Identified

The Fishery has "been the subject of comprehensive biological research and management since 1963,"[67] and thus, the problems have been assessed on an ongoing basis.

Since 1949,

> the commercial fishery has been documented through industry landing records and detailed research logbooks since 1963. Additional information from puerulus settlement (1968 onwards), observer programmes (starting 1971), and fishery independent spawning stock surveys (1991 onward) have provided particularly detailed databases for the fishery.[68]

Further,

> An extensive science program supports the management of lobster fishing. We collect data on commercial and recreational catches and fishing activity, and carry out independent monitoring to look at the abundance of the breeding stock, puerulus settlement and environmental factors that may affect breeding success and survival.
>
> These programs enable our researchers to estimate catches up to four years ahead, and assess the impacts of changes in fishing technology and practice.
>
> For the past 20 years, recreational catch and fishing effort has been estimated from the results of an annual mail-based survey of fishers. Since 2000/01, telephone diary surveys have provided extra data. The trends generated by the data collected, with data on puerulus settlement, are used to predict recreational catch and effort in following seasons.[69]

Effectiveness of Process for Identifying Problem

It has been noted that:

> Consultation occurs between the Department and the commercial sector either through the Western Rock Lobster Council or the Annual Management Meetings convened by the Department through the Western Australian Fishing Industry Council. Consultation with Recfishwest and other interested stakeholders is conducted through specific meetings and the Department's website. Consultation with non-fisher stakeholders is undertaken in accordance with the Department's Stakeholder Engagement Guidelines.[70]

Thus, the process for identifying the problems appears to be effective.

Steps Taken to Address the Problem

The Fishery includes both recreational and commercial fishing. Steps taken to address threats of overfishing by recreational fishers include the fact that "recreational lobster fishers need a licence and there are also gear restrictions and size, bag and boat limits, as well as a closed season."[71] For commercial fishing, there are:

- Areas closed to fishing;
- Lobster size limits;
- Protection for any females in breeding condition;
- Controls on the type of gear used; and
- A limit on the catch for the whole fishery, known as Total Allowable Commercial Catch (TACC).[72]

The annual TACC has fluctuated from a low of 5500 tons, to approximately 6300 to 6600 tons, and could possibly go as high at 7000 tons.[73]

With respect to bycatch, for instance, "the move of the rock lobster fishery from an effort-controlled fishery to a catch quota fishery, coupled with significant effort reductions, will ensure the octopus catch in the WCRLMF remains a low proportion of the overall catch."[74]

Results

Some results from the Fishery are:

- The Fishery was "one of the first limited entry fisheries in the world"[75];
- In 2000, the Fishery became the first fishery in the world to be certified to the MSC. In 2017, the Fishery was recertified to the MSC for a fourth time[76]; and
- In 2016–17, "there were 878 direct jobs and 1558 indirect jobs including in direct fishing, processing, boat building and tourism."[77]

Another major result of the Fishery is that researchers can predict within 2% the stock of lobsters in four years' time. Every year a stock assessment is conducted to determine what future takes will do for profitability and for the stocks. Relatedly, the Fishery was one of the first in Australia to undertake independent sampling. Data for fisheries often comes from the fishers, which can be bias, such as when reporting whale entanglements and bycatch. In contrast, the Fishery's sampling has been standardized, as much possible to include the same bait, traps, locations, and the same lunar cycles. In addition to this sampling, there is a smart camera system deployed in commercial lobster pots and gives a snapshot of the local environment (e.g., the fish swimming around, pictures of seafloor, etc.), takes baseline data such as water temperature, and all of this is geo-referenced.[78]

There is also a long-term coral reef monitoring program. Likewise, "the main activities at the Abrolhos are commercial rock lobster potting and line fishing and recreational fishing and diving. The Department has a long-term coral reef monitoring program at the Abrolhos to detect potential impacts from human use and natural influences."[79]

Another major result of the Fishery is that the "major long-standing and most successful processor in the Western Australian fishery has historically been a large fishers' cooperative, which distributes its profits back to fishers and has had no incentive to accumulate fishing rights on its own behalf and compete with its shareholders.[80]

With a reduced TACC, the Fishery processers—with half the supply—have focused on marketing and processing changes. This includes a focus on the export market for live lobsters mainly to China and less so to Japan and Taiwan. The price at the dock has gone from about AUS\$18 per kilogram to approximately AUS\$70 per kilogram. This has been a benefit to fishers who no longer need to go to sea during storms or holidays.[81]

There is also a positive result of a rebounding whale population, including Southern right whales {*Eubalaena australis*} and one of the world's fastest recovery rates for humpback whales {*Megaptera novaeangliae*}.[82]

Challenges and How They Were Met

One challenge for the Fishery has been bycatch and entanglement:

> Whale entanglements increased with the move to quota management, peaking in 2013 with 17 reported entanglements in western rock lobster fishing gear. The significant social and ethical concerns around these entanglements prompted two collaborative Fisheries Research and Development Corporation (FRDC) research projects between industry and government. These led to the introduction of gear modifications in 2014, which has seen a reduction in entanglements.
>
> In what was considered to be a world first, mitigation measures required fishers to remove slack from pot ropes, reduce the number of floats, avoid having pots in clusters, and minimise fishing in know migration pathways.

Gear modifications are estimated to have reduced entanglements of whales in western rock lobster gear by about 66%, with two and four entanglements recorded in the 2015 and 2016 seasons respectively. However, with an increasing whale population migrating through the fishery each year, a continued collaborative approach between industry and government to mitigate these entanglements is critical.

Our Western Rock Lobster fishery continues to monitor, respond and innovate to reduce whale entanglements to as low as they can possibly be in the waters off Western Australia.[83]

There are more entanglements, in part, because there are more whales. To address these concerns, there is industry and government collaboration and researchers are looking at designing an inexpensive, satellite tracking buoy. Traditionally, fishing boats are not legally allowed to approach endangered species, such as whales. There are whale disentanglement teams, but such teams typically need sunlight and good weather conditions to help the whales. With a move toward a social license to fish concept, industry is stepping up, and in the future, all commercial fishing boats may be equipped with the tracking buoy that would enable a faster response by the government's whale disentanglement teams.[84]

Another challenge concerns the movement of such lobsters. Likewise, "there are a few invertebrate species capable of long-distance movement, for example spiny lobsters. Lobsters are also nocturnal, making visual surveys unproductive."[85] In fact, it is not uncommon for a western rock lobster to migrate up to 800 km (nearly 500 miles).[86]

The challenge of relatively low tax revenue and a high volume of the Fishery being exported has often been cited. For instance:

> The western rock lobster resource is currently worth over $5 billion, however the net return to the Western Australian Government through licence fees for this community resource is only $10 million per annum. Currently more than 95 per cent of commercially caught WA rock lobster is exported to China, meaning little flows into our local market for the enjoyment of Western Australians and tourists.[87]

There have also been challenges with stakeholders and consultations:

> Some of these controls also required frequent stock assessments, real-time monitoring of catch, and adjustments midseason to ensure the fishery achieved the nominal target catch. For these reasons, the TAE changes created significant industry debate, uncertainty, and costs which ultimately contributed to the Government's unilateral decision to simplify management by directly setting

a TAC and adopting ITQs. This occurred despite a lack of overall industry support at the time, although there had been industry consultation and debate in the preceding years.[88]

A potential challenge, which has not materialized, is concerns about a monopoly being formed:

However, despite the lack of specific controls on who could own lobster fishing licences, ownership outside of the Western Australian fishing community has not become a feature of the fishery to date. While licence ownership controls were not imposed, there was until recently a legislated requirement that lobster processing companies (also historically subject to limited licences) were majority (80%) Australian owned to control offshore ownership and the associated possibility for transfer pricing to minimize Australian taxes.[89]

Beyond Results

The Fishery appears to be sustainable. In addition to regular recertifications against the MSC, the Western Australia Government notes:

Commercial and recreational catch rates have been maintained near their record-high levels. Fishery-independent egg production indices at all sites are well above both threshold and long-term levels indicating that the biomass and egg production in all locations of the WCRLMF is at record-high levels since surveys began in the mid 1970s. The breeding stock is therefore considered sustainable-adequate.[90]

Yet, "in recent years, research has shown a drop in 'puerulus settlement' – the numbers of late larval-stage lobsters settling on inshore reefs – which can be used to predict catches up to four years ahead. We are continuing to focus on securing long-term sustainability."[91]

Regarding bycatch and entanglements, several measures have been taken:

Sea lion exclusion devices (SLEDs) have now been implemented for rock lobster pots near Australian sea lion breeding colonies. Demersal gillnet fishing effort in the West Coast Bioregion, which has historically been responsible for a very small number of sea lion captures, is now less than 10% of its peak level of the late 1980s. Regulated modifications to rock lobster fishing gear configuration during humpback and southern right whales' northerly winter migration have successfully reduced entanglement rates by more than 65% in recent years.[92]

However, regarding the overall sustainability of the Fishery, it is important to recognize:

> While the case study experience with ITE and particularly ITQ-based management suggests that both are capable of generating significant economic benefits, the longer term impacts of these control systems on ownership of the fishing rights, the flow of benefits from fishing, and the resulting manageability of lobster stocks are questions yet to be answered.[93]

Lessons Learned

A few important lessons learned are:

> The review suggests that early adoption of tradable 'access rights' similar to franchisee contracts in the case study fishery was a critical first step in gaining fisher support for management and the licence values created provide a useful indicator of fishery viability. The historical process where basic biological controls are enhanced with limited entry before developing sophisticated ITE or ITQ-based systems appears an effective and efficient pathway for fishery development with wider relevance.[94]

According to Dr. Simon de Lestang, research scientist with the Fisheries division of Western Australia's Department of Primary Industries and Regional Development, one important lesson learned is that there is no equilibrium in fisheries. Maximum sustainable yield and finding the sweet spot was an old way of thinking. This said, the environment is changing so much and in ways we do not fully understand. Increasing water temperatures is often reported, as well as the impacts of increased ocean acidification on shells; in addition, are changing food sources, currents, and availability of zooplankton.

A more conservative approach to fisheries management is a must. The resulting reduction in catches can be compensated. For instance, more effort can be put into marketing and best handling practices to reduce injury to the lobsters and to ensure lobsters are in peak health to survive the journey. Fisherfolk can make more money and have more confidence in their business due to the greater resiliency of the fishery.

Another important lesson learned for Dr. de Lestang is the importance of demystifying fisherfolk to the general consumer. In contrast to popular opinion, the lobster fishery is a big business with lots of smart, hardworking people in the industry. The lobsters are live caught, not aquaculture, and the MSC certification helps promote a constant evolution of the fishery. This said, if more people bought only MSC certified seafood, the oceans would be in a better place.[95]

Other Resources for Western Australia's MSC-Certified Rock Lobster Fishery

Government of Western Australia's Department of Primary Industries and Regional Development

- http://www.fish.wa.gov.au/Species/Rock-Lobster/Lobster-Management/Pages/default.aspx;
- 2019 Annual Report: https://dpird.wa.gov.au/sites/default/files/2019-10/DPIRD%20Annual%20Report%202019%20-%20PDF.pdf; and
- Resource Assessment Report Western Rock Lobster Resource of Western Australia: http://www.fish.wa.gov.au/Documents/wamsc_reports/wamsc_report_no_9.pdf

MSC

- https://www.msc.org/en-us/home/meet-the-wild-ones/western-australia-rock-lobster; and
- http://rock-lobster-stories.msc.org/

The Lobster Newsletter

- http://www.fish.wa.gov.au/Species/Rock-Lobster/Pages/The-Lobster-Newsletter.aspx

Recfishwest

- https://recfishwest.org.au/

Western Australian Fishing Industry Council

- https://www.wafic.org.au/

Western Rock Lobster Council

- https://www.westernrocklobster.org/

Case Study #3: Rare's Fish Forever Work in the Philippines and Indonesia

Introduction

While many focus on addressing the impacts from large-scale, commercial fisheries, it is important to understand that:

> Although the production (in tonnes of fish) of small-scale fisheries is similar to large-scale fisheries, the former employ most of the fishers and fish workers, have lower production costs and less bycatch. Furthermore, small-scale fisheries provide food (animal protein) and cash to purchase goods and services to poor people worldwide, and aquatic environments can be thus considered as 'banks in the water' that support local economics.[96]

Rare, a global nonprofit organization, has developed the Fish Forever Program, which is being used throughout the Philippines and Indonesia (Fig. 14.3).

Fish Forever is under the umbrella of coastal fisheries reform using fully protected areas to protect essential fish habitat and to both replenish and sustain fish production:

> The idea of Fish Forever was born through three major realizations: that coastal fisheries were largely unmanaged and in decline; that coastal communities were facing an existential crisis impacting the foundation of their economy, food security, culture and wellbeing; and that the most widely-used management tool in coastal waters — Marine Protected Areas (MPAs) — were struggling to be effective, given a lack of community support and fisher compliance, among other factors.[97]

OurFish is an application applied to the program and a signature program of Rare, which is also applied to the Fish Forever Program, is the use of Pride Campaigns to shift key behaviors. The OurFish app is a specific tool used by fish buyers to record their transactions between local fishers—as a business management tool—to manage their business and to collate their taken data. This data (i.e., how many people are fishing, how many fish are taken, etc.) are then reported to local governments and management councils.

The umbrella campaign is to elevate the discussion of how to reform fisheries and connecting this back to communities through being a responsible fisher, a responsible fishing community, participating in the management of its fishery, and abiding by rules (i.e., not fishing in protected areas). There are

Fig. 14.3 Map of the Philippines and Indonesia (*Credit* Peter Hermes Furian)

also tactical campaigns, such as how to drive fisher registration and not just due to legal requirements, but to take pride in their role as a fisher and to supply data. Essentially, Rare is helping to build the social movement.[98]

One of the reasons behind establishing Fish Forever is that limited financing goes toward marine conservation, particularly toward small-scale fisheries. As estimated by Rare approximately "0.5% of all foundation grant-making goes to marine conservation, and an independent study Rare commissioned estimates that between 5-12% of that is directed to small-scale fisheries relevant projects; further, small-scale fisheries related projects makes up less than 0.5% on average of development finance institutions' portfolios."[99] With respect to funding, "since 2012, public funders, including Germany's International Climate Initiative (BMUB- IKI) and Blue Action Fund, Sweden's International Development Cooperation Agency, the GEF, the World Bank/Nordic Development Fund and Nordic Climate Facility, and USAID have committed $27.5 million to Fish Forever."[100]

Rare works throughout Mesoamerica and the Pacific Islands, but its programs in the Philippines and Indonesia are more mature. In the Philippines, it is a slightly different context as the government has already devolved governance to municipal authorities to manage at the local level (i.e., manage fisheries, protection efforts, etc.). Likewise, Rare has been working with local governments and community groups. There is a marvelous networking system set up and the government support is great. For instance, within the national government, managing coastal zones is written into their national development plan through 2022. Further, Rare has approximately 40 staff in the Philippines, a couple of offices, and is building a training hub. In Indonesia, there are two parallel models with the provincial level and under the provincial level is the sub-districts level, which Rare is helping to build up and align.[101]

Identify the Problem

In the Philippines, which is similar to Indonesia:

> A combination of rapid population growth, growing fishing effort, the introduction of destructive fishing practices, and weak fisheries governance, along with coastal habitat loss and degradation through development and pollution, has extracted a heavy toll on the health of coral reefs, seagrass beds and mangrove forests impacting the productivity of the country's fisheries.[102]

This said, "coral reefs grow slowly and die quickly. Global coral cover has declined precipitously, with Indonesia and the Pacific losing 50% of live coral cover over the last 40 years."[103]

One such problem is bleaching as "a three-year global coral bleaching event that began in 2014 was the longest and most damaging event of its kind on record. It followed another severe bleaching event in 2010, leaving little time for recovery."[104]

There are also problems after building reserves and with the structures associated with having a system in place to collect data in perpetuity. Such data can inform how many fish are there, how to protect critical habitat, and how to manage the marine reserve. Rare recognized early that data was very important to measure impact and upon joining Rare, Dr. Steve Box brought the OurFish app. The app's largest footprint was in Honduras and Myanmar and is being rolled out in the Philippines, Indonesia, Mozambique, and Brazil. The problem that OurFish is attempting to resolve is the need for fisheries data, especially data "from the beach." This data is the record of landing data, essentially the transaction between the fisher and the buyer. Community based buyers are the first point of sale and quite often, the fisher is connected to one buyer. From a behavior point of view, the problem is people are not accustomed to writing things down in a logbook. At the end of the day, they want to sell the fish and go home. Thus, you need to train them to undertake a new behavior that they do not want to do. Dr. Box found that buyers are accustomed to writing everything down. For instance, they write down receipts. The buyer might have lent the fishermen money to fish and then the buyer would pay the fishermen the balance when they come back to the shores.[105]

Similarly:

Small-scale fishers catch for subsistence, sale in local markets, recreation and other non-commercial purposes. It's been estimated that half of global catch for human consumption comes from small-scale fisheries, and the communities where such fishing takes place are some of the most dependent on fishing as a trade and food source. Historically, those communities, largely concentrated on the coasts of developing countries, have lacked low-cost, user-friendly means of tracking daily catch.[106]

Another problem is that many MPAs are top-down:

The reasons for small-scale fisher opposition to top-down MPAs are commonly threefold. First, exclusionary MPA policies can lead to unintended social and economic consequences {…}. Second, top-down MPAs have often resulted in the further marginalization of vulnerable groups and populations from marine governance processes {…}. Third, the conservation goals, and thus legitimacy, of top-down MPAs is often not well communicated and therefore questioned or not understood by local fishers.[107]

Thus, community engagement is essential.

Why the Problem Is Important

The Philippines' Tubbataha Reef National Park is a World Heritage Marine Programme, numerous sites throughout the Philippines are listed among the top fifty bioclimatic units (i.e., Palawan, Mindanao and Negros) and the Philippines is one of seven key countries for the Coral Reef Rescue Initiative.

Similarly, Indonesia's Komodo and Ujung Kulon National Parks are World Heritage Marine Programmes, Indonesia's Raja Ampat and Kalimantan are among the world's coral reef "bright spots,"[108] numerous sites throughout Indonesia are listed among the top fifty bioclimatic units (i.e., North Sumatra & Aceh, and Riau Islands), and Indonesia is also one of seven key countries for the Coral Reef Rescue Initiative.

How Problem Was Identified

The Fish Forever Program has taken steps to identify and monitor the problems, including:

- In Indonesia and the Philippines, "Fish Forever conducted more than 1,200 fish and 1,200 coral transects and counted more than 689,000 individual fish to measure the ecological responses to Fish Forever"[109];
- "Fish landings monitoring across the three countries {of Brazil, Indonesia, and the Philippines} documented nearly 56,000 individual fishing trips, recording over 674,000 kilos of fish. Catch data were collected for all target species in Indonesia and the Philippines but were limited to five primary target species in Brazil"[110]; and
- Rare "completed more than 15,000 Knowledge Attitude and Practice (KAP) and household surveys at the beginning and end of the Fish Forever campaigns to measure the social responses to Fish Forever."[111]

Effectiveness of Process for Identifying Problem

Dr. Steve Box often heads straight for the beach and spends time on the boats of fishermen to better understand their reality. As described by Rare:

> These days, he {Dr. Box} starts each project by going fishing with the local fishers. 'Sitting in their boats, watching the way they work and listening to the stories they tell' he says. Many of their stories reveal the same conflict: the risk inherent in relying on fishing in a time of widespread fisheries decline.[112]

Thus, the process of identifying the problems appears quite effective.

Steps Taken to Address the Problem

Rare's three steps for saving nature with behavioral science are:

- 1. Generate collective demand for change (such as through using the Fish Game);
- 2. Coordinate shifting behavior (such as through public pledges); and
- 3. Strengthen the norm (such as through voluntary sea patrols, maintain buoys marking reserve areas, etc.).[113]

One of the steps taken in Indonesia and the Philippines is to develop a data collection system, which Dr. Box has been running for a long time. The first one with cell phones are often the buyers, who are oftentimes a bit wealthier than the fishers. Thus, the buyers are often the initial focus for the data collection system. This said, why train a semiliterate fisher who does not want to use a logbook anyway? Yet, the benefit to the fisher is that a sort of income statement is created and sent back to the fisher via a text or WhatsApp message showing how much money they received. The fisher now has formal, documented income and they could get a loan, demonstrate collateral, and/or be eligible for community financing. Rare is now looking at how to pay fishers through mobile applications instead of cash. Likewise, the rural economy is often informal and does not touch banks, so the government does not have a good handle of the size of the local economy nor a handle on fishery statistics, which are not very good. Implementing data collecting systems, like OurFish, can solve these small-scale fisheries issues (i.e., provide social protections and social funds) and link into digital banking, which are transformational changes in the context of coastal fisheries.[114]

As further explained:

Steve Box and his team from the Smithsonian developed a digital tool to address the challenge. Called OurFish, it's a reporting system that taps into something fishers already carry around — their phones. OurFish is an Android app that consolidates three important pieces of information: the person fishing, the fish species being caught, and from where the fish originate. Designed to work with fisher registration cards that fishers obtain when they register with their local fisheries management department, these apps quickly scan fishers' information from QR codes on the back of their registration cards when they head out and come back in each day. When fishers come back with their catch, buyers weigh and identify fish stock and input the data into the system, which operates using cloud-based computing. OurFish is now actively used in Myanmar and Honduras, and has just gone live in Palau.[115]

Additional steps include:

- In 2011, "Rare, in partnership with National Geographic, launched the first-ever Solution Search, a crowdsourcing competition designed to identify the best examples of community-led solutions to the global challenge of coastal overfishing. 'Bright spots' sourced from the Turning the Tide for Coastal Fisheries competition helped to outline the basic elements of Fish Forever, and in 2012, Rare assembled a network of experts and institutions to identify overfishing's drivers and design and broadly prototype a sustainable solution for the world's coastal fishers and fisheries."[116]
- Rare "partnered with UCSB's Sustainable Fisheries Group and EDF to design and launch Fish Forever. {…} The partnership then transitioned to a less formal, collaborative relationship in 2016. While all three organizations continue as thought partners in advancing global coastal fishery reform, Rare now directs Fish Forever with ongoing scientific support from UCSB-SFG."[117]
- "In-country Fish Forever staff trained and mentored campaign managers to design and implement the campaigns. {…} In the Philippines and Indonesia, most campaign managers were selected from local or national government authorities responsible for Marine Protected Areas (MPAs) or other area-based management systems."[118]
- "To operationalize the program, Rare set up an office in Brazil with technical expertise and bolstered existing offices in Indonesia and the Philippines with additional expertise. Country teams were organized into cross-geography management groups that helped to crystallize best practices, spread them across the program and translate them into local action".[119]
- Before Fish Forever, "Rare had implemented Pride campaigns in Indonesia and the Philippines focused on MPAs (2010–14). The experience, trust and capacity gained in this work laid the foundation for Fish Forever in these two countries {…}.[120]

Results

To date, Rare's Fish Forever Program has achieved numerous results. For instance, the July 2018 Fish Forever Report:

> {…} describes the results of 41 Fish Forever sites, representing over 250 communities across Brazil, Indonesia and the Philippines. It is the first opportunity to analyze the past five years of design (2012–14) and implementation

(2014–17). Using a comprehensive monitoring and evaluation protocol, the report synthesizes information from three country learning reports, 2,400 in-water surveys of coral reefs, 15,000 individual and household surveys, and the landing records from nearly 56,000 fishing trips — and represents the work of 70 Rare staff and 80 partner organizations who have committed the time of more than 557 global staff to this project.[121]

This said, Dr. Box is excited to see how these ideas are catching on and spreading. Likewise, a social movement is building where the practices are not being done at just one community, but are spreading out as communities are networking with other communities and with governments. The key is how to do all of this at scale that makes ecological sense. For instance, in Southeast Sulawesi (Indonesia), Rare is reaching 10,000 fishers and working with 215 communities, 22 councils, 11 district governments, and one provincial government across 360,000 hectares of coastal waters. The behaviors are being reinforced and taking root. Each community has a specific area of reserve they are responsible to manage and the provincial government has adopted this approach as a plan for how to manage coastal waters. Instead of a competitive problem, this is becoming a cooperative stance. The same is occurring in the Philippines, where Rare is working on coordination with local governments and convened a gathering with 350 mayors to talk about coastal fisheries reforms and to bring sustainable production to municipal waters. This level of coordination—to underpin food security and well-being—is a major deal.

Important results from the data presented in the Fish Forever July 2018 Report suggests that the Fish Forever Program is working:

Ecologically, fish are recovering — fish biomass is increasing, both inside and outside no-take reserves; Socially, communities are empowered — social resilience, pride and livelihoods are improving; 51 legal and functional management bodies were established across the 41 sites; 63 managed access areas were built or strengthened, encompassing nearly 600,000 hectares of coastal waters with 27,000 hectares secured in fully protected reserves; and strengthened policies and governance provide a clear path to scale.[122]

Another important result is the development of the Fish Forever Toolkit.

The resulting Fish Forever Toolkit, built iteratively throughout the program, enabled fisheries data collection, fisheries management body development, fisher and community goal-setting and management plan development, and a process for implementing the Fish Forever approach. The Tool kit included: Pride curriculum; Fish Forever Curriculum and E-courses; Fisheries landscape and goal setting; managed access with reserves design; and adaptive fisheries assessment and management.[123]

In July 2018, Indonesia "officially designated the Dampier Strait as a protected TURF (Territorial Use Rights in Fisheries) network. At 211,000 hectares and encompassing 19 villages, the Dampier Strait network is the largest comprehensive TURF-Reserve network in the world."[124]

A further result is the incubation of the Meloy Fund, which is now a wholly owned subsidiary of Rare.[125]

Challenges and How They Were Met

There are many challenges facing the marine environment and more specifically Rare's work throughout Indonesia and the Philippines. It is important to note that:

> Marine systems tend to recover, even from severe pressure, if marine habitat is not irreversibly destroyed. Full recovery can require a decade or more. Marine systems are highly dynamic; recovery tends to be uneven, subject to temporary setbacks, and difficult to attribute to specific causes or triggers. Given this timeframe and the lack of consistent feedback, it is difficult to design and maintain a new system of steadily improving management effectiveness.[126]

This said, one main challenge is how do you sustain the behavioral change long enough so the ecology can bounce back. You need to make long-term, sustainable change. You can mobilize people, give them a sense of direction, get them to act, and have them feel things are changing. However, how do you sustain this sense of purpose for the ecology to bounce back and for the fisheries to recover? You need to make sure to sustain the political will and harness the feedback loop.[127]

One issue facing fisheries management is that "women face systemic and cultural gender-based biases and discrimination in accessing key fisheries resources and in being involved in their management."[128]

There are also challenges with data.

> Most data are based on a three year period. Given the natural variability of fish stocks and marine systems in general and the typically slow speed of fishery recovery, this is insufficient time to establish causality between management action and ecological outcome in analytically irrefutable terms. While the results are statistically relevant by themselves, they do not yet provide proof of the efficacy of Fish Forever's management interventions.[129]

Further, "most of the ecological and social outcome data do not have control groups from non-intervention sites, and this is due simply to time and resource constraints."[130]

There are challenges related to building local leadership, identifying the solution set, both ending at the same end point, and whether management capacity can be built. Another real challenge is the spatial scale, such as trying to build a province model. For instance, how does provincial governments devolve to village councils? What information do communities need to make their decisions? Can we get enough fisheries registered and can we get the data? Ultimately, fishers and the buyers of fish can be managed as household enterprises and for government officials, it is important they see what is really happening in their coastal fisheries.[131]

An overarching challenge as noted by Dr. Steve Box is that in coastal fishing communities:

'people have all the normal day-to-day issues that life brings us, but with a layer of uncertainty about the future that makes things even more complicated,' Steve says. 'Not knowing how their income would change in the future, not being able to make long-term plans, never sure if when they went out that day, they would come back with nothing or something. Understanding that perspective became ingrained in how I look at developing conservation solutions. A focus on how to identify opportunities to reduce that uncertainty, how to work with humility and compassion, knowing that life is difficult and conservation action will often mean making difficult decisions for families and communities — as they decide to trade off short-term gains against long-term sustainability. It's much easier said than done, especially if your uncertainty about what the future has in store has always been high.'[132]

Related to this challenge is the challenge of financial market inclusion. Small-scale fisheries are largely in the informal economy and do not touch the formal, financial community. Thus, there are no formal loans and no credit, and fishers can be susceptible to predatory loans and get caught in a debt trap. Further, the informal nature sets a limit for what price you can sell your fish. The challenge is how do you formalize small-scale fisheries and enable local fishing communities to not only capture wealth, but to build wealth and financial resiliency over time in order to overcome the stress associated with financial uncertainty.[133]

With COVID-19, one of the key challenges that is becoming apparent is that in terms of a crisis, people go home. One of the things we have been talking about is how to stop environmental problems from becoming a humanitarian crisis. If fisheries collapse, the source of high-quality protein goes with it. Likewise, national food supplies and food security based off small-scale fisheries is underestimated, because the data is lacking. If rural economies—whether they are underpinned by agriculture or fisheries—collapse, people move and you get a lot of urban migration. Thus, the challenge is how do you help these communities safeguard these safety nets.[134]

Beyond Results

Fish Forever appears sustainable.

> Fish Forever was designed as a 'mass prototype' model: rather than focus efforts on maximizing testing of individual site-level pilots, Rare chose to test multiple pilots simultaneously. The goals were to understand not only whether managed access with reserves would work at a site, but also if it could 1) make dozens work simultaneously and 2) better understand the key ingredients needed to scale the approach.[135]

Rare is also spreading the word. Rare "launched the Financing Small-Scale Fisheries dialogue at Our Ocean 2017 in Malta, and will expand on this initial dialogue during Our Ocean 2018 in Indonesia and Our Ocean 2020 in Palau. Rare also provided expert input on the creation of the Blue Action Fund {…}."[136]

It is stated that "in two to three years, Rare will have made available a world-class, robust data set on coastal fisheries reform, based on internationally accepted monitoring, evaluation and reporting standards — an essential pre-condition for broad adoption."[137]

The next steps for Fish Forever are:

- Including women and empowering them as decision-makers;
- Enhancing fishing households' financial inclusion and providing financial identity;
- Building resilient coastal communities and climate-smart reserve design;
- Creating proof points at provincial (subnational) scale;
- Stimulating investment in small-scale fisheries;
- Supporting participatory data collection and use; and
- Downscaling elements of The Guidelines for local government implementation.[138]

Rare set up and incubated the Moley Fund, which is an impact investing fund designed to invest in small and medium enterprises by providing capital for more sustainable fishing practices. The Fund overlaps geographically in the Philippines and Indonesia with Fish Forever.

The Fish Forever Program is also looking at blended funding such as development finance linked to government financing, including through municipal budgets and international funding. The key is how to scale up and few are pursuing small-scale coastal fisheries. Likewise, the real trick is how to get the capital down through the layers to the beach, and to figure out what these transfer mechanisms will look like.[139]

Furthermore, "if this progress continues, within five years {as of approximately 2018}, Rare will have supported 10 of the world's highest marine biodiversity and fish-dependent countries on a path to nationwide coastal fishery reform."[140]

Lessons Learned

According to Fish Forever July 2018 Report, there are several lessons learned:

- The Fish Forever approach works under a variety of settings;
- The program needs to build in greater flexibility and patience for empowering communities to co-manage their fisheries;
- Community engagement is central to change and sustainability;
- Peer-to-peer networks increase demand for the approach. Likewise, the increasing number of peer-to-peer networks, such as mayors' networks and fisher associations and federations, have intensified demand for the Fish Forever approach;
- Subnational (provincial) engagement and support are essential to scale;
- Reserve networks and connectivity in network design are needed to optimize both governance and ecology;
- Alternative livelihoods and value chain enhancements must be carefully planned and correctly sequenced; and
- Any new country launch must be contingent on the availability of sufficient financial, operational and political resources.[141]

Dr. Steve Box, who has worked on fisheries his entire professional life, is starting to see trends. An important lesson learned is that the ability to affect change is striking, particularly if you can set a clear path, show communities they do not have to continue to work as they were (i.e., in part by laying out logical arguments), and enable local communities and governments to take ownership of these sets of solutions. These same pathways are starting to work in Indonesia and the Philippines, to Honduras and Mozambique. While these are very different places—socially, politically, culturally, geographically, and linguistically—there are very similar ways to solve problems. Despite all the differences, there is a common thread and that is to mobilize local communities to help steward their natural resources. Anyone can point out the differences, but the key is to find the similarity and pull on those threads to make those local solutions a global solution. Getting to the right elevation and pulling some of the same levers to replicate in other places—whether it be Guatemala, the Pacific Islands, or Brazil—no one really thought that level of similarity approach was practicable.[142]

Other Resources for Rare's Fish Forever Work in the Philippines and Indonesia

An Ocean Mystery: The Missing Catch

- https://www.livingoceansfoundation.org/outreach/films/an-ocean-mystery/

Bloomberg Podcasts on Small Scale Fisheries

- Part 1: https://www.bloomberg.org/blog/follow-the-data-podcast-why-are-small-scale-fisheries-such-a-big-deal/; and
- Part 2: https://www.bloomberg.org/blog/community-based-conservation-local-approach-global-impact/

CNN's Interview with Rare's Rocky Sanchez Tirona

- http://nine.cnnphilippines.com/videos/2018/06/14/Profiles-Rocky-Sanchez-Tirona.html

FAO's Voluntary Guidelines on Securing Sustainable Small-Scale Fisheries in the Context of Food Security and Poverty Eradication

- www.fao.org/3/a-i4356en.pdf

OurFish, Available via Google Play

- https://play.google.com/store/apps/details?id=com.tellybug.fishapp&hl=en

Rare's Work in the Philippines and Indonesia

- http://www.rare.org/program/fish-forever-in-indonesia/; and
- http://www.rare.org/program/philippines/

Saving Nature with Behavioral Science, Dr. Erik Thulin's TEDx Cambridge Talk

- https://www.tedxcambridge.com/talk/saving-nature-with-behavioral-science/; and
- Also see: https://behavior.rare.org/cooperative-behavior-adoption/

"Smokey the Bear Should Come to the Beach: Using Mascot to Promote Marine Conservation" paper by Daniel Hayden and Benjamin Dills

- https://journals.sagepub.com/doi/10.1177/1524500414558126

Case Study #4: Blue Ventures' Work with Locally Managed Marine Areas in Madagascar

The author would like to especially thank Ms. Nicola Bassett from Blue Ventures, who wrote the vast majority of the following case study.

Introduction

Madagascar's coastal waters and littoral zone up to 25 meters from the high-tide point are in the public domain (i.e., owned and governed by the Malagasy State). Locally managed marine areas (LMMAs) are a concept/approach rather than a formal legal framework, and the areas are designated/managed under a number of different legal and governance frameworks. The most formal of these is a gazetted Marine Protected Area under the national Code des Aires Protegees (CoAP).

The management of these MPAs, in most cases, is formally delegated in a co-management arrangement between the government (i.e., who maintains responsibility for enforcement) and the NGOs and community-based organizations representative of local users. Management responsibilities and use rights for local users are defined within an approved management and social safeguards plan; in most cases, use of the resources in the area is limited to local, small-scale users.[143]

In mangrove areas, forestry legislation allows for the delegation of management and sustainable use rights to community-based associations under the "GELOSE" act, which is also being piloted in the local management of fisheries resources.[144]

At the heart of the frameworks mentioned above, however, is a customary rule system known as "dina," which is a legally recognized social contract that allows communities to set and apply rules within their own community. This system has been applied to natural resources since the 1990s, and in most protected areas and management transfer systems, is the main mechanism for the community to ensure local compliance with management plans.[145]

Many LMMAs are not formally designated as MPAs, or management transfers, but are governed by dina alone. This is effective at the local level, but is limited in that the area is not formally gazetted and there is no formal transfer of exclusive use rights to local people. The dina can also not be applied to users who are not part of the local communities, for example, an industrial fleet or migrant fishers.[146]

Blue Ventures, a UK-based charity, works throughout Madagascar, including Velondriake. Velondriake is located in Southwest Madagascar and was initially set up using only local dina, led by a local association known as the Velondriake Association (VA)[147] (Figs. 14.4 and 14.5).

Velondriake was then formally recognized as a protected area in 2015, under co-management between the VA and Blue Ventures (BV) as delegated managers of the protected area, with oversight and support from government services. The overall management objectives and rules were designed in the Velondriake management plan through wide participation and consultation, while the management rules are implemented through a local dina, which has been ratified by regional courts to become a local by law.[148]

Madagascar's LMMAs, which are co-managed with BV, are supported by a mixture of funding sources depending on available resources in the area. The majority of funding for BV's support comes from philanthropic and institutional donors in the form of project grants. The more developed LMMAs, such as Velondriake, are now generating their own sources of income through tourism and fees for alternative livelihood support. New approaches such as blue carbon payments may soon provide additional sources of income to the

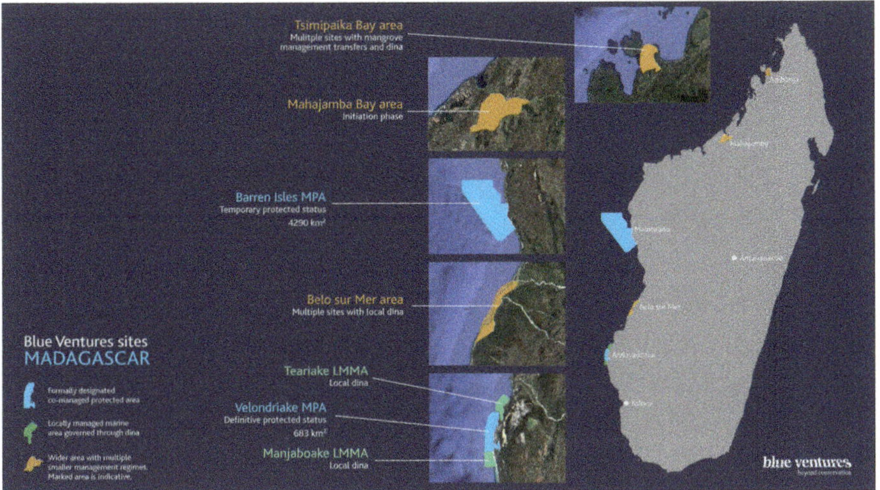

Fig. 14.4   Map of Blue Blue Ventures' Work in Madagascar (*Credit* Blue Ventures)

community groups.[149] Likewise, "to ensure the long-term financial sustainability of these LMMAs, BV is working to develop a variety of mechanisms including marine ecotourism programmes, eco-certifications for sustainable fisheries, and payment for ecosystem services such as mangrove REDD+."[150]

Identify the Problem

The majority of southern Madagascar is arid and infertile, and lacks industry and infrastructure. This results in local communities being extremely dependent on marine resources. However, the region's coastal ecosystems are rapidly degrading as a result of overfishing and the use of destructive fishing practices. The population of settled fishing communities has also grown significantly in the region, adding further pressure on the resources. A lack of access to key health services, such as family planning, along with the lack of access to education and alternative livelihoods, has also contributed to high birth rates and dependency on fishing.[151]

Another problem is that promoting rule enforcement can be a challenge with LMMAs. The Velondriake's dina was developed via a bottom-up process and allows for rules to be applied at the village level without escalation to a higher authority. However, there can be a reluctance to punish infractions due to complex webs of social, family and political ties within the community and a desire to avoid conflict. Thus, promoting ownership of the LMMA dina

Fig. 14.5 Map of Velondriake in Madagascar (*Credit* Blue Ventures)

and the application of rules within the community remains a challenge, especially when compared with traditional dina (for example, concerning theft or domestic violence) which are often applied with little hesitation.[152]

A further problem is that governing associations often have a limited ability to prevent rule breaking by outsiders. For example, the VA has problems with industrial fishing vessels that have acted with impunity within the Velondriake LMMA, and beach seine fishing has proven difficult to control in the north of the LMMA. Outside of formally designated MPAs, LMMAs also lack exclusive use rights to the resources in the area, meaning they are often competing with industrial fishers such as the national trawling fleet. Fisher communities lack the power to counter outsiders and depend on state authorities to provide "back-up" law enforcement. However, the local and state government agencies are severely under resourced and often do not view fisheries as a top priority.[153] Similarly, members of the Velondriake LMMA also have difficulty applying the LMMA *dina* against migrant fishers that pass through the area or settle temporarily, in part due to problems of communicating the LMMA zoning and associated rules.[154]

Lastly, the LMMAs throughout Madagascar are chronically underfunded (i.e., lack sustainable financing and are dependent on NGO support) and suffer from capacity and resource challenges.[155]

Why the Problem Is Important

The reason the aforementioned problems are important is because effective marine and fisheries governance and management measures need to be in place in order for local communities to safeguard their local marine resources for future generations.

Communities with a shared, highly valued resource, such as a fishing ground, are often at risk of falling prey to the 'tragedy of the commons.' Fishers often will not limit their catch as they cannot be sure another competing fisher will reduce their catch and thus, catch all the available fish. However, involving the entire community in planning the management of their resources helps to educate and engage them in resource management. This encourages the local communities to take ownership of the management, implementation, and regulation of the LMMA. Ensuring the rules of the LMMA are enforced gives the area a higher chance of success and acts as a catalyst to broad scale change.[156]

Within the Velondriake LMMA, there are a range of sustainable use management practices in place. For example, there is a 3.2 km^2 area of no-take zones covering a variety of habitats (e.g., coral reefs and mangroves) where all extractive activities are prohibited, areas of temporary closure, and also areas of sustainable mangrove harvest.[157]

Velondriake also forms part of a network of LMMAs along Madagascar's southwest coast, set up using different forms of governance. To the north and south, BV has also supported communities to set up LMMAs using *dina* (although not formally gazetted as protected areas), with the aim of increasing connectivity with neighboring reefs and mangroves to help buffer the Velondriake LMMA. Other NGOs are also supporting similar sustainable use MPAs further south along the coast.[158]

Furthermore, Madagascar is listed as among the top fifty most important bioclimatic units,[159] and is one of seven key countries identified by the Coral Reef Rescue Initiative.[160]

How Problem Was Identified

Community leaders from the 32 villages within the Velondriake LMMA make up the Velondriake Association (VA). The VA has an annual, general assembly where they meet to assess activities over the past year, discuss perspectives, and plan for the future. Association members are employed on a per diem basis to conduct outreach and promote community participation across the LMMA. Problems and issues within the LMMA are therefore raised with community leaders, who then bring and discuss these issues with the VA. BV supports the group, working with them to facilitate discussions and develop management strategies.[161]

Members of the VA are also part of a wider, nationwide network of LMMA managers called MIHARI. They organize learning exchanges and regular forums at regional and national levels, providing invaluable opportunities for LMMA managers to explore common issues and develop collaborative solutions.[162]

Effectiveness of Process for Identifying Problem

The process of identifying the problems appears quite effective.

For instance, BV and the Wildlife Conservation Society (WCS) supported the village of Andavadoaka, located within Velondriake LMMA, to implement a periodic fisheries closure for octopus in 2004, the first marine resource management initiative in the area. The closure led to visible results in very short time scales, sparking neighboring villages to follow suit and get started on fisheries management initiatives. The VA was formed out of these initial meetings and by 2006, 24 neighboring villages along 40 km of coastline had joined together to establish a dynamic and locally appropriate management strategy to improve fisheries.[163]

This has further catalyzed a national movement to develop LMMAs around Madagascar's coasts. The broader grassroots LMMA network, called MIHARI, has since grown from this movement, and is supporting and connecting over 200 communities in developing LMMAs collaboratively around Madagascar's coastline.[164]

Steps Taken to Address the Problem

BV "has published a framework for setting up an LMMA, which can be found in the online toolkit. An outline of the four keys stages is below:

- 1. Initial assessment—gathering as much information as possible about the resource problem and potential solutions, identifying stakeholders and engaging with them to ensure their support and involvement;
- 2. Planning and design—detailed consideration of the design of the LMMA and planning for the management approaches that will be used. The impacts of management actions should be considered to allow for monitoring and adapting (Theory of Change), and the legal framework will be decided to establish legal protection. Emphasis is placed on using a community-based participatory approach;
- 3. LMMA implementation—a management committee and plan are put together, in order to provide structure and accountability. The rules and regulations decided in the planning stage are established and enforced. As such, potential sources of conflict should be identified as well as methods for conflict resolution. A monitoring plan is put together to identify indicators and methods, and those responsible for monitoring. A process and plan is put in place to get demarcation of LMMA zones and the MPA legally gazetted by the government; and
- 4. Ongoing management—LMMA management is an adaptive process of learning and improvement. This often includes continuous ecological and social monitoring of the LMMA and the communities in order to identify issues, measure success and adapt the management practices in place."[165]

Results

As noted by BV, there have been a lot of results, such as:

- BV's LMMA programme focuses on three zones along Madagascar's west coast, which together include over 75 communities, a combined coastal population of more than 40,000 people, and a total marine area of almost 6000 km^2;
- Creation of the Barren Isles protected area, the largest LMMA in the Indian Ocean;
- Creation of Velondriake, the first LMMA in Madagascar to embark on registration as a nationally-recognised protected area;
- Expansion of the LMMA model to communities beyond Velondriake, inspiring and guiding the creation of large-scale LMMAs throughout Madagascar;
- Establishment of over 250 community-managed temporary fishing closures at sites around Madagascar, based on a model for community-based fisheries management first developed in Velondriake; and
- Development of the largest community-based monitoring programme for artisanal sea turtle and shark fisheries in the western Indian Ocean.[166]

As previously mentioned, the success of the initial fisheries closure in 2004, led to the development of a national movement to implement LMMAs around Madagascar's coasts and a nationwide network of LMMA communities (MIHARI) was established. This network has generated a community of individuals involved in local marine management, developing a platform for sharing experiences, knowledge exchange, and championing the rights of small-scale fishers. As a result of this movement within small-scale fishing communities, there has been a push toward policy reform and several fisheries laws have been reformed to reflect the experiences and insight gained from local management efforts.[167]

Periodic, locally managed fishery closures have demonstrated a positive impact on fishery catches and villages fishery income.[168,169]

Challenges and How They Were Met

With respect to challenges, "in recent years Madagascar's unique and fragile marine environment has experienced severe degradation from climate change and overfishing, which threaten the natural resilience of the ecosystems upon which the region's vulnerable communities depend. These fisheries are critical to the livelihoods, cultures and food security of rapidly growing coastal populations. However, declining catches, rapid population growth and a lack of livelihood alternatives have pushed small-scale fishers into deepening poverty, forcing communities to adopt increasingly unsustainable fishing practices to support dwindling catches."[170]

Beyond Results

According to BV, the results are still modest, but plans are in place to expand ecological protection and improve governance. However, the results have been realized against a worsening background trend, with government crises, frequent governmental changes, and decreases in law and order.[171] LMMA and MPA management still needs significantly more support from governments and in funding in order to be sustainable in the long term.[172]

Nevertheless, the Velondriake LMMA is a pioneer in the region, delivering benefits beyond its boundaries, and stimulating the broad replication of LMMAs across Madagascar and overseas. This said, Madagascar now has at least 65 LMMAs and the underlying octopus periodic fisheries closure model has been replicated in a diversity of countries including Comoros, Indonesia, Mauritius, Mayotte, Mexico, Mozambique and Tanzania.[173]

Furthermore:

Next steps for the octopus 'no take' model in the reefs of Southwest Madagascar is to potentially undergo a Marine Stewardship Council certification, an eco-certification granted to potentially increase the value of the octopus. Certified eco-label products can potentially bring in a higher return on octopus, particularly when selling in international markets around the world. This is in discussion with the fishermen in Velondriake, with potential pitfalls of consistent monitoring, surveys, and field management that are difficult to complete without sufficient technical capacity.[174]

Lessons Learned

According to Nicola Bassett of Blue Ventures, it is important to recognize the collaborations between fisher communities and an international NGO. While communities can be effective and efficient managers of their natural resources, they may need outside technical and financial support. Academic literature places great emphasis on "community-based" management, whereas all of BV's supported LMMAs are a co-managed initiative that heavily depend on NGO support. A failure to acknowledge the critical role of outside agencies in these initiatives hampers our ability to understand these systems and muddies the debate about the role of local users in the conservation of natural resources, and the effectiveness and sustainability of community-based interventions.[175]

Although BV does not engage in a formal, adaptive management process with LMMA association members, the management of LMMAs is able to be frequently adjusted to adapt to changing conditions or unexpected outcomes, due to the permanent presence of BV and frequent communications with community members. Adjustments are also made in conjunction with data from ecological, fisheries, and socio-economic monitoring programs. For example, the opening of the first octopus periodic fisheries closure attracted 1,300 fishers from across the region, greatly diminishing per capita yields for Andavadoaka villagers. Discussions over subsequent months led to decisions that all villages should implement their own periodic fisheries closures and that only residents of participating villages should be permitted for fish on the opening day.[176]

Another important lesson learned is that it is important to get communities initially involved with conservation via short-term fisheries management measures, to ensure understanding, trust, and confidence. Only once the results have been demonstrated and the communities have bought into the process and concept of conservation should strategies move on to more ambitious conservation measures.[177]

Lastly, BV has sought to build a diversified portfolio of incentive-driven models, adopting an entrepreneurial approach to fund both its own operations and those of LMMA associations (e.g., the VA), while developing alternative incomes to fishing for local resource users. In addition to fisheries management, this includes holothurian (sea cucumber) and seaweed mariculture, along with community homestay ecotourism developments. However, while the diversification of funding streams has increased resilience and helps LMMAs appeal to a broader range of donors, it has not served to reduce dependence on donors.[178]

Other Resources for Blue Ventures' Work with Locally Managed Marine Areas in Madagascar

Blue Ventures

- https://blueventures.org/;
- LMMA Factsheet: https://bjyv3zhj902bwxa8106gk8x5-wpengine.netdna-ssl.com/wp-content/uploads/2015/10/BV-LMMA-Factsheet-2015.pdf;

- LMMA Toolkit and Open Data Kit: https://blueventures.org/conservation/toolkits/; and
- LMMA Toolkit for communities in Kenya and Mainland Tanzania: https://bjyv3zhj902bwxa8106gk8x5-wpengine.netdna-ssl.com/wp-content/uploads/2019/04/BV_LMMA-Guide-FINAL-LO-RES.pdf

<u>MIHARI (Madagascar Locally Managed Marine Area Network)</u>

- https://mihari-network.org/

<u>PHE (Population-Health-Environment) Network</u>

- https://phemadagascar.org/

Financial Analysis

The following financial analysis will look at return and risk.

Return

Sustainably certified commodities and blue procurement models may have the potential to generate revenues greater than their traditional counterparts. For example, there might be premiums offered for products such as ASC or MSC-certified seafood. In addition to price premiums, products adopting a blue procurement model may gain access to new markets.

Risk

Despite the potential, there are risks associated with sustainably certified commodities and blue procurement.

Business Risk

There are several business risks attributable to fishing quotas:

> The most common objections to ITQs are that they often lead to a concentration of ownership of fishing quotas. However, this issue can be addressed by

placing limits on the total quota that can be owned by any single company or group of related companies (based on the same principles as the anti-monopoly or antitrust laws in many countries). Another objection to ITQs is that they can result in unearned windfall profits when the original quota holders sell their quotas. This can be addressed by taxing the profits of the quota holders who have paid nothing for their original quotas.[179]

A related business risk is enforcement of quotas:

{...} The most common sanctions for violating (i.e., exceeding) quota limits are to impose a fine or to reduce the violator's quotas for the following year(s). In cases of repeated violations, an individual's or company's fishing license(s) can be revoked or criminal penalties can be imposed. ITQ systems may work best in places where the total number of fishing operators is relatively small (e.g., New Zealand), and where there is a tradition of respect for the law and effective law enforcement. Sometimes even in countries that are well known for widespread corruption, quota systems may still be able to work effectively at the municipal or community level. Indeed, some form of IFQs or catch limits is one of the bases for many traditional systems of customary fishing rights.[180]

Ecological attributes must also be taken into account:

In any fishing quota system, the issue of ecological uncertainty also has to be addressed. The agency administering the quota system must be able to measure the current stocks of particular fish species, which vary from year to year based on ecological factors; and then calculate the total allowable (i.e., environmentally sustainable) catch for each target species. The agency must also determine the most appropriate size and geographical boundaries of whatever fisheries management units serve as the basis for allocating quotas.[181]

Another risk concerns profitability:

So far, few sustainable fisheries projects have reached a stage where sufficient profitability can be assured. Several factors can explain it: the complexity of fisheries management, which often involves many stakeholders and regulatory bodies; the difficulty in making firm projections of fisheries' recovery and resulting improved profitability; and the difficulty in making long-term, illiquid investments in developing countries.[182]

The World Benchmarking Alliance (WBA) and its Seafood Stewardship Index, which measures companies' performance against the UN SDGs:

{…} identified some examples of best practice, {but} it found that the long-term sustainability of the seafood industry is being hampered by a lack of oversight of operations and supply chains, impacting measures to eliminate illegal, unreported and unregulated (IUU) fishing, protect ecosystems and respect human rights and working conditions. IUU fishing represents up to 26 million tonnes of fish caught annually. WBA said it is clear that companies are committed to excluding IUU fish in their operations but only about a third of the companies can prove they have specific mechanisms in place to reduce IUU risks in their supply chain.

Strategic Risk

Strategic risks include choosing the right suppliers and choosing the best certification standards. This said, certification standards can be revised and updated over time. Getting certified can also be expensive and time consuming. Part of the expense and time could be attributable to the independent audits and the lack of information (i.e., templates, peers' lessons learned, etc.). In addition, some certified products may not be able to fetch a premium price and/or there could be a long payback period.

Reputation Risk

There was a Marine Aquarium Council (MAC), which no longer exists, but which was trying to certify that cyanide was not used to catch marine aquarium fish. However, there were many challenges and because the testing was faulty, MAC had a hard time definitively proving no cyanide was used and the program's cost was prohibitively expensive.[183]

Liquidity Risk

A liquidity risk occurs if cash is unable to be generated in a timely fashion or inventory is unable to be sold.

Operational Risk

There are numerous operational risks associated with sustainable commodities and blue procurement.

Legal and Regulatory Risk

There can be legal and regulatory risks associated with inconsistent practices and the lack of transparency with regard to the distribution of fishing licenses, quotas, and rights. For instance:

- Who should get the fishing license or quota? Just local communities? Local communities and foreign fleets?
- How large of a catch is allowed by the quotas?
- How long will the quotas last? One year? Five years? Are quotas renewable? If so, for how long and for how many times?
- Are the quotas divisible and transferable? Is it okay for someone to apply for the quota, only to on sell it to someone else?[184]

It is also important to note that the application forms for fishing licenses are sometimes difficult to understand and to fill out as some fisherfolk are illiterate.[185]

Illegally sourced goods or fraudulent papers—such as landing records—claiming legally sourced products can lead to legal and regulatory risk. In addition, there is risk of an increase in royalties or increase in the price of commercial fishing licenses for exported products. This situation may occur whether or not the commodities were produced in a sustainable or non-sustainable fashion.

Credit Risk

Credit risk includes default risk, bankruptcy risk, downgrade risk, and settlement risk.

Market Risk

Market risk includes interest rate risk, equity price risk, foreign exchange risk, and commodity price risk.

Risk, Return, Time (Horizon), Taxes, Liquidity, Legal and Unique (RRTTLLU)

Risk and Return

Please see above for the risk and return associated with blue procurement models.

Time Horizon

The time horizon for investments in sustainable commodities and blue procurement models could be short-term or long-term commitments.

Taxes

There are likely a variety of taxes, such as sales taxes and employment taxes. There can also be tax holidays and competition between countries regarding tax policy.

Liquidity

Sustainably sourced and produced, blue economy commodities are likely to continue having liquidity risks in global markets due to insufficient supply.

Legal

Legal considerations associated with blue procurement models will vary from country to country.

Unique

One unique aspect of blue procurement is that it can be difficult to discern the source of one's inputs due to the global trade of commodities. While this is not really an issue for ecotourism or debt-for-nature conversions, this issue presents a unique situation for companies and countries involved in the global trade of commodities, whether it be seafood or marine ornamental fish.

Policy Analysis

The following policy analysis will look at: defining the problem; establishing goals; selecting a policy; implementing a policy; and evaluating the policy.

Defining the Problem

There are policies that encourage blue procurement models and policies that prohibit unsustainable activities, such as illegal fishing.

Establishing Goals

The Aichi Biodiversity Targets of the CBD are an important collection of twenty targets designed to outline and establish international conservation goals. More specifically:

> Target 6: By 2020 all fish and invertebrate stocks and aquatic plants are managed and harvested sustainably, legally and applying ecosystem based approaches, so that overfishing is avoided, recovery plans and measures are in place for all depleted species, fisheries have no significant adverse impacts on threatened species and vulnerable ecosystems and the impacts of fisheries on stocks, species and ecosystems are within safe ecological limits.[186]

Similarly, SDG #14, "Life Below Water," and particularly targets 14.4 and 14.6 state:

> 14.4: By 2020, effectively regulate harvesting and end overfishing, illegal, unreported and unregulated fishing and destructive fishing practices and implement science-based management plans, in order to restore fish stocks in the shortest time feasible, at least to levels that can produce maximum sustainable yield as determined by their biological characteristics
>
> 14.6: By 2020, prohibit certain forms of fisheries subsidies which contribute to overcapacity and overfishing, eliminate subsidies that contribute to illegal, unreported and unregulated fishing and refrain from introducing new such subsidies, recognizing that appropriate and effective special and differential treatment for developing and least developed countries should be an integral part of the World Trade Organization fisheries subsidies negotiation.[187]

Selecting a Policy

With respect to sustainable fisheries, there are several frameworks, including the Seafood Stewardship Index:

> The World Benchmarking Alliance (WBA) said its Seafood Stewardship Index is the first detailed ranking of these keystone companies based on their sustainability efforts. It presents an overall ranking based on the weighted sum of the results in five key areas:
>
> - Governance and management of stewardship practices;
> - Stewardship of the supply chain;
> - Ecosystems;
> - Human rights and working conditions; and
> - Local communities.[188]

There is also the MSC and ASC-certified seafood standards.

Implementing a Policy

According to the Seafood Stewardship Index:

> Setting leading examples are Labeyrie Fine Foods and Parlevliet & Van der Plas, who share detailed risk assessment tools to monitor specifically for IUU risks in their supply chains.
>
> Meanwhile, 22 of the 30 companies have human rights commitments in place. But they have yet to turn their commitments into procedures. For example, only 20% could demonstrate that they have a remediation mechanism in place.
>
> Overall, Thai Union Group tops the WBA Seafood Stewardship Index 2019 ranking. The company, known for its John West and Chicken of the Sea brands, stands out with robust environmental and social commitments, targets and activities on which it reports publicly.
>
> Second and third in the ranking, respectively, are Mowi, a Norwegian integrated salmon farming company formerly known as Marine Harvest, and Thai-based conglomerate Charoen Pokphand Foods, both demonstrating strengths across most measurement areas. Mowi scored highest on transparency with a strong sustainability strategy detailing how it manages the impacts of its farming operations. Charoen Pokphand Foods, one of the largest shrimp producers in the world, demonstrates strong human rights commitments and group-level sustainability strategies and targets.[189]

Regarding fishing nations, the U.S. turnaround in fisheries management is the "great untold story." This includes improvements in ground fisheries on the West Coast, along with red snapper and grouper in the Gulf of Mexico.[190] As detailed in Charles Clover's book, *The End of the Line: How Overfishing Is Changing the World and What We Eat*, the Alaskan pollock, New Zealand hoki, and Iceland fisheries were relatively well managed.

In addition, Belize's fisheries include a managed access system covering 100% of its waters and fishers had to join management councils. Belize is practicing adaptive management. Further, while the initial focus was on the main commercial species of lobsters and conch, Belize is now starting to manage finfish.[191]

Furthermore, in 2017, Indonesia "became the first nation to make its fishing fleet visible to the world via GFW {Global Fishing Watch}. Since then, Peru, Chile, Panama, Mexico, Namibia, and Costa Rica have either published their vessel data to the GFW map or have publicly committed to do so."[192]

Evaluating the Policy

When evaluating policies, it is important to understand that good quality, climate resilient management of fisheries can help sustain fish populations and increase productivity (in terms of food for people), while improving ocean structure and resilience. Likewise, ocean systems are fairly resilient and can come back in 7–10 years' time.[193]

Yet, around the world, fisheries are typically managed for fisheries only, and often managed as species by species. This said, one must ask whether it is reasonable to expect developing countries to build top-down governance systems with expensive, scientific monitoring, and assessment on a stock by stock basis.

It is also important to understand that the overall productivity shift of fisheries is moving toward the poles. Likewise, the tropics are going to see a significant reduction in fisheries productivity. This begs the questions, what do you do with emerging arctic fisheries? Should decarbonation, wildlife, and biodiversity—which could be more valuable than fisheries—be factored in? Also, what does climate mean for the entire world, and for fisheries, when species move and productivity shifts? There are some regional advisory committees, like for the North Sea and Baltic Sea, which use a more flexible access system for countries.

Nevertheless, it is not clear whether the future oceans will sustain past productivity rates and a serious problem that needs to be addressed is what to do with higher productivity for some stocks and less productivity for others.[194]

Future Outlook for Instrument

One encouraging future outlook for blue procurement is that the International Organization for Standardization (ISO) developed the world's first international standard (ISO 20400: 2017) for sustainable procurement, which was published in April 2017.[195]

Another encouraging outlook is that it is likely consumers will demand more sustainable products. According to study conducted by GlobeScan and MSC:

> One of the major findings of the study is that seafood consumers are increasingly demanding independent verification of sustainability claims in supermarkets (70 percent in 2018, compared to 68 percent in 2016). In fact, independent labeling is particularly important to consumers buying health supplements and fish oils (76 percent), and pre-packed fresh fish (75 percent). We also learned that 70 percent of seafood consumers in North America say that they would like to hear more from companies about the sustainability of their seafood.[196]

There are several supply chain tracking mechanisms and related tools applicable to reducing tropical deforestation, such as Trase,[197] and there are also publicly announced no-deforestation supply chain commitments which are tracked by Supply Change.[198] Forest 500, a project of the Global Canopy Programme, "ranks 350 of the biggest companies in forest-risk supply chains and the 150 biggest investors in these companies."[199] In the future, there is a chance these organizations will expand their coverage to marine commodities or new organizations could fill this current gap. Likewise, Global Fishing Watch (GFW), which is now available on Bloomberg Terminals, will hopefully help to:

1. Improve insurance products for coastal infrastructure, including coral reefs and mangroves that act as natural barriers against extreme weather.
2. Highlight coral reefs' contribution to tourism. The map identifies where reefs are bleached or experiencing strong fishing pressures, information that can be used by the tourism industry to protect assets on which it heavily relies.

3. Contribute to environmental, social and governance (ESG) analysis between land and sea. The new map view allows subscribers to combine Bloomberg Terminal data and locations of interest with a range of other fishing and environmental data sources, as investors consider and map out a broad range of ESG factors.
4. Understand fishing activity. For instance, fishing vessels catch enough tuna to contribute more than $42 billion to the global economy annually. The interactive map visualises where commercial fishing is taking place, what types of fishing gear are used, and in which countries boats are registered.[200]

With the emergence of zero deforestation supply chain commitments from many of the world's largest private and publicly traded companies, including Burger King, Cargill, McDonald's, and Nestlé,[201] one should ponder whether such a movement will arise for sourcing seafood. Similarly, there are jurisdictions and companies beginning to implement terrestrial Produce & Protect models. What about developing a similar structure for fisheries? This said, Planet Tracker's Seafood Tracker Initiative "investigates the impact that financial institutions have in financing global wild-catch fisheries and seafood trade. {Its} aim is to align capital markets with sustainable fisheries management."[202]

There may also be more investment firms, such as Aqua-Spark,[203] that become active in the blue procurement space.

Other Resources on Blue Procurement Models

Aquaculture Stewardship Council (ASC)

- https://www.asc-aqua.org/

Chatham House's resourcetrade.earth

- https://resourcetrade.earth/data

Chefs for Oceans

- https://chefsforoceans.com/

EDF's Seafood Selector

- http://seafood.edf.org/

FAO's Code of Conduct for Responsible Fisheries

- http://www.fao.org/3/v9878e/v9878e00.htm

FAO's The State of World Fisheries and Aquaculture

- http://www.fao.org/state-of-fisheries-aquaculture

FishBase

- https://www.fishbase.us/search.php

FishChoice

- www.fishchoice.com

Global Fishing Watch

- http://globalfishingwatch.org/

Greenpeace's Sustainable Seafood Work

- https://www.greenpeace.org/usa/2018-supermarket-seafood-ranking/

Marine Stewardship Council (MSC) and MSC's Track a Fishery

- https://www.msc.org/home; and
- https://fisheries.msc.org/en/fisheries/

Monterey Bay Aquarium's Seafood Watch

- https://www.seafoodwatch.org/

NEPCon Sourcing Hub

- http://beta.nepcon.org/sourcinghub

NEPCon's Report: Fake Documents: How to Spot Them and What to do About Them

- http://www.nepcon.org/newsroom/fake-documents-how-spot-them-and-what-do-about-them

Planet Tracker

- Aquaculture: https://planet-tracker.org/tracker-programmes/soft-commodities/aquaculture/;
- Seafood: https://planet-tracker.org/tracker-programmes/soft-commodities/seafood/; and
- Soft Commodities: https://planet-tracker.org/tracker-programmes/soft-commodities/

Principles for Investment in Sustainable Wild-Caught Fisheries

- http://www.fisheriesprinciples.org/

Sea Around Us

- http://www.seaaroundus.org/

Sea of Opportunity: Supply Chain Investment Opportunities to Address Marine Plastic Pollution

- http://plasticreport.vulcan.com/wp-content/uploads/2017/02/Sea-of-Opportunity_Full_Final-Digital.pdf

SeaChoice

- www.seachoice.org

SeaWeb

- https://seaweb.org/

Script, Soft Commodities Risk Platform

- https://www.script.finance/

Sustainability Policy Transparency Toolkit (SPOTT)

- https://www.spott.org/

Sustainable Tropical Fisheries and Aquaculture

- https://www.jcu.edu.au/tropical-fisheries-and-aquaculture

The Sustainability Consortium

- https://www.sustainabilityconsortium.org/

Towards A Blue Revolution: Catalyzing Private Investment in Sustainable Aquaculture Production Systems

- https://www.nature.org/content/dam/tnc/nature/en/documents/TNC_ EncourageCapital_TowardsABlueRevolution_FINAL.pdf

Vibrant Oceans

- https://www.bloomberg.org/program/environment/vibrant-oceans/#pro blem

World Benchmarking Alliance's (WBA) Seafood Stewardship Index

- https://seafood.worldbenchmarkingalliance.org/

Notes

1. FSC. "About Us." Accessed December 27, 2019. https://fsc.org/en/page/abo ut-us.
2. Sampson, R. Neil. "The Sustainable Forestry Initiative Program: Seven Years of Sustainable Forestry." *FAO*. Accessed December 27, 2019. http://www.fao. org/3/XII/0700-A1.htm.
3. MSC. "Our History: MSC Timeline." Accessed August 7, 2019. https://www. msc.org/about-the-msc/our-history.
4. Balmford. *Wild Hope*. 173.
5. CDP. "Staff: Paul Dickinson." Accessed November 23, 2016. https://www. cdp.net/en/info/staff.

6. ASC. "Our History." Accessed December 31, 2019. https://www.asc-aqua.org/about-us/history/.

7. Balmford. *Wild Hope.* 175.

8. Martin. *On the Edge.* 82.

9. Roos, Gina. "CDP Launches Water Disclosure Project." *Environmental Leader.* November 20, 2009. Accessed April 13, 2020. https://www.enviro nmentalleader.com/2009/11/cdp-launches-water-disclosure-project/.

10. Forest Trends. "New Site Tracks Corporate Action on Deforestation." March 25, 2015. Accessed April 15, 2017. http://forest-trends.org/releases/uploads/Supply%20Change_PR_.pdf.

11. Marchant, Christopher. "Fishing and Sustainability Data Added to Bloomberg Terminals." *Environmental Finance.* October 24, 2019. Accessed December 31, 2019. https://www.environmental-finance.com/content/news/fishing-and-sustainability-data-added-to-bloomberg-terminals.html.

12. World Benchmarking Alliance. "Press Release: WBA Launches First of Its Kind Global Benchmark Ranking Seafood Companies on Their UN SDG Impact." October 23, 2019. Accessed December 31, 2019.https://assets.sea food.worldbenchmarkingalliance.org/app/uploads/2019/10/WBA-SeafoodSt ewardshipIndex-publication2019-pressrelease.pdf.

13. Bloomberg. "Data on Global Fishing Activity and Ocean Ecosystems Now Available on Bloomberg Terminal." October 24, 2019. Accessed December 31, 2019. https://www.bloomberg.com/company/press/data-on-global-fishing-act ivity-and-ocean-ecosystems-now-available-on-bloomberg-terminal/.

14. Pothering, Jessica. "BlackRock Asks Companies to Report Climate Risks." *Impact Alpha.* March 15, 2017. Accessed April 13, 2020. http://impactalpha. com/blackrock-asks-companies-to-report-climate-risks/.

15. Spergel, Barry and Melissa Moye. "Financing Marine Conservation: A Menu of Options." 49.

16. Ibid. 49.

17. CDP. "Water Program." Accessed December 27, 2019. https://www.cdp.net/en/water.

18. Global Fishing Watch. "Home." Accessed December 27, 2019. https://global fishingwatch.org/.

19. Sea Around Us. "Home." Accessed December 27, 2019. http://www.seaaro undus.org/.

20. ASC. "Certification Update: January 2020." January 2020. Accessed January 17, 2020. https://mailchi.mp/a8327114877e/xr162vrjvq-2684685.

21. Jones, Robert and Jason Scott. "Catalyzing the Blue Revolution: How Investors Can Turn the Tide on Aquaculture." *TNC.* May 8, 2019. Accessed April 13, 2020. https://www.nature.org/en-us/what-we-do/our-insights/perspe ctives/how-investors-can-turn-the-tide-on-aquaculture/.

22. Sheppard (n 79), 207; Wabnitz (n 86), 19. In: Goodwin. *International Environmental Law and the Conservation of Coral Reefs.* 218.

23. FAO. "GLOBEFISH—Information and Analysis on World Fish Trade." February 6, 2017. Accessed April 13, 2020. http://www.fao.org/in-action/glo befish/news-events/details-news/en/c/469648/.

24. MSC. "Working Together for Thriving Oceans: The MSC Annual Report 2018–19." Accessed December 27, 2019. https://www.msc.org/docs/default-source/default-document-library/about-the-msc/msc-annual-report-2018-2019.pdf?sfvrsn=e37c6f59_7. 21.

25. Ibid. 32.

26. Reksodihardjo-Lilley, Gayatri. Interviewed by Brian McFarland. August 2019.

27. Ibid.

28. Ibid.

29. Swanson, Shannon Switzer. Interviewed by Brian McFarland. January 2020.

30. Reksodihardjo-Lilley, Gayatri. Interviewed by Brian McFarland. August 2019.

31. Ibid.

32. Allen, G.R and Donaldson, T.J. 2007. *Pterapogon kauderni. The IUCN Red List of Threatened Species* 2007: e.T63572A12692964. http://dx.doi.org/10.2305/IUCN.UK.2007.RLTS.T63572A12692964.en. Downloaded on 12 December 2019.

33. COP14 Prop. 19, 2; Second FAO Ad Hoc Expert Advisory Panel for the Assessment of Proposals to Amend Appendices I and II of Cites Concerning Commercially Exploited Aquatic Species, 93; COP14 Com. I Rep. 10 (Rev. 1), 2. In: Goodwin. *International Environmental Law and the Conservation of Coral Reefs.* 227.

34. Reksodihardjo-Lilley, Gayatri. Interviewed by Brian McFarland. August 2019.

35. Ibid.

36. Ibid.

37. Ibid.

38. Ibid.

39. Ibid.

40. Ibid.

41. Ibid.

42. Swanson, Shannon Switzer. Interviewed by Brian McFarland. January 2020.

43. Reksodihardjo-Lilley, Gayatri. Interviewed by Brian McFarland. August 2019.

44. Ibid.

45. Maidenhead Aquatics. "About Us: Meet the team." Accessed August 28, 2019. https://www.fishkeeper.co.uk/about-us.

46. Reksodihardjo-Lilley, Gayatri. Interviewed by Brian McFarland. August 2019.

47. FishWise. "Safeway and Fair Trade Launch World's First Fair Trade Certified Seafood." December 20, 2017. Accessed August 28, 2019. https://fishwise.org/human-rights/safeway-and-fair-trade-launch-worlds-first-fair-trade-certif ied-seafood/.

48. Reksodihardjo-Lilley, Gayatri. Interviewed by Brian McFarland. August 2019.

49. Ibid.

50. Swanson, Shannon Switzer. Interviewed by Brian McFarland. January 2020.

51. Reksodihardjo-Lilley, Gayatri. Interviewed by Brian McFarland. August 2019.
52. Ibid.
53. Ibid.
54. Ibid.
55. Ibid.
56. Western Rock Lobster Council. "Map of the Fishery." Accessed March 30, 2020. https://www.westernrocklobster.org/the-western-rock-lobster/map-of-the-fishery/.
57. Penn, J.W. et al. "A Review of Lobster Fishery Management." i23.
58. de Lestang, Simon. Interviewed by Brian McFarland. March 2020.
59. Ibid. i27.
60. Ibid. i24.
61. Gaughan, D.J. et al. (eds). 2019. Status Reports of the Fisheries and Aquatic Resources of Western Australia 2017/18: The State of the Fisheries. 24.
62. Ibid. 31.
63. Ibid. 30.
64. Government of Western Australia. "Rock lobster." Accessed March 25, 2020. http://www.fish.wa.gov.au/Species/Rock-Lobster/Pages/default.aspx.
65. Government of Western Australia. "Lobster Commercial Fishing." Accessed March 25, 2020. http://www.fish.wa.gov.au/Species/Rock-Lobster/Pages/Lobster-Commercial-Fishing.aspx.
66. Gaughan, D.J. et al. (eds). 2019. Status Reports of the Fisheries and Aquatic Resources of Western Australia 2017/18: The State of the Fisheries. 30.
67. Penn, J.W. et al. "A Review of Lobster Fishery Management." i23.
68. Ibid. i23.
69. Government of Western Australia. "Lobster Management." Accessed March 25, 2020.http://www.fish.wa.gov.au/Species/Rock-Lobster/Lobster-Management/Pages/default.aspx.
70. Gaughan, D.J. et al. (eds). 2019. Status Reports of the Fisheries and Aquatic Resources of Western Australia 2017/18: The State of the Fisheries. 31.
71. Government of Western Australia. "Lobster Management." Accessed March 25, 2020. http://www.fish.wa.gov.au/Species/Rock-Lobster/Lobster-Management/Pages/default.aspx.
72. Ibid.
73. de Lestang, Simon. Interviewed by Brian McFarland. March 2020.
74. Gaughan, D.J. et al. (eds). 2019. Status Reports of the Fisheries and Aquatic Resources of Western Australia 2017/18: The State of the Fisheries. 45.
75. Ibid. 28.
76. Ibid.
77. Government of Western Australia. "WA's Western Rock Lobster: Industry Growth Plan to Create WA Jobs." Accessed March 25, 2020. https://www.mediastatements.wa.gov.au/MediaDocuments/Lobster%20fact%20sheet.pdf.
78. de Lestang, Simon. Interviewed by Brian McFarland. March 2020.

79. Gaughan, D.J. et al. (eds). 2019. Status Reports of the Fisheries and Aquatic Resources of Western Australia 2017/18: The State of the Fisheries. 26.

80. Penn, J.W. et al. "A Review of Lobster Fishery Management." i31.

81. de Lestang, Simon. Interviewed by Brian McFarland. March 2020.

82. Ibid.

83. Western Rock Lobster Council. "Sustainable: Marine Stewardship Council." Accessed August 7, 2019. https://www.westernrocklobster.org/sustainability/.

84. de Lestang, Simon. Interviewed by Brian McFarland. March 2020.

85. Nowlis and Friedlander. "Research Priorities and Techniques." In: Sobel and Dahlgren. *Marine Reserves*. 200.

86. de Lestang, Simon. Interviewed by Brian McFarland. March 2020.

87. Government of Western Australia. "WA's Western Rock Lobster: Industry Growth Plan to Create WA Jobs." Accessed March 25, 2020. https://www.mediastatements.wa.gov.au/MediaDocuments/Lobster%20fact%20sheet.pdf.

88. Penn, J.W. et al. "A Review of Lobster Fishery Management." i25.

89. Ibid. i28.

90. Gaughan, D.J. et al. (eds). 2019. Status Reports of the Fisheries and Aquatic Resources of Western Australia 2017/18: The State of the Fisheries. 29.

91. Government of Western Australia. "Lobster Management." Accessed March 25, 2020. http://www.fish.wa.gov.au/Species/Rock-Lobster/Lobster-Management/Pages/default.aspx.

92. Gaughan, D.J. et al. (eds). 2019. Status Reports of the Fisheries and Aquatic Resources of Western Australia 2017/18: The State of the Fisheries. 25.

93. Penn, J.W. et al. "A Review of Lobster Fishery Management." i33.

94. Ibid. i32–i33.

95. de Lestang, Simon. Interviewed by Brian McFarland. March 2020.

96. Hallwass et al. "Could Payment for Environmental Services Reconcile Fish Conservation with Small-Scale Fisheries in the Brazilian Amazon." In *Economic Incentives for Marine and Coastal Conservation*, edited by Essam Yassin Mohammed. 157.

97. Rare. "Stemming the Tide of Coastal Overfishing: Fish Forever Program Results 2012–2017: Full Report, July 2018." 10.

98. Box, Steve. Interviewed by Brian McFarland. November 2018.

99. Rare. "Stemming the Tide of Coastal Overfishing: Fish Forever Program Results 2012–2017: Full Report, July 2018." 44.

100. Ibid. 45.

101. Box, Steve. Interviewed by Brian McFarland. November 2018.

102. Lomboy, Christopher et al. "Building Household Economic Resilience to Secure a Future for Near Shore Fishers in the Philippines." 1.

103. Rare. "Stemming the Tide of Coastal Overfishing: Fish Forever Program Results 2012–2017: Full Report, July 2018." 32.

104. Ibid. 32.

105. Box, Steve. Interviewed by Brian McFarland. November 2018.

106. Irby, S.H. "He Dreams in Data: Fisheries Innovator Steve Box Joins Rare as Its New Vice President of Fish Forever." *Rare*. March 28, 2017. Accessed March 25, 2020. https://rare.org/story/he-dreams-in-data/.

107. Weigel, Jean-Yves et al. "Marine Protected Areas and Fisheries: Bridging the Divide." 203.

108. Cinner, Joshua E. et al. "Bright Spots Among the World's Coral Reefs." *Nature*. (2016). https://doi.org/10.1038/nature18607.

109. Rare. "Stemming the Tide of Coastal Overfishing: Fish Forever Program Results 2012–2017: Full Report, July 2018." 30.

110. Ibid. 30.

111. Ibid. 30–31.

112. Irby, S.H. "He Dreams in Data: Fisheries Innovator Steve Box Joins Rare as Its New Vice President of Fish Forever." *Rare*. March 28, 2017. Accessed March 25, 2020. https://rare.org/story/he-dreams-in-data/.

113. Thulin, Erik. "Saving Nature With Behavioral Science." TEDx Cambridge. Accessed April 7, 2020. https://www.tedxcambridge.com/talk/saving-nature-with-behavioral-science/.

114. Box, Steve. Interviewed by Brian McFarland. November 2018.

115. Irby, S.H. "He Dreams in Data: Fisheries Innovator Steve Box Joins Rare as Its New Vice President of Fish Forever." *Rare*. March 28, 2017. Accessed March 25, 2020. https://rare.org/story/he-dreams-in-data/.

116. Rare. "Stemming the Tide of Coastal Overfishing: Fish Forever Program Results 2012–2017: Full Report, July 2018." 9.

117. Ibid. 10.

118. Ibid. 14.

119. Ibid. 15.

120. Ibid. 16.

121. Ibid. 7.

122. Ibid. 7.

123. Ibid. 18.

124. Rare. "Indonesia Designates World's Largest TURF Network." July 12, 2018. Accessed April 9, 2020. https://rare.org/story/indonesia-designates-worlds-largest-turf-network/#.XAAhW2hKjIU.

125. Meloy Fund. "About the Fund." Accessed April 13, 2020. https://www.meloyfund.com/about.

126. Rare. "Stemming the Tide of Coastal Overfishing: Fish Forever Program Results 2012–2017: Full Report, July 2018." 9.

127. Box, Steve. Interviewed by Brian McFarland. March 2020.

128. Rare. "Stemming the Tide of Coastal Overfishing: Fish Forever Program Results 2012–2017: Full Report, July 2018." 24.

129. Ibid. 31.

130. Ibid. 31.

131. Box, Steve. Interviewed by Brian McFarland. November 2018.

132. Irby, S.H. "He Dreams in Data: Fisheries Innovator Steve Box Joins Rare as Its New Vice President of Fish Forever." *Rare*. March 28, 2017. Accessed March 25, 2020. https://rare.org/story/he-dreams-in-data/.
133. Box, Steve. Interviewed by Brian McFarland. March 2020.
134. Ibid.
135. Rare. "Stemming the Tide of Coastal Overfishing: Fish Forever Program Results 2012–2017: Full Report, July 2018." 19.
136. Ibid. 45.
137. Ibid. 53.
138. Ibid. 50–52.
139. Box, Steve. Interviewed by Brian McFarland. November 2018.
140. Rare. "Stemming the Tide of Coastal Overfishing: Fish Forever Program Results 2012–2017: Full Report, July 2018." 53.
141. Ibid. 47–49.
142. Box, Steve. Interviewed by Brian McFarland. March 2020.
143. Bassett, Nicola. Email message to author. June 18, 2019.
144. Ibid.
145. Ibid.
146. Ibid.
147. Ibid.
148. Ibid.
149. Ibid.
150. Blue Ventures. "Locally Led Marine Conservation." Accessed April 9, 2020. https://blueventures.org/conservation/community-conservation/.
151. Bassett, Nicola. Email message to author. June 18, 2019.
152. Ibid.
153. Ibid.
154. Ibid.
155. Ibid.
156. Ibid.
157. Ibid.
158. Ibid.
159. Beyer et al. "Risk-Sensitive Planning for Conserving Coral Reefs Under Rapid Climate Change." 3.
160. WWF. "Coral Reef Rescue: Resilient People, Coral Reefs, and Their Future in a 1.5° World." Side Event at WWF Pavilion—The Ambition Emergency Room—At the Conference of the Parties (COP25). Madrid, Spain. December 11, 2019.
161. Bassett, Nicola. Email message to author. June 18, 2019.
162. Ibid.
163. Ibid.
164. Ibid.
165. Ibid.

166. Blue Ventures. "Locally Led Marine Conservation." Accessed January 24, 2020. https://blueventures.org/conservation/community-conservation/.
167. Bassett, Nicola. Email message to author. June 18, 2019.
168. Oliver, Thomas A. et al. "Positive Catch & Economic Benefits of Periodic Octopus Fishery Closures: Do Effective, Narrowly Targeted Actions 'Catalyze' Broader Management?" *PLOS ONE*. June 17, 2015. https://doi.org/10.1371/journal.pone.0129075.
169. Blue Ventures. "Marine Management Pays." Accessed April 8, 2020. https://discover.blueventures.org/marine-management-pays/#1.
170. Bassett, Nicola. Email message to author. June 18, 2019.
171. USAID. "Back on the Path to Democracy." October 15, 2019. Accessed April 9, 2020. https://www.usaid.gov/madagascar/back-on-the-path-to-democracy.
172. Bassett, Nicola. Email message to author. June 18, 2019.
173. Ibid.
174. Iyer, Venkat et al. *Finance Tools for Coral Reef Conservation.* 48.
175. Bassett, Nicola. Email message to author. June 18, 2019.
176. Ibid.
177. Ibid.
178. Ibid.
179. Spergel, Barry and Melissa Moye. "Financing Marine Conservation: A Menu of Options." 50.
180. Ibid, 50.
181. Ibid, 50.
182. Vertigo Lab. "Innovations for Coral Finance, ICRI Publication." 41.
183. Swanson, Shannon Switzer. Interviewed by Brian McFarland. January 2020.
184. Janssen et al. "Allocation of Fishing Rights to Support Local Fishermen in South Africa's Western Cape." In *Economic Incentives for Marine and Coastal Conservation*, edited by Essam Yassin Mohammed. 120–135. Emerton, Lucy. "Using Valuation to Make the Case for Economic Incentives: Promoting Investments in Marine". In *Economic Incentives for Marine and Coastal Conservation*, edited by Essam Yassin Mohammed. New York: Earthscan from Routledge, 2014.
185. Ibid.
186. CBD. "Aichi Biodiversity Targets." Accessed September 9, 2019. https://www.cbd.int/sp/targets/.
187. SDGs Knowledge Platform. "Sustainable Development Goal 14." Accessed April 8, 2020. https://sustainabledevelopment.un.org/sdg14.
188. Cripps, Peter. "Seafood Stewardship Index Launched." *Environmental Finance*. October 23, 2019. Accessed December 31, 2019. https://www.environmental-finance.com/content/news/seafood-stewardship-index-launched.html.
189. Ibid.
190. Rader, Doug. Interviewed by Brian McFarland. April 2020.
191. Ibid.

192. Marchant, Christopher. "Fishing and Sustainability Data Added to Bloomberg Terminals." *Environmental Finance*. October 24, 2019. Accessed December 31, 2019. https://www.environmental-finance.com/content/news/fishing-and-sustainability-data-added-to-bloomberg-terminals.html.
193. Rader, Doug. Interviewed by Brian McFarland. April 2020.
194. Ibid.
195. ISO. "ISO 20400:2017: Sustainable Procurement—Guidance." Accessed December 27, 2019. https://www.iso.org/developing-sustainably.html.
196. Sustainable Brands. "Study: Consumers Want to Trust the Fish They Are Eating." Accessed April 8, 2020. https://sustainablebrands.com/read/marketing-and-comms/study-consumers-want-to-trust-the-fish-they-are-eating.
197. Trase. "About Trase." Accessed December 27, 2019. http://trase.earth/about.html.
198. Supply Change. "Profiles." Accessed December 27, 2019. http://supply-change.org/profiles/.
199. Forest 500. "Rankings." December 27, 2019. http://forest500.org/rankings.
200. Marchant, Christopher. "Fishing and Sustainability Data Added to Bloomberg Terminals." *Environmental Finance*. October 24, 2019. Accessed December 31, 2019. https://www.environmental-finance.com/content/news/fishing-and-sustainability-data-added-to-bloomberg-terminals.html.
201. Supply Change. "Profiles." Accessed February 9, 2017. http://supply-change.org/profiles/.
202. Planet Tracker. "Seafood." Accessed April 8, 2020. https://planet-tracker.org/tracker-programmes/oceans/seafood/.
203. Aqua-Spark. "Home." Accessed April 8, 2020. https://www.aqua-spark.nl/.

15

Blue Bonds and Seascape Bonds

Introduction

Green bonds, particularly blue bonds applied to the marine sector, are a newly emerging conservation finance instrument. As explained in the workshop report, *Unlocking Forest Bonds*, and which can be applied to blue bonds, "to access the deepest pools of capital managed by institutional investors, forests bonds will need to be simple, transparent, comparable and liquid, and must hold an investment-grade credit rating."[1]

Historical Overview

The first widely recognized green bonds were issued in 2007 and 2008 by the World Bank[2] and European Investment Bank.[3]

Other notable dates throughout the history of green, blue, and seascape bonds include:

- 2006: International Finance Facility for Immunisation (IFFIm) issues its inaugural bonds.[4]
- 2009: The Climate Bonds Initiative is established.[5]
- 2012: TNC's Conservation Notes are first issued.[6]
- 2014: First corporate green bond is issued, which used the Green Bond Principles and was subjected to a second opinion.[7]
- 2014 (October): The Climate Bonds Initiative's Phase I Technical Working Group is established for the Land Use Sector.[8]

© The Author(s) 2021

B. J. McFarland, *Conservation of Tropical Coral Reefs*,

https://doi.org/10.1007/978-3-030-57012-5_15

- 2014 (December): Credit Suisse's Nature Conservation Notes are issued.[9]
- 2015: Rhino Impact Bonds, "an innovative financing mechanism for site-based rhinoceros conservation," has funding approved for project implementation by the GEF.[10]
- 2016: Luxembourg launches first "100% green" exchange.[11]
- 2016 (May): "The San Francisco Public Utilities Commission (SFPUC) has become the first organisation to issue a green bond certified under the Climate Bonds Water Criteria."[12]
- 2017 (March): The "world's first green bond index-linked exchange-traded fund (ETF)" is listed.[13]
- 2017 (July): The second Climate Certified Water Bond was issued by the City of Cape Town, South Africa.[14]
- 2018 (April): DC Water "successfully issued $300 million in tax-exempt, fixed rate bonds, which includes $100 million designated as (Series A) green bonds for Clean Rivers projects and $200 million in (Series B) bonds for general capital improvements."[15]
- 2018 (April): The "first green bond to be certified under the Marine Renewable Energy Criteria of the Climate Bonds Standard" is issued by ABN AMRO.[16]
- 2018 (August): The World Bank "prices the first Sustainable Development Bond to raise awareness for water and ocean resources."[17]
- 2018 (October): The first blue bond for the use towards "sustainable fisheries and marine projects" was launched for the Seychelles.[18]
- 2019 (January): The Nordic Investment Bank (NIB) priced and launched "a 'landmark' blue bond to help protect and rehabilitate the Baltic Sea. The proceeds of the SEK2 billion ($200 million) inaugural 'Nordic-Baltic Blue Bond' will be used to support the Bank's lending to projects such as the modernisation of a waste water plant in Finland, and the redevelopment of water locks in Stockholm to help deal with extreme weather and rising sea levels."[19]
- 2019 (February): Atlanta, Georgia's Department of Watershed Management "announces first publicly-issued environmental impact bond."[20]
- 2019 (April): The World Bank launches "a Sustainable Development Bond to draw attention to the challenge of plastic waste pollution in the ocean."[21]
- 2019 (June): Chile issues the first sovereign green bond from the Americas.[22]
- 2019 (September): Enel, an Italian energy company, "inks 'world-first SDG-linked' bond" that links "its coupon to the company's achievement of a renewable energy generation target in line with the UN SDGs."[23]

- 2019 (September): The European Bank for Reconstruction and Development (EBRD) "issued what it says is the first bond to exclusively finance climate resiliency projects. It raised $700 million from the issuance of the five-year 'climate resilience' bond. It has a coupon of 1.625%. It was issued at 99.466% for a yield of 1.737% – 9.2 basis points over the five-year Treasury."[24]

Mechanisms of Instrument

As explained by the WWF, "bonds are debts issued by governments, companies, and other institutions as a way to raise funds. The debt must be repaid by the seller over a specified time period. U.S. government bonds must be issued in one of three ways: as general obligation bonds, which are repaid from future tax revenues; as special revenue bonds, which are repaid from revenues generated by specific projects being financed; and as bonds that are hybrids of both."[25]

The green bond industry generally involves bond issuers and investors, along with the supporting infrastructure of standards and verifiers. Furthermore, there is at least one green bond exchange- traded fund (ETF) and the Luxembourg Green Exchange, which offers green bonds, is a member of the Sustainable Stock Exchanges Initiative, which has also been established.[26] Furthermore, the Climate Bonds Initiative's Climate Bonds Standard has a process for the certification of a green bond that includes both a pre-issuance certification and a post-issuance certification.[27]

A few specific types of bonds that are also relevant to the blue bonds discussion are development impact bonds, social impact bonds, and environmental impact bonds.

Development Impact Bonds (DIBs)

DIBs are often structured for development projects that require significant upfront costs, with declining costs overtime. Such development projects could include hydroelectric facilities and wind power projects. This model could be applied to watershed protection payments. For instance, a freshwater offtake arrangement could be structured with downstream users, and then as long as revenues exceed ongoing costs, the profits could be used to pay back the DIBs.[28]

Social Impact Bond (SIBs)

SIBs are often structured with a government as the offtaker, and thus, the government pays for the outcome of a particular social good. For instance, instead of a more traditional approach where a government provides upfront funding for the training of people released from prison, the government could issue a SIB and just pay for the measured outcomes (e.g., number of people released from prison who obtain employment, a degree, or vocational training).[29]

With SIBs, "governments only end up paying for successful programs that achieve intended outcomes, so they have an incentive to support innovation."[30]

Environmental Impact Bonds (EIBs)

EIBs are also often structured with a government as the offtaker, and similarly, the government pays for the outcome of a particular environmental good.

Another somewhat similar financing mechanism is a resilience bond. The "concept of a resilience bond lies in the recognition that upfront investments in programs that foster resilience, will help to reduce larger expenses in the event of a catastrophic event."[31] Similarly, there are catastrophe bonds which "works in the following way: if the disaster does not occur, the investors receive both principal and interest; if the disaster does occur, the principal is used to pay for the cost of recovery, and the investors receive only the interest."[32]

Size of Instrument

According to the Climate Bonds Initiative's (CBI) report entitled, *Green Bonds: The State of the Market 2018*, there was an estimated USD$167.6 billion in green bonds that were issued in 2018 and which met "the CBI green bond database screening criteria."[33] Furthermore, the cumulative issuance of global green bonds, since 2007, was estimated at USD$521 billion.[34]

In 2018, the top five green bond issuers were from:

- 1. The United States (USD$34.1 billion, market share of 20%, and 63 total issuers);
- 2. China (USD$32 billion, market share of 18%, and 69 total issuers);
- 3. France (USD$14.2 billion, market share of 8%, and 12 total issuers);

- 4. Germany (USD$7.6 billion, market share of 5%, and 14 total issuers); and
- 5. The Netherlands (USD$7.4 billion, market share of 4%, and 6 total issuers).[35]

As of mid-2018, outstanding climate-aligned green bonds from developed markets was estimated at USD$489 bilion from 253 climate-aligned issuers and the top five sectors were:

- 1. Transport (USD$257 billion outstanding);
- 2. Energy (USD$128 billion outstanding);
- 3. Water (USD$68 billion outstanding);
- 4. Land Use (USD$29 billion outstanding); and
- 5. Waste (USD$4 billion outstanding).[36]

On a smaller note, the World Bank "raised more than US$660 million in Sustainable Development Bonds highlighting the critical role of ocean and water resources."[37]

Introduction to Case Study

The following case study will examine the Seychelles' Blue Bond. The Seychelles has undertaken several innovative financing schemes, including the Seychelles' Climate Adaptation and Impact Investment Debt Swap. The Seychelles' Blue Bond is the first blue bond with funds ring-fenced to fund "sustainable fisheries and marine projects." The USD$25 million financing arrangement includes the bond issuance, along with assistance from the World's Bank International Bank for Reconstruction and Development and the GEF.

Case Study #1: The Seychelles' Blue Bond

Introduction

The Seychelles, located off the coast of East Africa and relatively close to Madagascar, is comprised of 115 islands and manages an EEZ of approximately 1.4 million square kilometers[38] (Fig. 15.1).

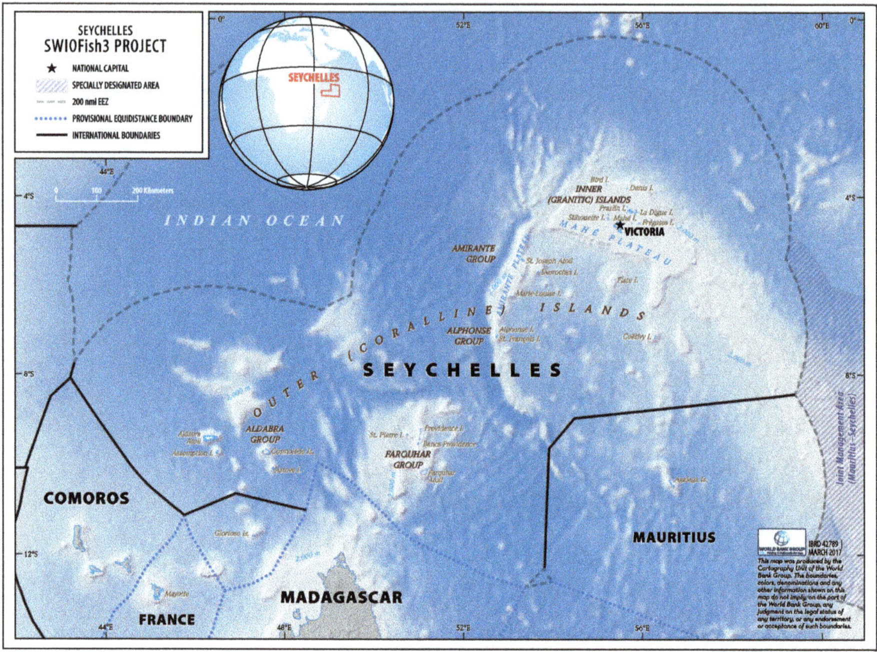

Fig. 15.1 Map of the Seychelles (*Credit* Cartography Unit, The World Bank)

The Seychelles has undertaken several innovative financing schemes. As previously mentioned, the Seychelles participated in the first of its kind Climate Adaptation and Impact Investment Debt Swap. Then in September 2017:

> The World Bank approved a bond of over US$25 million for the Republic of Seychelles, 'to improve the conservation of its marine resources and expand seafood value chains' {...}

In financial terms, this blue bond will work as follow:

1. A US$15 million bond issuance, supported by a US$5 million loan and a US$5 million guarantee;
2. This will later be supplemented with a US$5 million grant from GEF and a US$5 million non-grant instrument;
3. The total funds entering Seychelles to allocate to the project will thus be US$25 million (plus US$10 million in the form of a guarantee and a loan to repay the coupon in the early years).

Two entities in the Seychelles will work jointly: SeyCCAT will be issuing grants, while the Development Bank of Seychelles will be issuing concessionary loans to eligible organisations and businesses.

The principal goal of this strategy, in complementarity with the DFNS {debt-for-nature swap} funds, is 'to achieve the Blue economy strategy of the country, by diversifying the local economy, creating high value jobs, ensuring food security, while sustainable managing and protecting marine resources.' This World Bank project is part of the third phase of the South West Indian Ocean Fisheries Governance and Shared Growth Program (SWIOFish) that aims to increase the economic, social and environmental benefits from sustainable fisheries.

SeyCCAT and Seychelles representatives are also thinking to extent [sic, extend] the scope of financeable projects, as some non-marine projects may have indirect benefits for coral ecosystems (e.g. waste and wastewater treatment in coastal areas, climate change mitigation projects, etc.).[39]

Similarly, issued in 2018:

The Seychelles Blue Bond is a $15 million debt issue, with a loan guarantee from the International Bank for Reconstruction and Development (IBRD, World Bank). The bond issue is coupled with an additional $10 million of funding, consisting of a $5 million grant from the Global Environment Facility and $5 million in low interest loans from the IBRD. The full $25 million will go to two implementing entities – the Seychelles Conservation and Climate Adaptation Trust (SeyCCAT) and the Development Bank of the Seychelles. The economic development programs are designed to promote a healthy and sustainable fishing industry, thus generating increased tax revenue, which will enable the Seychelles to repay the bond.[40]

As explained by Rob Weary of TNC, the government itself issued the USD$15 million blue bond. Approximately USD$3 million will come to SeyCCAT over six years, half of this will go toward the management plan. The remaining USD$12 million will go to a revolving fund through the Development Bank of the Seychelles.[41] The Development Bank of Seychelles will then "on-lend to eligible projects."[42]

Furthermore, "Standard Chartered acted as placement agent for the bond and Latham & Watkins advised the World Bank as external counsel. Clifford Chance acted as transaction counsel."[43]

Identify the Problem

The Seychelles has a very specific country background, which in the relatively recent past, evolved out of its fiscal crisis of 2008–2009 and which effectively led to the bankruptcy of the country. While the Seychelles have been able to gradually reduce debt—in part due to the debt-for-nature conversion—the Seychelles still had difficulty borrowing money from the capital markets.[44]

Why the Problem Is Important

The problems are important because the Seychelles has been a strong leader in maritime protection.[45] Further, this is the first blue bond with funds ring-fenced to fund "sustainable fisheries and marine projects."

How Problem Was Identified

The problems were in part identified by the World Bank and the GEF, which assisted with the blue bond.[46]

Effectiveness of Process for Identifying Problem

The process for identifying the problems was effective as the World Bank Group has a standardized approach that involves two critical, publicly available documents:

- 1. The Systematic Country Diagnostics Report, which discusses a country's key development challenges; and
- 2. The Country Partnership Framework Report.

Both of these documents are agreed to by the host country and are approved by the World Bank's board of directors. Such documents essentially approve the areas of engagement and the overall strategy. Furthermore, there is a triangulation of:

- 1. The challenges identified;
- 2. The host government's priorities; and
- 3. The institutional comparative advantages.

Steps Taken to Address the Problem

The debt-for-nature conversion came out of the 2008–2009 fiscal crisis and helped to build up an institutional infrastructure to undertake ocean protection work. The blue bond tapped into this existing infrastructure, namely

SeyCCAT, and helped the Seychelles to overcome some of the financial constraints.[47]

Results

There were several notable results from the Seychelles blue bond. The "$15 million deal was privately placed with Calvert Impact Capital, Nuveen, and Prudential. Its primary goal is to assist the implementation of fisheries management through economic diversification and sustainability, through loans issued by the Development Bank of Seychelles."[48] Likewise, "it was sold in a private placement to three US-based investors – Nuveen, the asset management arm of TIAA, Prudential Financial and Calvert Impact Capital – which each bought $5 million of the notes."[49]

It is important to understand that people often get excited before there are actual results on the ground. Thus, while funding has been mobilized, the process has just begun for selecting the first projects. Likewise, there will be two disbursement windows, with the first for NGO grants and then second, for providing financing with attractive terms for private sector projects.[50]

Challenges and How They Were Met

While there are overall financing constraints for the blue economy, the blue bond did not face challenges associated with lack of investor interest and an undersubscription because the World Bank worked with a few institutional investors.

The challenges were more based around how to get the "biggest bang for the buck." Likewise, if the Seychelles sought to borrow commercially, they would not have received attractive rates; rather, how do you leverage finite public resources to bring in private money?[51] To address this challenge, the bond is backed with a "$5 million loan from the GEF and also received a partial guarantee of $5 million from the World Bank to help reduce the effective interest rate of the coupon from 6.5% to 2.8%. An unspecified donation from the Rockefeller Foundation covered the transaction costs."[52]

Although "the issue has no form of external assessment, Benoit Bosquet, practice manager for environment and natural resources at the World Bank, said the projects to which the proceeds are disbursed 'will be subject to World Bank policies and procedures'."[53]

Beyond Results

As reported by *Environmental Finance*:

> The bond issuance is seen as an important development for sustainable finance. It is being trumpeted as a good example of 'blended finance', which sees public and private capital combine to promote development goals, and it is hoped that it could help catalyse a market for 'blue bonds' to help promote the protection of the oceans.[54]

Thus, ideally, the Seychelles blue bond will be a model for sustainability and hopefully, other countries will follow suit with similar blue bonds.

Lessons Learned

One important lesson learned is that it is critical to have a broad coalition of stakeholders and one must understand what all the different partners want. This is especially important in a small, developing country such as the Seychelles. For instance, with a blue bond issuance, you need the finance ministry involved, but you also need other responsible ministries such as the ministries of fisheries, the environment, tourism, or another ministry that is responsible for coastal zone protection (i.e., such as the public ministry).[55]

Other Resources on the Seychelles' Blue Bond

World Bank Group's Country Partnership Framework for the Seychelles

- https://projects.worldbank.org/en/projects-operations/country-strategies#1; and
- http://documents.worldbank.org/curated/en/300111532057432476/pdf/SEYCHELLES-CPF-06222018.pdf

World Bank Group's Systematic Country Diagnostic for the Seychelles

- https://projects.worldbank.org/en/projects-operations/country-strategies#3; and
- http://documents.worldbank.org/curated/en/191181499447495374/pdf/Seychelles-SCD-FINAL-23Jun17-06282017.pdf

Financial Analysis

The following financial analysis will look at return and risk.

Return

The returns for blue bonds generally need to meet the market return. The challenge is how to structure the blue bond so that it meets this market return, while also delivering on its blue credentials. The use of proceeds, in general, needs to be ring-fenced and used specifically for eligible blue projects.

Risk

Business Risk

On-the-ground projects are often too small and lack both the business and financial expertise to attract large-scale financing from the issuance for blue bonds as well as from impact investors.

Strategic Risk

One strategic risk to the blue bond industry is that there are "multiple taxonomies that exist to define the impacts of green bonds {which} is blocking the flow of capital and slowing market growth."[56]

Furthermore, the "lack of size, diversification and awareness are among the main challenges facing the fledgling green bond market, a poll of fund managers has revealed. {...} But the most common gripe was the size of the market, which was cited by 10 of the 22 respondents. There were some $175 billion of issues in 2017, meaning it is still a fraction of the overall fixed income market."[57] Likewise, "another major potential stumbling block for blue finance is whether there is a big enough investment universe across public and private asset classes to attract institutional investors, or whether large-scale public financing is still needed as a guarantee."[58]

Reputation Risk

One reputation risk to the blue bond industry is the perception that some bonds lack additionality.

> According to an extract from the report {by the 2° Investing Initiative}: 'Issuers of green bonds usually invest in both green and brown projects. On average, green bond issuers invest more in renewables, but because of their existing assets, they will be slightly less invested in renewables than the market as a whole in 2022, and largely misaligned with the IEA 2°C target. Thus, green bond investors end up being less exposed to green projects.' {…} 'Nothing forces the issuers to deviate from their 'business as usual' investment plans when issuing green bonds,' it added.[59]

Similarly, "the People's Bank of China will step up scrutiny of green bonds to improve transparency and ensure the capital raised is channelled into projects that bring environmental benefits," to address issues of potential greenwashing.[60]

Another reputational risk, "{…} just as with other types of environmental instruments, there are difficulties in defining the sorts of projects and investments that should be eligible as blue finance, and there is plenty of debate around what is truly blue."[61] For instance:

> {…} there is a whole range of sectors beyond aquafarming and cleaning water that are part of the blue economy. These include investments in offshore wind, which has a huge, growing and highly investable services industry with logistics businesses, tourism and particularly eco-tourism, which is a growing area, he adds. Wave and tidal power technologies also form part of the blue economy. Yet, more controversial, is whether shipping, particularly fuel efficiency in shipping, should be included in blue financing.[62]

Similarly, some countries are issuing green bonds, while simultaneously financing environmental destructive practices through other means.[63] A similar reputation risk is greenwashing, which could result from the lack of consistent approaches by issuers in identifying the main issues to report on. This risk can then be compounded by sector expertise of second-party opinion providers. These issues are currently being addressed by the Climate Bonds Initiative.[64]

Liquidity Risk

The liquidity risk associated with a blue bond will depend on the characteristics of the bond. If the bond is unique, it will have less liquidity. If the bond is more "vanilla" (i.e., standardized), it will have greater liquidity.

However, due to the relatively small size of the blue bond market when compared to the larger fixed income market, the overall liquidity of the blue bond market is small.

Operational Risk

There are operational risks if the use of proceeds does not deliver on the stated objectives of the blue bond. There are also operational risks associated with designing, structuring, and commercializing the blue bond. In addition, there may be operational risks associated with meeting a blue bond certification standard, along with satisfying third-party reviewers and other stakeholders who may be critical such as NGOs and host governments.

Legal and Regulatory Risk

One regulatory risk associated with blue bonds is novel risk, as these bonds are perceived as new, even though they use time-tested structures.

Credit Risk

Credit risk includes default risk, bankruptcy risk, downgrade risk, and settlement risk.

This said, credit risk to consider is that:

The finance raised from selling a bond and the revenue generated to pay it back can either be held on the financial accounts of the issuing institution (on balance sheet) or in a legally independent entity (off balance sheet). Combined with the choice of revenue generating mechanism, the choice of institutional arrangements has important implications for the risk to both the bond investor and the bond issuer.[65]

Market Risk

Market risk includes interest rate risk, equity price risk, foreign exchange risk, and commodity price risk. Potential market risks that could lead to a "{…} possible dip in the green bond market could be triggered by prolonged market volatility and emerging market deleveraging. A cut in green bond supply could also be unleashed by excessive regulation and green bond cannibalization,"[66] {which is} "the replacement of green bonds with social or sustainability bonds."[67]

Risk, Return, Time (Horizon), Taxes, Liquidity, Legal and Unique (RRTTLLU)

Risk and Return

See above for the risk and return associated with green, blue, and seascape bonds. Essentially, it is important to provide investors with an attractive risk-adjusted return.

Time Horizon

One aspect to consider with respect to the time horizon is that the use of proceeds from blue bonds goes to projects and programs which may terminate before or after the bond funding time horizon. This creates a time horizon mismatch within the bond structure.

Taxes

From the investors' point of view, taxes will likely be comparable to traditional bonds. If you are able to construct and issue a bond with an attractive tax structure (i.e., such as U.S. municipal bonds), then this will influence (i.e., increase) investor demand.

Liquidity

One component of liquidity will depend on the issuer. For instance, a company issuing a blue bond with an investment grade will likely have greater liquidity, than a country issuing a blue bond with a lower or nonexistent investment grade. Due to the relatively small size of the blue bond market compared to the larger fixed income market, the overall liquidity of the blue bond market is small.

Legal

Legal considerations associated with blue and seascape bonds will vary from country to country. While such bonds will likely have very similar legal requirements as traditional bonds, there may be some unique legal considerations related to the certification of the bond's blue projects.

Unique

One unique aspect is that there could be a funding mismatch over the time horizon. In addition, there could be unique concerns about the coupon payments, such as payments made from the sale of blue carbon offsets.

Policy Analysis

The following policy analysis will look at: defining the problem; establishing goals; selecting a policy; implementing a policy; and evaluating the policy. This said, blue bonds depend on a wide variety of policies to be in place.

Defining the Problem

Choosing a country to make an investment via the issuance, or purchase, of a blue bond depends on factors such as the overall ease of doing business, level of corruption, rule of law, and whether there is sufficient demand for blue bonds backed by blue projects.

Establishing Goals

Clear public policies allow value to be ascribed to a potential issuer (e.g., allow revenues from the sale of blue carbon credits, for example, to be used to pay back investors). This, in turn, can incentivize investors to buy such bonds.[68] In addition, public policies that help to promote an overall ease of doing business, minimize corruption, and promote a strong rule of law would be helpful for blue bonds.

Selecting a Policy

As of April 2020, only the Seychelles has issued a blue bond. Regarding selecting a policy for future blue bonds, the global rankings of the top twenty countries with the largest coral reefs for ease of doing business, according to the World Bank as of July 2019, were as follows[69]:

- #8: U.S.
  - #16: enforcing contracts
  - #36: trading across borders
  - #50: protecting minority investors
  - #53: starting a business
- #9: U.K.
  - #15: protecting minority investors
  - #19: starting a business
  - #30: trading across borders
  - #32: enforcing contracts
- #15: Malaysia
  - #2: protecting minority investors
  - #33: enforcing contracts
  - #48: trading across borders
  - #122: starting a business
- #18: Australia
  - #5: enforcing contracts
  - #7: starting a business
  - #64: protecting minority investors
  - #103: trading across borders
- #32: France
  - #1: trading across borders
  - #12: enforcing contracts
  - #30: starting a business
  - #38: protecting minority investors
- #73: Indonesia
  - #51: protecting minority investors
  - #116: trading across borders
  - #134: starting a business
  - #146: enforcing contracts
- #115: Solomon Islands
  - #98: starting a business
  - #110: protecting minority investors
  - #156: enforcing contracts
  - #160: trading across borders
- #139: Maldives
  - #71: starting a business
  - #125: enforcing contracts
  - #132: protecting minority investors
  - #155: trading across borders
- #77: India
  - #7: protecting minority investors
  - #80: trading across borders
  - #137: starting a business
  - #163: enforcing contracts
- #92: Saudi Arabia
  - #7: protecting minority investors
  - #59: enforcing contracts
  - #141: starting a business
  - #158: trading across borders
- #94: Vanuatu
  - #110: protecting minority investors
  - #132: starting a business
  - #136: enforcing contracts
  - #147: trading across borders
- #101: Fiji
  - #79: trading across borders
  - #97: enforcing contracts
  - #99: protecting minority investors
  - #161: starting a business
- #108: Papua New Guinea
  - #89: protecting minority investors
  - #140: trading across borders
  - #143: starting a business
  - #173: enforcing contracts
- #144: Tanzania
  - #64: enforcing contracts
  - #131: protecting minority investors
  - #163: starting a business
  - #183: trading across borders

- #118: Bahamas
  - #84: enforcing contracts
  - #105: starting a business
  - #132: protecting minority investors
  - #161: trading across borders
- #120: Egypt
  - #72: protecting minority investors
  - #109: starting a business
  - #160: enforcing contracts
  - #171: trading across borders
- #124: Philippines
  - #104: trading across borders
  - #132: protecting minority investors
  - #151: enforcing contracts
  - 166: starting a business

- #150: Marshall Islands
  - #75: trading across borders
  - #75: starting a business
  - #103: enforcing contracts
  - #180: protecting minority investors
- #160: Micronesia
  - #61: trading across borders
  - #170: starting a business
  - #185: protecting minority investors
  - #184: enforcing contracts
- #189: Eritrea
  - #103: enforcing contracts
  - #164: trading across borders
  - #174: protecting minority investors
  - #187: starting a business

The global rankings by Transparency International of the Corruption Perceptions Index 2018[70] for the top twenty countries with the largest coral reefs, were as follows:

- #11 (Tied): U.K.
- #13: Australia
- #21: France
- #22: U.S.
- #29: Bahamas
- #58: Saudi Arabia
- #61 (Tied): Malaysia
- #64 (Tied): Vanuatu
- #70 (Tied): Solomon Islands
- #78 (Tied): India

- #89 (Tied): Indonesia
- #99 (Tied): Philippines
- #99 (Tied): Tanzania
- #105 (Tied): Egypt
- #124 (Tied): Maldives
- #138 (Tied): Papua New Guinea
- #157: Eritrea
- Unranked: Fiji
- Unranked: Marshall Islands
- Unranked: Micronesia

The global rankings by the UNDP of the 2017 Human Development Index (HDI),[71] which combines life expectancy, education, and per capita income indicators, for the top twenty countries with the largest coral reefs were as follows:

1. #3: Australia (HDI of 0.939)
2. #13: U.S. (0.924)
3. #14: U.K. (0.922)
4. #24: France (0.901)
5. #39 (Tied): Saudi Arabia (0.853)
6. #54: Bahamas (0.807)
7. #57: Malaysia (0.802)
8. #92 (Tied): Fiji (0.741)
9. #101 (Tied): Maldives (0.717)

11. #113 (Tied): Philippines (0.699)
12. #115: Egypt (0.696)
13. #116 (Tied): Indonesia (0.694)
14. #130: India (0.640)
15. #131: Micronesia (0.627)
16. #138: Vanuatu (0.603)
17. #152: Solomon Islands (0.546)
18. #153: Papua New Guinea (0.544)
19. #154: Tanzania (0.538)

10. #106 (Tied): Marshall Islands (0.708)   20. #179: Eritrea (0.440)

The World Justice Project has designed a rule of law ranking which primarily utilizes nine factors (e.g., Constraints on Government Powers, Absence of Corruption, Open Government, Fundamental Rights, Order and Security, Regulatory Enforcement, Civil Justice, and Criminal Justice, and Informal Justice).[72] Of the 126 countries assessed, the following global rankings are for the top 20 countries with the largest coral reefs:

1. #11: Australia (Overall Score of 0.8)
2. #12: U.K. (0.8)
3. #17: France (0.73)
4. #20: U.S. (0.71)
5. #39: Bahamas (0.61)
6. #51: Malaysia (0.55)
7. #62: Indonesia (0.52)
8. #68: India (0.51)
9. #90: Philippines (0.47)
10. #91: Tanzania (0.47)
11. #121: Egypt (0.36)
12. Unranked: Eritrea (N/A)
13. Unranked: Fiji (N/A)
14. Unranked: Maldives (N/A)
15. Unranked: Marshall Islands (N/A)
16. Unranked: Micronesia (N/A)
17. Unranked: Papua New Guinea (N/A)
18. Unranked: Saudi Arabia (N/A)
19. Unranked: Solomon Islands (N/A)
20. Unranked: Vanuatu (N/A)

Implementing a Policy

The Seychelles' blue bond helps to fund the SeyCCAT Blue Grants Fund, which is "governed by the Board of Directors, supported by a finance committee and grants committee, with the day-to-day leadership provided by the CEO and Executive team."[73]

Evaluating the Policy

It is difficult to evaluate the policies surrounding the Seychelles' blue bond, as the issuance is still relatively new.

Future Outlook for Instrument

The recent trend for green bonds, which may foreshadow the case for blue bonds, has seen tremendous growth and this trend is likely continue in the years ahead. For instance, "HSBC expects a boost in supply of green, social and sustainability (GSS) bonds in 2019, with as much as $250 billion in total GSS bonds, up from $199.5 billion in 2018."[74] This said, "as markets stabilise, supply growth could be spurred by factors including emerging market green bond supply and by the impacts of climate change becoming visible {…}."[75] Similarly, "cumulative issuance of green bonds, since the market began in 2007, has reached $800 billion which means 'we should reach the $1 trillion market next summer {2020}'."[76] Yet:

Given the current trend of green bonds (not yet designed for conservation purposes) and blue bonds (still to be issued and not designed specifically for coral ecosystems but for broader marine issues), a new kind of ad hoc bond should be developed to better address the specific financing challenges facing coral conservation: coral bonds. This needs to be designed and developed in a different way, enabling it to address inherent coral conservation targets while giving a return on investment different from classic financial market bonds. The framework of its application needs to be regulated to make sure that only sustainable projects are funded and to avoid possible misuse of the bond. It is also important to determine the kind of funds and partners (taxes, private funds, international institutional back-up, etc.) that are most likely to be mobilized through the coral bond.[77]

It is also important to note that the issuance of sukuk bonds could increase and this could specifically increase coral reef financing in Indonesia or Malaysia.

In the ASEAN region, the ASEAN Green Bond Standards (AGBS) catalysed green bond issuance in 2018. The momentum has carried over into 2019, with one new issuer from the Philippines becoming a repeat issuer in short order and The Republic of Indonesia coming to market with a second sovereign sukuk.

ASEAN is an association of 10 countries and six of them have seen green bond issuance: Indonesia, Malaysia, the Philippines, Singapore, Thailand and Vietnam. AGBS is supported by the ASEAN Capital Markets Forum, a cooperation platform for the ten financial sector regulators, and provides a common framework for green bond issuance, which should facilitate market transparency.

The world's first green sukuk was issued from Malaysia's Tadau Energy (MYR250m/USD58m), and the first green sovereign sukuk by the Republic of Indonesia (USD1.25bn).[78]

This said,

Indonesia has the largest green bond market in ASEAN, with 39% of total regional issuance. Singapore and Malaysia follow with shares of 35% and

19% respectively. Issuance from the 6 ASEAN markets is dominated by non-financial corporates (30% of the regional total).

{…} Island nations Fiji and Seychelles have both issued a sovereign green bond, with the Republic of Seychelles coming to market with the first sovereign blue bond. A blue bond finances projects that support marine reserves and sustainable fisheries, which also classify as green assets. These two issuances are particularly significant for island nations as they are particularly exposed to longer-term climate-related risks, such as rising sea levels and extreme weather.[79]

In addition:

Australia's green bond market is set to reach record highs this year, after A$3.9 billion ($2.6 bln) worth of new bonds were issued in the first half of the year, putting it on track to beat last year's A$6 bln, according to a new Climate Bonds Initiative report. As much as 43% of the funds were raised for green building, with energy getting 25% and transport 24%. Australia remains the third biggest green bond market in the Asia-Pacific and China and Japan, and the tenth largest in the world.[80]

Another thought, related to Australia, is whether a pool of farms could back a green or blue bond that would be used to implement on-farm improvements and which would ultimately help offshore reefs.[81]

Resilience bonds are, "at this stage, purely conceptual. However, with the achievement of the necessary scale and engagement by a multilateral or private actor working in concert with governments, there is potential for resilience bonds to provide both a source of reef finance and an investment vehicle for conservation-minded impact investors."[82]

Then again:

Following the Seychelles issue, TNC is looking to enter the market by raising $1.6 billion in blue bonds for ocean conservation. It will do this by acquiring debt from countries with junk status, ideally for $0.80 in the $1. The difference in what it borrows and the price it pays for the debt will be spent on protecting four million square kilometres of coastline.

The NGO's plan is to refinance and restructure a portion of a coastal nation's sovereign debt with the help of proceeds of the blue bonds and credit enhancement from a public institution, such as the World Bank or Overseas Private Investment Corporation (OPIC). As this will lead to lower interest rates and longer repayment periods for the sovereign's debt, the nation is asked, in return, to pledge to protect at least 30% of its near-shore ocean areas.

To ensure that the nation follows through with its commitment, TNC will use a trust fund—which is independent of the government and fully controls the funding—to make the payments.

Other issuers of blue bonds include the Nordic Investment Bank (NIB), which earlier this year raised SEK2 billion ($200 million) with a blue bond to protect and rehabilitate the Baltic Sea, in response to market demand.

The Nordic-Baltic Blue Bond targets investments in water projects, such as wastewater treatment, stormwater and flood defences, and protection of water resources and marine ecosystems.[83]

As of April 2019, TNC had "already secured more than $23 million in funding from various donors out of the $40 million TNC ultimately requires to unlock $1.6 billion toward marine conservation,"[84] and could work with twenty countries over the next five years on blue bonds. There are also other nonprofit organizations exploring blue bonds, such as EDF's potential environmental impact bonds for financing wetlands restoration.[85]

The future of green bonds, and possibly blue bunds, could involve more electronic trading.[86]

In addition, there are several countries such as New Zealand and Fiji which have issued, or are considering issuing climate-related bonds, which could have implications for the blue economy.[87]

Furthermore, there have been additional, unique bond developments in 2020 such as Central Arkansas Water (CAW) "preparing an innovative $31 million US municipal green bond, believed to be the first to buy and protect forests in order to secure clean drinking water"[88] and the creation of 'nature performance bonds'.[89]

Other Resources on Blue Bonds and Seascape Bonds

Blue Forest Conservation

- http://blueforestconservation.com/#aboutus

Carbon Yield

- www.carbonyield.org

Climate Bond Initiative's Climate Bonds Standard and Taxonomy

- https://www.climatebonds.net/standard/download; and
- https://www.climatebonds.net/standard/taxonomy

Climate Bond Initiative's Marine Renewable Energy Page

- https://www.climatebonds.net/standard/marine

Instiglio

- http://www.instiglio.org/en/impact-bonds/

International Capital Market Association's Green Bond Principles

- http://www.icmagroup.org/Regulatory-Policy-and-Market-Practice/green-bonds/

KPMG International: Gearing up for Green Bonds: Key Considerations for Bond Issuers

- https://assets.kpmg/content/dam/kpmg/pdf/2015/03/gearing-up-for-green-bonds-v1.pdf

Levering ecosystems: A business-focused perspective on how debt supports investments in ecosystem services; a Report by Clarmondial, Climate Bonds Initiative, and Credit Suisse

- https://www.credit-suisse.com/media/assets/corporate/docs/about-us/responsibility/banking/levering-ecosystems.pdf

Planet Tracker's Sovereign Bonds Programme

- https://planet-tracker.org/tracker-programmes/food-and-land-use/sovereign-bonds/

Sustainability Bond Database by Environmental Finance

- http://www.bonddata.org/

World Bank's Green Bond Issuances to Date

- http://treasury.worldbank.org/cmd/htm/GreenBondIssuancesToDate. html; and
- http://treasury.worldbank.org/en/about/unit/treasury/ibrd/ibrd-green-bonds

Notes

1. Cranford et al. 2011. *Unlocking Forests Bond: A High-Level Workshop on Innovative Finance for Tropical Rainforests, Workshop Report*. WWF Forest & Climate Initiative, Global Canopy Programme and Climate Bonds Initiative. Accessed April 11, 2017. https://www.climatebonds.net/files/uploads/2011/10/FBWork shop_report_web_A.pdf.
2. World Bank. "World Bank Green Bonds: Green Bond Issuances to Date." Accessed December 13, 2016. http://treasury.worldbank.org/cmd/htm/GreenB ondIssuancesToDate.html.
3. Climate Bonds Initiative. "History." Accessed December 13, 2016. https:// www.climatebonds.net/market/history.
4. International Finance Facility for Immunisation. "International Finance Facility for Immunisation Issues Inaugural Bonds." November 7, 2006. Accessed April 11, 2017. http://www.iffim.org/library/news/press-releases/2006/international-finance-facility-for-immunisation-issues-inaugural-bonds/.
5. LinkedIn. "Climate Bonds Initiative: About Us." Accessed April 11, 2017. https://www.linkedin.com/company-beta/1032198/.
6. TNC. "Conservation Note Fact Sheet." Accessed March 29, 2017.https://www. nature.org/about-us/conservation-note-fact-sheet-2016.pdf.
7. Climate Bonds Initiative. "Green Bond Labels and Standards Green Bond Labels and Standards Webinar." June 2016. Accessed April 7, 2017. https:// www.climatebonds.net/files/files/Green%20Bond%20Labels%20and%20Stan dards%2010-06-2016.pdf. 14.
8. Ibid. "Land Use." Accessed April 6, 2017. https://www.climatebonds.net/sta ndard/land-use.
9. Credit Suisse. "Sustainable Products & Services: Conservation Finance: Our Role in Conservation Finance." Accessed March 29, 2017.https://www.credit-suisse.com/us/en/about-us/responsibility/banking/sustainable-products-services. html.
10. The GEF. "Rhino Impact Bonds: An Innovative Financing Mechanism for Site-Based Rhinoceros Conservation." Accessed April 6, 2017. https://www. thegef.org/project/rhino-impact-bonds-innovative-financing-mechanism-site-based-rhinoceros-conservation.
11. Environmental Finance. "Luxembourg Launches First '100% Green' Exchange." Last modified September 28, 2016. https://www.environme

ntal-finance.com/content/news/luxembourg-launches-first-100-green-exchange. html.

12. Kidney, Sean. "San Francisco Public Utilities Commission (SFPUC) Issues World's First 'Climate Certified' Water Bond: A \$240 m Boost for Sustainable Water & US Muni Market." *Climate Bonds Initiative.* May 18, 2016. Accessed June 1, 2018. https://www.climatebonds.net/2016/05/san-francisco-public-uti lities-commission-sfpuc-issues-world's-first-'climate-certified'.

13. Ali, Hamza. "World's First Green Bond ETF Launches." March 1, 2017. *Environmental Finance.* Accessed April 6, 2017. https://www.environmental-finance. com/content/news/worlds-first-green-bond-etf-launches.html.

14. City of Cape Town, Media Office. "Green Pays: City's R1 Billion Bond a Resounding Success in the Market." July 11, 2017. Accessed April 13, 2020. http://www.capetown.gov.za/media-and-news/Green%20pays%20City.

15. DC Water and Sewer Authority. "DC Water Announces New Bond Offering." April 17, 2018. Accessed April 13, 2020. https://www.dcwater.com/dc-water-announces-new-bond-offering.

16. Climate Bonds Initiative. "Green Bond Fact Sheet: ABN AMRO." February 19, 2018. Accessed December 3, 2018. https://www.climatebonds.net/files/files/2018-04%20NL_ABN%20AMRO.pdf.

17. World Bank. "World Bank Prices First Sustainable Development Bond to Raise Awareness for Water and Ocean Resources." August 30, 2018. Accessed December 16, 2019. https://www.worldbank.org/en/news/press-release/2018/08/30/world-bank-prices-first-sustainable-development-bond-to-raise-awaren ess-for-water-and-ocean-resources.

18. Wahlén, Catherine Benson. "Seychelles Launches First Blue Bond." International Institute for Sustainable Development. 6 November 2018. Accessed 7 August 2019. https://sdg.iisd.org/news/seychelles-launches-first-blue-bond/.

19. Cripps, Peter. "NIB Prices Inaugural Blue Bond." *Environmental Finance.* January 25, 2019. Accessed December 31, 2019. https://www.environmental-finance.com/content/news/nib-prices-inaugural-blue-bond.html.

20. City of Atlanta Department of Watershed Management. "New Release: City of Atlanta Department of Watershed Management Announces First Publicly-Issued Environmental Impact Bond." February 21, 2019. Accessed April 13, 2020. https://www.atlantawatershed.org/first-publicly-issued-enviro nmental-impact-bond/.

21. World Bank. "World Bank Launched Bonds to Highlight the Challenge of Plastic Waste in Oceans." April 3, 2019. Accessed December 16, 2019. https://www.worldbank.org/en/news/press-release/2019/04/03/world-bank-launches-bonds-to-highlight-the-challenge-of-plastic-waste-in-oceans.

22. Caminha, Mariana. "Chile: First Sovereign Green Bond in the Americas: Strong Investor Demand: More to come says Finance Minister Larraín." *Green Bonds Initiative.* June 21, 2019. Accessed December 16, 2019. https://www.climatebonds.net/2019/06/chile-first-sovereign-green-bond-americas-strong-investor-demand-more-come-says-finance.

23. Environmental Finance. "Enel Inks 'World-First SDG-Linked' Bond." Accessed September 9, 2019. https://www.environmental-finance.com/content/news/enel-inks-world-first-sdg-linked-bond.html.

24. Hurley, Michael. "EBRD Issues World-First 'Climate Resilience' Bond." *Environmental Finance.* September 20, 2019. Accessed December 31, 2019. https://www.environmental-finance.com/content/news/ebrd-claims-worldfirst-climate-resilience-bond.html.

25. WWF. *Guide to Conservation Finance.* 43.

26. Environmental Finance. "A Progress Report on the Luxembourg Green Exchange." February 13, 2017. Accessed April 7, 2017. https://www.environmental-finance.com/content/analysis/a-progress-report-on-the-luxembourg-green-exchange.html.

27. Climate Bonds Initiative. "How to Get Certified." Accessed April 6, 2017.https://www.climatebonds.net/standards/certification/get-certified.

28. Garbaliauskas, Romas. Interviewed by Brian McFarland. July 2017.

29. Ibid.

30. Iyer, Venkat et al. *Finance Tools for Coral Reef Conservation.* 36.

31. Ibid. 34.

32. Ibid. 35.

33. Climate Bonds Initiative. "Green Bonds: The State of the Market 2018." Accessed December 27, 2019. https://www.climatebonds.net/files/reports/cbi_gbm_final_032019_web.pdf. 2.

34. Ibid. 2.

35. Ibid. 2.

36. Ibid. 11.

37. World Bank. "Investors Partner with World Bank to Highlight Vital Role of Ocean and Fresh Water Resources." November 25, 2018. Accessed December 16, 2019. https://www.worldbank.org/en/news/press-release/2018/11/25/investors-partner-with-world-bank-to-highlight-vital-role-of-ocean-and-fresh-water-resources.

38. So Seychelles. "The Geography of the Seychelles." Accessed April 9, 2020. http://www.seychelles.org/seychelles-info/geography-seychelles.

39. Vertigo Lab. "Innovations for Coral Finance, ICRI Publication." 47.

40. Iyer, Venkat et al. *Finance Tools for Coral Reef Conservation.* 34.

41. Weary, Rob. Interviewed by Brian McFarland. February 2019.

42. Cripps, Peter. "First Sovereign Blue Bond Launched." *Environmental Finance.* October 29, 2018. Accessed December 31, 2019. https://www.environmental-finance.com/content/news/first-sovereign-blue-bond-launched.html.

43. Ibid.

44. von Uexkull, Erik. Interviewed by Brian McFarland. January 2020.

45. Ibid.

46. Ibid.

47. Ibid.

48. Walsh, Joe. "Investing in the Oceans." *Environmental Finance.* October 21, 2019. Accessed December 31, 2019. https://www.environmental-finance.com/content/analysis/investing-in-the-oceans.html.

49. Cripps, Peter. "First Sovereign Blue Bond Launched." *Environmental Finance.* October 29, 2018. Accessed December 31, 2019. https://www.environmental-finance.com/content/news/first-sovereign-blue-bond-launched.html.

50. von Uexkull, Erik. Interviewed by Brian McFarland. January 2020.

51. Ibid.

52. Walsh, Joe. "Investing in the Oceans." *Environmental Finance.* October 21, 2019. Accessed December 31, 2019. https://www.environmental-finance.com/content/analysis/investing-in-the-oceans.html.

53. Cripps, Peter. "First Sovereign Blue Bond Launched." *Environmental Finance.* October 29, 2018. Accessed December 31, 2019. https://www.environmental-finance.com/content/news/first-sovereign-blue-bond-launched.html.

54. Ibid.

55. von Uexkull, Erik. Interviewed by Brian McFarland. January 2020.

56. Hurley, Michael. "Proliferation of Standards Is Impediment to Green Bond Market, Says EIB." *Environmental Finance.* October 30, 2017. Accessed December 31, 2019. https://www.environmental-finance.com/content/news/proliferation-of-standards-is-impediment-to-green-bond-market-says-eib.html.

57. Wilson, Molly. "The Growing Pains of Green Bond Funds." *Environmental Finance.* August 31, 2018. Accessed December 31, 2019. https://www.environmental-finance.com/content/analysis/the-growing-pains-of-green-bond-funds.html.

58. Walsh, Joe. "Investing in the Oceans." *Environmental Finance.* October 21, 2019. Accessed December 31, 2019. https://www.environmental-finance.com/content/analysis/investing-in-the-oceans.html.

59. Hurley, Michael. "NGO Attacks 'Non-Additional' Green Bond Market." *Environmental Finance.* March 12, 2018. Accessed December 31, 2019. https://www.environmental-finance.com/content/news/ngo-attacks-nonadditional-green-bond-market.html.

60. Chen, Kathy and Stian Reklev. "China to Scrutinise Green Bonds Over 'Greenwash' Concerns." *Carbon Pulse.* March 9, 2018. Accessed December 30, 2019. https://carbon-pulse.com/48695/.

61. Walsh, Joe. "Investing in the Oceans." *Environmental Finance.* October 21, 2019. Accessed December 31, 2019. https://www.environmental-finance.com/content/analysis/investing-in-the-oceans.html.

62. Ibid.

63. Razzouk, Assaad. "Green Bonds Do More Harm Than Good." *Eco-Business.* March 6, 2018. Accessed April 13, 2020. https://www.eco-business.com/opinion/green-bonds-do-more-harm-than-good/.

64. Havemann, Tanja. Email message to author. January 8, 2017.

65. Ibid. 7.

66. Johansson, Elena K. "HSBC predicts up to $250bn in green, social and sustainability bonds in 2019." *Environmental Finance*. January 10, 2019. Accessed December 31, 2019. https://www.environmental-finance.com/content/news/hsbc-predicts-up-to-$250bn-in-green-social-and-sustainability-bonds-in-2019.html.

67. Ibid.

68. Havemann, Tanja. Email message to author. January 8, 2017.

69. World Bank Group. "Doing Business: Economy Rankings." Accessed July 23, 2019. http://www.doingbusiness.org/rankings.

70. Transparency International. "Corruption Perceptions Index 2018." Accessed July 23, 2019. https://www.transparency.org/cpi2018.

71. UNDP. "Human Development Data (1990–2017)." Accessed July 23, 2019. http://hdr.undp.org/en/data#.

72. World Justice Project. "Rule of Law Index 2019." Accessed July 24, 2019. http://data.worldjusticeproject.org/#table and https://worldjusticeproject.org/our-work/research-and-data/wjp-rule-law-index-2019.

73. SeyCCAT. "Our Structure." Accessed April 8, 2020. https://seyccat.org/about-us/.

74. Johansson, Elena K. "HSBC Predicts Up to $250bn in Green, Social and Sustainability Bonds in 2019." *Environmental Finance*. January 10, 2019. Accessed December 31, 2019. https://www.environmental-finance.com/content/news/hsbc-predicts-up-to-$250bn-in-green-social-and-sustainability-bonds-in-2019.html.

75. Ibid.

76. Cooper, Graham. "Green Bond Market on Track to Hit $1trn Next Summer, Says SEB." *Environmental Finance*. October 21, 2019. Accessed December 31, 2019. https://www.environmental-finance.com/content/news/green-bond-market-on-track-to-hit-$1trn-next-summer-says-seb.html.

77. Vertigo Lab. "Innovations for Coral Finance, ICRI Publication." 48.

78. Climate Bonds Initiative. "Green Bonds: The State of the Market 2018." Accessed December 27, 2019. https://www.climatebonds.net/files/reports/cbi_gbm_final_032019_web.pdf. 13.

79. Ibid. 13.

80. Carbon Pulse. "CP Daily: Tuesday August 27, 2019." August 27, 2019. Accessed December 30, 2019. https://carbon-pulse.com/81105/.

81. Bentley, James. Interviewed by Brian McFarland. October 2019.

82. Iyer, Venkat et al. *Finance Tools for Coral Reef Conservation*. 64.

83. Walsh, Joe. "Investing in the oceans." *Environmental Finance*. October 21, 2019. Accessed December 31, 2019.

84. The Fish Site. "A Massive Boost for Marine Conservation." April 17, 2019. Accessed April 8, 2020. https://thefishsite.com/articles/a-massive-boost-for-marine-conservation.

85. EDF. "Environmental Impact Bonds: Financing for Wetlands Restoration." Accessed April 13, 2020. https://www.edf.org/ecosystems/environmental-impact-bonds-financing-wetlands-restoration.

86. Wirz, Matt. "Green-Bond Trading Is Going Electronic." *Wall Street Journal*. January 28, 2020. Accessed April 3, 2020. https://www.wsj.com/articles/green-bond-trading-is-going-electronic-11580212808.

87. World Bank. "Fiji Issues First Developing Country Green Bond, Raising $50 Million for Climate Resilience." October 17, 2017. Accessed April 8, 2020. https://www.worldbank.org/en/news/press-release/2017/10/17/fiji-issues-first-developing-country-green-bond-raising-50-million-for-climate-resilience; Also see: Scoop. "Auckland Council Exploring Green Bonds." March 7, 2018. Accessed April 8, 2020. https://www.scoop.co.nz/stories/AK1803/S00169/auckland-council-exploring-green-bonds.htm.

88. Lester, Ahren. "First green bond to secure drinking water by buying forests proposed." *Environmental Finance*. October 7, 2020. Available: https://www.environmental-finance.com/content/news/first-green-bond-to-secure-drinking-water-by-buying-forests-proposed.html.

89. Environmental Finance. "Call for 'nature performance bonds'." 2020. Available: https://www.environmental-finance.com/content/news/call-for-nature-performance-bonds.html.

16

Additional Solutions

Introduction

Alongside the financing and implementation of activities aimed at conserving tropical coral reefs, there are a variety of additional solutions including, but are not limited to: reducing discarded fishing nets; undertaking coral restoration activities; and tackling marine plastic pollution.

Historical Overview

A few notable dates include:

- 2003: "Designed by Rinkevich and located near Eilat's northern beach in the Red Sea, the world's first midwater floating nursery propagated eight thousand coral fragments. Created within the last twenty years, this nursery was the first attempt of its kind to raise thousands of coral colonies at a single site."[1]
- 2006: "An even newer technique for coral propagation, the midwater rope nursery, was installed in the Philippines in 2006".[2]
- 2013: Bureo is founded by Ben Kneppers, David Stover, and Kevin Ahearn.[3]

© The Author(s) 2021

B. J. McFarland, *Conservation of Tropical Coral Reefs*,

https://doi.org/10.1007/978-3-030-57012-5_16

Mechanisms of Instrument

As the case studies will highlight, there are a wide range of additional solutions. For instance, coral restoration activities can be from land-based or sea-based operations and such restoration activities are done for a variety of reasons:

> Corals are propagated in a wide range of sites and for a variety of reasons, of which restoration for conservation purposes is only one. Corals are propagated for public education exhibits in aquariums and zoos, for profit in the ornamental trade, for research and biomedicine, and for the mariculture industry – a brand of aquaculture developed in the late nineteenth country that involves the cultivation of marine organisms in saltwater. As a tool in conservation management, cultivated corals can be transplanted back onto the reef, usually onto areas damaged by ship or storm, in order to improve coral recovery.[4]

Another solution is reducing discarded fishing nets and tackling marine plastic pollution by utilizing the discarded nets as inputs into consumer goods.

Introduction to Case Studies

The following case studies will examine Bureo's reuse of discarded fishing nets, the Coral Restoration Foundation™ (i.e., the world's largest restoration program), and Coral Vita (i.e., the world's largest, land-cased coral restoration initiative).

Case Study #1: Bureo's Reuse of Discarded Fishing Nets

Introduction

Bureo, founded in 2013 by Mr. Ben Kneppers, Mr. David Stover, and Mr. Kevin Ahearn, is an entrepreneurship that has essentially setup an innovative supply chain program called Net+Positiva. Net+Positiva involves the collection of discarded (i.e., end-of-life) fishing nets, converts such fishing nets into raw materials, and then these raw materials are incorporated into a wide range of products (upcycling) including water bottle cages for Trek bicycles and Jenga game sets.

Bureo's current operations are primarily focused in Chile, Peru, and Argentina, with several potential expansion opportunities throughout other South American countries and in Asia. Bureo works with both smaller, artisanal fisherfolk and with the largest, commercial fisherfolk.

With the commercial fishing industry, the firms already have the infrastructure in place to manage the volumes of discarded fishing nets. Thus, for every kilo of discarded fishing net that are donated by the commercial firms to Bureo, Bureo donates money to local community projects. In exchange, the commercial firms receive free, certified end-of-use recycling and the firms receive positive public relations as a result of the local community projects that are sponsored in their name. To help facilitate this process, Bureo trains a team of locals who collect the commercial nets and these locals are paid for each kilo they collect in the manner they were taught.[5]

With the artisanal fisheries, the local fisherfolk are directly paid for their discarded nets. There is a local collection manager in each community, who is the intermediary between the artisanal fisherfolk and the regional collection manager. When the local collection manager has a full truckload, a regional manager collects the load.[6]

Whether working with the artisanal or commercial fisheries, Bureo is ultimately trying to set up microentrepreneurs and train how to collect discarded fishing nets, how to clean the nets, sort the nets, and to properly pack the nets. A regional team collects all the nets and once a cargo load is full, the shipment is sent to Bureo's recycling partners. These recycling partners convert the discarded nets into raw materials and Bureo sells the raw materials to companies.[7]

Bureo's experience is to set up the programs, develop the local infrastructure, provide training, and find the people to do the work, along with growing the networks of fisherfolk and facilitating partnerships.[8]

While such an operation may not appear to directly benefit the conservation of coral reefs, it is important to understand that discarded fishing nets, if left to the currents, could end up harming coral reefs and other marine life.

Identify the Problem

The main problem that Bureo is focused on is the widespread prevalence of ghost gear, particularly discarded fishing nets. In fact, "18 billion pounds of plastic enter our waterways every year creating An Ocean of Plastic. Discarded fishing nets make up an estimated 10% of plastic in the ocean, and have been found to be four times more harmful than all other forms of plastic pollution.

Bureo's programs provide a tangible solution to prevent this harmful material from entering our oceans."[9]

There is also a laundry list of additional problems in Chile. For instance, there are numerous deeper problems associated with education, poor infrastructure, income inequality, and the lack of incentives to properly dispose of discarded fishing nets. There is also constantly changing quotas for catch permits. Similarly, there are lots of complications related to the commercial fishing industry overinfluencing the Chilean Government to receive higher fish quotas, especially vis-à-vis artisanal fishers. Likewise, there is controversy surrounding who is getting the right to fish and what is the proper level. This, in turn, is slowly wiping out the trade of artisanal fisherfolk and a lot of children do not want to go into fishing as their parents did. Yet, the artisanal fisherfolk can be the least sustainable, as people are often not teaching the local artisanal fisherfolk how to do things more sustainably. Case-in-point, collecting algae is a popular local business, but is not being done in a sustainable fashion.[10]

Furthermore, farm-raised salmon fishing in Chile contributes a lot of plastic pollution and water pollution. In Peru, there are a whole host of additional problems such as the access to clean drinking water in remote fishing villages.[11]

Why the Problem Is Important

Ghost gear is a major problem facing the oceans. As presented by the Global Ghost Gear Initiative, it was estimated in 2009 by UNEP and the FAO, that at least 640,000 tons of fishing gear is annually abandoned, lost, or discarded into the ocean. This ghost gear is the result of strong currents and storms, getting snagged on underwater features, improper fishing methods or gear, intentional disposal, and either the lack of, or expensive, disposal facilities. Furthermore, it is estimated that 5–30% of global harvestable fish stocks are annually killed by ghost gear and approximately 70% of all macro-plastics in the ocean are made up of ghost gear.[12]

How Problem Was Identified

The founders of Bureo had a diversity of experiences prior to establishing Bureo. Mr. Kneppers was a sustainability consultant, Mr. Stover was a financial consultant, and Mr. Ahearn was a design engineer. Having met in Australia while surfing, the founders often discussed how to connect their passion with their skill sets and focus on addressing the issues of ocean pollution, particularly plastics, as a result of what they witnessed while surfing. Mr. Knepper's work took him to Chile to examine the wild-caught fisheries in Chile and discarded fishing nets. With a backdrop of pristine Chilean coastlines, Mr. Kneppers reflected back on their prior conversations.[13]

Effectiveness of Process for Identifying Problem

The process followed by Bureo for identifying the problems appears to be effective. To further think through and assess the problems, the founders took an academic approach and conducted lots of informational interviews with firms such as 5 Gyres and Net-Works, including a site visit to Net-Works' operations in the Philippines in 2017. Ultimately the solution, as formulated by Bureo, was that you need to stop the problem of ocean plastic pollution at the source. As Mr. Kneppers explained, "if your tub is overflowing, are you going to stop the tap or grab a mop? You need to stop the tap."[14]

Steps Taken to Address the Problem

There are several steps taken by Bureo to address the widespread prevalence of ghost gear, particularly discarded fishing nets. This includes:

- One major step has been setting up the recycling program which "{…} provides collection points that empower people to dispose of their fishing nets responsibly, diminishing the amount of harmful materials that enter our oceans. After the nets are collected, they are washed, shredded and melted into pellets, which are then injected into molds that form Bureo's products. To date, Bureo's programs have recycled over 500,000 kg of discarded materials and 15,000,000 square feet of fishing nets"[15]; and
- Bureo also provides education. This said, "Bureo's products not only reduce plastic pollution in our oceans, they also work to educate people about waste in general. Net Positiva educates fishermen and locals in Chile first-hand about the correct way to recycle, and the amazing things that can come from it. By creating fun products out of materials that most people disregard, Bureo shows the entire world how to think outside the box and value the seemingly invaluable. Most importantly, they show us that it's possible to combine sustainability and education with passion, and that making an impactful difference can be fun and creative."[16]

Results

In a relatively short period of time, Bureo has notched some impressive results. For instance:

> Creating a positive impact since 2013, over 500,000 kg of NetPlus materials have been collected to date from 50 participating fisheries in South America. Aligned with end of life incentives, these collections have financed community projects supporting sustainable development in coastal communities. In addition, Bureo provides contributions to nonprofit partners through the 1% For The Planet Network, is a certified B Corporation, and remains actively involved in the global fight against plastic pollution.[17]

Bureo has established a global distribution via retailers from Japan, Australia, Chile, and to both coasts of the United States.[18] In addition, Bureo has launched product partnerships with several high-profile firms including Trek and Patagonia,[19] and across a wide range of products including water bottle cages, surfboard fins, skateboards, and sunglasses. Other products include the "fishnet flyer," the Humanscale "Smart Ocean chair," and "Jenga ocean."[20] Bureo has been working with Patagonia for over three years on R&D to discover new techniques regarding how to incorporate raw materials produced from discarded fishing nets into more products. In July 2019, this process was to become public with an announcement at Outdoor Retailer. This will be a huge result, as the process will enable firms to incorporate the raw materials into a lot more products and simultaneously, the raw material volumes are increasing.[21]

Bureo has also established partnerships with the Global Ghost Gear Initiative.

Challenges and How They Were Met

As Mr. Kneppers explained, the general answer is that every day is a challenge, because Bureo is really about setting up an entirely new supply chain in a different country. Mr. Kneppers and his colleagues need(ed) to be there, in person, to see things through. It was a challenge to originally get fisherfolk onboard and to train workers on how to sort, clean, and pack the discarded fishing nets. However, once Bureo got through the first big shipments and the local people got paid, it sort of clicked with the local fisherfolk, and the process started to work like clockwork. It is a fairly straightforward supply chain, but that is easier said than done.[22]

Another challenge is that nylon and high-density polyethylene plastic nets are the best nets for fishing. While some other organizations dive down to collect discarded fishing nets, this is not what Bureo is doing with its operations. Likewise, it is not cost-effective for Bureo to go out to the ocean to collect the discarded nets because it is hard to get a quality discarded net once they are floating on the seas due to the nets getting contaminated or degraded as a result of seawater and/or exposure to sunlight.[23]

Beyond Results

The results, thus far, for Bureo appear to be sustainable. The company has pivoted to a more B2B (business-to-business) model and started focusing more on the sale of raw materials to businesses. Simultaneously, Bureo will need to scale up the collection of raw materials (i.e., discarded fishing nets). Bureo is now covering all of Chile, launched in Peru with WWF in 2019, and will be expanding to all of Uruguay and Argentina by 2020. While logistics and tariffs make it difficult, Bureo will look to expand into Southern Brazil and the next big push will be into Asia where well over 50% of the wild-caught fisheries are located.[24]

In addition, all of Bureo's products, to date, have been "injection molded, solid products which do not generate microplastics."[25] Furthermore, Bureo's goal is to increase the annual quantity of discarded fishing nets that are removed, while generating significant funding for local community projects.[26]

Lessons Learned

According to cofounder Mr. Ben Kneppers and related to the first challenge, one major lesson learned is that you need to understand that really nobody wants to get done what you are doing as bad as you, so it is going to be on your shoulders to get it done.[27]

There is also a lesson learned, which could be a cultural difference, related to for example, getting a meeting, talking through all the issues, and then coming back weeks later and no one has done anything. As Mr. Kneppers explained, everything is at stake and unfortunately, he needs to be a pest and constantly follow-up on items.[28]

Lastly, another lesson learned is that it is one thing to setup a program that does some cool and flashy beach cleanups, and it is an entirely other thing to do a long-term, sustainable program. Bureo did a lot of flashy things in the beginning, but are now keeping their heads down to focus on making this program last, to provide a solution of raw materials, and ultimately help get nets out of the water. Likewise, the top priority is to make sure you are doing something that will last and now, Bureo could care less about tweets and Facebook likes.[29]

Other Resources on Bureo

Bureo

- https://bureo.co/

The Global Ghost Gear Initiative

- https://www.ghostgear.org/; and
- https://www.ghostgear.org/projects/2018/11/21/qf8ta90ssp85rbkpd6cqh ls0w2mnfi

Net-Works

- https://net-works.com/

Ocean Conservancy and its Trash Free Seas Alliance

- https://oceanconservancy.org/; and
- https://oceanconservancy.org/trash-free-seas/plastics-in-the-ocean/trash-free-seas-alliance/

Ocean Recovery Alliance

- https://www.oceanrecov.org/

Patagonia

- https://www.patagonia.com/blog/2016/02/a-different-path-living-in-sou thern-chile-with-bureo-co-founder-ben-kneppers/

Projects of the Global Ghost Gear Initiative (GGGI)

- https://www.ghostgear.org/projects

<u>World Animal Protection</u>

- https://www.worldanimalprotection.us/

<u>WWF</u>

- https://www.wwf.org.pe/?343290%25252FJoint-efforts-to-keep-over-50-tons-of-discarded-fishing-nets-from-entering-the-Peruvian-sea=

Case Study #2: Coral Restoration Foundation™, the World's Largest Restoration Program

Introduction

The Coral Restoration Foundation™ (CRF™), founded by Mr. Ken Nedimyer, is headquartered in Key Largo, Florida and is now the world's largest coral reef restoration organization in the world. CRF™ primarily works in the Florida Keys and "{…} consists of five nurseries that together comprise the largest coral nursery effort in the world, with half a million corals propagated for transplantations as of 2017"[30] (Fig. 16.1).

CRF™ was originally "founded in response to the wide-spread loss of the dominant coral species on the Florida Reef Tract."[31] By "enlisting his family and a small, dedicated staff, Nedimyer founded CRF™ in 2007 and began working to build up stock of staghorn and elkhorn (*Acropora palmata*) coral."[32] Focusing on both science and education, the CRF™ is now working primarily with staghorn (*Acropora cervicornis*), elkhorn (*Acropora palmata*), Boulder star coral (*Orbicella annularis*), and mountainous star coral (*Orbicella faveolata*).

Fig. 16.1 Map of Where CRF™ Works in the Florida Keys (*Credit* CRF™)

With respect to funding, CRF™ is funded by a combination of:

- U.S. federal grants;
- Local grants from Florida municipalities;
- Grants from nonprofit and philanthropic organizations;
- Corporate sponsorships; and from
- Individual donors.[33]

In addition, it is important to note that CRF's™ "work is only possible through the participation of a large number of volunteers. {In fact}, just under 50% of CRF's™ work in the Florida Keys is performed by volunteers, totaling over 15,000 hours annually."[34]

Identify the Problem

The coral reefs of the "Caribbean, including the Florida Reef Tract, have lost upwards of 95% of their coral cover over the last 40 years. In short, it has been a 'death by a thousand little cuts' over a sustained period of time that has led to the degradation of Florida's reefs."[35]

For instance, "high concentrations of fertilizers, pesticides, herbicides, and other exotic chemicals flowing primarily from large agricultural developments in south Florida, especially sugar plantations, coupled with decades of diversion and manipulation of traditional sources of freshwater, are believed to be responsible for the wholesale collapse of the bay's ecosystem."[36]

Regarding the Florida Keys, "from the 1970s onwards increasing development and tourism throughout the Florida Keys caused a dramatic decline of populations of *Acroporid* corals along the length of the Florida Reef Tract."[37] While there are wide range of subfactors responsible for this decline on coral reef cover, two disease outbreaks in the 1980s—attributed to poor water quality and specifically sewage runoff—were highly significant:

1. A disease event that wiped out local populations of long spine sea urchins, which are important algal grazers; and
2. The white pox disease which is linked to the lack of a sewage system in the Florida Keys, was responsible for 90% mortality of elkhorn (*Acropora palmata*) and staghorn (*A. cervicornis*) corals.[38]

Additional problems facing the Florida Reef Tract include, but are not limited to:

- Nutrient runoff from agricultural operations (i.e., primarily citrus, sugarcane, etc.);
- Overfishing (i.e., both amongst sportsmen and commercial fisheries);
- Coastal development (i.e., housing, transportation and tourism infrastructure, etc.);
- Lack of mooring balls and poor boating practices contributing to anchor damage;
- Direct damage to corals by divers and snorkelers;
- Other pollutants, including oxybenzone which is a common ingredient in traditional sunscreens; and
- The increased frequency of large-scale coral bleaching events attributed to global climate change.[39]

It is "suspected that the continuous, long-term pressure of many of these factors, resulted in significant decreases in the populations of reef-building corals (primarily the *Acroporid* corals, i.e., staghorn and elkhorn corals) in the Florida Keys. With the diminished populations, the coral reef ecosystems are reaching a tipping point and degrading to an alternative stable state that is devoid of the coral reef system's original complexity."[40]

Today, the Florida Keys' coral reefs continue to face significant local and global stressors that can negatively impact hard coral populations. This said, "the reefs of the Florida Keys are still under significant pressure from human activity."[41] For instance:

- Water quality is still not optimal with sewage outlets along Florida's coast continuing to discharge effluent into the ocean, and agricultural runoff remains an issue;
- Divers, snorkelers, anglers, and boaters visit the reefs in enormous numbers every year, with associated impacts; and
- Global climate change, including significant temperature fluctuations (i.e., leading to coral bleaching) and the corresponding increase in ocean acidification, along with the increase in both storm frequency and strength will continue to impact the Florida Key's coral reefs.[42]

Furthermore:

In 2014, an alarming new coral disease broke out in the northern part of the Florida Reef Tract. It quickly became known as the Stony Coral Tissue Loss Disease (SCTLD). Its cause is unknown but it affects more than twenty different species of stony coral, including brain, pillar, star, and starlet corals. This multi-year outbreak of SCTLD is unique due to its large geographic range, extended duration, rapid progression, high rates of mortality, and the number of species affected.[43]

CRF™, along with other NGOs and combined "with local education and government action, are attempting to address some of the local problems."[44]

Why the Problem Is Important

These aforementioned problems are important because the Florida Reef Tract is "home to almost 1,400 species of marine plants and animals, including more than 40 species of stony corals and 500 species of fish. This is the world's third largest barrier reef and comprises an ecologically significant community providing critical ecosystem services."[45]

The Florida Reef Tract "directly supports the regional economy, and underpins the culture of life in south Florida. As the Florida Reef Tract (FRT) declines, it loses its ability to provide ecosystem services that are significant on local, regional, and national levels."[46] More specifically, NOAA "suggests that coral reefs in south-east Florida have an asset value of USD$8.5 billion, generating USD$4.4 billion in local sales, USD$2 billion in local income, and 70,400 full and part-time jobs."[47]

The FRT is critical to commercial and recreational fishing interests in the Caribbean. [48] In addition, the FRT is "the most visited dive destination globally, which drives the tourism industry of the Florida Keys. In 2018, tourism in the Florida Keys generated USD$1.8 billion in total economic impact. Divers have been visiting these reefs in droves since the 1960s when recreational scuba diving first took off."[49] Yet, as the reef "has degraded and as fish populations have seen sharp declines, the Keys is no longer considered a premier dive destination."[50] This said, Fodor's Travel recently listed the Florida Reef Tract on its "No List 2020."[51]

The FRT is also an "important coastal barrier, dissipating wave energy during storm events which reduces damage to coastal infrastructure. Damage from storm surges during events such as hurricanes can cost state and federal governments billions of dollars. As the reef degrades, it is increasingly unable to dissipate this wave energy, resulting in more extreme damage to homes, business, and essential services."[52]

In addition to the "economic value and ecosystem services that the Florida Reef Tract provides, the FRT is the corner stone of the culture of south Florida. Many local people recognize that this reef has intrinsic value as a beautiful ecosystem, and its decline is felt on a personal level by people who have watched it degrade over the past decades."[53]

How Problem Was Identified

The problems impacting the Florida Reef Tract were identified through a variety of ways.

With respect to CRF™, its "founder, Ken Nedimyer, witnessed the decline in coral cover along the Florida Reef Tract, firsthand. At the time Ken was a commercial fisherman to the pet trade." The idea for the Foundation actually:

> {…} began with a random event - in 2003, Mr. Nedimyer noticed three small staghorn corals had settled on some of the 'live rock' he sold to the pet trade. These were 'juveniles' and the result of sexual reproduction. Over a period of months, they survived and continued to grow. Then, adopting techniques from the aquarium trade and researchers, he tried to 'fragment' these corals. Fragmenting corals is the process of splitting a coral colony to produce small sections of coral, each of which will continue to grow – analogous to plant 'clippings.' Not only did the new coral fragments survive, but after experimenting with different ways to grow the coral fragments, he developed the 'coral tree' growing technique.[54]

Another way the problem was identified was that "in 2006, the two species of *Acroporid* corals that were disappearing from the Florida Reef Tract (*A. cervicornis* and *A. palmata*) were listed under the U.S. Endangered Species Act. {Then} in 2008, these two species were assessed and also added to the IUCN Red List of Endangered Species with a listing of Critically Endangered."[55]

Effectiveness of Process for Identifying Problem

The process for identifying the problem was very effective, in part because the "identification of the problem was obvious and widely shared by the general public (i.e., those familiar with the area), the scientific community, and the local/national governmental organizations charged with protecting marine resources." For example, in 2006, the U.S. Government placed the major reef-building corals (*Acropora cervicornis* and *A. palmata*) on the Endangered Species List, which subsequently catalyzed the formation of a recovery plan for the species.[56]

Steps Taken to Address the Problem

There are numerous steps taken to address the problems facing the Florida Reef Tract.

Generally speaking, the steps of the CRF™ involve coral reproduction, the establishment of Coral Tree™ nurseries, and then the outplanting of corals (i.e., coral restoration). As described by the CRF™, there are two issues that need to be addressed:

- 1. The recovery of natural populations of hard-coral species that create coral reefs; and
- 2. The mitigation of factors that decimated the hard-coral populations.[57]

CRF™ "works on the former: increasing the size of hard-coral populations." [58] Likewise, CRF™:

> began developing techniques for growing and outplanting Acroporid corals, based on methods common in the aquarium industry. These techniques take advantage of the ways in which corals naturally reproduce asexually, through fragmentation. The Coral Restoration Foundation™ invented the Coral Tree™ after developing and studying multiple iterations of nursery designs, including block nurseries and string nurseries. The Coral Tree™ proved to be the fastest, easiest, and most cost-effective way of growing Acroporid corals. This method is now widely used by other restoration groups around the world. Outplanting methods currently employed by Coral Restoration Foundation™ are also the result of years of research and development.[59]

It is important to note that in 1990, the Florida Keys National Marine Sanctuary and Protection Act was enacted.[60] This said, "thanks to permits obtained from the Florida Keys National Marine Sanctuary, CRF was also able to collect clippings and broken fragments of coral from the remaining wild populations of staghorn and elkhorn in south Florida."[61] More specifically, the CRF™ was established as a 501(c)(3) nonprofit organization and then was "able to work with the coral species and setup the nurseries required various government approvals."[62] Likewise:

> Since the primary corals we work with (*Acropora cervicornis* and *A. palmata*) are on the US Endangered Species and CITES lists, permits are required to work with these animals. These are issued by the U.S. Fish and Wildlife Service, specifically the Florida Fish and Wildlife office. These permits allow the species to held by the organization. Additionally, these permits set limits on where the outplants can be located as well as the number of individuals. As CRF™ grew and was able to demonstrate responsible management practices, the state agency has worked with CRF™ to increase the number of species, the number of outplants, and the locations.[63]

Furthermore:

> Coral restoration efforts in the state of Florida are conducted under several state and federal agencies including: Department of Environmental Protection (DEP), Army Corps of Engineers (AOCE), Florida Fish and Wildlife Conservation Commission (FWC), and the Florida Keys National Marine Sanctuary (FKNMS). Specifically, in the Florida Keys, the Foundation holds operational permits from FKNMS, ACOE, and FWC. Permitted activities include the creation of nurseries, maintenance of coral species within, outplanting activities, research activities, and monitoring requirements.[64]

Results

Since its founding in 2007, CRF™ has achieved many notable results. Since 2007, the CRF™ "has returned more than 100,000 corals to the reefs of the Florida Keys and we have now restored at least 5,000 square meters of coral reef across 8 sites in the Florida Keys."[65] Since 2012, the CRF™ has "outplanted more than 74,000 corals out onto the Florida Reef Tract,"[66] and is "currently working with 323 coral genotypes across 10 species."[67] This said, the Coral Restoration Foundation™ now manages the largest Coral Tree nurseries in the world,[68] and now "manage the world's largest genetic ark and database of coral genotypes."[69]

The CRF™ has a "total of more than 740 Coral Trees™ and are capable of raising more than 45,000 corals to a 'reef ready' size every year."[70] Many of these corals have been "observed to spawn in around two to three years after being returned to the wild."[71] This said, CRF's™ "restoration sites are essentially ecological stepping stones, refugia of biodiversity scattered along the reef that will be able to 'seed' the rest of the reef with life."[72]

Going forward:

> In the next three years, CRF™ will be scaling up its restoration work and focusing on returning a huge volume of corals to three reef sites: Pickles, Carysfort, and Sombrero. At Carysfort, CRF™ will be restoring 27,300 m^2, at Pickles CRF™ will be restoring 40,900 m^2, and at Sombrero CRF™ will be restoring 23,650 m^2. The total area that CRF™ will be working to restore over these three years is 91,850 m^2 - an area that is about the size of 17 American football fields.[73]

Other notable results to date include:

- In 2009, CRF™ "made history as the Wellwood corals were the first documented nursery-raised corals to spawn."[74] CRF™ made history again "with the first ever nursery-raised elkhorn restoration in July of 2012."[75]
- In 2016, CRF™ was "part of a Florida Keys-wide effort in collaboration with Summerland Marine Lab, Florida Aquarium, FWC, FKCC, and Mote Marine Lab to safeguard the remaining pillar coral genotypes present on the Florida Reef Tract."[76]

Furthermore, "for his work with CRF™, Nedimyer has been given the Wyland ICON Award and named a CNN Hero and a Disney Worldwide Conservation Hero."[77]

In 2020, *Mission: Iconic Reefs* was launched, which will see NOAA and its Florida Keys National Marine Sanctuary "intent to raise almost $100 million to restore seven iconic coral reef sites off the Florida Keys."[78] CRF™ is a partner in the initiative.

Challenges and How They Were Met

There were many challenges faced by the CRF™'s restoration work. This includes:

- 1. Proving to the academic community that the coral restoration work can be done at ecologically meaningful scales;
- 2. Educating funders, permitting agencies, academics, and the general public about the following matters:

  - A. "Survivorship of individual corals for large outplanted populations and natural populations is not necessarily the same as survivorship of individual corals in a small-sample research study."[79]
  - B. "Just because one genetic strain of coral appears to grow better then another, this does not mean that we should exclusively outplant that particular genetic strain. The notion of a 'super-coral' is inherently flawed and we should work to maintain as much genetic variation in the population as possible, applying best-practice of population genetics, as we would with any terrestrial species."[80]
  - C. "While hard-coral loss of the Great Barrier Reef appears to be due to climate change, hard-coral in the Caribbean and in Florida specifically is

the result of many factors over many years. There are still reasons to be hopeful and continue the work in spite of the presence of climate change and other global factors, since the corals continue to grow and thrive in offshore nurseries: it is not a coral growth problem, but a population size problem."[81]

- 3. Ensuring survival of corals both during and after hurricane events;
- 4. Fundraising—this includes the challenge of "raising funds to do the work,"[82] and the challenge of "educating people on the appropriate metrics for the project. Specifically, that per-coral costs are misguided and that one should evaluate the cost to restore a certain size of reef."[83]
- 5. Developing new approaches to restoration that allow for increased scalability;
- 6. Raising awareness of the issues and threats to coral reefs amongst the local, regional, national, and international community; and
- 7. Mitigating stressors including poor water quality and other human impacts.[84]

To address these challenges, CRF™ has undertaken many steps. In order to help prove to the academic community that the work can be done at ecologically meaningful scales, the CRF™ showed, through its work, "that this work could be scaled up to ecologically relevant areas. Additionally, the CRF™ partnered with various academics to study the results of its work, yielding a series of peer-reviewed papers showing the benefits of the method."[85]

With respect to educating stakeholders, the CRF™ has:

{…} been working with scientists to better inform various agencies that survivorship results for large outplanted populations (and natural populations) are not the same as survivorship for research studies, and in fact there is no reason to suspect they should be. CRF™ approached this by looking at historical data, as well as proposing new methods for monitoring survivorship, such as using photomosaics. The principal discrepancy is that after 2-3 years, many corals will 'fuse' forming a single larger thicket. Thus, if you are tracking individual corals, after 1-2 years they can no longer be reliably tracked. Additionally, the biology of the species is that they can fragment naturally, yielding new colonies as part of the project area (i.e., something that is desired). {Furthermore}, CRF™ has worked extensively with academics to help introduce best-management practices for populations. This includes a large, collaborative study whereby we sequenced multiple genetic strains of corals to see how related they were. Moreover, we drew attention to NOAA's Acropora Recovery Implementation Plan which highlighted the need for genetic diversity across the reef area.[86]

CRF™ is continuously educating people that coral restoration work is not futile due to climate change and other regional/global factors. If "they were the primary drivers of coral loss in the Florida Keys, then they would affect the corals in our nurseries – and they do not. In fact, the corals in our nurseries not only grow, but thrive, indicating that it is not a global problem, but rather a local problem related to the population size on the existing reefs. By outplanting many corals back onto a reef, we can move the system back to a healthy stable state."[87]

Regarding the challenge of ensuring coral survival as a result of hurricane events, it should be noted that "hurricanes are naturally occurring phenomena. Part of the reason hurricanes are so devastating now is that the natural break created by the reef line is gone. By outplanting corals in sufficient densities, we start to reform that reef break."[88]

Fundraising is always a challenge. As explained by the CRF™:

> As with any new technology, the initial costs of development are high. As technology moves from proof of principal to scalable implementation, costs can be driven down through efficiencies. We try to educate people that the cost of getting a reef-ready coral for outplanting is minimal, the majority of the cost is in the labor to outplant the coral to the reef. By increasing efficiencies, we have been able to continuously lower this cost. Moreover, by utilizing the volunteers we have further reduced the cost, as well as developed a model where restoration can be effective in other parts of the work with lower labor costs. {In addition}, although we receive funding from federal and local sources, the majority of our funding comes from private individuals and foundations. The US Government recognizes the need for protection of these species and the Florida Reef Tract, but has been unable to secure funding at the scale of the Australian Government. As a result, most of the funding for restoration work comes through private charitable efforts.[89]

Other challenges in addition to the scalability of coral restoration and the claims that restoration is not based off science, is the need for selective breeding to ensure corals are resistance to future scenarios of global climate change and ocean acidification. In addition, there is the challenge of having the ability to properly tag and keep track of corals, and there are claims that coral restoration activities take away funding from coral reef conservation:

> Reducing the cost and increasing the scale of production are important preconditions for making restoration into a feasible conservation enterprise in the Anthropocene. Still, critics of coral restoration repeatedly refer to its high price. According to one study, gardening one square meter of corals can cost up to $500. Others criticize the limited geographic scale of coral restoration. A recent

fact sheet from a Florida nursery sets a goal of restoring one thousand acres, or 1.5 square miles, of reef in the Florida Keys within the next ten years; this, when the entire Florida Reef tract is six hundred square miles and the Great Barrier Reef – the world's largest – is more than two hundred times that size.[90]

Furthermore, there are challenges associated with permitting as "there was a restoration permit. The problem was that nobody had ever done that before: there were no standards, no protocols, no best practices, nothing. It just hadn't been done, at least in Florida or in the National Marine Sanctuary Program."[91] Likewise:

> Nurseries and other restoration initiatives that have been obtaining take permits from NOAA were especially concerned that their fragmentation and outplanting operations (i.e., transplanting corals from nurseries onto reefs) would be hindered or at least stalled. From their perspective, stronger legal protections would actually result in lower levels of restoration and could thus end up harming rather than benefitting the corals.[92]

Beyond Results

The work of the CRF™ appears sustainable. Ultimately, the CRF's™ focus is on "restoring genetically diverse wild coral populations. In recent years, scientific collaborations have helped to ensure that corals returned to the wild are as genetically diverse as possible. Collaborations around coral spawning have also resulted in the breeding and rearing of new coral genotypes."[93]

In order for the CRF's™ work to be successful in the long term, the following are needed:

- 1. Stressors (noted above such as sewage and agricultural runoff) need to be mitigated;
- 2. Increased protection for restoration sites;
- 3. Increased funding for restoration efforts;
- 4. Ongoing collaborations in the restoration and scientific community; and
- 5. Further developments in outplanting techniques to allow corals to be returned to the wild in larger quantities.[94]

To this end, the success of the CRF's™ Coral Tree™ nurseries, has led to the growing of more corals than CRF™ can outplant. The CRF's™ team is now developing new outplanting methods that will allow the CRF™ to return the corals its growing to the wild at a much larger scale.[95] Alongside these developments, "in the coming year, it is our goal to sequence every coral and make this information publicly accessible through a genetic database that will be used as the focal point for all genetically -conscious coral restoration."[96]

CRF™ is also seeking to expand its work through training other coral restoration practitioners:

> {…} We train people to do restoration in other Caribbean locations. We provide the material and teach them how to do everything. We've helped locals set up nonprofits in some of these different islands. First, we go in and show them how to do it, where to find the corals, how to handle them, where to set up the nurseries, how to take care of them, how to plant the corals, and all that. Then we work with them on how to run a nonprofit to support this work. We've gone to different places like Bonaire, Curaçao, Jamaica, and Roatán. We've also been approached by people in the Philippines and Indonesia – but we're just not quite ready to go there.[97]

CRF™ believes that the outcomes of its work are sustainable and moreover, there is mounting evidence in the scientific literature. This said:

> We know that the hard-corals grow in our offshore nurseries. Consequently, we do not believe that the factors leading to the decline of the hard-corals in the Florida Keys are systematic. Moreover, we maintain an 'ark' of multiple strains of each coral type, thus demonstrating the populations' resiliency. In our nurseries, the corals survive, and even grow large enough to exhibit spawning behavior. Additionally, corals that are placed on the reef – post outplanting – also survive, continue to grow, and spawn. Overtime, we also see that these outplanted corals serve as refugia for other reef-dwelling species, thus indicating that the ecosystem services are starting to come back. By outplanting large numbers of corals representing multiple species and multiple genotypes per species, we believe that the population has the potential to be resilient in the face of ongoing threats. {Furthermore}, a recent set of publications (some in pre-print stage) review the history of CRF's outplanting and conclude that our work has made a positive impact on the local coral reef communities.[98]

With respect to continuing the work, CRF™ notes that:

> We hope that the work can continue. As stated above, we know what work must be done and now have demonstrated its success. This is acknowledged in the openness and collaboration of state and federal agencies to support this work through permits. However, the work is relatively expensive to perform. As a consequence, the long-term risk is predominately funding. Presently, a large portion of CRF's funding comes from individual and foundation donors, with nearly half of the work performed by volunteers. The federal government is supportive of reef restoration work through NOAA, but the funds available

are only a fraction of what is needed to address the entire Florida Reef Tract. Local government involvement (i.e., Monroe county) has been very supportive and applies a portion of its tourism bed tax to restoration efforts; but, this too only covers a small portion of the reef tract. What is needed is either a significant, material change in scalability to the work, or a significant investment in restoration similar to that being made by the Australian Government to address coral loss on the Great Barrier Reef.[99]

Lessons Learned

According to a white paper put together by CRF™, a few key lessons learned were:

- 1. The delineation of coral nursery stock as either production or preservation. This allowed for greater capacity in our nurseries to produce outplant stock and maintain genetic diversity.
- 2. The focus on continuous, incremental improvements to efficiency in order to increase scalability. By making consistent, additive changes (e.g. maintaining the base of elkhorn colonies as nursery stock or pre-placement of outplant clusters by an experienced diver) we have become more efficient in our daily dive activities.
- 3. The development of standardized data systems and collection processes, which has been essential to supporting the influx of large amounts of diverse information from multiple sources. CRF has created new standard operating procedures (SOPs) that encompass data collection, training, and processing of SOPs across nursery, outplanting, and monitoring activities. Combined, these advancements have resulted in an infrastructure platform and systematic approach that can support increased growth and scalability.[100]

Other Resources on Coral Restoration Foundation™

Coral Restoration Foundation™

- 2018 Annual Report: https://docs.wixstatic.com/ugd/60a969_7648b6cca ada407ebdcb04d6b64fee3f.pdf;
- Coral Chronicles: https://www.coralrestoration.org/get-involved and https://us10.campaign-archive.com/home/?u=6f00e3fbf0a16180aa857 4a73&id=efef6cffec;
- REEFocus Blog: https://www.coralrestoration.org/reefocus-blog; and

- White papers: https://www.coralrestoration.org/white-papers

Drury, Crawford et al. "Genomic Patterns in *Acropora cervicornis* show Extensive Population Structure and Variable Genetic Diversity." *Ecology and Evolution.* 7.16 (2017): 6188–6200.

- https://doi.org/10.1002/ece3.3184

Flint, Mark and John T. Than. "Potential Spawn Induction and Suppression Agents in Caribbean *Acropora cervicornis* Corals of the Florida Keys." *PeerJ.* 4.e1982 (2016).

- https://peerj.com/articles/1982/

Hein, Margaux Y. et al. "Coral Restoration: Socio-Ecological Perspectives of Benefits and Limitations." *Biological Conservation.* 229 (2018):14–25.

- https://doi.org/10.1016/j.biocon.2018.11.014

Johnsen, Emilie C. "Toxicological Effects of Commercial Sunscreens on Coral Reef Ecosystems: New Protocols for Coral Restoration." *Nova Southeastern University* (2018).

- https://nsuworks.nova.edu/cgi/viewcontent.cgi?article=1335&context=cnso_stucap

Ladd, Mark C. et al. "Density Dependence Drives Habitat Production and Survivorship of *Acropora cervicornis* Used for Restoration on a Caribbean Coral Reef." *Frontiers in Marine Science.* 3.261 (2016):1–14.

- https://www.frontiersin.org/articles/10.3389/fmars.2016.00261/full

Lohr, Kathryn E. and Joshua T. Patterson. "Intraspecific Variation in Phenotype among Nursery-Reared Staghorn Coral *Acropora cervicornis* (Lamarck, 1816)." *Journal of Experimental Marine Biology and Ecology.* 486 (2017): 87–92.

- https://www.sciencedirect.com/science/article/pii/S0022098116301885?via%3Dihub

Miller et al. "Reef-Scale Trends in Florida *Acropora spp.* Abundance and the Effects of Population Enhancement." *PeerJ.* 4(1689).e2523 (2016).

- https://peerj.com/articles/2523/

Pausch et al. "Impacts of fragment genotype, habitat, and size on outplanted elkhorn coral success under thermal stress." *Marine Ecology Progress Series.* 592 (2018).

- https://doi.org/10.3354/meps12488

Rodriguez-Casariego, Javier A. et al. "Coral epigenetic responses to nutrient stress: Histone H2A.X phosphorylation dynamics and DNA methylation in the staghorn coral Acropora cervicornis." *Ecology and Evolution.* 8.23(2018):12193–12207.

- https://doi.org/10.1002/ece3.4678

Strand, Matt, et al. "The Colorado Coralition: Engaging Landlocked Youth in the Restoration of Staghorn Coral in the Florida Keys, USA." IUCN/SSC Reintroduction Specialist Group & Environment Agency. (2018).

- Also See: https://www.youtube.com/watch?v=JPpux8OKQtI

van Woesik et al. "Macroalgae reduces survival of nursery-reared Acropora corals in the Florida reef tract." *Restoration Ecology.* 26.3(2017):563–569.

- https://doi.org/10.1111/rec.12590

Case Study #3: Coral Vita, the World's Largest, Land-Based Coral Restoration Initiative

Introduction

Coral Vita, based near the city of Freeport on the island of Grand Bahamas in the Bahamas archipelagos, is the world's largest, land-based coral restoration initiative (Fig. 16.2).

Fig. 16.2 Map of Grand Bahama Restoration Sites (*Credit* Coral Vita)

Coral Vita's unique approach involves micro-fragmenting. Essentially, Coral Vita collects coral fragments from the reefs, brings the fragments to the farm, and gives them a safe space to grow and mature. This said, you can either take no more than 10% from living corals or you can collect from what are called "corals of opportunity" (i.e., basically broken off corals as a result of anchor damage or from a storm, or corals that are lying on the sand and not yet dead). In addition, you can also grow corals sexually, which is something Coral Vita is looking into, but right now, Coral Vita is focused on asexual reproduction. It is important to note that the micro-fragmenting process allows one to almost cut down to an individual polyp and when they breakup, they fuse back together—this technique is revolutionary and will increase the speed of restoration up to 50x.[101]

Identify the Problem

The essential problem being addressed by Coral Vita is that:

There is great concern that the high rates, magnitudes, and complexity of environmental change are overwhelming the intrinsic capacity of corals to adapt and survive. Although it is important to address the root causes of a changing climate, it is also prudent to explore the potential to augment the capacity of reef organisms to tolerate stress and to facilitate recovery after disturbances.[102]

In addition to the degradation and loss of coral reefs, the traditional approach of small-scale, ocean-based, grant-based coral restoration is seen as insufficient when compared to the scale of the problem.

Furthermore, Caribbean spiny lobster (*Panulirus argus*), queen conch (*Strombus gigas*), and the Nassau grouper (*Epinephelus striatus*) "{...} account for approximately 90 percent of the value of Bahamian fisheries."[103] If the coral reefs are lost and not restored, these species and the Bahamian fisheries will collapse.

Why the Problem Is Important

As explained by Coral Vita, "sadly, coral reef health is collapsing around the world. As reefs die, these ecological wonders and the critical benefits they provide for people disappear. The threats posed from this crisis – from wildlife loss, to homes underwater, to climate refugees – matters to everyone everywhere."[104] For instance, the Bahamas, including Grand Bahama, Bimini, and Abacos, are among the top fifty most important bioclimatic units.

How Problem Was Identified

The founders, Mr. Sam Teicher and Mr. Gator Halpern, originally met at Yale University. Their interest in scaling up reef conservation and identifying the problems with coral reefs can be traced back to the fact that Mr. Teicher has been a diver since the age of thirteen and Mr. Halpern grew up in San Diego, California and was always passionate about the oceans. Over time, Mr. Halpern was more academic-focused and Mr. Teicher was more policy-focused with Mr. Teicher having worked on a reef restoration project in Mauritius with his friend's organization.[105]

Effectiveness of Process for Identifying Problem

It appears the process for identifying problems has been effective.

Steps Taken to Address the Problem

Coral Vita is using a large-scale, land-based reef restoration approach that will not be solely (i.e., or not mainly) dependent on grant financing. As previously mentioned, traditional approaches are too dependent on donations and grants. Thus, one of the first steps was for Coral Vita to establish itself as a for-profit entity.

Next, Coral Vita looked for a place that needed coral restoration and a location for their first operation. After looking at the British Virgin Islands and the Dominican Republic, Coral Vita met Mr. Rupert Hayward, the Executive Director of the Grand Bahama Port Authority. Likewise, there is a large-scale development lease with a vision for sustainability and Coral Vita was offered a great plot of land for next to nothing, so Coral Vita chose Grand Bahama as its first location.[106]

Coral Vita is effectively helping to build the market of land-based coral restoration. To this end, Coral Vita positioned itself well with coral reef scientists, practitioners, and investors.[107] For instance, Coral Vita is in the process of becoming an Affiliate Member of the Global Island Partnership.[108]

With respect to the restoration process,

> Coral Vita restores reefs by growing and transplanting resilient corals. We restore reefs using a proven 5-step process:
>
> - Assess target reef's health and water quality conditions;
> - Create a restoration plan, determining which coral species should be grown, how many, and where they will be planted to revive the reef;
> - Raise these corals using accelerated growth and improved resiliency methods in our land-based farms;
> - Install the corals into the reef once they are sufficiently mature; and
> - Monitor the reef's progress after installation.[109]

It is also important to note that:

> unlike ocean-based projects, land-based farms can use breakthrough techniques that allow for faster growth of more diverse array of corals. By using microfragmenting - a process pioneered by the Mote Marine Laboratory - we can accelerate coral growth up to 50× natural rates. Many coral species that serve

as critical building blocks for reefs (such as Brain or Great Star corals) grow too slowly to be feasible for restoration projects using ocean-based nurseries. Now, Coral Vita can grow these corals in months rather than decades. {In addition}, land-based coral farming enables the usage of assisted evolution techniques to improve coral resiliency to changing oceanic conditions that threaten reef health.[110]

Results

Coral Vita has accomplished a lot to date. This includes site analysis and site selection, lining up advisors, and Coral Vita initially raised USD\$2 million for reef restoration. Most of the investors were angel investors (i.e., high-net worth individuals), but there were some Silicon Valley venture capitalists and the investment time horizon is approximately 10–15 years, as opposed to the more traditional seven years.[111]

In addition to working with concerned citizens, communities, corporations, and foundations, Coral Vita has also developed "tailor-made restoration plans" for: resorts and ecotourism operators; governments and marine park managers; and damage repair and mitigation banking.

For instance, with resorts and ecotourism operators:

Our 'Coral Cabañas' offer a dual-purpose experience. As we grow corals to restore your resort's reef, guests and their families will also have the opportunity to visit the Cabaña and learn about coral farming and threatened reefs. They can sponsor coral restoration on their behalf, and even help plant corals alongside our team. If your resort has its own dive shop, we can also arrange for the creation of a coral restoration dive certification program. {…} By investing in restoration, forward-thinking hotels like the Four Seasons Seychelles and Buddy Dive Resort Bonaire are already capitalizing on local projects as prime marketing tools that protect their reef assets.[112]

For governments and marine park managers:

Direct government investment, such as funds designated in the American Recovery and Reinvestment Act, help to both revive reefs and create local jobs. As well, there are growing opportunities for innovative conservation finance mechanisms. From the new COAST initiative by the World Bank and partners to use reef restoration as a way to reduce insurance premiums to the Caribbean Biodiversity Fund to the Seychelles debt-for-adaptation swaps to the Palau Green Fee, governments are finding ways to invest in climate adaptation. A local Coral Vita farm gives your nation the opportunity to protect its reefs and the people who depend on them.[113]

With respect to damage repair and mitigation banking:

> There is a growing need for corals to repair damaged sites, as reefs are regularly impacted by dredging projects, ship groundings, and reckless behavior. Existing ocean-based nurseries are limited in the scalability and coral diversity necessary to service large-scale restoration needs. Coral Vita provides the opportunity for mitigation projects to obtain the corals needed at-scale to restore the vitality of damaged reefs. {…} Damaging corals is illegal in many countries, and there are penalties or tools for financing restoration projects. In some areas, offenders who damage reefs get off without penalty- but local communities and industries are still affected. In the United States, mitigation banking has proved to be an effective tool for damage caused to wetlands. The State of Hawai'i is now piloting a bank to offset coral losses, proposing 'to maintain or improve the quantity and condition of coral reefs at banks sites and provide high quality compensatory mitigation that yields ecologically successful and sustainable results.' Such a model can prove effective for countries in need of effective compensatory mechanisms to restore reefs.[114]

Challenges and How They Were Met

There are several challenges with the traditional, ocean-based coral reef restoration operations:

> Ocean-based nurseries are appealing for small-scale restoration projects because they can be assembled at low cost and support fast-growing branching species like Staghorn and Elkhorn corals. Unfortunately, such nurseries don't offer an effective and scalable means for executing large-scale restoration projects, especially since they are almost entirely grant-funded. Coral Vita grows corals in land-based farms, which provide crucial advantages for the company and reef stakeholders.[115]

Further:

> Farming corals on land further reduces risks that ocean-based nurseries are exposed to, such as storms, boating accidents, or warming events. Also, ocean-based farming limits the type of corals to the site-specific location of the nursery. Growing corals on land allows us to produce a range of native corals which meet the needs of a particular site, as Coral Vita can tailor water conditions that replicate those of restoration site. Sooner rather than later, Coral Vita's land-based farming offers a viable solution to counter large-scale degradation.[116]

Another challenge was that as of April 2019, Coral Vita had not yet done any coral reef restoration; although, this will soon be accomplished.[117]

Sadly, a major challenge facing the Bahamas and Coral Vita's operations is the recovery from Hurricane Dorian, a devasting hurricane that hit the Bahamas in September 2019 and which to date, was "{…} the strongest known tropical system to impact the Bahamas."[118]

Beyond Results

Coral Vita is already looking into new expansion areas because Coral Vita's goal is to have reef restoration facilities set up "everywhere." Likewise, because of CITES, Coral Vita cannot export corals from Grand Bahama to different countries. Furthermore, Coral Vita will have a tourism and education component. This will help to diversify revenue and to increase awareness. For instance, tourists would pay to visit the coral restoration farm, people can volunteer to participate in planting the corals back onto the reefs, and people can also donate to adopt the corals.[119]

As highlighted on Coral Vita's website:

Through this restoration process, you can protect coral reefs and the value they create for you. Benefits from a restoration project include:

- Increased marine life abundance and fisheries production;
- Enhanced tourism experiences and attractions;
- Improved coastal property values and protection; and
- Preservation of unique biodiversity.[120]

Furthermore, this will be the first large-scale, land-based, commercial-based coral reef restoration and Coral Vita, which is fortunate to have very smart, passionate people involved, would like to scale up to millions, if not billions of corals.[121] As stated by Mr. Teicher, "hopefully we are not the only ones doing it."[122]

Lessons Learned

According to co-founder Mr. Sam Teicher, there are several lessons learned thus far. First, island time is real and things take longer, so one should have contingency plans. Second, one should always integrate the local communities as much as possible. For instance, fisherfolk can become part of the local dive team and scientists from the host country can be incorporated into the project. Third, it is an important goal to make these coral restoration facilities both a tourism attraction and an educational facility, in addition to being a commercial operation.[123]

Other Resources on Coral Vita

Bahamas National Trust

- https://bnt.bs/

"Building Coral Reef Resilience Through Assisted Evolution," by Madeleine J. H. van Oppen, James K. Oliver, Hollie M. Putnam, and Ruth D. Gates

- https://www.pnas.org/content/pnas/112/8/2307.full.pdf

Coral Vita

- http://www.coralvita.co/

Global Island Partnership

- http://www.glispa.org/

"Growing Coral Larger and Faster: Micro-Colony-Fusion as a Strategy for Accelerating Coral Cover," by Zac H. Forsman, Christopher A. Page, Robert J. Toonen, and and David Vaughan

- https://www.ncbi.nlm.nih.gov/pmc/articles/PMC4614846/pdf/peerj-03-1313.pdf

The Caribbean Challenge Initiative

- https://www.caribbeanchallengeinitiative.org/

Future Outlook for Instrument

The future of additional solutions will likely involve the maturation of approaches and the development of new business models which focus on coral conservation, coral restoration, marine pollution cleanup, and fisheries management.

Marine pollution cleanups could follow the work of 4ocean,[124], Bracenet,[125] and Bureo. For example, "Dell has begun shipping its XPS 13 in materials made in part from plastic pollution from the ocean. The company spent about 18 months doing a detailed assessment and validation, followed by a pilot, based on using ocean plastic in a way that is cost-effective and commercially scalable."[126]

There are a wide range of river cleanup projects and improvements to wastewater treatment plants to improve water quality and lessen the offshore impact on coral reefs.

There are also a wide range of climate change mitigation projects from the energy efficiency, renewable energy, and forestry sectors which can help to stabilize global sea surface temperatures and reduce the trend of ocean acidification. Furthermore, some of these climate change mitigation projects in the forestry space are focused on blue carbon and mangrove forestry restoration.

Restoration work, both similar to and beyond to the work of the CRF™ and Coral Vita, is needed. For instance, SEA LIFE helped to build reef spiders made from rebar and attached pieces of coral to them in the Indo-Pacific region. Over time, the reef spiders can be lifted up and the fish can be harvested for the aquarium trade. As the corals grow ever larger, eventually one can no longer move the reef spiders, and they are left behind. Mars International, which has a large presence in the pet industry, has a lot of marine scientists and they did one reef restoration study like the reef spiders rebar project. However, the neighboring island destroyed their reef spiders, possibly out of spite.[127]

There will also likely be more artificial reef projects. There is a distinction between intentional artificial reefs (e.g., the USS Kittiwake in Grand Cayman, Cayman Islands) and unintentional artificial reefs (e.g., WWII wreckage at Chuuk Atoll /Truk Lagoon or the SS Yongala off Queensland, Australia). Likewise, there are artificial reef projects all around the world, from artificial reefs in Bahraini waters off Saudi Arabia to reconstituted lumps of rock off the coast of Florida.[128] Regarding artificial reefs, "{…} a nagging question remains: Do they simply concentrate existing fish or do they add living space and so increase the amount of fish a habitat can support? {…} Sometimes they do, sometimes they don't. But even if artificial reefs did provide some uplift in fish stocks, is that a good enough reason to fill the sea with junk?"[129] Likewise, tires are not good for artificial reefs because of potential disintegration and release of toxins.

Other Resources on Additional Solutions

Coral Gardeners

- https://www.coralgardeners.org/

Coral Gardening

- https://www.imdb.com/title/tt5807770/plotsummary?ref_=tt_ov_pl

Coral Restoration Consortium

- http://crc.reefresilience.org/

Coral Restoration Toolkit

- https://restoreyourcoast.org/

International Coastal Cleanup

- https://oceanconservancy.org/trash-free-seas/international-coastal-cleanup/

Reef Resilience Network

- https://reefresilience.org/

The Coral Gardener

- https://www.imdb.com/title/tt3207618/plotsummary?ref_=tt_ov_pl

The Society for Ecological Restoration

- https://www.ser.org/

Trash Free Seas Alliance

- https://oceanconservancy.org/trash-free-seas/plastics-in-the-ocean/trash-free-seas-alliance/

Notes

1. Braverman. *Coral Whisperers.* 118.
2. Ibid.
3. Kneppers, Ben. Interviewed by Brian McFarland. June 2019.
4. Braverman. *Coral Whisperers.* 111.
5. Kneppers, Ben. Interviewed by Brian McFarland. June 2019.
6. Ibid.
7. Ibid.
8. Ibid.
9. Bureo. "Bureo Collection." Accessed June 17, 2019. https://bureo.co/pages/bureo-collection2.
10. Kneppers, Ben. Interviewed by Brian McFarland. June 2019.
11. Ibid.
12. Giskes, Ingrid. "Ghost Fishing Gear: The Global Problem and the Global Solution." Online Presentation, Co-Coordinated by OCTO and NatureServe, Online. 10 April 2018. https://www.openchannels.org/webinars/2018/ghost-fishing-gear-global-problem-and-global-solution.
13. Kneppers, Ben. Interviewed by Brian McFarland. June 2019.
14. Ibid.
15. EarthHero. "Bureo." Accessed September 13, 2019. https://earthhero.com/brands/bureo/.
16. Ibid.
17. Bureo. "Bureo Collection." Accessed June 17, 2019. https://bureo.co/pages/bureo-collection2.
18. Bureo. "Our Retail Partners." Accessed June 17, 2019. https://bureo.co/pages/bureo-retailers.
19. Bureo. "Bureo Collection." Accessed June 17, 2019. https://bureo.co/pages/bureo-collection2.
20. Bureo. "Shop." Accessed June 17, 2019. https://bureo.co/collections/bureo.
21. Kneppers, Ben. Interviewed by Brian McFarland. June 2019.
22. Ibid.
23. Ibid.
24. Ibid.
25. Ibid.
26. Ibid.
27. Ibid.
28. Ibid.
29. Ibid.
30. Braverman. *Coral Whisperers.* 120–121.
31. CRF™. "About Us." Accessed December 30, 2019. https://www.coralrestoration.org/about.

32. CRF™. "Our Origins." Accessed December 30, 2019. https://www.coralrest oration.org/about.

33. Grainger, Alice and Scott Winters. Email message to author. November 26, 2019.

34. Ibid.

35. Ibid.

36. Earle. *Sea Change*. 227.

37. Grainger, Alice and Scott Winters. Email message to author. November 26, 2019.

38. Ibid.

39. Ibid.

40. Ibid.

41. Ibid.

42. Ibid.

43. Ibid.

44. Ibid.

45. Ibid.

46. Ibid.

47. NOAA. "Coral Reefs Support Jobs, Tourism, and Fisheries." Accessed December 30, 2019. https://floridakeys.noaa.gov/corals/economy.html.

48. Grainger, Alice and Scott Winters. Email message to author. November 26, 2019.

49. Ibid.

50. Ibid.

51. Fodor's Editors. "Thirteen Places to Reconsider in the Year Ahead." Fodor's Travel. November 13, 2019. Accessed December 30, 2019. https://www.fod ors.com/news/photos/fodors-no-list-2020.

52. Grainger, Alice and Scott Winters. Email message to author. November 26, 2019.

53. Ibid.

54. Ibid.

55. Ibid.

56. Ibid.

57. Ibid.

58. Ibid.

59. Ibid.

60. Sobel and Dahlgren. *Marine Reserves*. xii.

61. CRF™. "About Us." Accessed September 19, 2019. https://www.coralrestora tion.org/about.

62. Grainger, Alice and Scott Winters. Email message to author. November 26, 2019.

63. Ibid.

64. Ibid.

65. Ibid.

66. CRF™. "Restoration." Accessed September 19, 2019. https://www.coralrestoration.org/restoration.
67. Ibid.
68. Grainger, Alice and Scott Winters. Email message to author. November 26, 2019.
69. CRF™. "Science." Accessed September 19, 2019. https://www.coralrestoration.org/science.
70. Grainger, Alice and Scott Winters. Email message to author. November 26, 2019.
71. Ibid.
72. Ibid.
73. Ibid.
74. CRF™. "Restoration." Accessed September 19, 2019. https://www.coralrestoration.org/restoration.
75. Ibid.
76. Ibid.
77. CRF™. "About Us." Accessed September 19, 2019. https://www.coralrestoration.org/about.
78. Monroe County Tourist Development Council. "Florida Keys Sanctuary Announces $100 Million Coral Restoration Programme." January 6, 2020. Accessed April 13, 2020.
79. Grainger, Alice and Scott Winters. Email message to author. November 26, 2019.
80. Ibid.
81. Ibid.
82. Ibid.
83. Ibid.
84. Ibid.
85. Ibid.
86. Ibid.
87. Ibid.
88. Ibid.
89. Ibid.
90. Braverman. *Coral Whisperers.* 119–120.
91. Ibid. 142–143.
92. Ibid. 164.
93. Grainger, Alice and Scott Winters. Email message to author. November 26, 2019.
94. Ibid.
95. Ibid.
96. CRF™. "Restoration." Accessed September 19, 2019. https://www.coralrestoration.org/restoration.
97. Braverman. *Coral Whisperers.* 150.

98. Grainger, Alice and Scott Winters. Email message to author. November 26, 2019.

99. Ibid.

100. Levy, J., K. Ripple, and R. S. Winters. 2018. Lessons Learned for Increased Scalability for in Situ Coral Restoration Efforts [White paper]. Retrieved April 8, 2020, from Coral Restoration Foundation. https://www.coralrestoration.org/white-paper-lessons-learned.

101. Teicher, Sam. Interviewed by Brian McFarland. April 2019.

102. Van Oppen et al. 2015. Building Coral Reef Resilience Through Assisted Evolution. Proceedings of the National Academy of Sciences of the United States of America. Vol. 112. No. 8. www.pnas.org/cgi/doi/10.1073/pnas.1422301112. 1.

103. Dahlgren. "Bahamian Marine Reserves—Past Experience and Future Plans. In: Sobel and Dahlgren. *Marine Reserves.* 270.

104. Coral Vita. "Why Does Coral Reef Health Matter to You?" Accessed April 13, 2020. https://www.coralvita.co/why-reefs.

105. Teicher, Sam. Interviewed by Brian McFarland. April 2019.

106. Ibid.

107. Ibid.

108. Global Island Partnership. "Members." Accessed December 30, 2019. http://www.glispa.org/participate/members.

109. Coral Vita. "What We Do." Accessed August 13, 2019. http://www.coralvita.co/what-we-do.

110. Ibid. "Coral Farming." Accessed December 30, 2019. http://www.coralvita.co/coral-farming.

111. Teicher, Sam. Interviewed by Brian McFarland. April 2019.

112. Coral Vita. "Resorts & Eco-Tourism Operators." Accessed August 15, 2019. http://www.coralvita.co/eco-tourism.

113. Ibid. "Governments & Marine Park Managers." Accessed August 15, 2019. http://www.coralvita.co/governments-and-marine-parks.

114. Ibid. "Damage Repair & Mitigation Banking." August 15, 2019. http://www.coralvita.co/damage-repair.

115. Ibid. "Coral Farming." Accessed August 15, 2019. http://www.coralvita.co/coral-farming.

116. Ibid.

117. Teicher, Sam. Interviewed by Brian McFarland. April 2019.

118. National Weather Service. "Hurricane Dorian—September 6, 2019." September 6, 2019. Accessed December 30, 2019. https://www.weather.gov/mhx/Dorian2019.

119. Teicher, Sam. Interviewed by Brian McFarland. April 2019.

120. Coral Vita. "What We Do." Accessed August 15, 2019. http://www.coralvita.co/what-we-do.

121. Teicher, Sam. Interviewed by Brian McFarland. April 2019.

122. Ibid.

123. Ibid.
124. 4ocean. "Home." Accessed December 30, 2019. https://4ocean.com/.
125. Bracenet. "Home." Accessed December 30, 2019. https://bracenet.net/en/.
126. Nastu, Jennifer. "Dell Says 'Ocean Plastic' Packaging Is Cost-Effective, Scalable." *Environmental Leader*. June 19, 2017. Accessed December 30, 2019. https://www.environmentalleader.com/2017/06/dell-says-ocean-plastic-packaging-cost-effective-scalable/.
127. Wright, James. Interviewed by Brian McFarland. January 2018.
128. Ibid.
129. Roberts. *The Ocean of Life*. 290.

17

The Future of Coral Reef Conservation Finance

With USD\$52 billion[1] spent annually on conservation projects—not just coral reef conservation—and approximately USD\$8.2 billion invested by the private sector in sustainable food and fiber production, habitat conservation, and water quality and quantity protection projects from 2004 to 2015,[2] there is still a long way to go to reach the estimated USD\$200 to USD\$300 billion needed on an annual basis to finance the world's ecosystems.[3] There is clearly a funding gap when it comes to conserving the world's tropical coral reefs.

One of the main barriers to conservation enhancement is that the funds currently needed to achieve effective, lasting conservation significantly exceed the available funds. An average US\$270 million commitment per year for marine conservation was recorded between 2010 and 2016. Meanwhile, the funds necessary to achieve the United Nations Convention on Biological Diversity target of having 20% of the ocean protected are estimated at between US\$4 and \$8 billion per year in management costs only.[4]

There are solutions. As this book has outlined, there are numerous financial and strategic solutions to conserve the world's remaining tropical coral reefs. Essentially what is needed right now and going into the future is, in no particular order:

- Clean up plastic pollution and reduce plastic usage;
- Protect adjacent land;
- Restoration of degraded areas;
- Utilize technology;

© The Author(s) 2021
B. J. McFarland, *Conservation of Tropical Coral Reefs*,
https://doi.org/10.1007/978-3-030-57012-5_17

- Mitigation of global climate change;
- Conservation of large seascapes;
- Investable, scalable projects for the private sector;
- Development of new financing structures;
- Public funds used to leverage private financing;
- Education, reduced public apathy, and the next generation inspired;
- A new economic system that values natural capital;
- Environmental justice; and
- Effective public policies.

Similarly, these are also the reoccurring themes throughout the case studies' lessons learned.

Clean Up Plastic Pollution and Reduce Plastic Usage

Reducing marine pollution, particularly plastics, is essential to the health of the world's oceans and coral reef ecosystems. Everyone, from individuals to companies and governments, can be part of this solution. This could include packing lunch in reusable containers instead of single-use plastics, bringing reusable bags to the grocery stores, declining the use of straws, or even to lobby for "common single-use plastics {to} carry a cigarette-style warning label."[5]

Numerous governments and companies are moving to eliminate or reduce plastic usage whether through "no straw initiatives" or municipalities charging a plastic bag tax. Governmental examples around the world include:

- Boston, Massachusetts has banned plastic bags[6];
- London, United Kingdom announced a plan to install "public water fountains to tackle plastic waste" and to "increase accessibility of tap water"[7];
- Chile adopted a new law to ban the use of plastic bags[8];
- The EU's European Parliament approved a law that will ban single-use plastics such as straws, plates and cutlery starting in 2021[9];
- Malaysia plans to ban all single-use plastic by 2030[10];
- Narendra Modi, the prime minister of India, has pledged "to remove all single-use plastics by 2022"[11];
- New Zealand banned single-use plastic bags in 2019[12];

- Thailand, "in a bid to stem the flow of imported rubbish coming into Southeast Asia following China's waste ban, {…} has announced a plan to halt all imports of plastic waste by 2021"[13];
- Taiwan introduced a "plan to outlaw plastic waste by 2030"[14]; and
- China, the world's most populated country and "one of the world's biggest plastic producers, has unveiled a new plan to phase out single-use plastics," by the end of 2020.[15]

Similarly, many companies are taking action such as:

- Sandals Resorts International planned to eliminate single-use plastic straws and stirrers as of November 2018[16];
- American Airlines is eliminating plastic straws from all its flights and lounges[17];
- Marriott International planned to eliminate plastic straws and stirrers by July 2019[18];
- Starbucks plans to eliminate plastic straws from its stores by 2020, which is expected to be "more than 1 billion straws from the company's 28,000 stores annually"[19];
- Adidas produced five million pairs of shoes made from recycled plastic in 2018 and had a 2019 goal to produce another eleven million such pairs[20];
- McDonald's "restaurants will switch from plastic straws to paper ones in the UK and Ireland" in 2019[21];
- Mondelēz has committed to making all of its packaging recyclable by 2025[22];
- Eight international sporting bodies, including the International Olympic Committee, have "signed up to a major UN campaign to reduce single-use plastics"[23];
- In September 2018, Tesco launched "reverse vending machines for plastic bottles"[24];
- Unilever's line of "PG tips tea bags will now be plastic-free and 100% biodegradable"[25];
- Ekoplaza, a Dutch supermarket, as "result of a campaign from UK-based charity A Plastic Planet, which works to persuade supermarkets to lessen the amount of plastic they use to stock everyday food items, opened the world's first plastic-free supermarket aisle in Amsterdam"[26];
- Evian, the water product line from Danone, "will use 100% recycled plastic bottles by 2025"[27];
- Lego, the iconic toy company, "has announced plans to stop the production of plastic blocks by 2030" and the new "Lego shapes will alternatively

be made from plant-based materials in an attempt to reduce plastic waste"[28];

- PepsiCo, one of the world's largest consumer goods companies, has "signed a deal with Loop Industries to introduce 100 per cent recycled plastic packaging"[29];
- Coca-Cola launched the 'World Without Waste' campaign and "has pledged to collect and recycle the equivalent of 100 percent of its packaging worldwide, and manufacture its bottles using at least 50 percent recycled plastic by 2030"[30]; and
- The National Geographic has launched a major new initiative to raise awareness about single-use plastics.[31]

Consumers can pressure these companies, such as the world's largest contributors to plastic pollution (i.e., Coke, PepsiCo and Nestlé),[32] to adopt more sustainable forms of packaging and product design. To reduce plastic waste from straws, consumers can stop using straws, or use reusable straws like the foldable FinalStraw.[33]

In addition to reducing marine pollution, we need to clean up the pollution that is already in the marine environment. There are many groups that are tracking marine pollution, such as the Australian Marine Debris Database,[34] and many groups organizing trash pickups such as Let's Do It![35] and Trash Hero.[36]

Furthermore, there are also alternative uses being sought. Take for example:

- The company Neste is "exploring the use of plastic waste as a raw material for fuel"[37];
- India could construct buildings from recycled plastics as researchers "demonstrated that replacing 10 per cent of sand in concrete with plastic waste could simultaneously solve a national sand shortage and reduce the growing amounts of rubbish on the streets"[38];
- American Express is developing a new credit card that is made from recycled ocean plastic[39];
- French designers, Lilian and Fred, designed a watch called "Awake," that is "made from plastic waste, recycled stainless steel and is powered by solar energy"[40]; and
- Dell, "HP and Ikea join a consortium of 10 firms that aims to turn the equivalent of 1.2 billion single-use plastic water bottles into products and packaging by 2025."[41]

Protect Adjacent Land

As stated by WWF:

The protection of land in coastal areas, of islands, and even of offshore underwater property (in countries whose legal systems permit this) can be effective ways of protecting marine biodiversity against land-based threats such as:

- industrial and urban pollution and sewage;
- agricultural runoff from fertilizer and pesticides;
- construction and development of docks, marinas, coastal roads, hotels, and housing near the shore or in ecologically sensitive areas; and
- other human activities that disturb nesting birds, turtles and other wildlife along the shore and in the ocean.[42]

This is why work such as the Palmyra Atoll purchase, the conservation easement structured for Pez Maya in the Sian Ka'an Biosphere Reserve, and the restoration of Lady Elliot Island in the Southern Great Barrier Reef are so important.

Restoration of Degraded Areas

As highlighted in Chapter 16, *Additional Solutions*, coral restoration—despite concerns of its cost and scalability—is necessary. Yet, tensions remain:

In a nutshell, I would offer that such tensions go deeper than personal and professional disagreements. They signify the ongoing, and arguably intensifying, rift between those conservation scientists who still assume that it is possible to use traditional conservation tools (chiefly the removal of adverse local human impacts) to allow natural systems to restore themselves to a prior state, and others who hold that even with the deleterious impacts removed, natural systems will not return to a prior state because their environment has fundamentally altered.[43]

For example,

the Australians have been historically and culturally more inclined toward traditional preservation models such as securing marine protection areas, while their colleagues in the Caribbean have focused more strongly on restoration and other intensive management strategies. Currently, however, the Australians are reassessing their approach, as corals have suffered a serious death toll in their region, too.[44]

As explained by Dr. Petra Lundgren:

> I think there will be two avenues, one that is focused on the smaller scale, where people power matters, where it is about saving 'your' reef, or your patch. Where people (or tourist operators, or communities that rely on a patch of reef) accept that they will have to keep restoring through coral garden type activities for a long time (until the climate stabilizes and coral reefs begin to recover by themselves again).
>
> Then there will be the massive, hugely expensive, engineering efforts, that will combat loss through every mean possible. These interventions will include solar radiation management, assisted migration/evolution, large scale aquaculture, and industrial sized deployment technologies.
>
> They both play a critical role, and as most coral reefs exist in developing countries, it is really important that we, in the developed world, do not lose sight of the smaller scale, and the possibility of people power and affordable technological solutions when we do our large R&D programs.
>
> It is not only about 'our' reefs, it has to be about the coral reefs of the world, and there will never be one solution that fits them all. Policy, strategy, finance, it is so vastly different in different countries, and there will be as many variations as there are management authorities (or lack thereof).[45]

In addition to coral restoration, whether it be land-based or sea-based operations, there needs to be the restoration of complete ecosystems and not just the replanting of corals. For instance, the Healthy Reefs for Healthy People Initiative recognizes that you need corals to be replanted, but the macroalgae is a problem and putting herbivores back can help. Thus, the Initiative is looking at ecosystem services, working with coral nurseries, and putting more herbivores (e.g., crabs and parrotfish) back on the reefs. More specifically, the Initiative is raising hundreds of crabs up to ten centimeters at a farm and then releasing them back onto the reefs.[46] According to Tom Moore, a "balanced, strong, diverse population is the goal" of coral restoration.[47]

Utilize Technology

The role of technology will hopefully better assist humanity with coral reef conservation and restoration. This could take the form of more accurately mapping and monitoring coral reef cover, more effectively enforcing fishing regulations, better marine spatial planning, and educating the general public about the wonders of life under the sea.

Technology is advancing fast:

New technologies are helping to overcome human limitations – acoustic probes, satellite surveillance, sensors, lasers, innovative diving systems, remotely operated vehicles, autonomous unmanned and manned submersibles, and computers to store, analyze, and communicate the vast amount of data acquired.[48]

Similarly, there are:

advances in research ships and scientific platforms, diving technologies, climb-aboard submersibles, remote and autonomous underwater vehicles, side-scan sonar, multibeam and airborne surveillance systems, miniaturized tracking devices, and space-based remote sensing systems.[49]

For instance, drones can be used to monitor for illegal fishing vessels and for monitoring beach litter.[50] Reef Design Labs is using 3D printers for coral restoration,[51] the Ocean Cleanup's The Interceptor is an autonomous, solar powered plastic collector being deployed to rivers,[52] and researchers are developing ways to monitor coral reef health with satellites.[53]

Social media and mobile phones, which have the ability to reach literally billions of people around the world, can be a powerful force for conservation. For instance:

This spring an initiative called Apps for Earth, saw 100 percent of revenues from 27 apps at the Apple Store go to WWF over a ten day period that ended on April 24th. This campaign not only raised awareness it generated more than $8 million to support the WWF's conservation work.[54]

New technologies will also be applied. For example, Paso Pacifico, an NGO based in California with conservation projects in Nicaragua, developed fake turtle eggs made from a 3D printer and implanted with a GPS tracking device to catch poachers.[55]

Mitigation of Global Climate Change

Without mitigating global climate change, the conservation of tropical coral reefs will be only more difficult, if not impossible. The largest global sources of greenhouse gas emissions are fossil fuel combustion for power generation and transportation, along with land-use change as a result of agriculture and tropical deforestation. Thus, to mitigate global climate change, the world needs to quickly and drastically:

- Increase energy efficiency;
- Increase renewable energy generation;
- Reduce tropical deforestation;
- Increase forest restoration; and
- Electrify transportation.

Conservation of Large Seascapes

The need to conserve large seascapes will help mitigate the impacts on biodiversity from global climate change, reduce the impact of border effects and activities in the buffer zone, and provide habitat for wildlife to reproduce. Likewise, "a recent meta analysis of 87 MPAs found that conservation benefits increased exponentially with the accumulation of five attributes: no take, well enforced, old (>10 years), large (<100 km^2) and isolated by deep water or sand."[56] This is why places such as the Chagos Archipelago, the Great Barrier Reef, the Phoenix Islands, and the Seychelles are vitally important.

In addition, key lessons for effectively bridging the divide between biodiversity conservation and fisheries sustainability goals in the context of MPAs include: creating spaces and processes for engagement, incorporating fisheries in MPA design and MPAs into fisheries management, engaging fishers in management, recognizing rights and tenure, coordinating between agencies and clarifying roles, combining no-take-areas with other fisheries management actions, addressing the balance of costs and benefits to fishers, making a long-term commitment, creating a collaborative network of stakeholders, taking multiple pressures into account, managing adaptively, recognizing and addressing trade-offs, and matching good governance with effective management and enforcement.[57]

Investable, Scalable Projects for the Private Sector

Additional funding is currently available from the private sector, but projects and programs need to be investable and scalable. As noted by the *Analysis of International Funding for the Sustainable Management of Coral Reefs and Associated Coastal Ecosystems* summary report:

Many projects are relatively small. Of projects for which detailed funding information was available (294 of 314 projects), 120 received up to USD 100,000 and a further 92 up to USD 1 million, with a total combined value of USD 40.4 million. In contrast, {only} 33 projects received in excess of USD 10 million each, collectively accounting for USD 1.72 billion of the total funding commitments since 2010.[58]

Furthermore, the Caribbean-Pacific Alliance's Blue Challenge is looking for investable marine initiatives to help take them to market such as by upgrading production capacity, by developing better traceability, or by implementing better management structures.[59]

Ecosystem Marketplace nice summarized a conservation finance report developed by Credit Suisse and the McKinsey Center for Business and Environment, which[60]:

> {…} presents a few concrete ideas that we deem to be scalable, repeatable, and investable. A toolkit suggests two distinct approaches. One involves applying the same proven project management approach across a single project type, financing it through plain equity or debt vehicles, and scaling it up. The other involves bundling heterogeneous projects linked by a common feature – a national park, for instance – and creating a common risk-and-return sharing vehicle. Interviews suggested that risk is often a greater concern to potential investors than the financial return or even the conservation impacts of a project.[61]

This is important advice from the private sector.

Develop New Financing Structures

New financing structures need to be developed. For instance, what if megacities such as Manila or Jakarta used a real estate transfer tax or water usage surcharge to protect upstream watersheds, which in turn benefited offshore reefs? What if the Conservation Notes or the Nature Conservation Notes came down to a minimum investment of say USD$500 or USD$1000 for smaller investors?

As summarized by Ecosystem Marketplace, a conservation finance report developed by Credit Suisse and the McKinsey Center for Business and Environment noted that,[62] an "incubation workshop with 15 NGOs also revealed some fresh ideas: a marine protected area bond, insurance payments for risk mitigation, and a 'substitutes fund' that would support research and development of substitutes for destruction-inducing commodities such as meat, palm oil, or diamonds."[63] Similarly, the Conservation Finance Alliance recently announced the winners of its first call for proposals for the CFA Incubator. Of the five financed projects, the following four relate to reefs, watersheds, and coastal restoration:

- The "Investment in Coral Reef for Coastal Protection" by Barbados Reef Keepers—Coenostrum Inc.;
- The "Community Reef and Forest Bank" by GreenFi Systems Limited (East Africa);
- The "Kenyir For Life – Protecting the Kenyir Watershed through Sustainable Financing Mechanisms" by Rimba (Malaysia); and
- The "Biodiversity Lending Instruments for Coastal Restoration and Protection" by Natrx, Inc. (USA, Global).[64]

The CFA Incubator also announced a virtual incubator with mentorship support for:

- The "Blue Buyout Fund" by the IUCN – Conservation Capital (Africa);
- "Adapting PACE to finance conservation" by Possible Planet (U.S.);
- "Retrieving ghost fishing gear in Mexico: a circular economy solution to plastic pollution" by an NGO to be created (Mexico);
- "Building a conservation due diligence standard" by Conservation Alpha (Global); and
- "HORUS - A boat sharing marketplace for diving tour operators," by Conictus (Egypt, Global).[65]

Further, as noted by the International Coral Reef Initiative, "other forms of financial engineering may also merge in the future (e.g. Park Bonds, Project Finance for Permanence."[66]

Public Funds Used to Leverage Private Financing

Public funds can be used to de-risk investments for tropical coral reef conservation, which, in turn, helps leverage private financing. This was demonstrated with Mirova's Althelia Sustainable Ocean Fund.

Education, Reduced Public Apathy, and the Next Generation Inspired

As explained by Dr. Sylvia Earle, "far and away the greatest threat to the sea and to the future of mankind is ignorance."[67] Related to this is the pressing question of how to reduce "environmental fatigue, environmental despair and apathy, and environmental melancholia."[68] This "apparent indifference also may be related to the widely held view that the ocean is so vast and resilient that there is little reason to worry, either about what is put in or about what is taken out."[69]

Thus, we need much more education and public awareness, which can come in a variety of forms. For instance, "so far, more than eight thousand people in twenty-seven countries, mostly women, have collaborated to crochet displays of coral reefs."[70] Take as another example:

> Across the far-flung islands of the western Pacific, where an innovative conservation NGO called Rare broadcasts a popular radio soap opera dramatizing social and environmental issues, a quarter of listeners report that it's prompted them to stop eating sea turtle eggs, and almost a fifth say they've adopted family planning as a result of the show.[71]

As explained by Dr. Melanie McField, the Founder and Director of the Healthy Reefs for Healthy People Initiative, one thing we all tend to do is put our head down and do the work, but we need to take more time to speak together and collaborate. We need to collaborate upfront, in the middle, and at the end. We cannot just share our results at the end, because things would have been improved if groups talked to each other throughout the process. Belize is a pretty good example where there are nationally focused working groups and a diverse audience—from coral monitoring experts, spawning and aggregation experts, to park managers—that comes together. Essentially, we need to make time for sharing and we need to improve how we do things at a faster rate by sharing.[72]

We also need to mobilize and not just scientists and practitioners, but others such as Greta Thunberg.[73] Whether this mobilization occurs around the dire situation facing our tropical coral reefs and species extinction, or mobilization around the positive actions being taken each day, such as highlighted by the Ocean Optimism Movement, we need to rise to the challenges that lie ahead.[74] Similarly, there are numerous organizations around the world that are dedicated to inspiring the next generation. This includes Heirs to Our Oceans,[75] Jr Ocean Guardians,[76] and Bye Bye Plastic Bags,[77] along with the Kids Can Save The Planet documentary series.[78]

Further, there are countless engagement and volunteer opportunities.

Develop a New Economic System that Values Natural Capital

On a general level, there is an increased understanding or awareness of natural capital. For instance, 10–15 years ago, people would not know what you are talking about. While public citizens may still be a bit clueless, people in the private sector are starting to get it and many local communities and Indigenous Peoples have always gotten it. The challenge is translating the awareness of our dependence on natural capital (i.e., for society in general including companies and consumers) and the benefits we receive from ecosystems (e.g., seafood, water, storm protection, and medicine) into a more sustainable system of incentives for conservation (i.e., redirect resources to these areas that are providing all these benefits).[79]

As explained by Dr. Melanie McField, the private sector is going to realize that if they do not get on board, they are going to lose more than if they do not get on board. From Cozumel to Tulum, a tremendous amount of money has been generated from tourism and what built the tourism industry, in large part, are the offshore coral reefs. Yet, the dive industry and hotels do not pay for the conservation and restoration of coral reefs. In Belize, tourists pay a departure tax, but that is it. In Mexico, it is a little different, but a lot of the marine conservation finance is in the form of philanthropic dollars from the United States and Europe, and a lot of time is spent fundraising. In contrast, the private sector should do more. There is a mantra from companies about sustainability and those genuinely committed need to come together and promote their efforts in the media. Those other companies, whose business as usual for them is a reef keeping their hotel or restaurants protected, need to get on board.[80]

Furthermore, the climate movement is getting some momentum from the We Are Still In coalition,[81] to the RE100 movement which is a collection of the largest companies pledging to support 100% renewable power.[82]

Environmental Justice

The rights of local communities and Indigenous Peoples need to be recognized. From the perspectives of Latin America, Sub-Saharan Africa, and SIDS, there is little they can do to address the global climate crisis and yet, these countries are often the most at risk. Here arises a lack of environmental justice, as the larger developed countries of the United States and Europe, along with Brazil, Russia, India, China, and Indonesia, are adding far more GHG emissions to the atmosphere. Similarly, when pollution washes ashore on an atoll in Belize, such plastic pollution, for example, is often not coming from Belize, but largely from passing ships or neighboring countries. Furthermore, many local communities are being impacted by such pollution (e.g., poor sewage systems, plastic pollution washing ashore, etc.) and by immediate development pressures.[83]

There is an ongoing need to recognize the traditional livelihoods and customary fishing grounds of local communities and indigenous peoples. This said:

> A growing number of different user groups are seeking greater access to coastal resources, leading to overcrowding, ecosystem service degradation and community dissatisfaction. It often involves different user groups with diverse rights such as harvest rights (fish extraction), use rights (tourist permits and passive recreation), conservation rights (the right to conserve threatened species) and management rights with different degrees of exclusivity. This makes fishing and property rights very complex.[84]

It is vitally important that conservation initiatives are not done at the expense of the world's poorest and marginalized communities. Thus, social and environmental safeguards must be built into coral restoration and conservation initiatives. These social safeguards include, but not limited to:

- Free, Prior and Informed Consent (FPIC);
- Respect for women, children, and minority rights;
- No child, slave, or forced labor;
- Respect for local traditions, local knowledge, and intellectual property rights; and
- Equitable distribution of benefits, income, and risks.

A few of the biodiversity safeguards that should be considered when investing in, or undertaking, the development of coral reef conservation and/or coral reef restoration projects, include the prevention of:

- Coral nursery/coral farm monocultures;
- The introduction of genetically modified organisms (GMOs);
- The introduction of invasive species; and
- Biopiracy, particularly related to bioprospecting.

These biodiversity safeguards, along with their related social safeguards, should be regularly monitored, reported on, and then independently verified against third-party certification standards, if applicable.

Effective Public Policies

Politicians have not been held accountable by voters, to the extent needed, to address the issues impacting tropical coral reefs. While there is still a good degree of public awareness about the environment in general, a lot more needs to be raised as a serious political issue. Such policies need to be implemented at local, subnational, national, international, and company levels.

Other Resources

Business & Human Rights Resource Centre and Corporate Human Rights Benchmark

- https://www.business-humanrights.org/; and
- https://www.corporatebenchmark.org/

ConservationDrones.org

- https://conservationdrones.org/

Institute for Human Rights and Business

- https://www.ihrb.org/

Integrated Biodiversity Assessment Tool

- https://www.ibatforbusiness.org/about

<u>TED Talks on Coral Reefs, Conservation, and Technology</u>

- https://www.ted.com/talks/lian_pin_koh_a_drone_s_eye_view_of_cons ervation;
- https://www.ted.com/topics/coral+reefs; and
- https://reefbites.wordpress.com/2018/10/15/5-ted-talks-on-coral-reefs-that-will-wow-you-in-less-than-15-minutes/

<u>World Overview of Conservation Approaches and Technologies (WOCAT)</u>

- https://www.wocat.net/

Notes

1. Huwyler, Fabian et al. "Conservation Finance: From Niche to Mainstream: The Building of an Institutional Asset Class." 3.
2. Hamrick, Kelley. "State of Private Investment in Conservation 2016." vii.
3. Huwyler, Fabian et al. "Conservation Finance: From Niche to Mainstream: The Building of an Institutional Asset Class." 3.
4. Vertigo Lab. "Innovations for Coral Finance, ICRI Publication." 1.
5. Cooper, Rachel. "New Study Finds Public in Favour of Cigarette-Style Plastic Warnings." *Climate Action*. November 8, 2018. Accessed April 13, 2020. http://www.climateaction.org/news/new-study-finds-public-in-favour-of-cigarette-style-plastic-warnings.
6. Climate Action Programme. "Boston Bans Plastic Bags to Reduce Pollution." December 21, 2017. Accessed December 30, 2019. http://www.climateaction.org/news/boston-bans-plastic-bags-to-reduce-pollution.
7. Climate Action Programme. "London Installs Public Water Fountains to Tackle Plastic Waste." January 23, 2018. Accessed December 30, 2019. http://www.climateaction.org/news/london-installs-public-water-fountains-to-tackle-plastic-waste?utm.
8. Climate Action Programme. "Chile Adopts New Law to Ban the Use of Plastic Bags." May 31, 2018. Accessed June 1, 2018. www.climateactionprogramme.org/news/chile-adopts-new-law-to-ban-the-use-of-plastic-bags.
9. Watkiss, Camilla. "EU Finally Seals Ban on Single-Use Plastics by 2021." *Climate Action*. March 28, 2019. Accessed November 6, 2019. http://www.climateaction.org/news/eu-finally-seals-ban-on-single-use-plastics-by-2021.
10. Zein, Zafirah. "Malaysia to Ban Single-USE Plastic." *Eco-Business.com*. September 26, 2018. Accessed October 16, 2018. https://www.eco-business.com/news/malaysia-to-ban-single-use-plastic/.

11. Wentworth, Adam. "India Pledges to Remove All Single-Use Plastics by 2022." *Climate Action*. June 8, 2018. Accessed December 6, 2018. www.climateaction. org/news/india-pledges-to-remove-all-single-use-plastics-by-2022.

12. Ministry for the Environment. "Single-Use Plastic Shopping Bags Are Banned in New Zealand." July 1, 2019. Accessed December 30, 2019. https://www. mfe.govt.nz/waste/single-use-plastic-shopping-bags-banned-new-zealand.

13. Zein, Zafirah. "Thailand to Ban Plastic Waste Imports by 2021." *Eco-Business.com*. October 17, 2018. Accessed November 2, 2018. https://www.eco-business.com/news/thailand-to-ban-plastic-waste-imports-by-2021/.

14. Wentworth, Adam. "Taiwan Introduces Plan to Outlaw Plastic Waste by 2030." *Climate Action*. February 27, 2018. Accessed December 30, 2019. http://www.climateaction.org/news/taiwan-introduces-plan-to-outlaw-plastic-waste-by-2030?utm.

15. Cooper, Rachel. "China to Phase Out Single-Use Plastics." *Climate Action*. January 21, 2020. Accessed January 24, 2020. www.climateaction.org/news/china-to-phase-out-single-use-plastics.

16. Hermes, Jennifer. "Sandals Resorts Joins Ranks of Straw Abolishers." *Environmental Leader*. September 24, 2018. Accessed October 16, 2018. https://www.environmentalleader.com/2018/09/178142/.

17. Wentworth, Adam. "American Airlines Switches Away from Plastic Straws to Cut Waste." *Climate Action*. July 12, 2018. Accessed December 6, 2018. www.climateaction.org/news/american-airlines-switches-away-from-plastic-straws-to-cut-waste.

18. Danigelis, Alyssa. "Marriott International Eliminating Plastic Straws Next Year." *Environmental Leader*. July 18, 2018. Accessed October 16, 2018. https://www.environmentalleader.com/2018/07/marriott-international-plastic-straws/.

19. Danigelis, Alyssa. "Starbucks Vows to Ditch Plastic Straws by 2020." *Environmental Leader*. July 13, 2018. Accessed October 16, 2018. https://www.environmentalleader.com/2018/07/starbucks-plastic-straws/.

20. Adidas Group. "Adidas to Produce More Shoes Using Recycled Plastic Waste in 2019." January 21, 2019. Accessed April 13, 2020. https://www.adidas-group.com/en/media/news-archive/press-releases/2019/adidas-to-produce-more-shoes-using-recycled-plastic-waste/.

21. Danigelis, Alyssa. "McDonald's Begins Transition Away from Plastic Straws." *Environmental Leader*. June 15, 2018. Accessed October 16, 2018. https://www.environmentalleader.com/2018/06/mcdonalds-transition-plastic-straws/.

22. Cooper, Rachel. "Mondelēz Commits to Making All Packaging Recyclable by 2025." *Climate Action*. October 18, 2018. Accessed November 2, 2018. www.climateaction.org/news/mondeleez-commits-to-making-all-packaging-recyclable-by-2025.

23. Wentworth, Adam. "Major Sporting Bodies Make 'Biggest Commitment Ever' to Cut Plastic Waste." *Climate Action*. June 5, 2018. Accessed December 6, 2018. http://www.climateaction.org/news/major-sporting-bodies-make-biggest-commitment-ever-to-cut-plastic-waste.

24. Cooper, Rachel. "Tesco Launches Reverse Vending Machines for Plastic Bottles." *Climate Action.* September 27, 2018. Accessed December 30, 2019. http://www.climateaction.org/news/tesco-launches-reverse-vending-mac hines-for-plastic-bottles.
25. Wentworth, Adam. "PG Tips Tea Bags Will Now Be Plastic-Free and 100% Biodegradable." *Climate Action.* March 1, 2018. Accessed December 30, 2019. http://www.climateaction.org/news/pg-tips-tea-bags-will-now-be-plastic-free-and-100-biodegradable?utm.
26. Wentworth, Adam. "World's First Plastic-Free Supermarket Aisle Opens in Amsterdam." *Climate Action.* March 1, 2018. Accessed December 30, 2019. http://www.climateaction.org/news/worlds-first-plastic-free-aisle-opens-in-amsterdam.
27. Climate Action Programme. "Evian Will Use 100% Recycled Plastic Bottles by 2025." January 23, 2018. Accessed December 30, 2019. http://www.climateac tion.org/news/evian-will-use-100-recycled-plastic-bottles-by-2025.
28. Cooper, Rachel. "Lego to Ban Plastic Blocks by 2030." *Climate Action.* September 3, 2018. Accessed December 30, 2019. http://www.climateaction. org/news/lego-to-ban-plastic-blocks-by-2030.
29. Cooper, Rachel. "PepsiCo Signs Deal to Introduce 100% Recycled Plastic Packaging." *Climate Action.* October 15, 2018. Accessed December 30, 2019. http://www.climateaction.org/news/pepsico-signs-deal-to-introduce-100-recycled-plastic-packaging.
30. Climate Action Programme. "Coca-Cola Launches 'World without Waste' Campaign." January 24, 2018. Accessed December 30, 2019. www.climateac tion.org/news/coca-cola-launches-world-without-waste-campaign.
31. Ibid. "National Geographic Launches Major Campaign to Fight Single-Use Plastics." May 17, 2018. Accessed June 1, 2018. www.climateactionprogr amme.org/news/national-geographic-launches-major-campaign-to-fight-single-use-plastics.
32. Hicks, Robin. "World's Biggest Plastic Polluters Revealed." *Eco-Business.com.* October 9, 2018. Accessed October 16, 2018. https://www.eco-business.com/news/worlds-biggest-plastic-polluters-revealed/.
33. FinalStraw. "Home." Accessed June 1, 2018. https://finalstraw.com/.
34. Australian Marine Debris Database. "Home." Accessed August 29, 2018. http://amdi.tangaroablue.org/.
35. Let's Do It! "Overview." Accessed August 29, 2018. https://www.letsdoitworld. org/about/overview/.
36. Trash Hero. "What We Do." Accessed August 29, 2018. https://trashhero.org/what-we-do/.
37. Cooper, Rachel. "Plastic Waste to Be Used for Fuel." *Climate Action.* August 31, 2018. Accessed December 30, 2019. http://www.climateaction.org/news/plastic-waste-to-be-used-for-fuel.
38. Cooper, Rachel. "India Could Construct Buildings from Recycled Plastic." *Climate Action.* September 14, 2018. Accessed December 30, 2019. http://

www.climateaction.org/news/india-could-construct-buildings-from-recycled-pla
stic.

39. Wentworth, Adam. "American Express to Launch Credit Card Made from Ocean Plastic." *Climate Action.* June 12, 2018. Accessed December 30, 2019. http://www.climateaction.org/news/american-express-to-launch-credit-card-made-from-ocean-plastic.

40. Cooper, Rachel. "New Solar-Powered Watch Is Made from Recycled Plastic." *Climate Action.* August 24, 2018. Accessed December 30, 2019. http://www.climateaction.org/news/new-solar-powered-watch-is-made-from-recycled-plastic.

41. Hicks, Robin. "HP Joins Rival Dell's Consortium to Turn Ocean-Bound Plastic Trash into Products." *Eco-Business.* October 23, 2018. Accessed December 30, 2019. https://www.eco-business.com/news/hp-joins-rival-dells-consortium-to-turn-ocean-bound-plastic-trash-into-products/.

42. Spergel, Barry and Melissa Moye. "Financing Marine Conservation: A Menu of Options." 42.

43. Braverman. *Coral Whisperers.* 5.

44. Ibid. 5.

45. Lundgren, Petra. Email message to author. March 24, 2020.

46. McField, Melanie. Interviewed by Brian McFarland. January 2020.

47. Moore, Tom. Interviewed by Brian McFarland. December 2019.

48. Earle. *The World Is Blue.* 197.

49. Helvarg. *50 Ways to Save the Ocean.* 141.

50. Martin, Cecilia, Stephen Parkes, Qiannan Zhang, Xiangliang Zhang, Matthew F. McCabe, and Carlos M. Duarte. 2018. "Use of Unmanned Aerial Vehicles for Efficient Beach Litter Monitoring." *MarXiv.* May 16, 2018. https://doi.org/10.1016/j.marpolbul.2018.04.045.

51. Klinges, David. "A New Dimension to Marine Restoration." *Eco-Business.* August 28, 2018. Accessed April 13, 2020. https://www.eco-business.com/news/a-new-dimension-to-marine-restoration/.

52. The Ocean Cleanup. "The Interceptor." Accessed April 13, 2020. https://theoceancleanup.com/rivers/.

53. Asner, Greg and Clare LeDuff. "Reefscape: A Global Reef Survey to Build Better Satellites for Coral Conservation." *Leonardo DiCaprio Foundation.* 2019. Accessed April 13, 2020. https://www.leonardodicaprio.org/reefscape-a-global-reef-survey-to-build-better-satellites-for-coral-conservation/.

54. The Green Market Oracle. "Apple's Apps for Earth Initiative Raised Awareness and Funds for WWF." Last modified June 22, 2016. http://www.thegreenmarketoracle.com/2016/06/apples-apps-for-earth-initiative-raised.html.

55. Peters, Adele. "These Fake GPS-Implanted Sea Turtle Eggs Will Help Track Poachers." *Fast Company.* November 10, 2017. Accessed December 30, 2019. https://www.fastcompany.com/90445110/the-best-and-worst-brand-designs-of-the-decade.

56. Rotjan et al. "Establishment, Management, and Maintenance of the Phoenix Islands Protected Area." 298.

57. Weigel, Jean-Yves. "Marine Protected Areas and Fisheries: Bridging the Divide." 199–200.
58. UN Environment et al. "Analysis of International Funding for the Sustainable Management of Coral Reefs and Associated Coastal Ecosystems." 4.
59. Hernandez, Marcello. "Caribbean-Pacific Alliance." *Conservation Finance Alliance*. February 25, 2019. Accessed March 16, 2020. https://www.conservationfinancealliance.org/webinars-1/2019/2/25/caribbean-pacific-alliance.
60. Huwyler, Fabian et al. "Conservation Finance: From Niche to Mainstream: The Building of an Institutional Asset Class."
61. Ecosystem Marketplace. "The Carbon Chronicle." January 26, 2016. Accessed April 15, 2017.
62. Huwyler, Fabian et al. "Conservation Finance: From Niche to Mainstream: The Building of an Institutional Asset Class."
63. Ecosystem Marketplace. "The Carbon Chronicle." February 25, 2016. Accessed April 15, 2017.
64. CFA. "CFA Incubator: 2020 Call for Proposals Results." Accessed April 13, 2020. https://www.conservationfinancealliance.org/incubator.
65. Ibid.
66. Vertigo Lab. "Innovations for Coral Finance, ICRI Publication." 2017. http://vertigolab.eu/wp-content/uploads/2018/04/rapport-innovation-for-coral-finance.pdf. 71.
67. Earle. *Sea Change*. XXI.
68. Braverman. *Coral Whisperers*. 46.
69. Earle. *Sea Change*. XVIII.
70. Braverman. *Coral Whisperers*. 247.
71. Balmford. *Wild Hope*. 15.
72. McField, Melanie. Interviewed by Brian McFarland. January 2020.
73. Ibid.
74. Ocean Optimism. "Home." Accessed April 13, 2020. http://www.oceanoptimism.org/.
75. Heirs to Our Oceans. "Home." Accessed June 1, 2018. https://www.heirstoouroceans.com/about/.
76. Jr Ocean Guardians. "Home." Accessed June 1, 2018. https://www.jroceanguardians.org/.
77. Bye Bye Plastic Bags. "Home." Accessed June 1, 2018. http://www.byebyeplasticbags.org/.
78. Rock Island Media. "Kids Can Save The Planet." Accessed June 1, 2018. http://www.kidscansavetheplanet.com/.
79. Settelmyer, Scott. Interviewed by Brian McFarland. February 2017.
80. McField, Melanie. Interviewed by Brian McFarland. January 2020.
81. We Are Still In. "Home." Accessed April 13, 2020. https://www.wearestillin.com/.
82. RE100. "Home." Accessed April 13, 2020. http://there100.org/.
83. McField, Melanie. Interviewed by Brian McFarland. January 2020.
84. Mohammed (Ed.). *Economic Incentives for Marine and Coastal Conservation*. 7.

18

Concluding Thoughts About Conservation Finance

Tropical rainforests and tropical coral reefs have a lot in common. They are mostly located between the Tropics of Cancer and Capricorn and are often located in "developing" countries. These ecosystems provide homes to the world's greatest biodiversity on Earth and are dependent upon by billions of people. Yet, both ecosystems are being severely degraded, and at times lost forever, due in large part to the patterns of human consumption and human development.

Such loss is due to the expanding agricultural and livestock frontier into tropical forests, which results in the increase in greenhouse gas emissions and nutrient runoff, and which directly impacts offshore coral reefs. While tropical forests are more often privately owned than tropical coral reefs, there are innovative project types and innovating financial instruments to promote their conservation.

Challenges in the conservation of tropical rainforests and coral reefs are similar. For instance, there are funding shortfalls and monitoring these ecosystems—which can still stretch vast areas such as the Amazon Rainforest and Congo Basin or the Great Barrier Reef and the Mesoamerican Barrier Reef—remains difficult.

Ocean finance and the creation of MPAs is behind terrestrial finance and terrestrial protected areas. For instance, the iconic Yellowstone National Park was established in 1872,[1] yet it was not until the 1970s, when "Australia's Great Barrier Reef Marine Park Authority began by zoning broad areas of the 2,414-kilometer-long (1,500 mile) reef system for various uses, including about 3 percent of the area where fishing was restricted."[2] The first terrestrial debt-for-nature swap was conducted in 1987,[3] while one of the first

© The Author(s) 2021   **707**
B. J. McFarland, *Conservation of Tropical Coral Reefs*,
https://doi.org/10.1007/978-3-030-57012-5_18

debt-for-nature swaps related to the marine environment, the Seychelles' Climate Adaptation and Impact Investment Dept Swap, was completed in 2015.[4] Similarly, the first widely recognized green bonds were issued in 2007 and 2008 by the World Bank[5] and European Investment Bank.[6] The first "blue bond" with funds ring-fenced to fund "sustainable fisheries and marine projects" was launched for the Seychelles in 2018.[7]

While it is true that billions of dollars are needed for conservation, as asked by David de Rothschild, "If I can build a boat made entirely of materials that are fully recyclable materials and cross the ocean, why can't we build everyday household items that are cradle to cradle, rather than cradle to grave?"[8]

In addition, volunteers in India planted a record 66 million trees in just 12 hours,[9] while in China, "some 200 million Chinese users have downloaded an app by financial services provider Ant Financial that helps them make carbon-friendly decisions in everyday life."[10] Why not scale this level of volunteerism to clean up marine pollution and to assist with large-scale coral restoration?

It is vitally important that we reduce deforestation on land and the degradation of our world's coral reefs, otherwise, we will lose a tremendous amount of biodiversity and the means to feed billions of people on earth. As noted by Dr. Doug Rader, "if we do our jobs, with our intellection and financial power together, there can be reefs for our grandchildren and our great grandchildren despite climate change. We must get through the bottleneck – more MPAs, better fisheries management, restore reefs in places, find heat resilient combinations of algae and corals, etc. – and on the other side of the bottleneck, we could have vast productive corals."[11] Similarly, "recovery rates across studies suggest that substantial recovery of the abundance, structure and function of marine life could be achieved by 2050, if major pressures—including climate change—are mitigated."[12]

I will conclude with the following, paraphrased story told by a minister at the WWF Coral Reef Rescue Initiative Event at COP25 in Madrid: My daughter wants to become a marine biologist when she grows up. My wife and I discussed that if we, as humanity, do not get our act together, then there will be no marine biologists in the future. Instead, our daughter will be a marine archeologist.

Please do everything in your power, no matter how great or how small, to make a positive impact on our world's last remaining tropical ecosystems, which harbor the world's greatest biodiversity and whose ecosystem services, all of humanity depend.

Notes

1. National Park Service. "Yellowstone National Park: Frequently Asked Questions: History." Accessed November 23, 2016. https://www.nps.gov/yell/learn/historyculture/history-faqs.htm.
2. Ibid. 246.
3. WWF. *Guide to Conservation Finance*. 41.
4. SeyCCAT. "Seychelles' Conservation and Climate Adaptation Trust (SeyCCAT): Achieving Conservation Through Innovative Finance and Creative Collaborations." Accessed August 7, 2019. https://seyccat.org/wp-content/uploads/2017/09/infographic_talking_points_bat_web-1.pdf.
5. World Bank. "World Bank Green Bonds: Green Bond Issuances to Date." Accessed December 13, 2016. http://treasury.worldbank.org/cmd/htm/GreenBondIssuancesToDate.html.
6. Climate Bonds Initiative. "History." Accessed December 13, 2016. https://www.climatebonds.net/market/history.
7. Wahlén, Catherine Benson. "Seychelles Launches First Blue Bond." International Institute for Sustainable Development. November 6, 2018. Accessed 7 August 2019. https://sdg.iisd.org/news/seychelles-launches-first-blue-bond/.
8. Earle. *The World Is Blue*. 110.
9. Climate Action. "Record 66 Million Trees Planted in 12 Hours in India." July 4, 2017. http://www.climateactionprogramme.org/news/record-66-million-trees-planted-in-12-hours-in-india.
10. Carbon Pulse. "CP Daily: Tuesday May 30, 2017." May 30, 2017. http://carbon-pulse.com/35047/.
11. Rader, Doug. Interviewed by Brian McFarland. April 2020.
12. Duarte, C.M., et al. "Rebuilding Marine Life." *Nature*. https://doi.org/10.1038/s41586-020-2146-7.

Epilogue By Peter Gash

It is with pleasure that I have been asked to write this epilogue for Brian McFarland's book "Conservation of Tropical Coral Reefs: A Review of Financial and Strategic Solutions." This is a tremendous science-based document and Brian has created a great body of work here which I am sure you will agree covers critically important areas for our planet's future.

However, my epilogue is based on Australian history and my message of hope for our beautiful country.

As I turn the pages, I find myself pondering much about which he has referred.

Captain James Cook is credited with discovering the Eastern Australian Coast in 1770 and European settlement came to the Australian continent in 1788 with the arrival of the first fleet out of England.

On 11 June 1770, Cook, his botanist Joseph Banks and fleet arrived on Cook's ship the Endeavour. They became the first known Europeans to reach the east coast of Australia.

Parts of the landscape those early settlers would have found must have surely seemed like the promised land; Australia exhibited the desired life they sought. Closeness to the sea, the East Coast, major river systems, expansive land, and unique wildlife. Australia was surely the land of milk and honey.

However, if you dared to venture inland, West of the Great Dividing Range or toward the centre of the continent, it must have seemed a veritable hell on earth.

Red dirt, extreme heat, lack of obvious freshwater, vegetation, and wildlife like nowhere else on earth.

© The Editor(s) (if applicable) and The Author(s) 2021
B. J. McFarland, *Conservation of Tropical Coral Reefs*,
https://doi.org/10.1007/978-3-030-57012-5

Living in close custodianship with this strange and foreign continent was Australia's Indigenous People, our Aboriginal Australians.

These remarkable people had come South over land bridges, probably from as far away as Africa, as much as 60,000 years or more ago. Sea level changes had subsequently cut the Great Southern Continent off from the rest of the earth's land masses and the Australian Aboriginal People evolved in a way that has proven to be extremely unique on this planet.

Aboriginals had learned to live in harmony with the diverse continent and became extraordinary custodians of this wild place.

They learned numerous hunting methods with what, to other races seemed strange.

Weapons, like the boomerang and the woomera, are remarkable hunting tools.

However, the most remarkable of all the hunting tools the Australian Aboriginals mastered was fire, and with the mastering of fire, they learned to manage the forests and unique vegetation on this great southern land in a special way.

Nature had already put fire into the landscape with lightning strikes initiated from the severe electrical thunderstorms that this hot, dry and barren landscape had been giving birth to almost daily in certain parts of the continent.

That consistent use of fire by nature and by our Aboriginal People sculpted the landscape and fostered adaptations to the native vegetation that would not be found elsewhere. Cook and his fleet found the Australian vegetation and wildlife was almost unbelievable. Platypus, kangaroos, wombats, and koalas to name the more well known, along with thousands of additional, unique species on this strange southern land.

European population and thinking had come to an Aboriginal dry and unique continent.

The next 250 years saw great change and great pain; agriculture, mining, settlement, population growth. Human domination—of this remarkable, but fragile environment.

Considering its small population, considered less than 1 million Aboriginal People at the time of European settlement, to now just reaching 25 million people.

Our land mass is similar to the North American Continent, however there are 15 people there for every one person in Australia.

The Aboriginal People have until recent times been treated in the most terrible manner and still to this day, we have much to answer for.

The landscape has been stripped of the vast and ancient trees and forests by the millions. The precious soils farmed till exhausted and rivers cleared of mangroves and turned into fast flowing water drains…no longer wildlife drains and water catchment sources for the vegetation on this dry continent.

But there was also something that almost cost Captain James Cook and his fleet their lives when they ran aground on Endeavour Reef in June 1770. Cook sailed the Endeavour directly into razor-sharp coral of the Great Barrier Reef on his journey of discovery.

The world's largest living structure on Earth and the only living structure that can be seen from space, the Great Barrier Reef.

Over 2000 kms (1200 miles) long and home to thousands of marine species in the water and terrestrial species on its hundreds of various Islands.

An enormous mass stretching the East Coast of Australia, causing havoc for sailors trying to cross before the time of charts and navigation aids.

Today, the Great Barrier Reef is a monument of life forms, beauty and variety. Thousands on thousands of different species of marine and terrestrial life call this Great Barrier Reef home. But it is under extreme pressures. It faces uncertainty due to man made threats and climate change.

The real challenge has now begun for mankind.

Together we must minimise our footprint, not only on our continent but the whole planet.

We can… we will… because we must!

The Australian continent and more particularly, the Great Barrier Reef, have become poster children for the struggles that are going on today, between the Dark Greens (conservationists) and the Dark Browns (coal, oil, gas believers).

The answer will almost certainly sit somewhere in the middle…. when we all realise that we must work together and stop the blaming, finger pointing, and buck passing.

It's time we accept that we are all, yes all of us, are responsible for the current condition of our environment. It's time we join together to find ways to learn, share knowledge, collaborate and accept that we don't have all the answers.

Science learning from farmers and knowledge gained from land carers.

Farmers learning from scientists, miners, developers, fishermen, and tourism operators.

We are all in this together, the real-life Noah's Ark and if the ark sinks, we all go down together.

For us on Lady Elliot Island, located on the Southern tip of Australia's World Heritage Listed Great Barrier Reef, we have taken up the baton, we aren't waiting for someone else to tell us what direction to travel.

It comes from the heart…it's in our DNA. On Lady Elliot Island Eco Resort, we are dedicated to reducing our environmental footprint. Our solar energy program and sustainable initiatives have all been designed to mitigate climate change, our goal is to be 100% renewable by 2020 and we are on track to achieve that goal!

A quote on the initial pages of this book resonated with me "knowing is the key to caring, and with caring there is hope that people will be motivated to take positive actions"—Dr. Sylvia A. Earle.

This quote is what Lady Elliot Island Eco Resort is all about. Our goal is to help people fall in love with the Great Barrier Reef in the hope that they will then be motivated to protect it and all it's amazing inhabitants.

Implementing sustainability initiatives on a remote, off-grid Island is an extreme challenge. We use education as a way to pass on our message of conservation and climate action. If our guests take home just one climate action, we have done our job!

We believe that environmental sustainability and financial sustainability must go hand in hand.

Each and every one of us can make a difference, no matter how small.

Reducing our carbon footprint is an essential step to ensure that, as caretakers of this planet, we do not live beyond our means.

Buy locally, eat seasonally, plant trees, compost your food waste, buy ethically sourced products and recycle, to name a few.

Educate and encourage your neighbours, sisters, brothers, mothers, fathers, local members and government to take climate action.

Climate change is almost certainly the biggest challenge, as a collective, we will face in our lifetime.

If there is one message I could leave, is to use your voice, be persistent in your messages, and have faith in the ability of humanity to pull together in the face of crisis.

—Peter Gash
Chief Caretaker and Custodian
Lady Elliot Island Eco Resort
March 2020

Bibliography

Ali, Saleem H. (Ed.). *Peace Parks: Conservation and Conflict Resolution*. Cambridge, MA: The MIT Press, 2007.

Althelia Ecosphere. "White Paper: Investing for Impact and Value in the Marine Environment: The Sustainable Ocean Fund." June 6, 2016. Royal Society, London. https://archive.globallandscapesforum.org/wp-content/uploads/docs/London-white-papers/London-2016-WhitePaper-Investing-for-impact-and-value.pdf.

Bahadur, Jay. *The Pirates of Somalia: Inside Their Hidden World*. New York: Vintage Books, A Division of Random House, Inc., 2012.

Balmford, Andrew. *Wild Hope: On the Front Lines of Conservation Success*. Chicago: The University of Chicago Press, 2012.

Bennett, Genevieve and Franziska Ruef. "Alliances for Green Infrastructure: State of Watershed Investment 2016." December 2016. *Ecosystem Marketplace*. https://www.forest-trends.org/wp-content/uploads/2017/03/doc_5463.pdf.

Beyer, Hawthorne L., Emma V. Kennedy, Maria Beger, Chaolun Allen Chen, Joshua E. Cinner, Emily S. Darling, C. Mark Eakin, Ruth D. Gates, Scott F. Heron, Nancy Knowlton, David O. Obura, Stephen R. Palumbi, Hugh P. Possingham, Marji Puotinen, Rebecca K. Runting, William J. Skirving, Mark Spalding, Kerrie A. Wilson, Sally Wood, John E. Veron, and Ove Hoegh-Guldberg. "Risk-Sensitive Planning for Conserving Coral Reefs Under Rapid Climate Change." *Conservation Letters*. Vol. 11 (June 5, 2018): e12587. https://doi.org/10.1111/conl.12587.

Bigue, Marcel, Manuel Bravo, Oswaldo Rosero, Rich Arnold, Carmen Revenga, and Jeremy Rude. "Nearshore Artisanal Fisheries Enforcement Guide." September

© The Editor(s) (if applicable) and The Author(s) 2021
B. J. McFarland, *Conservation of Tropical Coral Reefs*,
https://doi.org/10.1007/978-3-030-57012-5

18, 2015. http://wildaid.org/wp-content/uploads/2017/09/Nearshore-Artisanal-Fisheries-Enforcement-Guide_0.pdf.

Bos, Melissa, Robert L. Pressey, and Natalie Stoeckl. "Marine Conservation Finance: The Need for and Scope of an Emerging Field." *Ocean & Coastal Management.* Vol. 114 (2015). 116–128. http://dx.doi.org/10.1016/j.ocecoaman.2015.06.021.

Braverman, Irus. *Coral Whisperers: Scientists on the Brink.* Berkeley, CA: University of California Press, 2018.

Buckley, Ross P. "Debt-for-Development Exchanges: An Innovative Response to the Global Financial Crisis." *University of New South Wales Law Journal.* Vol. 32 (2009). 620–645. https://papers.ssrn.com/sol3/papers.cfm?abstract_id=1583197.

Bull, Joseph W., K. Blake Suttle, Ascelin Gordon, Navinder J. Singh, and E.J. Milner-Gulland. "Biodiversity Offsets in Theory and Practice." *Oryx.* Vol. 47. Issue 3. 369–380. https://doi.org/10.1017/S003060531200172X.

Carr, Archie. *The Windward Road: Adventures of a Naturalist on Remote Caribbean Shores.* New York: Alfred A. Knopf, 1955.

Carson, Rachel. *The Sea Around Us.* New York: Oxford University Press, 1950.

Carson, Rachel. *The Edge of the Sea.* Boston: A Mariner Book, Houghton Mifflin Company, 1998.

Chao, N.L., P. Petry, G. Prang, L. Sonneschein, and Tlusty, M.F. (Eds.). *Conservation and Management of Ornamental Fish Resources of the Rio Negro Basin, Amazonia, Brazil—Project Piaba.* Editoria da Universidade do Amazonas, Manaus, Brazil.

Charteris, Mickey. *Caribbean Reef Life: A Field Guide for Divers (Third Edition).* Minneapolis, MI: Mill City Press, 2017.

Cinner, Joshua E., Cindy Huchery, M. Aaron MacNeil, Nicholas A.J. Graham, Tim R. McClanahan, Joseph Maina, Eva Maire, John N. Kittinger, Christina C. Hicks, Camilo Mora, Edward H. Allison, Stephanie D'Agata, Andrew Hoey, David A. Feary, Larry Crowder, Ivor D. Williams, Michel Kulbicki, Laurent Vigliola, Laurent Wantiez, Graham Edgar, Rick D. Stuart-Smith, Stuart A. Sandin, Alison L. Green, Marah J. Hardt, Maria Beger, Alan Friedlander, Stuart J. Campbell, Katherine E. Holmes, Shaun K. Wilson, Eran Brokovich, Andrew J. Brooks, Juan J. Cruz-Motta, David J. Booth, Pascale Chabanet, Charlie Gough, Mark Tupper, Sebastian C. A. Ferse, U. Rashid Sumaila, and David Mouillot. "Bright Spots Among the World's Coral Reefs." *Nature* (2016). https://doi.org/10.1038/nature18607.

Clark, Story. *A Field Guide to Conservation Finance.* Washington, DC: Island Press, 2007.

Clover, Charles. *The End of the Line: How Overfishing Is Changing the World and What We Eat.* Berkeley, CA: University of California Press, 2006.

Cousteau, Jacques-Yves, with Frédéric Dumas and James Dugan. *The Silent World.* New York: Ballantine Books, 1950.

Crew, Bill. *Port Douglas: An Environmental Playground.* Port Douglas, QLD, Australia: Odyssey Bound Pty Ltd., Date Unknown.

Daily, Gretchen C. and Katherine Ellison. *The New Economy of Nature: The Quest to Make Conservation Profitable.* Washington, DC: Island Press, 2002.

Daley, Ben. *The Great Barrier Reef: An Environmental History.* New York: Routledge, 2014.

Darling, Emily S., Tim R. McClanahan, Joseph Maina, Georgina G. Gurney, Nicholas A. J. Graham, Fraser Januchowski-Hartley, Joshua E. Cinner, Camilo Mora, Christina C. Hicks, Eva Maire, Marji Puotinen, William J. Skirving, Mehdi Adjeroud, Gabby Ahmadia, Rohan Arthur, Andrew G. Bauman, Maria Beger, Michael L. Berumen, Lionel Bigot, Jessica Bouwmeester, Ambroise Brenier, Tom C. L. Bridge, Eric Brown, Stuart J. Campbell, Sara Cannon, Bruce Cauvin, Chaolun Allen Chen, Joachim Claudet, Vianney Denis, Simon Donner, Estradivari, Nur Fadli, David A. Feary, Douglas Fenner, Helen Fox, Erik C. Franklin, Alan Friedlander, James Gilmour, Claire Goiran, James Guest, Jean-Paul A. Hobbs, Andrew S. Hoey, Peter Houk, Steven Johnson, Stacy D. Jupiter, Mohsen Kayal, Chao-yang Kuo, Joleah Lamb, Michelle A. C. Lee, Jeffrey Low, Nyawira Muthiga, Efin Muttaqin, Yashika Nand, Kirsty L. Nash, Osamu Nedlic, John M. Pandolfi, Shinta Pardede, Vardhan Patankar, Lucie Penin, Lauriane Ribas-Deulofeu, Zoe Richards, T. Edward Roberts, Ku'ulei S. Rodgers, Che Din Mohd Safuan, Enric Sala, George Shedrawi, Tsai Min Sin, Patrick Smallhorn-West, Jennifer E. Smith, Brigitte Sommer, Peter D. Steinberg, Makamas Sutthacheep, Chun Hong James Tan, Gareth J. Williams, Shaun Wilson, Thamasak Yeemin, John F. Bruno, Marie-Josée Fortin, Martin Krkosek, and David Mouillot. "Social-Environmental Drivers Inform Strategic Management of Coral Reefs in the Anthropocene." *Nature Ecology and Evolution.* Vol. 3 (2019, August 12). 1341–1350. Accessed February 28, 2020. https://doi.org/10.1038/s41559-019-0953-8.

Deacon, Robert T. and Dominic P. Parker. "Encumbering Harvest Rights to Protect Marine Environments: A Model of Marine Conservation Easements." *The Australian Journal of Agricultural and Resource Economics.* December 22, 2008. Accessed March 14, 2020. https://doi.org/10.1111/j.1467-8489.2007.00429.x.

Dela Cruz, Dexter W. and Peter L. Harrison. "Enhanced Larval Supply and Recruitment Can Replenish Reef Corals on Degraded Reefs." *Scientific Reports.* Vol. 7 (October 11, 2017). 13985. https://doi.org/10.1038/s41598-017-14546-y.

Donofrio, Stephen, Patrick Maguire, William Merry, and Steve Zwick. "Financing Emission Reductions for the Future: State of the Voluntary Carbon Markets 2019." *Ecosystem Marketplace.* December 2019. Available at: https://www.ecosystemmarketplace.com/carbon-markets/.

Duarte, Carlos M., Susana Agusti, Edward Barbier, Gregory L. Britten, Juan Carlos Castilla, Jean-Pierre Gattuso, Robinson W. Fulweiler, Terry P. Hughes, Nancy Knowlton, Catherine E. Lovelock, Heike K. Lotze, Milica Predragovic, Elvira Poloczanska, Callum Roberts, and Boris Worm. "Rebuilding Marine Life." *Nature.* Vol. 580 (2020). 39–51. https://doi.org/10.1038/s41586-020-2146-7.

Earle, Sylvia A. *Sea Change: A Message of the Oceans.* New York: Fawcett Columbine, 1995.

Earle, Sylvia A. *The World Is Blue: How Our Fate and the Ocean's Are One.* Washington, DC: National Geographic, 2009.

Ebbesmeyer, Curtis and Eric Scigliano. *Flotsametrics and the Floating World: How One Man's Obsession with Runaway Sneakers and Rubber Ducks Revolutionized Ocean Science.* New York: Harper, 2010.

Fitzpatrick, Richard. *Shark Tracker: Confessions of an Underwater Cameraman.* Kensington, NSW, Australia: NewSouth Publishing, 2016.

Food and Agriculture Organization of the United Nations. 2018. "The State of World Fisheries and Aquaculture 2018—Meeting the Sustainable Development Goals." Rome. Licence: CC BY-NC-SA 3.0 IGO. http://www.fao.org/3/I95 40EN/i9540en.pdf.

Forsman, Zac H., Christopher A. Page, Robert J. Toonen, and David Vaughan. "Growing Coral Larger and Faster: Micro-Colony-Fusion as a Strategy for Accelerating Coral Cover." *PeerJ*. Vol. 3 (2015). e1313. https://doi.org/10.7717/peerj. 1313.

Gaughan, D.J., Molony, B. and Santoro, K. (Eds). 2019. Status Reports of the Fisheries and Aquatic Resources of Western Australia 2017/18: The State of the Fisheries. Department of Primary Industries and Regional Development, Western Australia. http://www.fish.wa.gov.au/Documents/sofar/status_rep orts_of_the_fisheries_and_aquatic_resources_2017-18.pdf.

Gómez-Baggethun, Erik, Rudolf de Groot, Pedro L. Lomas, and Carlos Montes. "The History of Ecosystem Services in Economic Theory and Practice: From Early Notions to Markets and Payment Schemes." *Ecological Economics*. Vol. 69. Issue 6 (April 2010). 1209–1218. https://doi.org/10.1016/j.ecolecon.2009. 11.007.

Goodwin, Edward J. *International Environmental Law and the Conservation of Coral Reefs*. New York: Routledge of Taylor & Francis Group, 2011.

Government of Seychelles. "Conservation and Climate Adaptation Trust of Seychelles Act." 2015. Accessed May 15, 2017. https://www.seylii.org/sc/sc/legisl ation/Act%2018%20of%202015%20Conservation%20and%20Climate%20A daptation%20Trust%20of%20Seychelles%20Act%2C%202015.pdf.

Great Barrier Reef Marine Park Authority 2019. Annual Report 2018–19. GBRMPA, Townsville. http://www.gbrmpa.gov.au/about-us/corporate-inf ormation/annual-report.

Greenlaw, Linda. *The Hungry Ocean: A Swordboat Captain's Journey*. New York: Hyperion, 1999.

Halpern, Benjamin S., Shaun Walbridge, Kimberly A. Selkoe, Carrie V. Kappel, Fiorenza Micheli, Caterina D'Agrosa, John F. Bruno, Kenneth S. Casey, Colin Ebert, Helen E. Fox, Rod Fujita, Dennis Heinemann, Hunter S. Lenihan, Elizabeth M. P. Madin, Matthew T. Perry, Elizabeth R. Selig, Mark Spalding, Robert Steneck, and Reg Watson. "A Global Map of Human Impact on Marine Ecosystems." *Science*. Vol. 319 (February 15, 2008). https://doi.org/10.1126/science. 1149345.

Hamrick, Kelley. *State of Private Investment in Conservation 2016: A Landscape Assessment of an Emerging Market.* Washington, DC: Ecosystem Marketplace, December 2016. http://www.forest-trends.org/documents/files/doc_5474.pdf#.

Hastings, Jesse G., Michael K. Orbach, Leah B. Karrer, and Les Kaufman. "Lessons Learned from the Marine Management Area Science Program: Insights for Global Conservation Science Programs." *Coastal Management.* Vol. 43 (2015). 189–215. https://doi.org/10.1080/08920753.2015.1005538.

Hayden, Daniel and Benjamin Dills. "Smokey the Bear Should Come to the Beach: Using Mascot to Promote Marine Conservation." *Social Marketing Quarterly.* Vol. 21. Issue 1 (2015). 3–13. https://doi.org/10.1177/1524500414558126.

Helvarg, David. *50 Ways to Save the Ocean.* San Francisco: Inner Ocean Publishing, 2006.

Hoegh-Guldberg, O. et al. *Reviving the Ocean Economy: The Case for Action—2015.* WWF International, Gland, Switzerland., Geneva, 60 pp. 2015. https://www.worldwildlife.org/publications/reviving-the-oceans-economy-the-case-for-action-2015.

Hooker, Sascha K. and Leah R. Gerber. "Marine Reserves as a Tool for Ecosystem-Based Management: The Potential Importance of Megafauna." *BioScience.* January 2004. Vol. 54. Issue 1. https://doi.org/10.1641/0006-3568(2004)054[0027:MRAATF]2.0.CO;2.

Huwyler, Fabian, Jürg Käppeli, and John Tobin. *Conservation Finance: From Niche to Mainstream: The Building of an Institutional Asset Class.* New York: Credit Suisse, January 2016. https://www.credit-suisse.com/media/assets/corporate/docs/about-us/responsibility/banking/conservation-finance-en.pdf.

Huwyler, Fabian, Jürg Käppeli, Katharina Serafimova, Eric Swanson, and John Tobin. *Conservation Finance: Moving Beyond Donor Funding Toward an Investor-Driven Approach.* New York: Credit Suisse, January 2014. https://www.credit-suisse.com/media/assets/corporate/docs/about-us/responsibility/environment/conservation-finance-en.pdf.

Iyer, Venkat, Katy Mathias, David Meyers, Ray Victurine, and Melissa Walsh. *Finance Tools for Coral Reef Conservation: A Guide.* December 2018. https://www.icriforum.org/sites/default/files/50+Reefs+Finance+Guide.pdf.

Jambeck, Jenna R., Roland Geyer, Chris Wilcox, Theodore R. Siegler, Miriam Perryman, Anthony Andrady, Ramani Narayan, and Kara Lavender Law. "Plastic Waste Inputs from Land into Ocean." *Science.* Vol. 347. Issue 6223 (February 13, 2015). https://doi.org/10.1126/science.1260352.

Kikiloi, Kekuewa, Alan M. Friedlander, ʻAulani Wilhelm, Naiʼa Lewis, Kalani Quiocho, William ʼĀila Jr. & Sol Kahoʼohalahala. "Papahānaumokuākea: Integrating Culture in the Design and Management of One of the World's Largest Marine Protected Areas." *Coastal Management.* Vol. 45. Issue 6 (2017). 436–451. 2017. https://doi.org/10.1080/08920753.2017.1373450.

Legrand, Willly, Claudia Simons-Kaufmann, and Philip Sloan (Eds.). *Sustainable Hospitality and Tourism as Motors for Development: Case Studies from Developing Regions of the World.* New York: Routledge, Taylor & Francis Group, 2012.

Leopold, Aldo. "Conservation Economics." *Journal of Forestry*. Vol. 32. Issue 5 (May 1, 1934). 537–544.

Levitt, James N. (Ed.). *From Walden to Wall Street: Frontiers of Conservation Finance*. Cambridge, MA: Lincoln Institute of Land Policy, 2005.

Levitt, James N. (Ed.). *Conservation Capital in the Americas: Exemplary Conservation Finance Initiatives*. Cambridge, MA: Lincoln Institute of Land Policy, 2010.

Lomboy, Cristopher G., Flora Belinario, Robert Pomeroy, Joey Pedrajas, Raquel Sanchez Tirona, Steve Box, Paolo, Roberto Domondon, and Kristine Balbido-Ramirez. "Building Household Economic Resilience to Secure a Future for Near Shore Fishers in the Philippines." *Marine Policy*. November 6, 2018. https://doi.org/10.1016/j.marpol.2018.11.013.

Mackelworth, P.C., Y. Teff Seker, T. Vega Fernández, M. Marques, F.L. Alves, G. D'Anna, D.A. Fa, D. Goldborough, Z. Kyriazi, C. Pita, M.E. Portman, B. Rumes, S.J. Warr, and D. Holcer. "Geopolitics and Marine Conservation: Synergies and Conflicts." *Frontiers in Marine Science*. Vol. 6 (2019). 759. https://doi.org/10.3389/fmars.2019.00759.

Macy, Joanna and Chris Johnstone. *Active Hope: How to Face the Mess We're in Without Going Crazy*. Novato, CA: New World Library, 2012.

Madsen, Becca, Nathaniel Carroll, Kelly Moore Brands. "State of Biodiversity Markets 2009: Offset and Compensation Programs Worldwide." March 2010. *Ecosystem Marketplace*. Available at: https://www.forest-trends.org/publications/state-of-biodiversity-markets/.

McCalman, Iain. *The Reef: A Passionate History, The Great Barrier Reef from Captain Cook to Climate Change*. New York: Scientific American and Farrar, Straus, and Giroux, 2013.

Mohammed, Essam Yassin (Ed.). *Economic Incentives for Marine and Coastal Conservation: Prospects, Challenges and Policy Implications*. New York: Earthscan from Routledge, 2014.

Montgomery, Sy. *The Soul of an Octopus: A Surprising Exploration into the Wonder of Consciousness*. New York: Atria Paperback, 2016.

Myers, Norman. *The Sinking Ark: A New Look at the Problem of Disappearing Species*. Oxford: Pergamon Press, 1979.

NatureVest. "Seychelles Debt Restructuring for Marine Conservation and Climate Adaptation." 2018. Accessed January 10, 2018. http://www.naturevesttnc.org/business-lines/debt-restructuring/seychelles-debt-restructuring/?intc=nvest.header.projects.

Niner, Holly J., Ben Milligan, Peter J.S. Jones, and Craig A. Styan. "A Global Snapshot of Marine Biodiversity Offsetting Policy." *Marine Policy*. Vol. 81. 368–374. https://doi.org/10.1016/j.marpol.2017.04.005.

Obura, D., S.D. Donner, S. Walsh, S. Mangubhai, Rotjan R. Living Document. "Phoenix Islands Protected Area Climate Change Vulnerability Assessment and Management." Report to the New England Aquarium, Boston, USA. 35 pp. Updated January 18, 2016. http://www.phoenixislands.org/pdf/PIPA-CC-scoping-study-Jan-18-2016.pdf.

Oliver, Thomas A., Kirsten L. L. Oleson, Hajanaina Ratsimbazafy, Daniel Raberinary, Sophie Benbow, Alasdair Harris. "Positive Catch & Economic Benefits of Periodic Octopus Fishery Closures: Do Effective, Narrowly Targeted Actions 'Catalyze' Broader Management?" *PLOS ONE*. June 17, 2015. https://doi.org/10.1371/journal.pone.0129075.

Pascal, Nicolas, Angelique Brathwaite, Maxime Philip, and Melissa Walsh. "Impact Investing in Marine Conservation." *Duke Environmental Law & Policy Forum*. Vol. XXVIII (Spring 2018). 199. https://pdfs.semanticscholar.org/6d2b/e4faf94728ee781378054889fb4a41142eb3.pdf.

Penn, J. W., N. Caputi, and S. de Lestang. "A Review of Lobster Fishery Management: The Western Australian Fishery for Panulirus Cygnus, A Case Study in the Development and Implementation of Input and Output-Based Management Systems." ICES. *Journal of Marine Science*, Vol. 72. i22–i34.

Portman, Michelle E. and Yael Teff-Seker. "Factors of Success and Failure for Transboundary Environmental Cooperation: Projects in the Gulf of Aqaba." *Journal of Environmental Policy and Planning*. 31 January 2017. http://dx.doi.org/10.1080/1523908X.2017.1292873.

Prideaux, Bruce and Anja Pabel (Eds.). *Coral Reefs: Tourism, Conservation and Management*. New York: Routledge, 2018.

Rare. "Stemming the Tide of Coastal Overfishing: Fish Forever Program Results 2012–2017: Full Report, July 2018." https://rare.org/wp-content/uploads/2019/02/Fish-Forever-Full-Report-July-2018.pdf.

Roberts, Callum. *The Ocean of Life: The Fate of Man and the Sea*. New York: Penguin Books, 2012.

Roberts, Callum M., Colin J. McClean, John E. N. Veron, Julie P. Hawkins, Gerald R. Allen, Don E. McAllister, Cristina G. Mittermeier, Frederick W. Schueler, Mark Spalding, Fred Wells, Carly Vynne, and Timothy B. Werner. "Marine Biodiversity Hotspots and Conservation Priorities for Tropical Reefs." *Science*. 15 February 2002. Vol. 295. Issue 5558. 1280–1284. https://doi.org/10.1126/science.1067728.

Robinson, Michele. "Debt Restructuring Initiatives Paper for the Commonwealth Secretariat." June 2010. https://michelerobinson.net/yahoo_site_admin/assets/docs/Debt_Restructuring_Initiatives_paper_-_Final_Version_June_2010.43100215.pdf.

Rotjan, Randi, Regen Jamieson, Ben Carr, Les Kaufman, Sangeeta Mangubhai, David Obura, Ray Pierce, Betarim Rimon, Bud Ris, Stuart Sandin, Peter Shelley, U. Rashid Sumaila, Sue Taei, Heather Tausig, Tukabu Teroroko, Simon Thorrold, Brooke Wikgren, Teuea Toatu, and Greg Stone. "Establishment, Management, and Maintenance of the Phoenix Islands Protected Area." *Advances in Marine Biology*. Vol. 69 (2014). ISSN 0065-2881. http://dx.doi.org/10.1016.B978-0-12-800214-8.00008-6.

Selig, Elizabeth R., Will R. Turner, Sebastian Troëng, Bryan P. Wallace, Benjamin S. Halpern, Kristin Kaschner, Ben G. Lascelles, Kent E. Carpenter, and Russell A. Mittermeier. "Global Priorities for Marine Biodiversity Conservation." *PLoS*

ONE. Vol. 9. Issue 1 (2014). e82898. https://doi.org/10.1371/journal.pone.008 2898.

Seychelles Marine Spatial Planning Initiative. "Seychelles Marine Spatial Planning Initiative." 2014. Accessed January 17, 2018. http://seymsp.com/.

Seychelles Marine Spatial Planning Initiative. "Seychelles Marine Spatial Planning: On the Leading Edge of Marine Conservation and Climate Adaptation." 2015. Accessed January 22, 2018. https://www.cbd.int/doc/meetings/mar/soiom-2016-01/other/soiom-2016-01-seychelles-02-en.pdf.

Sheikh, Pervaze A. "Debt-for-Nature Initiatives and the Tropical Forest Conservation Act: Status and Implementation." *Congressional Research Service*. March 30, 2010. https://www.cbd.int/financial/debtnature/g-inventory2010.pdf.

Sheikh, Pervaze A. "Debt-for-Nature Initiatives and the Tropical Forest Conservation Act (TFCA): Status and Implementation." *Congressional Research Service*. Updated July 24, 2018. https://crsreports.congress.gov.

Silver, Jennifer J. and Lisa M. Campbell. "Conservation, Development and the Blue Frontier: The Republic of Seychelles' Debt Restructuring for Marine Conservation and Climate Adaptation Program." *International Social Science Journal*. September 13, 2018. Accessed October 1, 2018. https://doi.org/10.1111/issj.12156.

Sobel, Jack and Craig Dahlgren. *Marine Reserves: A Guide to Science, Design, and Use*. Washington, DC: Island Press, 2004.

Spalding, Mark, Lauretta Burke, Spencer A. Wood, Joscelyne Ashpolee, James Hutchison, and Philine zu Ermgassen. "Mapping the Global Value and Distribution of Coral Reef Tourism." *Marine Policy*. Vol. 82 (August 2017) 104–113. https://doi.org/10.1016/j.marpol.2017.05.014.

Spergel, Barry and Melissa Moye. "Financing Marine Conservation: A Menu of Options." January 2004. Washington, DC: WWF Center for Conservation Finance. http://awsassets.panda.org/downloads/fmcnewfinal.pdf.

Speth, James Gustave and Peter M. Haas. *Global Environmental Governance*. Washington, DC: Island Press, 2006.

Stone, Gregory S. and David Obura (Eds.). *Underwater Eden: Saving the Last Coral Wilderness on Earth*. Chicago: The University of Chicago Press, 2013.

The Nature Conservancy and EKO Asset Management Partners. *Investing in Conservation: A Landscape Assessment of an Emerging Market*. Arlington, VA: The Nature Conservancy, November 2014. https://thought-leadership-production.s3.amazonaws.com/2016/09/14/18/27/27/93df86a6-f916-47cc-8a54-0a818037bba1/InvestingInConservation_Report.pdf.

UN Environment, International Coral Reef Initiative, UN Environment World Conservation Monitoring Centre. "Analysis of International Funding for the Sustainable Management of Coral Reefs and Associated Coastal Ecosystems." 2018. Available at: www.wcmc.io/coralbrochure.

Van Oppen, Madeleine J.H., James K. Oliver, Hollie M. Putnam, and Ruth D. Gates. "Building Coral Reef Resilience Through Assisted Evolution." *Proceedings of the National Academy of Sciences of the United States of America.* Vol. 112. Issue 8 (2015). www.pnas.org/cgi/doi/10.1073/pnas.1422301112.

Vertigo Lab. "Innovations for Coral Finance, ICRI Publication." 2017. 80 p. http://vertigolab.eu/wp-content/uploads/2018/04/rapport-innovation-for-coral-finance.pdf.

Verutes, Gregory M., Katie K. Arkema, Chantalle Clark-Samuels, Spencer A. Wood, Amy Rosenthal, Samir Rosado, Maritza Canto, Nadia Bood, and Mary Ruckelshaus. "Integrated Planning That Safeguards Ecosystems and Balances Multiple Objectives in Coastal Belize." *International Journal of Biodiversity Science, Ecosystem Services & Management.* Vol. 13. Issue 3. 1–17. https://doi.org/10.1080/21513732.2017.1345979.

Warland, Linde and Axel Michaelowa. "Can Debt for Climate Swaps Be a Promising Climate Finance Instrument? Lessons from the Past and Recommendations for the Future." Perspectives GmbH. 5 November 2015. https://www.perspectives.cc/fileadmin/Publications/Can_debt_for_climate_swaps_be_a_promising_climate_finance_instrument_Warland_Linde__Michaelowa_Axel_2015.pdf.

Weigel, Jean-Yves, Kathryn Olivia Mannle, Nathan James Bennett, Eleanor Carter, Lena Westlund, Valerie Burgener, Zachary Hoffman, Alfredo Simão da Silva, Elimane Abou Kane, Jessica Sanders, Catherine Piante, Sukarno Wagiman, and Ashley Hellman. "Marine Protected Areas and Fisheries: Bridging the Divide." Aquatic Conservation: Marine and Freshwater Ecosystems. Vol. 24 (Suppl. 2) (2014). 199–215. https://doi.org/10.1002/aqc.2514.

Wilson, Edward O. *The Creation: An Appeal to Save Life on Earth.* New York: W. W. Norton, 2006.

Wohlleben, Peter. *The Secret Network of Nature: The Delicate Balance of All Living Things.* London, UK: Penguin Random House, 2017.

World Wildlife Fund's Center for Conservation Finance. "Commercial Debt-for-Nature Swaps." December 9, 2003. https://www.cbd.int/doc/external/wwf/wwf-commercial-swaps-en.pdf.

World Wildlife Fund. *Guide to Conservation Finance: Sustainable Financing for the Planet.* Washington, DC: World Wildlife Fund, 2009. http://awsassets.panda.org/downloads/wwf_guide_to_conservation_finance.pdf.

Zimmer, Carl. "The Partitioning of the Red Sea." *Science.* Vol. 293. Issue 5530 (July 29, 2001). 627–628. https://doi.org/10.1126/science.293.5530.627.

Index

GPSR Compliance
The European Union's (EU) General Product Safety Regulation (GPSR) is a set
of rules that requires consumer products to be safe and our obligations to
ensure this.

If you have any concerns about our products, you can contact us on

ProductSafety@springernature.com

In case Publisher is established outside the EU, the EU authorized
representative is:

Springer Nature Customer Service Center GmbH
Europaplatz 3
69115 Heidelberg, Germany

www.ingramcontent.com/pod-product-compliance
Ingram Content Group UK Ltd.
Pitfield, Milton Keynes, MK11 3LW, UK
UKHW020103090726
13644UKWH00003B/389